T0211385

Urban Flood Mitigation and Stormwater Management

Urban Flood Mitigation and Stormwater Management

James C. Y. Guo

CRC Press
Taylor & Francis Group
Boca Raton London New York

CRC Press is an imprint of the
Taylor & Francis Group, an **informa** business

CRC Press
Taylor & Francis Group
6000 Broken Sound Parkway NW, Suite 300
Boca Raton, FL 33487–2742

First issued in paperback 2019

CRC Press is an imprint of Taylor & Francis Group, an Informa business

No claim to original U.S. Government works

ISBN-13: 978-1-138-19814-2 (hbk)
ISBN-13: 978-0-367-88599-1 (pbk)

Library of Congress Cataloging-in-Publication Data
Names: Guo, James C. Y. (James Chwen-Yuan), author.
Title: Urban flood mitigation and stormwater management/James C.Y. Guo.
Description: Boca Raton, FL : CRC Press, [2017] | Includes bibliographical references and index.
Identifiers: LCCN 2016049390| ISBN 9781138198142 (hardback : alk. paper) | ISBN 9781315269917 (ebook)
Subjects: LCSH: Drainage--Design and construction. | Urban runoff--Management. | Flood control.
Classification: LCC TC970 .G84 2017 | DDC 628/.21--dc23
LC record available at https://lccn.loc.gov/2016049390

Visit the Taylor & Francis Web site at
http://www.taylorandfrancis.com

and the CRC Press Web site at
http://www.crcpress.com

Contents

Preface

Hydrology and hydraulics are the fundamental sciences to study the water environmental systems. Hydrology covers the basic understanding of the natural phenomenon about water circulations and movements, whereas hydraulic engineering emphasizes analyses, designs, and construction. In the United States, more than 80% of the population lives in urban areas. Therefore, applications of hydrology and hydraulics to human needs and safety in urban areas are critically important to the engineers and scientists who are dedicated to the predictions of floods and droughts, designs of water supply, flood mitigation, urban drainage, flood plain management, and water environmental protection. This book presents the latest developments of urban hydrology and hydraulic modeling techniques and design procedures for stormwater management using both the conventional and low-impact development (LID) approaches.

Stormwater is a natural resource that supports water supplies for urban needs, refreshes lakes and rivers, and recharges groundwater tables. It is important for an urban area to have a sound regional stormwater management planning that preserves the urban water environment and balances the pros and cons in every development project. Urban development leads to more pavements and impervious surfaces on the ground. Sharply increased storm runoff volumes and flows from urban areas change the spatial and temporal distributions of surface and subsurface runoff in the hydrologic cycle. The major negative impacts of urban development on storm runoff are (1) increase of peak flows (Q-problem) from flooding events and (2) increase of runoff volumes (V-problem) from frequent events. The former becomes a safety issue to the public, while the latter leads to the deterioration of water quality in the urban environment.

To mitigate the urbanization impacts, the first priority in developing an urban drainage system is to quickly collect and drain flooding water out of urban areas. The next is to improve the urban watershed to enhance the water quality for the purpose of preserving the water environment. Following such a thought, this book is structured first to cover the conventional urban drainage methods to design street hydraulic conveyance capacity, including inlets, gutter flows, sewer drains, and channels. An efficient drainage system not only achieves the first goal of quick removal of stormwater but also results in more concentrated flows that may unlawfully transfer flooding problems to downstream properties or water bodies. The second part of this book focuses on how to follow the drainage criteria to set up the allowable flow release and how to design a stormwater detention system to reduce the peak flows. Man-made basins and gardens are an effective method to recover stormwater storage capacity in an urban watershed. The more the detention and retention basins built in the neighborhood, the more the concerns about the public safety. Hydraulic structures must be equipped with safety measures, including flood gates, trash racks, fences, service roads, and flash flood warning signs. This book covers how to quantifiably determine the forces on the trash rack and how to design forebay, micropool, and diversion structures with concerns for public safety.

Under the mandate of the Federal Water Quality Act, the new concept of LID offers many infiltration-based approaches to improve conventional stormwater drainage systems. In this book, the conventional minor and major drainage system is further expanded into a 3M drainage system, or a new layer of micro drainage system is added for LID facilities. The 3M drainage system is a cascading overflow system that consists of three components: (1) *micro drainage system* such as rain gardens and pavers that intercept 3- to 6-month events, (2) *minor drainage system* such as storm drains and street inlets that collect 2- to 5-year peak runoff flows, and (3) *major drainage system* such as street gutters and channels that convey the 10- to 100-year peak runoff flows. In this book, the concept of urban *3M (micro, minor, and major) drainage system* dictates both new designs and retrofitting existing systems for urban renewal.

It is a challenge to retrofit existing drainage systems to be compatible with the 3M drainage system. Several examples are presented in this book to illustrate how to incorporate an LID infiltration basin into an existing detention basin that was built for the purpose of flood mitigation only and how to convert an orifice-weir outlet box into a perforated plate for a full-spectrum flow-release control. These examples may not be generalized for all similar cases, but they do provide some guidance to improve an existing drainage system with new functions for water quality enhancement.

The first 6 chapters in this book cover the hydrologic procedures for rainfall and runoff predictions, and the next 14 chapters focus on hydraulic designs of urban channel, culvert, street inlet, sewer drain, detention basin, retention basin, infiltration basin, LID designs, and stormwater modeling techniques by various routing methods. Hydrologic designs are often lengthy in calculations. Many design procedures presented in this book have been converted into Excel spreadsheets/books. All these efficient tools can be downloaded from the website www.udfcd.org at no cost.

This book is the summary of the author's 30-year research publications and class notes developed for graduate classes and senior design courses. More than 100 real-world design examples are used to illustrate the methods and procedures. Many of the design methods documented in this book have been adopted as the recommended design procedures by Denver, Las Vegas, and Sacramento metropolitan areas in the United States. For additional information, the reader can visit the website www.udfcd.org.

PowerPoint presentations and Excel computer models are provided via the downloads link at www.crcpress.com/9781138198142 which build on the exercises in the book.

Acknowledgments

The preparation of this book was supported by the Urban Watersheds Research Institute, and the author is grateful to the institute and its board of directors for that support. Ben Urbonas, president, and Ken MacKenzie, treasurer of the institute, provided comments and reviewed chapters and sections of this book. They also offered discussions on critical issues, new concepts, and public concerns related to stormwater management.

The Urban Drainage and Flood Control District (UDFCD) in Denver, Colorado, was the source of field data, photos, and maps that supplement the examples presented in this book. The website www.udfcd.org provides the software and Excel spreadsheets/books that all readers and users of the UDFCD's methods can access.

Author

James C. Y. Guo earned a PhD in water resources at the University of Illinois at Champaign-Urbana. Since 1982, he has been a teaching and research faculty at the University of Colorado Denver and serves as the director of the Hydrology and Hydraulics Graduate Program. Dr. Guo is a registered licensed professional engineer and actively participates in real-world design and planning projects. He has more than 30 years of experience in hydrology, hydraulics, water resources, and groundwater modeling. His approach is to apply the concept of system to the development of hydrologic methods for engineering designs and analyses. He also incorporated these new algorithms into practical design charts, procedures, and computer software models. Many of his publications have been adopted by the drainage manuals used in Colorado, Nevada, and California for stormwater designs.

Dr. Guo is the author of numerous technical articles, training notes, and conference proceedings, and is in demand as a speaker at technical meetings and training seminars in the areas of flood mitigation and stormwater management.

Chapter 1

Urban stormwater planning

One of the major tasks in urban development is to provide adequate stormwater drainage systems to preserve the water environment and to promote the public health and economic well-being of the region. Driven by the gravity, stormwater does not follow any man-made jurisdictional boundaries and regulations of water rights as it flows through depressions, gullies, and washes, seeking an ultimate terminus such as rivers, lakes, or oceans. An engineered design is an attempt to control the drainage of stormwater while at the same time maintaining the integrity of natural flow paths and existing legal and liability relationships arising from land ownerships. It takes a joint effort to achieve such a goal, starting from the local and heading to the regional. Stormwater planning for a region can affect all governmental jurisdictions and all parcels of property in the watershed. The major characteristics of stormwater require coordination among all entities involved and cooperation from both the public and the private sectors. *Regional Master Drainage Planning* (RMDP) is an approach that integrates both local and regional efforts to identify drainage conveyance and storage facilities based on hydrological optimization and cost minimization individually and collectively.

1.1 Drainage plan

An RMDP incorporates, insofar as possible, planning completed or undertaken by the local government or land developers. It sets forth the most currently effective structural and regulatory means for improving the existing flooding conditions within an area, taking into account the possible effects under future development conditions. An RMDP shall provide basic drainage information including

1. Locations of regional drainage facilities
2. Inflow and outflow information at all design points
3. Flooding problems and future improvements
4. Estimated costs for various alternatives

An RMDP can be modified and/or revised from time to time to reflect the changes desired by the local entities as long as the intent and integrity of the RMDP are not compromised.

Under the guidance of an RMDP, a set of *Local Drainage Plans* (LDPs) can be developed as elements to the RMDP required to preserve and to promote the general health, welfare, and economic well-being of the area. An LDP sets forth the site requirements for new developments and identifies the required local public improvements. All local flood mitigation facilities must be designed in a manner to collectively achieve the regional goals stated in the RMDP. If a local flood mitigation facility impacts other entities and/or regional flood-control facilities, then the local drainage planning must be coordinated with the affected entities.

When planning urban drainage facilities, certain underlying principles and design criteria provide directions for this effort. These principles are made operational through a set of policy statements stated in the local *stormwater* and urban drainage design *criteria*. The design criteria are developed with the support of local field data and facilitated by technical reviews, permitting approvals and construction inspections for all proposed designs and constructions within the watershed. Using a comprehensive approach by involving public and private sectors, all local and regional drainage facilities are provided as the watershed is being developed in such a manner that the entire watershed is protected from the preselected flood risk.

1.2 Doctrines for surface water drainage

Under the common enemy rule, all property owners may protect themselves from flood water, but they cannot make flood water more dangerous to their neighbors. Basically, the rule includes the concepts as follows:

1. Drainage problems should not be transferred from one location to another.
2. An upstream landowner can request a drainage easement over the downstream properties, but shall not unreasonably burden the downstream properties with increased flow rates or unreasonable changes to the natural waterway from upstream properties.
3. The downstream properties cannot block natural runoff through their site and must accept runoff from upstream properties.

In an urban area, the development process alters the historic or natural drainage paths and sets the possibilities to violate the aforementioned regulations. As a result, strict compliance with the abovementioned rules can produce drainage systems that become impractical or very costly to the general public. Therefore, the concept of *Reasonable Use of Drainage* was developed to design economic and efficient drainage systems within the limits of drainage laws. The concept of *Reasonable Use of Drainage* is defined for planning purposes to provide an economically and hydraulically efficient drainage system that is demonstrated by not adversely affecting downstream properties. Under the concept of *Reasonable Use of Drainage*, new developments are allowed to occur while preserving the rights of adjacent property owners. A stormwater drainage system is an integral part of the total urbanization process. The RMDP shall be included in the regional and local land use plans with the following considerations:

1. Multiple purpose land uses
   Drainage systems require space to accommodate their conveyance and storage functions. When the space requirements are considered, the provision for adequate drainage becomes a competing use for space along with other land uses. If adequate provision is not made in a land use plan for the drainage requirements, stormwater runoff will conflict with other land uses and may result in water damage and may impair or even disrupt the functioning of other urban systems. Therefore, often a stormwater detention system is designed with public access for picnic or sports.

2. **Multiple purpose resources**

   Stormwater runoff is a resource that has the potential of being utilized for different beneficial uses. These uses, however, must be compatible with adjacent land uses and applicable water laws.

3. **Water rights**

   A drainage design must be planned and constructed taking into consideration the existing water rights and applicable water laws. When the drainage system interferes with existing water rights, the value and use of the water rights are affected.

4. **Jurisdictional cooperation**

   Because drainage considerations and problems are regional in nature and do not respect jurisdictional boundaries, drainage planning must emphasize regional jurisdictional cooperation, unified standards, and similar drainage requirements in accomplishing the goals.

1.3 Design risk and consistency

The risk-based approach applies to the selection of design storm events, based on public perception, federal regulations, watershed physical characteristics, economics, and safety. Typically, the public perceives three types of storm events: *micro, minor, and major*. The *micro rainfall events* are frequent and small events that cause inconvenience. These events occur weekly and monthly. The *minor storm events* will fill up street gutters and slow down the traffic. These events occur once per several years. For instance, a minor storm event is defined by the City of Denver, CO, as the runoff rate with a magnitude that will, on an average, be exceeded once every 5 years. The *major storm events* are perceived as occurring infrequently and have a high potential to cause major damage to public property and possibly loss of life. For instance, in the City of Denver, a major storm event is defined as the magnitude of storm runoff that will statistically be exceeded once every 100 years. Without properly designed drainage facilities, the minor storms can also cause more damage and inconvenience than the public perception would allow. Therefore, facilities should be designed to minimize public inconvenience caused by minor storm events and protect public property and life from major storm events. The federal government has recognized the need to protect the general public from catastrophic damage and destruction associated with major storm events. This recognition has resulted in the issuance of floodplain regulations, mapping, and insurance using the 100-year storm event as the base flood.

Urban drainage design takes a risk-based approach. It is important to maintain consistency in selecting design events. In an urban area shown in Figure 1.1, all infiltration facilities for water quality enhancement shall be designed for the micro event of 3–6 months. Street gutters and storm sewers are sized to pass the 2- to 5-year storm event. During a major storm event (50- to 100-year event), the excess stormwater shall be spread into the traffic lanes in the street and then conveyed to the downstream receiving water body. An urban drainage system is designed to carry a cascading flow that will fill and then overtop the micro system into the minor system. When the stormwater reaches the street, the street gutters will be filled and then intercepted by the street inlets.

With a consistent underlying risk level, the relationship between the magnitude of storm runoff and tributary area can be established throughout the watershed. This practice warrants that the cross-sections of a drainage system are increased in size in the direction of flow.

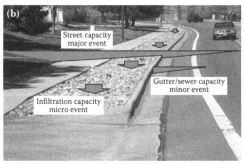

Figure 1.1 Cascading drainage system in urban area. (a) Minor–major drainage system and (b) micro–minor–major drainage system.

1.4 Common problems in urban drainage

The urbanization process generally increases runoff rates (Q-problem) and runoff volumes (V-problem) to downstream properties owing to the increases in impervious area, including more buildings, streets, and parking lots. The stormwater Q-problems are related to flood damage and public safety, whereas the V-problems are more an environmental issue related to stormwater quality.

Mitigation of Q-problems is generally accomplished through stormwater detention and/or retention facilities designed for extreme events such as 10- to 100-year events. As shown in Figure 1.2, stormwater detention is a viable method to reduce the postdevelopment peak flows. Temporarily storing stormwater runoff can significantly reduce downstream flood hazards as well as the sizes of sewers and channels required to safely convey the flood water in urban areas. A storage process also adds additional benefits to collect sediment and debris and to keep downstream channels cleaner and more efficient.

Mitigation of V-problems is often achieved with a cascading flow system to drain storm runoff generated from upstream impervious areas onto the downstream pervious areas. At the outfall point, a low-impact development (LID) device shall be installed, including porous paver, rain garden, grass swale, and infiltrating bed. An LID device is sized to capture the runoff volumes generated from frequent events such as 3- to 6-month events. An LID unit is structured with a storage layer on the surface and one or two layers of filtering media for subsurface water infiltration and filtering processes.

The urbanization process tends to transfer natural sheet flows into concentrated flows along property lines. These concentrated flows are usually generated by street gutters, storm sewers, and detention facilities. Concentrated flows released to an undeveloped downstream property can cause severe surface erosion. Mitigation of these point flows can be accomplished through energy dissipaters or flow spreaders as shown in Figure 1.2. As illustrated in Figure 1.3, urban developments also alter the natural flow paths. After the development, the streets intercept the surface runoff and relocate the flooding area from low points to the street intersections. When the outflow from an on-site drainage system does not return to its historic waterway, the flooding problem is transferred to a new location.

Figure 1.2 Drainage facilities in urban area. (a) Detention basin for peak flow reduction and (b) level spreader for infiltration bed.

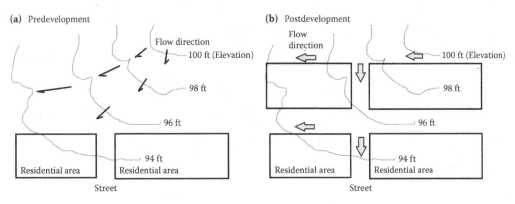

Figure 1.3 Development induced changes of flow path and flooding area. (a) Predevelopment flood flow pattern and (b) postdevelopment flood flow pattern.

Land development projects tend to achieve high density in land uses utilizing the most economic measures. Thus, floodplains become valuable if the low land areas can be reclaimed for development. The purpose of floodplain management is to provide guidance, conditions, and restrictions for development in floodplain areas while protecting the public's health, safety, welfare, and property from danger and damage. To provide impetus for proper floodplain management, the US government, acting through the *Federal Emergency Management Agency's National Flood Insurance Program (FEMA NFIP)*, has established regulations for development in floodplain areas. Compliance with these regulations allows property owners to obtain lower cost flood insurance premiums and/or eliminates the requirement for the owner to obtain flood insurance as a condition for obtaining government-supported loans. FEMA has adopted the 100-year flood (1% chance of annual occurrence) as *the base flood* for floodplain management purposes and delineates the *100-year floodplain* on their maps. For certain stream courses studied by FEMA by detailed methods, a floodway may also be depicted. The *floodway* is a portion of the floodplain and is defined as the channel itself plus any adjacent land areas that must be kept free from encroachment in order to pass the base flood without increasing water surface elevations by more than a designated height such as 1 ft in rural areas or 6 in. in urban areas.

In the mid-1980s, the *US Environmental Protection Agency* (US EPA) presented Congress with the results of the *Nationwide Urban Runoff Program* (NURP). The purpose of this study was to characterize the *quantity and quality of storm runoff* generated from urbanized areas. The results of this study showed that the process of urbanization decreases the quality and increases the quantity of storm runoff from a postdeveloped watershed. In 1987, the *Federal Clean Water Act* mandated additional regulations to control urban pollutants from entering the water environment through storm drainage facilities. These regulations are administered locally through the *National Pollutant Discharge Elimination System (NPDES) Stormwater Permitting Process*, resulting in many new designs under the concept of stormwater *Best Management Practices* (BMPs). After 2000, the concept of LID follows what we learned from stormwater BMPs to offer infiltration and filtering designs to manage urban storm runoff. The major challenge in stormwater LID designs is the selection of the proper design event. In lieu of design frequency, an urban stormwater system is composed of the *minor drainage system* designed for the 2- to 5-year events and the *major drainage system* that is able to pass the 50- to 100-year events. In the past two decades, stormwater filtering and infiltrating facilities (as shown in Figure 1.4) have been gradually added into the urban drainage systems as a *micro drainage system* that is often located upstream of the outfall point of a parking lot, business square, shopping mall, auto service site, residential subdivision, etc. These facilities appear to be grass-lined swales, depressed turf strips, vegetation beds, bush gardens, or shallow porous trenches. Stormwater from pavements will first drain onto a micro system that can only store up to the 3- to 6-month runoff events. The major functions of a micro system are twofold:

1. To infiltrate the stored water by the natural seepage process
2. To settle urban debris by the filtering process

The bypass design in a micro system allows the extreme events to overtop the infiltration basin into streets for quick collection.

Figure 1.4 LID facilities blended into street drainage system. (a) Streets without LID concept and (b) streets with LID concept.

1.5 Urban stormwater facilities

Stormwater is a natural resource for urban areas. It provides a renewal process to refresh lakes, streams, man-made greenbelts, and open space. Facing a random process, the challenge in stormwater management is to cope with the seasonal and uneven spatial distributions of rainfall and runoff amounts. A wet season produces floods, whereas a drought season creates a water shortage. Stormwater systems are vital to the reduction of flood damage and enhancement of the quality of urban water environments. As an overtopping system, an urban stormwater drainage system is composed of the following elements:

1. Watershed LID/Water quality facilities for volume reduction (V-problems)
2. Stormwater conveyance facilities through the waterway
3. Flood storage facilities at the outfall point for flow reduction (Q-problem)

Although these three systems are designed and built for different purposes, they jointly achieve the same goal that is to sustain a healthy and functional urban water environment.

1.5.1 Stormwater conveyance system

A stormwater conveyance facility is designed to collect and to pass the runoff flows on the street. Examples of stormwater conveyance facilities (Figure 1.5) include a roadside ditch, channel, street, storm sewer, and grass swale.

The primary design parameter for a conveyance facility is the peak flow rate for the design event. For instance, the cross-section of a flood channel is determined by Manning's equation as

$$V = \frac{K}{n} R^{\frac{2}{3}} \sqrt{S_o} \tag{1.1}$$

Figure 1.5 Street conveyance system. (a) Grass swale as roadside ditch and (b) trench for flow collection.

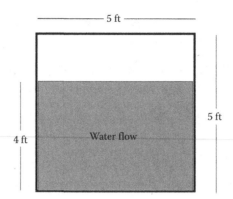

Figure 1.6 Flow in box sewer.

$$R = \frac{A}{P_w} \tag{1.2}$$

$$Q = VA \tag{1.3}$$

in which V = average velocity in [L/T], n = Manning's roughness coefficient, A = flow area in [L^2], P_w = wetted perimeter in [L], R = hydraulic radius in [L], S_o = slope of conveyance in [L/L], Q = discharge in [L^3/T], and $K = 1$ for the SI system or $K = 1.486$ for the English system. The Manning's roughness coefficient is a sensitive variable in Equation 1.1. The recommended values of Manning's roughness coefficient are 0.014 for concrete pipes, 0.016 for streets, 0.025 for corrugated pipes, 0.025–0.035 for natural swales, and 0.05–0.07 for floodplains.

EXAMPLE 1.1

A 5 ft × 5 ft box culvert (shown in Figure 1.6) carries a water flow 4 ft deep. Knowing that $n = 0.014$ and $S_o = 0.01$, determine the discharge, Q, at flow depth of 4 ft by Manning's equation.

Solution:

$A = 5 \times 4 = 20\,\text{ft}^2$

$P_w = 4 + 4 + 5 = 13\,\text{ft}$

$R = \dfrac{A}{P_w} = \dfrac{20}{13} = 1.54\,\text{ft}$

$V = \dfrac{1.486}{0.014} \times 1.54^{\frac{2}{3}} \sqrt{0.01} = 14.20\,\text{fps}$

$Q = 14.2 \times 20 = 284.3\,\text{cfs}$

1.5.2 Flood–control storage system

When the catchment is overly developed, the excess stormwater must be mitigated on the site before releasing it into the downstream properties. As a common practice, on-site flood mitigation must be provided after the development. A stormwater storage facility is an effective means by which the on-site stormwater can be temporarily stored and then gradually released at an allowable rate. Examples of stormwater storage facilities include detention basins, natural ponds, lakes, depression areas, and widened river floodplains. A stormwater detention basin is designed to have a large storage volume for controlling the extreme flood events. As illustrated in Figure 1.7, a large depressed area can be maintained as a garden area or open space that also serves as a designated area for storing flood water during a storm event.

The primary design parameter for a stormwater storage facility (Figure 1.8) is the storage volume. The cycle of loading and depletion of water volume is an unsteady process. At the beginning of an event, the storage volume continues to increase as long as the inflow is greater than the outflow. The increase of the storage volume means that there is increase of the water depth in the facility. Hydraulically, the deeper the water depth is, the greater the outflow will be. As soon as the outflow becomes greater than the inflow, the storage volume begins to deplete.

An unsteady flow is often numerically modeled by the finite difference technique. The continuous flow process is discretized by a series of time intervals, and the flow within each time interval is assumed quasi-steady. The average flow within the time interval

Figure 1.7 Detention basin for multiple land uses in urban area. (a) Basin as garden in dry days and (b) basin as storage facility in wet days.

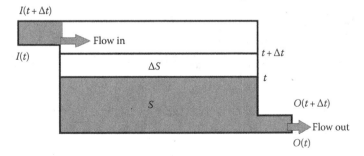

Figure 1.8 Volume balance through a storage facility.

represents the flow condition. As illustrated in Figure 1.8, the principle of continuity among inflow, outflow, and storage volume is described as follows:

$$I - O = \frac{\Delta S}{\Delta t} \tag{1.4}$$

or

$$\frac{I(t) + I(t + \Delta t)}{2} - \frac{Q(t) + O(t + \Delta t)}{2} = \frac{S(t + \Delta t) - S(t)}{\Delta t} \tag{1.5}$$

in which I = average inflow rate in $[L^3/T]$, O = average outflow rate in $[L^3/T]$, S = storage volume in $[L^3]$, ΔS = storage volume difference in $[L^3]$, and Δt = time interval in $[T]$. There are various units used in counting water volume. On top of common units in $[L^3]$, an acre-ft is equal to 43,560 ft^3, and a cfs per day (cfd) is equal to the flow rate in cubic feet per second (cfs) times 86,400 s. Similarly, a cms per day (cmd) is equal to the flow rate in cubic meter/second (cms) times 86,400 s.

EXAMPLE 1.2

A tank has a diameter of 10 ft (Figure 1.9). The initial water depth in the tank is 2 ft. The inflow to the tank is at a constant rate of 2 cfs. The outlet has a diameter of 6 in. Assuming that the outflow rate from the tank can be modeled by the orifice formula with an orifice coefficient of 0.65, determine the water depth after 5 min of operation.

Solution: According to the orifice formula, the release from the tank is calculated as

$$O = 0.65A\sqrt{2gH} = 0.65\frac{\pi(0.5)^2}{4}\sqrt{2.0 \times 32.2 \times H} = 1.03\sqrt{H}\ \text{cfs}$$

Beginning with 2 ft of water in the tank, the initial release is

$$O_1 = 1.03\sqrt{2.0} = 1.45\ \text{cfs}$$

At the end of the operation, the final release is

$$O_2 = 1.03\sqrt{H_2}\ \text{cfs}$$

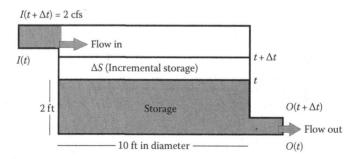

Figure 1.9 Water flowing through storage tank.

The volume in the tank is a function of water depth, and this is calculated as below:

$$S = H \times \frac{\pi \times 10^2}{4} = 78.50 H\, \text{ft}^3$$

The initial water volume in the tank is

$$S_1 = 78.50 \times 2 = 157.0\, \text{ft}^3$$

The final water volume in the tank is

$$S_2 = 78.50 H_2\, \text{ft}^3$$

With a time interval of $\Delta t = 5\, \text{min}$, the continuity principle states

$$\frac{2+2}{2} - \frac{1.45 + 1.03\sqrt{H_2}}{2} = \frac{78.50 H_2 - 157.0}{5 \times 60}$$

By trial and error, the final water depth is found to be

$$H_2 = 3.3\, \text{ft}$$

The corresponding release after 5 min is

$$O_2 = 1.87\, \text{cfs}$$

1.5.3 Watershed LID and quality enhancement system

The concept of LID provides a fundamental change in the watershed management regarding the land uses and flow systems. An LID drainage layout (as shown in Figure 1.10) is aimed at runoff volume reduction using cascading flow systems. At the outfall point, a *stormwater quality-control basin* (WQCB) is installed to enhance the water quality through the filtering and sedimentation processes. A WQCB is composed of a sand basin with an infiltrating bed for more efficient runoff volume reduction. A WQCB is typically small in size because it stores frequent events and has a bypass to release the

Figure 1.10 Example for WCQB. (a) WQCB at the outfall point and (b) LID units as runoff source control.

extreme events. The operation of a WQCB can be as long as 12–48 h. Of course, the longer the residence time, the higher the sediment trap ratio. The *stormwater runoff capture volume* (WQCV) for a WQCB is approximately equivalent to the street flush volume that is the minimum amount of water required to sweep urban streets in a catchment. For instance, a depth of 0.5 in. per catchment is considered by Boston, MA, and 1 in. per catchment is accepted by Tampa, FL, as the water quality runoff capture volume. A WQCB is designed to have a storage volume as

$$V_c = P_c \times A \qquad\qquad (1.6)$$

in which P_c = WQCV in [L] expressed in depth per catchment, V_c = water quality capture storage volume in [L^3], and A = watershed area in [L^2].

EXAMPLE 1.3

A WQCB is designed to capture the runoff volume of 1 in. from a catchment of 2 acres. The surface area of the basin is a 100 ft^2 × 100 ft^2. Determine the water depth in this basin (1 acre = 43560 sq feet).

Solution: For the given information, the storage volume is calculated as

$$V_c = P_c A = \frac{1}{12} \times 2 = 0.167 \, \text{acre-ft}$$

Considering the basin cross-sectional area of 100 ft × 100 ft, the water depth is

$$\text{Water depth} = \frac{V_c}{100 \times 100} = \frac{0.167 \times 43,560}{10,000} = 0.73 \, \text{ft}$$

1.6 Closing

For an urban area, the RMDP sets the regional level of flood protection, defines the design flows at all major design points, identifies the existing and future flood problems, recommends the mitigation measures, and reserves the land dedicated to flood-control facilities. All LDPs shall be incorporated into the RMDP to support the regional effort to preserve the water environment. From a site, the local drainage plan shall release flows no more than the allowable defined in the RMDP. Therefore, a LDP shall apply stormwater LID devices and water quality-control basins to maximize the on-site disposal of runoff volume. The RMDP outlines the best strategy for the regional stormwater detention to reduce the peak flow. The ultimate goal for the joint effort of local stormwater management and regional flood mitigation is to mimic the predevelopment watershed condition and to stabilize the flow patterns along the waterways.

1.7 Homework

Q1.1 A micro drainage system such as a rain garden shall be placed at the source of runoff. The minor system such as inlets and sewers is laid along the sidewalks. Street gutters are designed to carry the major storm event. Figure Q1.1 presents the runoff pattern through a residential site. Identify the following items in Figure Q1.1: (A) source of

Figure QI.I Runoff flow pattern through residential site.

runoff, (B) micro drainage system for runoff infiltration, (C) minor drainage system to pass the 2-year event, and (D) major drainage system to pass the 100-year event.

Bibliography

City and County of Denver, Colorado. (2010). *Urban Stormwater Drainage Criteria Manual*, Volumes 1, 2, and 3, Urban Drainage and Flood Control District, Denver, CO.

City and County of Sacramento, California. (1992). *Hydrologic Standards*, prepared by Brown and Caldwell Inc., Sacramento, CA.

City of Tucson, Arizona. (2010). *Standards Manual for Drainage Design and Floodplain Management*, prepared by Simons, Li, and Associates, Inc., Tucson, AZ.

Clark County, Nevada. (1999). *Hydrologic Criteria and Drainage Design Manual*, prepared by Montgomery Watson Inc., Las Vegas, NV.

Guo, J.C.Y. (1998). *Stormwater System Design*, Water Resources Publication, Littleton, CO.

Sheaffer, J.R., Wright, K.R., Taggart, W.C., and Wright, R.M. (1982). *Urban Storm Drainage Management*, Marcel Dekker, Inc., New York.

UDFCD. (2010). *Urban Stormwater Drainage Criteria Manual*, Volumes 1, 2, and 3, The Urban Drainage and Flood Control District, Denver, CO.

Chapter 2

Rainfall analysis

A rain gage is a basic unit installed in the field to collect precipitation information. The accuracy of the rainfall data collected at a rain gage depends on the wind speeds during the storm event and canopy effects around the rain gage. In practice, it is imperative that the raw data at a rain gage be examined and corrected before any further data analyses. Rainfall data collected from a rain gage network provide the database for regional rainfall statistical analyses. In the United States, there are published rainfall statistics available for hydrologic designs. This chapter presents the basic rainfall analytical methods that were developed to derive the rainfall statistics for flood flow predictions, including (a) *the time distribution* and *mass curve methods* for point rainfall analyses, (b) the *intensity–duration–frequency (IDF) curves* for design storm events, and (c) *rainfall temporal distributions* for stormwater numerical simulations.

2.1 Hydrologic cycle

Hydrologic cycle is the circulation of water through the atmosphere, lands, lakes, and oceans. Figure 2.1 indicates that the hydrologic cycle involves precipitation, surface runoff, groundwater, and evaporation. Precipitation takes different forms for raindrops in the atmosphere to be lifted up, to be condensed, and then to fall onto the ground.

During this process, raindrops experience interceptions by trees, bushes, and buildings before reaching the ground. On the ground, the raindrops first infiltrate into the soils and then fill up potholes and depressed areas. The infiltrating water will either percolate through the soil layers to reach the groundwater table or become subsurface runoff to drain into the nearby streams. During a dry season, the groundwater table is the source for base flows in rivers and lakes. *Rainfall excess*, also called *runoff depth*, is the amount of precipitation that survives the hydrologic losses and produces overland runoff toward streams, creeks, rivers, lakes, and oceans. Oceans and lakes are the major sources for evaporation, which is the process of converting the liquid water on the ground into moisture back to the atmosphere.

2.2 Formation of precipitation

Precipitation and evaporation are the important phenomena in the hydrologic cycle. Evaporation from ocean surfaces is the main source of moisture for precipitation. Approximately one-fourth of the total precipitation that falls on the continental areas is returned to the seas by direct runoff flows. In general, the location of a region with

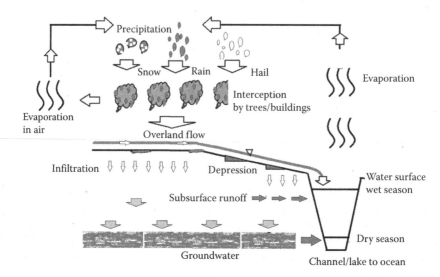

Figure 2.1 Illustration of the hydrologic cycle.

Table 2.1 Terminal velocities of raindrops at various sizes

Raindrop diameter (mm)	Terminal velocity (m/s)
0.51	2.06
1.00	4.03
1.50	5.41
2.00	6.49
3.00	8.10
4.00	8.83
5.00	9.10

respect to the atmospheric circulation system, latitude, and distance to a moisture source is the primary parameter for the climate. However, places near oceans may not have adequate precipitation, as evidenced by many desert lands near oceans. In addition to location, some mechanism is also required for precipitation to occur. Such a complicated process is briefly explained by the following three stages:

1. Under a nearly saturation condition in which a thermodynamic state of dew point is satisfied, condensation of air moisture may begin with the presence of freezing nuclei of various substances such as dust with a diameter of $0.1–1 \times 10^{-6}$ m.
2. Upon nucleation, the droplet grows and travels with air flows. It takes a number of collisions between the water particles before a raindrop grows to as large as 5 mm.
3. When a raindrop falls, its weight is balanced by the buoyancy and drag forces in the air. The *terminal velocity* of a raindrop tends to level off as the raindrop approaches the maximum size, because the raindrop's weight is balanced with the air resistance. Table 2.1 presents the terminal velocities for various sizes of raindrops.

2.3 Types of precipitation

Any hydrometeor is a type of precipitation due to the condensation of water vapor in the atmosphere including fog, drizzle, rain, frost, snow, hail, etc. Types of precipitation are classified by the mechanism of the lifting air flow that affects a large-scale cooling process in the atmosphere as follows:

1. *Convective precipitation* is caused by the rising of warmer, lighter air mass in colder, denser surroundings. As illustrated in Figure 2.2, this mechanism is induced by unequal heating at the ground surface and unequal cooling at the top of the air layer. This type of precipitation can have a high intensity and short duration.
2. *Cyclonic precipitation* (Figure 2.3) results from the lifting of air converting into a low-pressure area with a frontal or nonfrontal condition. Frontal precipitation results from the lifting of warm air on one side of a frontal surface over the colder, denser, and nearly stagnant air mass on the other side. *Warm-front precipitation* is formed

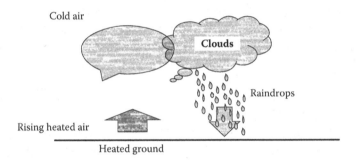

Figure 2.2 Convective precipitation.

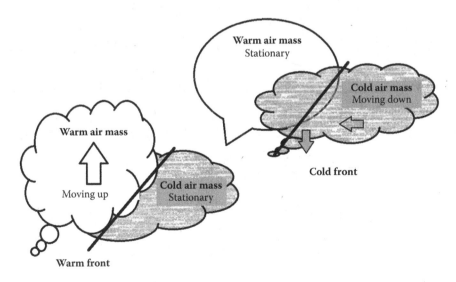

Figure 2.3 Cyclonic precipitation.

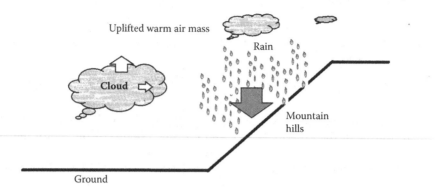

Figure 2.4 Orographic precipitation.

by the warm air moving upward over the colder air mass and is characterized by a large coverage in area and a light-to-moderate intensity. On the other hand, *cold-front precipitation* is formed by the warm air moving upward by the advancement of cold air mass. It exhibits as a long and less-intense shower.

3. *Orographic precipitation* results from warm and moist air mass lifted over mountain barriers, as illustrated in Figure 2.4. Condensation of warm air mass was forced by the cooling environment in higher layers. For instance, in the front range of the Rocky Mountains, the orographic influence is obvious.

2.4 Rainfall measurement

Many instruments and techniques have been developed for gathering information on various forms of precipitation. For the purpose of flood warning and forecasting, the total rainfall depth for an event and its temporal and spatial distributions are the most important parameters. These data can be directly measured by rain gages. The following are two basic types of rain gages:

1. A *tipping-bucket gage* (as shown in Figure 2.5) has a top orifice to collect rain-drops, a bucket to store the incremental rainfall amount, and a reservoir to store

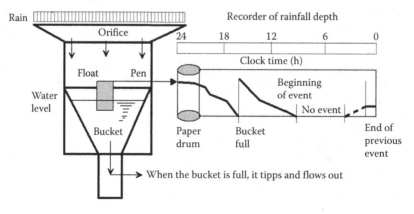

Figure 2.5 Tipping-bucket rain gage.

the accumulated rainfall volume. The rainfall caught by the orifice opening is first funneled into the bucket that will tip and pour the accumulated water into a reservoir when the depth reaches the capacity of the bucket. This process will trigger an electron pen to continuously register the accumulation of rainfall depth on a roll of paper loaded on the revolving drum.

2. A *weighting gage* weighs the rain or snow that falls into the gage orifice by a spring device. The increasing weight of the bucket continuously recorded on a chart indicates the accumulation of rainfall depth.

Operations of a rain gage during a storm involve many variables. The accuracy of precipitation measurements is subject to *operational error* and *interference error* (Curtis and Burnash, 1996). Operational errors of a rain gage are related to scale reading and instrumental errors, including evaporation from the receiver, adhesion on the funnel surface, inclination and size of the orifice, raindrop splash, etc. Interference errors are introduced by wind effects and vegetal covers. Inadvertent operational errors can be corrected or avoided, but the interference by winds at a gage site is inevitable. For instance, during the January 9–10, 1995 rain storms in Sacramento, CA, wind speeds ranged from 35 to 75 km/h continuously for several hours (Curtis and Humphrey, 1995). To reduce vegetal cover, it is preferred to have a higher orifice installed on a rain gage, although a higher orifice invites more wind effects to cause undercatch. In general, the rain gage is installed 1.0–1.5 m above the ground.

2.5 Empirical correction methods

As stated in *Field Manual for Research in Agriculture Hydrology* (Brakensiek et al., 1979), consistent rainfall records are the most significant input to a hydrologic analysis. Errors in rainfall measurements can cause serious problems in rainfall data reliability and consistency. Larson and Peck (1974) reported that the *rain undercatch* percentage for an unshielded rain gage increases at about *1.0% per every mile/h* or 2.2% per every m/s of wind speed. Kurtyka et al. (1953) suggested that two sets of undercatch rates be used: One is for *wind speed* (Table 2.2) and the other is for *gage height* (Table 2.3) (Gray, 1973; Sevruk, 1982).

Table 2.2 Undercatch under various wind speeds

Wind speed (mph)	Undercatch percentage for rain	Undercatch percentage for snow
5.00	6.00	20.00
10.00	15.00	37.00
15.00	26.00	47.00
25.00	41.00	60.00
50.00	50.00	73.00

Source: Kurtyka, J.C. et al., *Precipitation Measurement Study*, U.S. Army, Signal Corps of Engineering Laboratories, Ft. Monmouth, NJ, 1953.

Table 2.3 Rain catch percentages at various heights

Height of rain orifice	2.0 in.	6.0 in.	1.0 ft	5.0 ft	20.0 ft
Rain catch percentage	105.0	102.0	100.0	95.0	90.0

Source: Kurtyka, J.C. et al., *Precipitation Measurement Study*, U.S. Army, Signal Corps of Engineering Laboratories, Ft. Monmouth, NJ, 1953.

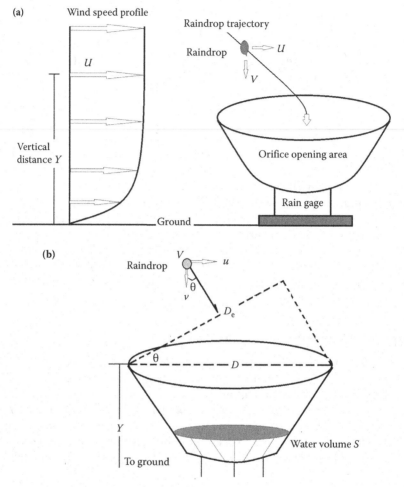

Figure 2.6 Illustration of rain gage under wind. (a) Wind speed around rain gage and (b) Rain-drop trajectory.

2.6 Rain undercatch

The trajectory of a raindrop approaching the orifice opening is important to the rain catch. Such an angle (as illustrated in Figure 2.6) depends on the raindrop's velocity components, as described below:

$$\overline{V} = ui + vi \tag{2.1}$$

$$\left|\overline{V}\right| = u^2 + v^2 \tag{2.2}$$

$$\theta = \tan^{-1}\left(\frac{v}{u}\right) \tag{2.3}$$

in which $\overline{V}$ = raindrop velocity in [L/T], u = horizontal velocity component in [L/T], v = vertical velocity component in [L/T], and θ = incoming angle. It is reasonable to assume that raindrops develop the terminal velocity, v, in the vertical direction when they approach the ground:

$$v = \left[\frac{4gd}{3C_d} \left(\frac{\rho_w}{\rho_a} - 1 \right)^{\frac{1}{2}} \right] \qquad (2.4)$$

where g = gravitational acceleration in $[L/T^2]$, d = diameter of raindrop in $[L]$, C_d = drag coefficient such as 0.67 for a 2-mm raindrop, ρ_a = density of air in $[M/L^3]$, and ρ_w = density of water in $[M/L^3]$. As shown in Table 2.1, the terminal velocity of approximately 6.0 mps is observed for the average size of raindrops of 2 mm in diameter (Chow et al., 1988).

The trajectory of a raindrop is also subject to the horizontal wind speed. The prevailing direction of air flow in a turbulent boundary layer is parallel to the ground, and its velocity profile can be described by (Guo, 2001):

$$U = m \log Y + n \qquad (2.5)$$

where U = wind speed in $[L/T]$ at vertical distance, Y is $[L]$, above the ground, and n and m = empirical parameters, depending on the turbulent flow velocity profile. As illustrated in Figure 2.7, to analyze the momentum exchange between the air flow (wind) and the raindrop, let us add "$-u$" to the entire flow field. In doing so, the raindrop becomes stationary, and its momentum exchange is caused by the drag force in the horizontal direction:

$$\rho_a A (U - u)^2 = C_d \frac{\rho_a u^2}{2} A \qquad (2.6)$$

where A = projected area of raindrop in $[L^2]$. Rearranging Equation 2.6 yields

$$u = \frac{1}{\left(\sqrt{0.5 C_d} + 1 \right)} U = KU \qquad (2.7)$$

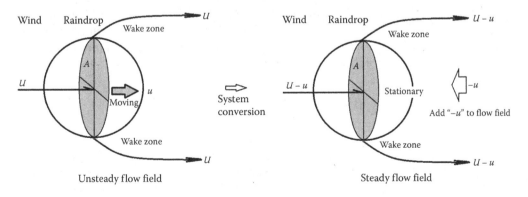

Figure 2.7 Exchange of momentum between wind and raindrop.

For instance, when $C_d = 0.67$, we have

$$K = \frac{1}{\left(\sqrt{0.5C_d} + 1\right)} = 0.63 \tag{2.8}$$

$$u = 0.63U \tag{2.9}$$

where K = horizontal velocity ratio.

Aided by Equations 2.2, 2.4, and 2.8, the velocity of a raindrop, $\vec{V}$, shortly above the rim of a rain gage can be estimated as

$$\vec{V} = ui + vj = KUi + vj \tag{2.10}$$

The raindrop speed, V in [L/T], is

$$V = |\vec{V}| = \sqrt{K^2U^2 + v^2} \tag{2.11}$$

In case that the rain gage orifice opening area is obstructed, the effective diameter, D_e in [L], as shown in Figure 2.6, of a rain gage orifice is defined as

$$D_e = (1-k)D\sin\theta \tag{2.12}$$

in which k = vegetal cover factor as a reduction to the orifice diameter due to vegetal coverage or other obstructions and D = diameter of gage orifice in [L]. A cover factor varies between zero for a clear condition and one for an entire coverage. As a result, the effective orifice area, A_e, in the direction perpendicular to the raindrop velocity is

$$A_e = \frac{\pi[(1-k)D\sin\theta]^2}{4} = A_o[(1-k)\sin\theta]^2 \tag{2.13}$$

in which A_o = opening area of rain gage orifice in [L^2]. The captured rainfall volume, V_c in [L^3], by the rain gage over a period of time is

$$V_c = \eta V A_e T_d \tag{2.14}$$

in which η = areal density of raindrops on the rain gage orifice and T_d = rainfall duration in [T]. Because raindrops do not form a continuous rate of flow through the orifice, the value of η reflects the intensity of rain intensity, heavy or light. The corresponding rainfall intensity, I in [L/T], over its duration, T_d, is

$$I = \frac{V_c}{A_o T_d} = \frac{\eta V A_e T_d}{A_o T_d} = \eta V \frac{A_e}{A_o} = \eta V \left(\frac{D_e}{D}\right)^2 = \eta V \left[(1-k)\sin\theta\right]^2 \tag{2.15}$$

The value of η can be calibrated by Equation 2.14 when rainfall intensity and raindrop velocity are measured. When a rain gage is free from wind and vegetal coverage effects, with $U = u = 0$, $k = 0$, and $\theta = 90°$, Equations 2.11 and 2.15 are reduced to

$$V_0 = v \tag{2.16}$$

$$I_0 = \eta v \tag{2.17}$$

Considering that V_0 and I_0 represent the measurements on the ground, the *rain catch rate*, R, at an elevated rain gage can be expressed as a ratio to Equation 2.17 as

$$R = \frac{I}{I_0} = \frac{V}{v}\left[(1-k)\sin\theta\right]^2 \tag{2.18}$$

If $k = 0$, Equation 2.18 is reduced to

$$R = \frac{V}{v}(\sin\theta)^2 \tag{2.19}$$

By definition, the *rain undercatch rate*, r, at a rain gage is

$$r = 1 - R \tag{2.20}$$

EXAMPLE 2.1

Consider a situation in which the wind speed varies between 8.0 feet per second (fps) at 1 ft above the ground and 16.5 fps at 10 ft above the ground. According to Equation 2.6, $m = 8.0$ and $n = 8.50$. Therefore, the wind velocity profile is described as

$$u = 8.0 \log Y + 8.5 \tag{2.21}$$

The horizontal velocity ratio between the air flow and raindrop is considered as 0.63 by Equation 2.9. The terminal velocity for a 2-mm raindrop is approximately 19.68 fps by Equation 2.4. Under this circumstance, the rain undercatch rates at various heights are estimated by Equation 2.20. As shown in Table 2.4, a rain gage at 5.0 ft above the ground will catch 91% of the actual rainfall amount. The experience of 14% rain catch reduction for the rain gage installed 20 ft above the ground is reproduced in this case. Table 2.4 shows good agreements between this case and Symon's data (Kurtyka et al., 1953; Curtis and Burnash, 1996).

Table 2.4 Rain capture rate for 2-mm raindrops

Vertical distance above ground, Y (ft)	Wind speed, U (fps)	Raindrop horizontal velocity, u (fps)	Total raindrop velocity, V (fps)	Raindrop incoming angle (°)	Rain capture rate, R	Symons field data (1880)
On ground	0.00	0.00	19.68	90.00	1.00	1.00
1.00	8.51	5.36	20.40	74.78	0.96	0.95
5.00	14.10	8.88	21.59	65.72	0.91	0.91
10.00	16.51	10.40	22.26	62.16	0.88	0.88
15.00	17.91	11.28	22.68	60.17	0.87	0.87
20.00	18.90	11.91	23.00	58.81	0.86	0.85

Table 2.5 Rain capture under vegetal and wind effects for 2-mm raindrops

Vertical distance, Y (ft)	Raindrop horizontal velocity, u (fps)	Total raindrop velocity, V (fps)	Raindrop incoming angle, θ (°)	Cover factor, k	Effective orifice diameter, D_e (ft)	Rain capture rate, R	Rain undercatch rate, I−R
0.00	0.00	19.68	90.00	0.00	1.00	1.00	0.00
1.00	5.36	20.40	74.78	0.10	0.96	0.78	0.22
1.00	5.36	20.40	74.78	0.20	0.96	0.62	0.38
5.00	8.88	21.59	65.72	0.00	0.91	0.91	0.09
5.00	8.88	21.59	65.72	0.10	0.91	0.74	0.26
5.00	8.88	21.59	65.72	0.20	0.91	0.58	0.42
5.00	8.88	21.59	65.72	0.50	0.91	0.23	0.77
5.00	8.88	21.59	65.72	0.75	0.91	0.06	0.94
5.00	8.88	21.59	65.72	1.00	0.91	0.00	1.00

Table 2.5 shows the analyses of vegetal coverage effects under the same wind speed profile as used in Table 2.4. It indicates that the rain catch rate of a gage at 5.0 ft above the ground will reduce to 58% when the gage has a 20% vegetal coverage. In comparison, a vegetal coverage has more impact on rain undercatch than wind speed. Table 2.5 is the comparison between Equations 2.19 and 2.18 for raindrops with a diameter of 2.5 mm. Again, good agreements are achieved.

2.7 Rainfall analysis

2.7.1 Continuous record

A rainfall record is continuous in time, as shown in Figure 2.8. How to define the beginning and the end of a rainfall event depends on the *minimum interevent time* used to separate events. A minimal interevent time is defined as the minimum period of time with no rain. Referring to Figure 2.8, using a minimal interevent time of 6 h, Groups A and B shall be considered as a single event, and so are Groups D and E.

Having a continuous record separated into individual events, the *event duration* is defined as the period of time from the beginning to the end of the event. An event rainfall volume is expressed in *depth per unit area*. The distribution of the incremental rainfall

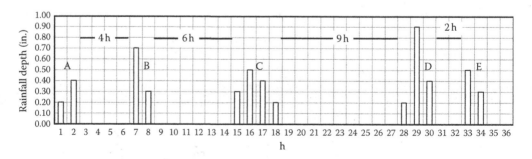

Figure 2.8 Continuous rainfall records.

depths (rainfall blocks) with respect to time is called *hyetograph* or rainfall distribution. A rainfall event is often recorded by the incremental amounts in time and then analyzed to obtain (1) *mass curve* and (2) *intensity–duration curve.*

2.7.2 Incremental rainfall distribution and mass curve

A *rainfall time distribution* represents the sequential incremental rainfall amounts recorded according to clock time. The *cumulative rainfall distribution* is the mass curve for the event that can be derived as

$$P(t) = \sum_{i=1}^{i=N} \Delta p(i\Delta t) \tag{2.22}$$

$$t = N\Delta t \tag{2.23}$$

in which $P(t)$ = cumulative rainfall depth in [L] at time t, $\Delta p(i\Delta t)$ = ith incremental rainfall depth or ith rain block, Δt = time interval such as 5 or 15 min, and N = total number of rainfall blocks.

2.7.3 Intensity–duration curve

A *rainfall time distribution* is a continuous record in time. In practice, we may divide a continuous record into segments using the concept of duration. Duration in rainfall analyses is a window width in time to pick the most intense amount in a rainfall event. For instance, a rainfall distribution is 30 min long. Considering a duration of 10 min, every continuous period of 10 min or two 5-min rainfall blocks is a segment. A total of five 10-min segments can be derived from this 30-min storm. Among these five 10-min segments, the most intense one is selected as the representative 10-min precipitation depth observed in this storm event. Similarly, for a duration of 20 min, a total of three 20-min segments can then be derived from this 30-min storm, and the most intense one is selected for further rainfall duration–depth analyses. Under the assumption that all segments are independent, an observed storm event can be dissected into many small segments to augment the database for conservative designs. A storm event can be converted from its time distribution into rainfall depth–duration (P–D) pairs. The plot of rainfall depth (P) versus duration (D) is termed P–D *curve*. A P–D curve is an increasing function of time, starting from the highest 5-min depth to the total depth for the entire event. It is important to understand that a rainfall distribution is plotted according to clock time, and a P–D curve is plotted using durations. The ratio of rainfall depth to duration gives the rainfall intensity. The average rainfall intensity is defined as

$$I = \frac{P}{T_d} \tag{2.24}$$

in which I = average intensity in in./h or mm/h, P = precipitation depth in inch or mm, and T_d = duration in hour. The plot of rainfall intensity (I) versus duration (D) is termed an I–D *curve*. An I–D curve is a decay curve with respect to duration, starting with the highest 5-min intensity to the event average intensity.

Table 2.6 Rainfall duration analysis

Clock time	Incremental rainfall depth (in.)	Cumulative depth (in.)	Duration (min)	Highest depth (in.)	Highest intensity (in./h)
(1)	(2)	(3)	(4)	(5)	(6)
16:00	0.00	0.00			
16:05	0.05	0.05	5.00	1.11	13.32
16:10	0.11	0.16	10.00	1.66	9.96
16:15	0.25	0.41	15.00	2.07	8.28
16:20	0.41	0.82	20.00	2.42	7.26
16:25	1.11	1.93	25.00	2.67	6.41
16:30	0.55	2.48	30.00	2.92	5.84
16:35	0.35	2.83	35.00	3.07	5.26
16:40	0.25	3.08	40.00	3.18	4.77
16:45	0.15	3.23	45.00	3.23	4.31
16:50	0.05	3.28	50.00	3.28	3.94
16:55	0.05	3.33	55.00	3.33	3.63
17:00	0.05	3.38	60.00	3.38	3.38

EXAMPLE 2.2

Table 2.6 presents a rainfall event recorded from 16:00 to 17:00. The total precipitation depth for this event is 3.38 in. for a period of 60 min. The mass curve for this event is derived in column 3. For this case, the highest 5-min rainfall depth is 1.11 in. observed at 16:25. The highest 10-min rainfall depth is the sum of 1.11 and 0.55 or 1.66 in. Similarly, the sum of the three blocks—1.11, 0.55, and 0.41 or a total of 2.07 in.—represents the 15-min rainfall depth for this case. Repeating the same procedure generates the *P–D* curve listed in column 5. The *P–D* curve can be converted into its *I–D* curve using Equation 2.24. For instance, the 5- and 10-min rainfall intensities are calculated as

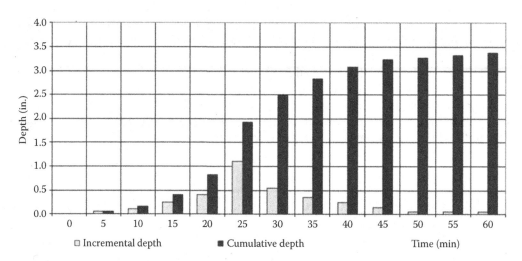

Figure 2.9 Time distributions for incremental and cumulative rainfall depths.

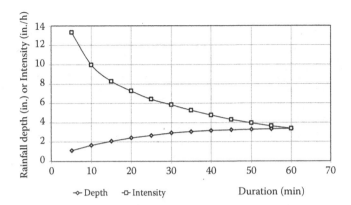

Figure 2.10 P–D and I–D curves.

$$I_5 = \frac{1.11}{5} \times 60 = 13.32\,\text{in./h}$$

$$I_{10} = \frac{1.66}{10} \times 60 = 9.96\,\text{in./h}$$

Repeating the above produces the I–D curve in column 6.

It is noticed that the mass curve (as shown in Figure 2.9) sharply increases before the peak and then becomes flatter as time increases. Figure 2.10 presents the P–D and I–D curves. Both the mass and the P–D curves appear as a cumulative. It is important to understand that the mass curve follows the clock time, while the P–D curve follows the window width in time that often starts from the peak and then symmetrically expands in both directions.

2.8 Design rainfall information

Rainfall data are available for the United States Weather Bureau in two forms: *Raw Data* and *Published Data*. Raw data are in forms of continuous records of rainfall events with various incremental time intervals. Raw data can be analyzed by the depth–duration–frequency approach and published in forms of isohyetal maps. Each map shows contours of equal precipitation depth for a specified duration and recurrence intervals. The precipitation–duration–frequency (PDF) information contains a series of generalized rainfall maps for durations from 5 min to 24 h and for return periods from 2 to 100 years.

2.8.1 Technical paper 40 (TP 40)

In 1961, *Weather Bureau Technical Paper No. 40 (TP 40)*, the results of previous Weather Bureau investigations of the precipitation–frequency regime of the conterminous United States were combined into a single publication. Investigations by the Weather Bureau during the 1950s had not covered the region between longitudes 90° and 105° W. *TP 40* has been accepted as the standard source for precipitation–frequency information in the United States since 1961. A sample of isohyetal maps is presented in Figure 2.11. *TP 40* includes isohyetal maps with return periods of 1, 2, 5, 10, 25, 50, and 100 years, and duration periods of 30 min, 1, 3, 6, 12, and 24 h.

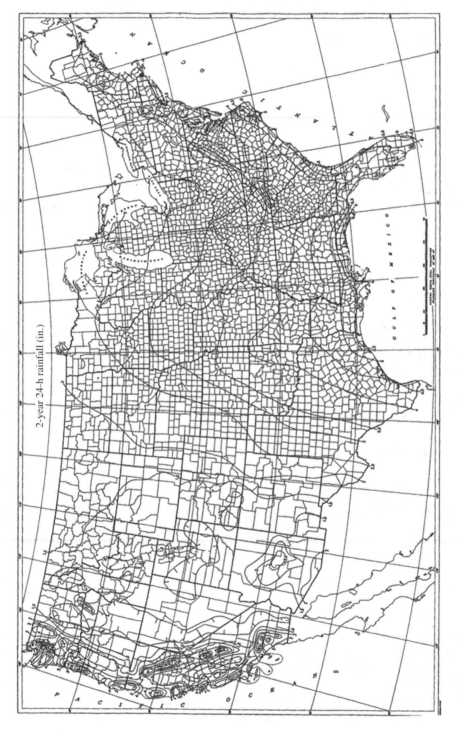

Figure 2.11 Sample of 2-year 24-h rainfall isohyetal map from TP 40.

2.8.2 Hydro 35

National Oceanic and Atmospheric Administration (NOAA) memorandum: National Weather Service HYDRO-35 was derived from *TP 40*, providing 5- to 60-min precipitation values for the eastern and central United States.

2.8.3 NOAA Atlas 2 for the 11 western states

Results presented in *TP 40* are reliable in relatively flat plains. Although the averages of point values over relatively large mountainous regions are reliable, the variations within such regions are not adequately defined. In the largest of these regions is the western United States, the topography of which plays a significant role in the incidence and distribution of precipitation. Consequently, the variations in precipitation–frequency values are actually greater than that portrayed in *TP 40*. In 1973, *Precipitation–Frequency Atlas of the Western United States, NOAA Atlas 2*, was published to refine the precipitation–frequency regime in mountainous regions of the 11 conterminous states west of approximately 103° W. Primary emphasis has been placed on developing generalized maps for precipitation of 6- and 24-h duration and for return periods of 2–100 years. The construction of isopluvial lines in mountainous regions has been done considering topography and its effect on precipitation in a general sense only. These investigations are intended to provide material for use in developing planning and design criteria of 2–100 years. Procedures have also been developed to estimate values for 1-h duration. Values for other durations can be estimated from the 1-, 6-, and 24-h duration values. Figure 2.12 is a sample map of *NOAA Atlas 2*. Table 2.7 summarizes the contents of Atlas 2 for the 11 western States.

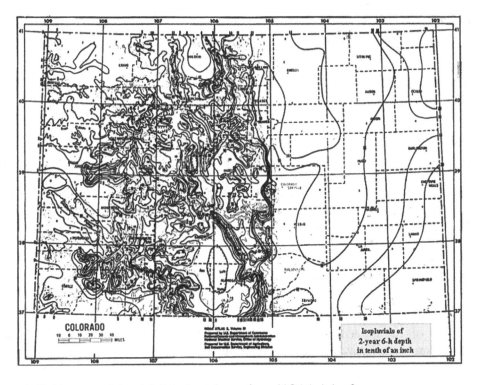

Figure 2.12 Two-year 6-h rainfall isohyetal map from NOAA Atlas 2.

Table 2.7 Design rainfall statistics National Weather Service Publications

A. *Durations to 1 day and return periods to 100 years*
   NOAA *Technical Memorandum NWS HYDRO-35 (1977)* "5–60-min Precipitation–Frequency for Eastern and Central United States"
   Technical Paper 40 covering 48 states (1961) recommended for 37 contiguous states east of the 105th meridian for durations of 2–24 h
   Technical Paper 42. Puerto Rico and Virgin Islands (1961)
   Technical Paper 43. Hawaii (1962)
   Technical Paper 47. Alaska (1963) NOAA Atlas

B. *Precipitation Atlas 2 for the 11 Western United States (1973)*
   Vol. I, Montana
   Vol. II, Wyoming
   Vol. III, Colorado
   Vol. IV, New Mexico
   Vol. V, Idaho
   Vol. VI, Utah
   Vol. VII, Nevada
   Vol. VIII, Arizona
   Vol. IX, Washington
   Vol. X, Oregon
   Vol. XI, California

C. *Durations from 2 to 10 days and return periods to 100 years*
   Technical Paper 49. 48 contiguous states (1964)
   Technical Paper 51. Hawaii (1965)
   Technical Paper 52. Alaska (1965)
   Technical Paper 53. Puerto Rico and Virgin Islands (1965)

D. *Probable maximum precipitation*
   Hydrometeorological Report 33. States east of the 105th meridian (1956)
   Hydrometeorological Report 36. California (1961)
   Hydrometeorological Report 39. Hawaii (1963)
   Hydrometeorological Report 43. Northwest States (1966)
   Hydrometeorological Report No. 49 Colorado and Great Basin Drainage
   Hydrometeorological Report No. 51 East of the 105th Meridian
   Technical Paper 38. States west of the 105th meridian (1960)

2.8.4 Probable maximum precipitation estimates

The probable maximum precipitation (PMP) presents the probable precipitation under the worst combination of meteorologic and hydrologic condition. *Hydrometeorological Report No. 51 (HMR 51)* was developed for the United States east of the 105th Meridian, and *Hydrometeorological Report No. 49 (HMR 49)* was developed for Colorado River and Great Basin Drainage. Reports applicable to other areas can be found in Table 2.7.

2.8.5 Continuous precipitation data

Under the concept of low-impact development (LID), hydrologic designs for extreme events need to be evaluated by all events observed on a long-term basis in order to detect the impact on the watershed regime. Usually, 15-min or 1-h continuous precipitation data series are recommended for continuous numerical simulations. The resultant flow frequency and duration curves serve as a basis to evaluate the performance of a hydraulic

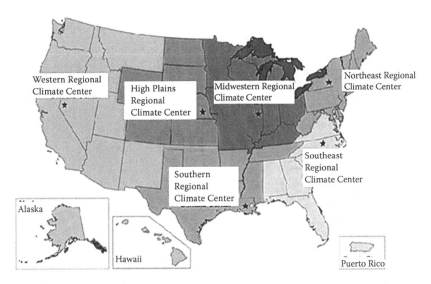

Figure 2.13 National Climate Center.

facility. Such long-term data records are provided by the National Climate Center (NCC). The service centers of NCC are presented in Figure 2.13.

EXAMPLE 2.3

Derive the PDF relationship for the City of Denver using the *NOAA Atlas 2, Volume 3*—Colorado.

1. **Geographic regions in the State of Colorado**
   The 11 US western States were separated into several geographic regions. The regions were chosen on the basis of meteorologic and climatologic homogeneity and are generally combinations of river basins separated by prominent divides. Four of these geographic regions are partially within the state of Colorado. They are shown in Figure 2.14 as
   Zone 1: South Platte, Republican, Arkansas, and Cimarron River Basins
   Zone 2: San Juan, Upper Rio Grande, Upper Colorado, and Gunnison River Basins and Green River Basin below confluence with the Yampa River
   Zone 3: Yampa and Green River Basins above confluence of Green and Yampa Rivers
   Zone 4: North Platte River Basins

2. **Derivation of PDF table**

   Step 1. Index precipitation depths
   The index precipitation values used in the *Atlas 2 Volume 3 for Colorado* include 2- and 100-year 6-h depths and 2- and 100-year 24-h depths. Based on the location of the project site, these precipitation index values can be obtained from the isohyetal maps, and then these serve as a basis to derive the 2- and 100-year 1-h precipitation values using the empirical formulas in Table 2.8.

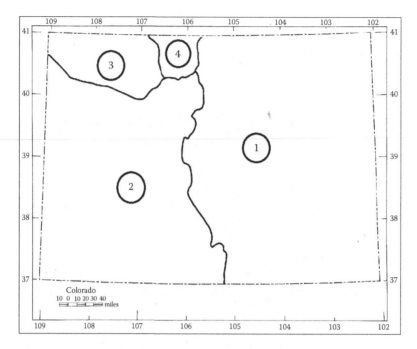

Figure 2.14 Geographic regions in the State of Colorado.

Table 2.8 Coefficients for precipitation formulas developed for Colorado

Zone	1 h						2 h		3 h	
	a	b	c	d	e	f	m	n	r	Q
1	0.218	0.709		1.897	0.439	−0.008	0.342	0.658	0.597	0.403
2	−0.011	0.942		0.494	0.755		0.341	0.659	0.569	0.431
3	0.019	0.711	0.001	0.338	0.670	0.001	0.250	0.750	0.467	0.533
4	0.028	0.890		0.671	0.757	−0.003	0.250	0.750	0.467	0.533

The Atlas 2 suggests the following empirical formulas be used to derive the precipitation values in Colorado:

2-year 1-h $\quad P_2^1 = a + b\left(\dfrac{P_2^6 \times P_2^6}{P_2^{24}}\right) + cZ$

(2.25)

100-year 1-h $\quad P_2^1 = d + e\left(\dfrac{P_{100}^6 \times P_{100}^6}{P_{100}^{24}}\right) + fZ$

(2.26)

2-h $\quad P^2 = mP^6 + nP^1$ for a specified frequency

(2.27)

3-h $\quad P^3 = rP^6 + qP^1$ for a specified frequency

(2.28)

Table 2.9 PDF table for Denver, CO

Return period (year)	Rainfall depth (in.)								
	5 min	10 min	15 min	30 min	1 h	2 h	3 h	6 h	24 h
	(1)	(2)	(3)	(4)	(5)	(6)	(7)	(8)	(9)
2	0.32	0.51	0.63	0.88	1.11	1.29	1.43	1.65	2.17
5	0.44	0.68	0.86	1.19	1.49	1.72	1.89	2.15	2.65
10	0.51	0.79	0.99	1.38	1.7	1.99	2.17	2.45	3.11
25	0.58	0.91	1.14	1.58	2.01	2.29	2.51	2.85	3.69
50	0.67	1.04	1.31	1.82	2.31	2.61	2.84	3.21	4.21
100	0.76	1.18	1.51	2.08	2.63	2.93	3.15	3.51	4.59

in which P_2^1 = 2-year 1-h precipitation values, P^2 = 2-h precipitation value, and Z = elevation in hundreds of feet. The City of Denver is located in Zone 1 with $Z = 55.0$ (at elevation of 5500 ft). The values in columns 8 and 9 of Table 2.9 are read off from the isohyetal maps in Volume 3 of Atlas 2.

Step 2. One-hour precipitation values

The 1-h 2-year precipitation depth is calculated as

$$P_2^1 = 0.218 + 0.709\left(1.65 \times \frac{1.65}{2.17}\right) = 1.11 \text{ in.}$$

The 1-h 100-year precipitation depth is calculated as

$$P_{100}^1 = 1.897 + 0.439\left(3.51 \times \frac{3.51}{4.59}\right) - 0.008 \times 55 = 2.63 \text{ in.}$$

Next, establish the straight line between the above two values on Figure 2.15. The 1-h precipitation values for other frequencies can be obtained and listed in column 5 of Table 2.9.

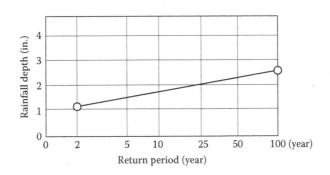

Figure 2.15 Monograph for 1-h values used in Volume 3 of Atlas 2.

Table 2.10 Ratio to 1-h values used in Volume 3 of Atlas 2

Duration (min)	5.0	10.0	15.0	30.0
Ratio to P_1	0.29	0.45	0.57	0.79

Table 2.11 IDF curve produced for Denver, CO

Return period (year)	Rainfall (inches)				Intensity (in./h)				
	5 min	10 min	15 min	30 min	1 h	2 h	3 h	6 h	24 h
	(1)	(2)	(3)	(4)	(5)	(6)	(7)	(8)	(9)
2	3.84	3.06	2.52	1.76	1.11	0.65	0.48	0.28	0.09
5	5.28	4.08	3.44	2.38	1.51	0.86	0.63	0.36	0.11
10	6.12	4.74	3.96	2.76	1.75	1.00	0.72	0.41	0.13
25	6.96	5.46	4.56	3.16	2.01	1.15	0.84	0.48	0.15
50	8.04	6.24	5.24	3.64	2.31	1.31	0.95	0.54	0.18
100	9.12	7.08	6.04	4.16	2.63	1.47	1.05	0.59	0.19

Step 3. Calculations of 2- and 3-h precipitation values
The following formulas for Zone 1 are used to calculate columns 6 and 7 in Table 2.9:

$$P^2 = 0.342P^6 + 0.658P^1 \text{ for a specified frequency} \tag{2.29}$$

$$P^3 = 0.597P^6 + 0.658P^1 \text{ for a specified frequency} \tag{2.30}$$

Step 4. Calculations of 5-, 10-, 15-, and 30-min precipitation values
The precipitation values with duration shorter than 1 h can be linearly related to the 1-h precipitation value by the ratios given in Table 2.10.

Ratios given in Table 2.10 are used to produce values in columns 1, 2, 3, and 4 in Table 2.11. Consider that the average rainfall intensity (in./h) is the ratio of precipitation (in.) to duration (h). Table 2.11 is the rainfall IDF curve produced from Table 2.9.

2.9 Seasonal variation

The maps in Atlas 2 are based on data for the entire year. In certain sections of the US West, precipitation is highly seasonal. Thus, rainy season precipitation–frequency values approach the annual values. In sections where the greatest annual n-hour precipitation amount may be observed in any season, seasonal precipitation–frequency maps would differ from those presented in this Atlas. In no case could the seasonal value be greater than the annual value. However, the seasonal values would be a certain percentage of the annual values, with the percentage varying according to the frequency of large storms during the season under investigation. Generalizations about the seasonal distribution of large storms can be obtained from U.S. *Weather Bureau Technical Paper No. 57.*

2.10 Area reduction

For a given rainfall duration and frequency, a value read from an isohyetal map is the amount of rainfall depth at the design point. For hydrologic designs, it is necessary to translate the point value into an area-averaged rainfall depth. The area covered under a storm is limited. It is implied that the entire watershed is not under a single storm. Second, the structure of a rain storm is dynamically decayed as it moves. Figure 2.16 is a recorded severe storm event at the City of Fort Collins, CO, on July 27, 1997. The rainfall depth decayed from its center of 14.5 in. over a distance of 5–10 miles. When modeling the runoff flow generated from an area of 100 mile^2 under this storm, the basic challenge is how to select the representative rainfall depth for the entire area.

The *depth-area reduction factor* (DARF) is an attempt to relate the average rainfall depth for the area covered under the storm to all the point values within the watershed. Generally, there are two types of depth-area reduction relations. The first is the *storm-centered* relation, as is the case shown in Figure 2.16, where the storm covers the entire watershed with its highest precipitation depth at the center of the area under the storm coverage. The second type is the *geographically fixed-area* relation, as is the case presented in Figure 2.17, where the storm is so displaced that only a portion of the storm covers the watershed. In comparison, the storm-centered rainfall data represent

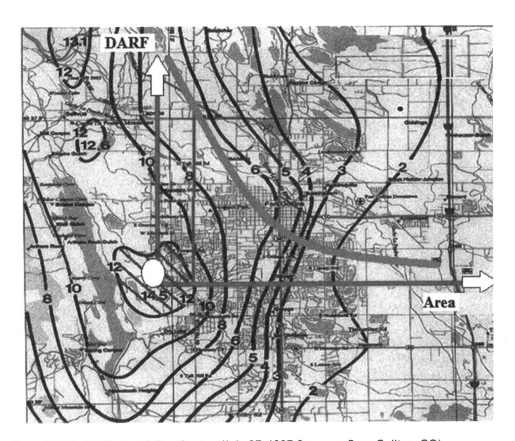

Figure 2.16 Rainfall spatial distribution (July 27, 1997 Storm at Fort Collins, CO).

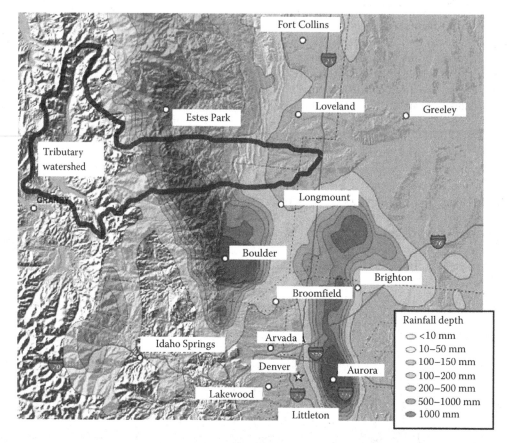

Figure 2.17 Storm coverage (September 25, 2013 Storm at Boulder, CO).

the decay profiles of discrete storms, whereas the fixed-area data are the statistical averages in which the maximum point values were not always at the center of the watershed. DARF is a statistical estimate of the area-averaged value using the point value at the center of the storm. DARF depends on the frequency and duration of the storm. Generally, the storm-centered relations are used for preparing estimates of *PMP*, whereas the geographically fixed relations are more suitable for watershed hydrologic studies using area-averaged precipitation–frequency values.

EXAMPLE 2.4

Figure 2.18 presents a set of contours of rainfall depths observed in a storm. The structure of this storm is identified with its highest depth at the center and decay curves as the distance increases from the center. Derive the DARF for this case.

Solution: The area-averaged rainfall depth is the ratio of the accumulated rainfall volume to the area under the storm. DARF is the ratio of the area-average to the central rainfall depth, which is the maximum value at the center of the storm.

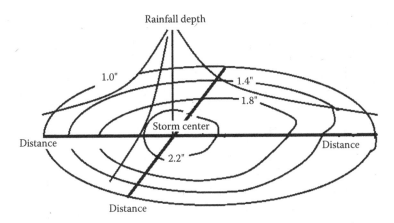

Figure 2.18 Spatial distribution of rainfall depths in storm.

Table 2.12 Derivation of DARF for $P_c = 2.2$ in.

ID for rainfall contour, j	Rainfall depth, P_j (in.)	Incremental area, ΔA_j (mile²)	Incremental volume, ΔV_j (in.-mile²)	Cumulative area, A_j (mile²)	Cumulative volume, V_j (in.-mile²)	Area depth, $\overline{P}_j$ (in.)	DARF
1	2.20	1.00	2.20	1.00	2.20	2.20	1.00
2	1.80	8.00	14.40	9.00	16.60	1.84	0.92
3	1.40	15.00	21.00	24.00	37.60	1.57	0.78
4	1.00	20.00	20.00	44.00	57.60	1.31	0.65

$$A_j = \sum_{j=1}^{j=N_j} \Delta A_j \tag{2.31}$$

$$V_j = \sum_{j=1}^{j=N_j} P_j \Delta A_j = \sum_{j=1}^{j=N_j} \Delta V_j \tag{2.32}$$

$$\overline{P}_j = \frac{V_j}{A_j} \tag{2.33}$$

$$R_j = \frac{\overline{P}_j}{P_c} \tag{2.34}$$

where N_j = jth contour of rainfall depth, A = cumulative area in $[L^2]$, V = cumulative rainfall volume in $[L^3]$, $\overline{P}$ = area-averaged depth in $[L]$, P_c = maximum rainfall depth in $[L]$ at the center of the storm, R = DARF, and Δ = incremental amount for the assigned variable and the subscript, j, represents the variables associated with the jth contour of rainfall depth. Table 2.12 is the summary of DARF calculations.

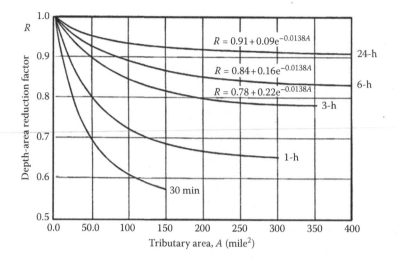

Figure 2.19 Precipitation depth-area reduction factor (DARF).

When working on a large watershed, the point rainfall depth will overestimate the runoff amount. It is necessary to apply the area-average value to the watershed. The following steps are recommended:

1. Estimate the point precipitation value at the centroid of the watershed.
2. Use Figure 2.19 to obtain a DARF required for the selected rainfall duration.
3. Multiply the point value by DARF to obtain the area-averaged rainfall depth.

DARF was derived as a decay curve for the selected duration:

$$R = \alpha + \beta A^{-\lambda} \tag{2.35}$$

where R = DARF, A = watershed area in square miles, and α, β, and λ are constants, as stated in Figure 2.19.

EXAMPLE 2.5

As illustrated in Figure 2.20, the 1-h 100-year design rainfall depth at the center of a watershed is found to be 2.6 in. Determine the areal rainfall depth for the entire watershed of 50 mile^2 under the design storm.

Solution: From Figure 2.19, DARF = 0.80 for an area of 50 mile^2 under 1-h event. Therefore, the areal precipitation depth, P_1, is

Area averaged $P_1 = 0.80 \times 2.60 = 2.08$ in.

For this case, the areal rainfall depth of 2.08 in. shall be used in the rainfall-runoff simulation study.

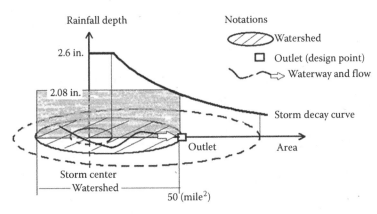

Figure 2.20 Area-averaged rainfall depth for 50 mile^2.

2.11 Design rainfall distribution

2.11.1 Twenty-four hour rainfall distribution curves

Hershield's study in 1962 led to a set of 24-h rainfall time distributions. Each rainfall distribution covers a period from clock time of 0:00 to 24:00 with the peak intensity at 12:00 or the center of the period of 24 h. The *Natural Resources Conservation Service* (NRCS) developed several Soil Conservation Service (SCS) 24-h rainfall curves to distribute the 24-h precipitation depths recommended by the National Weather Service's precipitation–frequency map. As shown in Table 2.13, these SCS synthetic 24-h rainfall distributions include Type I, Type IA, Type II, and Type IIA. The SCS 24-h distributions are expressed as $p(t)/P24$ ratios in which $p(t)$ is the cumulative amount at time, t, and $P24$ is the 24-h precipitation depth for any selected frequency. These distributions are characterized by a sharp rising curve with a major portion of design rainfall depth precipitated in the central 2 h. The SCS has also developed a nondimensional rainfall distribution using 6-h rainfall depth as listed in Table 2.13. The peak hours of the 6-h distribution is not so sharply rising as the 24-h distributions. Regional applicability of these curves is shown in Figure 2.21.

As discussed earlier, duration is the period of the burst that carries the highest intensity during the storm event. A synthetic 24-h rainfall distribution consists of bursts of 2-, 6-, and 12-h durations. A 6-h burst is the central 6 h from 9:00 to 15:00 around the peak time at 12:00, and so are the 2-h and 12-h bursts. Therefore, a 24-h rainfall curve also provides the most intense 2-, 6-, and 12-h time distributions. Figure 2.22 presents the concept of 2-, 6-, 12-, and 24-h rainfall distributions derived from the SCS Type I curve. Rainfall duration selected for hydrologic designs has to be compatible with the time of concentration of the watershed. It is critically important that a watershed is completely covered under the design storm. If not, then only a portion of the watershed is the tributary area. As a result, using the rainfall distribution shorter than the time of concentration of the watershed will underestimate the peak flow.

2.11.2 Two-hour design rainfall distributions

In the front range of the Rocky Mountain, the average size of watersheds is approximately 5–10 mile^2 because of the hilly condition. As a result, a 2-h design storm is long

Table 2.13 SCS rainfall distributions

Time (t) hour	$p(t)/P24$		Ratios		6-h curve $p(t)/P6$
	Type I	Type II	Type IA	Type IIA	
0	0	0	0	0	0
0.50	0.0080	0.0050	0.0025		0.0350
1.00	0.0170	0.0110	0.0050		0.0800
1.50	0.0260	0.0160	0.0075		0.1350
2.00	0.0350	0.0220	0.0100	0.0500	0.2300
2.50	0.0450	0.0280	0.0150		0.6000
3.00	0.0550	0.0350	0.0200		0.7050
3.50	0.0650	0.0410	0.0250		0.7800
4.00	0.0760	0.0460	0.0300	0.0750	0.8350
4.50	0.0870	0.0560	0.0500	0.1400	0.8800
5.00	0.0990	0.0630	0.0600	0.1600	0.9250
5.50	0.1220	0.0710	0.1000	0.1900	0.9650
6.00	0.1250	0.0800	0.7000	0.2200	1.0000
6.50	0.1400	0.0890	0.7500	0.2500	
7.00	0.1560	0.0980	0.7800	0.2750	
7.50	0.1740	0.1090	0.800	0.3650	
8.00	0.1940	0.1230	0.8200	0.4500	
8.50	0.2190	0.1330	0.8300	0.485	
9.00	0.2540	0.1470	0.8400	0.5250	
9.50	0.3030	0.1630	0.8500	0.5500	
10.00	0.5150	0.1810	0.855	0.575	
10.50	0.5830	0.2040	0.8600	0.6050	
11.00	0.6240	0.2350	0.8650	0.6250	
11.50	0.6540	0.2830	0.8850	0.6500	
12.00	0.6820	0.6630	0.8900	0.6750	
12.50	0.7050	0.7350	0.9000	0.6900	
13.00	0.7270	0.7720	0.9050	0.7100	
13.50	0.7460	0.8000	0.9100	0.7250	
14.00	0.7670	0.8200	0.9150	0.7400	
14.50	0.7840	0.8400			
15.00	0.8000	0.8540			
15.50					
16.00	0.8300	0.8800	0.9400	0.8000	
16.50	0.8440	0.8910			
17.00	0.8570	0.9020			
17.50					
18.00	0.8820	0.9200			
18.50	0.8930	0.9290			
19.00	0.9050	0.9370			
19.50	0.9160	0.9450			
20.00					
20.50	0.9360	0.9590			
21.00	0.9460	0.9650			
21.50	0.9550	0.9720			
22.00					
22.50	0.9740	0.9840			
23.00					
23.50	0.9920	0.9950			
24.00	1.0000	1.0000	1.0000	1.0000	

Figure 2.21 Applicability of SCS 24-h rainfall distributions.

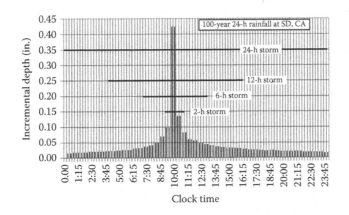

Figure 2.22 Duration of design storm.

enough to cover the entire watershed as a tributary area to the peak flow at the outlet. Using the 1-h rainfall depth, P_1, as the index precipitation, the IDF curve derived for the Denver area is

$$I(\text{in./h}) = \frac{28.5 P_1}{(10 + T_d)^{0.789}} \tag{2.36}$$

where I = rainfall intensity in in./h, P_1 = 1-h precipitation depth in inches (see Table 2.10), and T_d = rainfall duration in minutes from 5 min to 24 h. The value of P_1 represents the design frequency such as 2-, 5-, 10-, 50-, and 100-year. The 2-h rainfall distributions

Table 2.14 Design storm distribution used in the state of Colorado, USA

Time (t) (min)	$\Delta p(t)/P_1$				
	2 years	5 years	10 years	50 years	100 years
0.00	0.00	0.00	0.00	0.00	0.00
5.00	2.00	2.00	2.00	1.30	1.00
10.00	10.00	4.00	3.70	3.70	3.50
15.00	15.00	8.40	8.00	5.60	4.00
20.00	16.00	8.00	8.70	8.00	8.00
25.00	25.00	25.00	25.00	25.00	14.00
30.00	14.00	13.00	12.00	12.00	25.00
35.00	8.30	6.30	5.80	5.60	14.00
40.00	5.00	5.00	4.40	4.30	8.00
45.00	3.00	3.00	3.60	3.80	6.20
50.00	3.00	3.00	3.60	3.20	5.00
55.00	3.00	3.00	3.20	3.20	4.00
60.00	3.00	3.00	3.20	3.20	4.00
65.00	3.00	3.00	3.20	3.20	4.00
70.00	3.00	3.00	3.20	2.40	2.00
75.00	2.50	2.50	3.20	2.40	2.00
80.00	2.20	2.20	2.50	1.80	1.20
85.00	2.20	2.20	1.90	1.80	1.20
90.00	2.20	2.20	1.90	1.40	1.20
95.00	2.20	2.20	1.90	1.40	1.20
100.00	1.50	1.50	1.90	1.40	1.20
105.00	1.50	1.50	1.90	1.40	1.20
110.00	1.50	1.50	1.90	1.40	1.20
115.00	1.50	1.50	1.70	1.40	1.20
120.00	1.50	1.50	1.30	1.40	1.20

Note: The total percentage is 115% that is approximately the ratio of P_2/P_1.

are expressed as $\Delta p(t)/P_1$ in percentage versus time in Table 2.14. These design rainfall curves were derived from 73-year rainfall records. The total percentage of each rainfall distribution is 115%, which includes 100% of the design precipitation that occurs in the first 60 min and 15% of the design precipitation that occurs in the second hour. These rainfall distributions are recommended for the front range of the Rocky Mountain. In fact, these 2-h rainfall distributions are comparable to the sharp rise from 11:00 to 13:00 on the SCS Type IIA distribution.

2.12 Derivation of localized design rainfall distribution

The time distribution of an observed rainfall event can be converted into a dimensionless mass curve that plots the ratios of cumulative rainfall depth to total depth on the y-axis, and the ratios of cumulative t to event duration. Figure 2.23 presents five randomly selected storm events recorded at the Stapleton airport, Denver, CO. Comparisons with the SCS 24-h Type I and II curves suggest that the design rainfall curves are constructed using the low enveloping curve for the leading portion and the high enveloping curve for the tail portion, and then a sharp rise in between. The steeper the sharp rise is, the higher the peak discharge will be. Under the circumstance that the local rainfall data are inadequate, the conservative approach is to combine the low and high enveloping curves with a connection of sharp rise through the center of the rainfall duration.

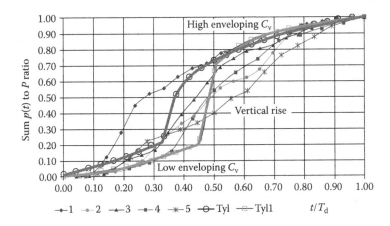

Figure 2.23 Normalized rainfall cumulative depth time distributions.

In comparison, the SCS Type I curve has a milder central rise. As expected, Type I rainfall curve produces less peak flows than the Type II curve. Therefore, Type I is recommended for the harbor, hilly city like San Diego, CA, whereas Type II is more suitable for the inland cities where the watersheds have a large tributary area (see Figure 2.21). Type I curve is also applicable to depict the distribution of winter storm, long and mild, whereas Type II is more recommended for thunderstorms that are short and intense.

EXAMPLE 2.6

As always, the basic challenge in the hydrologic design is not enough data. Table 2.15 presents the hourly incremental rainfall depths in two storm events observed on June 20 and November 23 in 2008 in Taiwan. Recommend a conservative design rainfall distribution.

Convert the two rainfall incremental depth time distributions into their mass curves. Normalize each mass curve by the total rainfall depth for the depth axis and the duration for the time axis. Plot these two normalized mass curves. Identify the low and high envelopes and the steepest connection. As shown in Figure 2.24, the solid line represents the most severe rainfall distribution that will produce the highest flood flow with 70% of the total rainfall amount blasted within 10% of the rainfall duration.

Table 2.15 Hourly incremental rainfall depths in inches for two storms in Taiwan

Date	Time (h)														
	1	*2*	*3*	*4*	*5*	*6*	*7*	*8*	*9*	*10*	*11*	*12*	*13*	*14*	*15*
June 20, 2008	0.00	0.10	0.25	0.75	1.00	1.30	0.50	0.25	0.20	0.15	0.10	0.00			
November 23, 2008	0.00	0.10	0.15	0.20	0.30	0.40	0.70	1.20	1.50	2.00	1.50	1.20	0.80	0.30	0.00

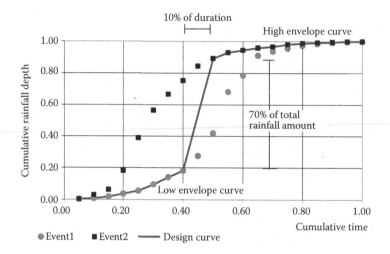

Figure 2.24 Conservative design rainfall distributions.

2.13 Closing

1. It is important to examine raw rainfall data and to make all necessary corrections before conducting rainfall duration and rainfall frequency analyses.
2. Design rainfall statistics for the United States are available on various rainfall technical reports that can be downloaded from the U.S. National Weather Service Web Sites.
3. Although there are many design rainfall distributions recommended for stormwater simulation studies, the proper design rainfall distribution for a specific project site shall be selected according to the design rainfall duration. As a rule of thumb, the rainfall duration shall be longer than the flow time through the waterway in the study area.
4. A design rainfall distribution is a composite curve that is composed of the leading and tailing envelop curves that are connected with a sharp rise. It does not require a large amount of rainfall events to derive the local design rainfall distribution. A thunderstorm shall be portrayed with a steep rise that carries 60%–70% of the total rainfall amount, whereas a winter storm shall have a mild rise carrying 40–50% of the total rainfall amount.

2.14 Homework

Q2.1 Observed incremental rainfall depths, $\Delta p(t)$ in Table Q2.1, were recorded, according to clock time, t. Your tasks are as follows:

1. Determine the cumulative rainfall time distribution, pairs of $[P(t), t]$.
2. Normalize the cumulative rainfall curve by the event duration, t/T_d, and total precipitation depth, $P(t)/P$.
3. Determine the duration–depth pairs for this case.
4. Convert the duration–depth to duration-intensity distribution.

Table Q2.1 Rainfall duration–depth analysis

Clock time, t (min)	Rainfall increment, $\Delta p(t)$ (in.)	Cumulative rain depth, P(t) (in.)	Normalized time, t/T_d	Normalized rainfall mass curve, P(t)/P	Duration, D (min)	Depth, P(D) (in.)	Intensity, I(D) (in./h)
5.00	0.01		0.08		5.00		
10.00	0.02		0.17		10.00		
15.00	0.04		0.25		15.00		
20.00	0.13		0.33		20.00		
25.00	0.22		0.42		25.00		
30.00	0.18		0.50		30.00		
35.00	0.10		0.58		35.00		
40.00	0.04		0.67		40.00		
45.00	0.02		0.75		45.00		
50.00	0.01		0.83		50.00		
55.00	0.01		0.92		55.0		
60.00	0.01		1.00		60.00		

Note: T_d = 60 min and P = 0.79 inches for this case.

Q2.2 At a rain gage, the wind speeds were measured at two points: 10 ft/s at 2 ft above the ground and 50 ft/s at 10 ft above the ground. Construct the wind speed profile for range of 1.0–15.0 ft above the ground. (Solution: $U = 10 \log y + 9$, y = vertical distance in feet).

Q2.3 Determine the terminal velocity for raindrops of 3 mm in diameter. (Solution: 8.2 m/s or 26.5 fps).

Q2.4 Consider the horizontal velocity ratio, $K = 0.65$. A rain gage is operated under the wind speed profile observed in *Q2.2*, and the raindrop terminal velocity determined in *Q2.3*. Complete Table Q2.4 for determining the rain undercatch.

Q2.5 The rainfall intensity–duration formula is given. We can construct the rainfall duration–intensity relationship for a range of duration from 5 to 50 min. The rainfall

Table Q2.4 Estimation of rain undercatch

Gage orifice above ground, Y (ft)	Wind speed at Y, U (fps)	Raindrop horizontal velocity, u = KU (fps)	Rainfall total velocity, V (fps)	Raindrop incoming angle, θ (°)	Value of $\sin(\theta)$	Vegetal cover percent, k (%)	Rain catch rate	Rain undercatch rate
1.00	9.00	5.85	27.21	77.58	0.98	0.00	0.98	0.02
3.00						0.00		
5.00	15.99	10.39	28.53	68.64	0.93	10.00	0.75	0.25
7.00						5.00		
9.00						2.00		
11.00						0.00		
13.00						0.00		
15.00						0.00		

Table Q2.5 Rainfall *I–D* curve and time distribution

Duration D (min)	I(D) (in./h)	P(D) (in.)	Δp(D) (in.)	Clock t (min)	ΔP(t) (in.)	P(t) (in.)	P(t)/P	t/T$_d$
5	8.75	0.73	0.73	5	0.11	0.11	0.045	0.1
10	6.97	1.16	0.43	10	0.15	0.26	0.106	0.2
15	5.85	1.46	0.30	15	0.23	0.38	0.155	0.3
20				20	0.43			
25				25	0.73			
30				30	0.30			
35				35	0.18			
40				40	0.13			
45				45	0.10			
50				50	0.09			

Note: T_d = 50 min and P = 2.44 inch for this case.

depth $P(D) = I(D) \times D$. The incremental rainfall depth, $\Delta p(D)$, is the difference between the two rainfall depths: $P(D)$ and $P(D + \Delta t)$.

$$I(\text{in.}/\text{h}) = \frac{74.1}{(10+D)^{0.789}} \text{ in which } D = \text{duration in minutes}$$

In order to reproduce the rainfall time distribution, we may consider the symmetric distribution with the highest $\Delta p(D)$ placed at the center of the rainfall event. Fill up Table Q2.5 to reproduce the time distribution for the rainfall event.

Q2.6 A storm event is mapped in Figure Q2.6. The circular contours represent the distribution of precipitation depth in inches decayed with respect to the radius in miles from the storm center. Construct the precipitation DARFs for this event.

Q2.7 Continue with Q2.6. The 24-h 100-year point rainfall depth at the project site is found to be 6.0 in. (A) For a tributary area of 40 mile², determine the area-average rainfall depth. (B) Distribute the rainfall depth onto the 24-h SCS Type II Rainfall Curve.

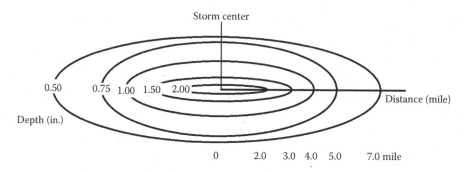

Figure Q2.6 Decay of precipitation depth versus storm coverage area.

Bibliography

Brakensiek, D.L., Osborn. H.B, and Rawls, W.J. (1979). *Field Manual for Research in Agricultural Hydrology*, Stock Number 011-000-03798-6, The Superintendent of Documents, U.S. Government Printing Office, Washington, DC.

Chow, V.T., Maidment, D.R., and Mays, L.W. (1988). *Applied Hydrology*, McGraw-Hill Publishing Company, New York.

Curtis, D.C., and Burnash, R.J.C. (1996). "Inadvertent Rain Gage Inconsistencies and Their Effect on Hydrologic Analysis," Proceedings of the 1996 California-Nevada ALERT Users Group Conference, Ventura, CA, May 15–17.

Curtis, D.C., and Humphrey, J.H. (1995). "Use of Radar-Rainfall Estimates to Model the January 9–10, 1995 Floods in Sacramento, CA," Presented at the 1995 Southwest Association of ALERT Systems Conference held in Tulsa, OK, October 25–27.

Frederick, R.H., Myers, V.A., and Auciello, E.P. (1977). "Five to 60-Minute Precipitation Frequency for the Eastern and Central United States," NOAA Technical Memorandum NWS HYDRO-35, National Weather Service, Silver Spring, MD.

Gray, D.M. (1973). *Handbook on the Principles of Hydrology*, Water Information Center, Water Research Building, Port Washington, NY.

Guo, J.C.Y. (2001). "Rain Undercatch Due to Wind and Vegetal Effects," *ASCE Journal of Hydrologic Engineering*, Vol. 6, No. 1, pp. 29–33.

Hamon, W.R. (1973). "Computing Actual Precipitation; Distribution of Precipitation in Mountainous Area," Volume 1. WMO Report, No. 362, World Meteorological Organization, Geneva, Switzerland.

Hanson, C.L., Johnson, G.L.M., and Rango, A. (1999). "Comparison of Precipitation Catch Between Nine Measuring Systems," *ASCE Journal of Hydrologic Engineering*, Vol. 4, No. 1, pp. 70–75.

HEC-1 Flood hydrograph Package. (1985). Published by the Hydrologic Engineering Center, U.S. Army Corps of Engineers, Davis, CA.

Hershfield, D.M. (1961). "Rainfall Frequency Atlas of the United States for Durations from 30 Minutes to 24 Hours and Return Periods from 1 to 100 Years," Technical paper 40, U.S. Department of Commerce, Weather Bureau, Washington, DC.

Hershfield, D.M. (1962). "Extreme Rainfall Relationships," *Journal of the Hydraulics Division American Society of Civil Engineers*, Vol. 88, No. HY6, pp. 73–92.

Hershfield, D.M. (1971). "Agricultural Research Service Precipitation Facilities and Related Studies," Report ARS 41-176, Agricultural Research Service, Washington, DC.

Kurtyka, J.C., Binks, V.M.B., and Buswell, A.M. (1953). "Precipitation Measurement Study," U.S. Army, Signal Corps of Engineering Laboratories, Ft. Monmouth, NJ, and Report of Investigation No. 20, State Water Survey Division, Urbana, IL.

Larson, L., and Peck, E.L. (1974). "Accuracy of Precipitation Measurements for Hydrologic Forecasting," *Water Resources Research*, Vol. 10, No. 4, pp. 857–863.

Liggett, J.A. (1994). *Fluid Mechanics*, McGraw-Hill, Inc., New York.

Michaud, J.D. (1994). "Effect of Rainfall-Sampling Errors on Simulations of Dessert Flash Floods," *Water Resources Research*, Vol. 30, No. 10, pp. 2765–2775.

Napor, V., and Sevruk, B. (1999). "Estimation of Wind-Induced Error of Rainfall Gauge Measurements Using a Numerical Simulation," *Journal of Atmospheric and Oceanic Technology*, Vol. 16, No. 5, pp. 450–464.

Sevruk, B. (1982). "Methods of Correction for Systematic Error in Point Precipitation Measurement for Operation Use," Operational Hydrology Report 12, World Meteorological Organization, Geneva, Switzerland.

Symons, G.J. (1866). *Notes on Some Results of Various Sets of Experimental Gages*, British Rainfall, UK.

Symons, G.J. (1880). *On the Amount of Rain Collected at Very Considerable Heights Above the Ground*, British Rainfall, UK.

Chapter 3

Watershed hydrology

A natural drainage network collects surface runoff from its watershed. A watershed is defined by its boundaries along the ridge lines. Within a watershed, all surface runoff flows will be collected into the waterways flowing toward the watershed's outlet. As shown in Figure 3.1, watersheds are formed in all shapes and sizes. They cross the boundaries among states and nations. The watershed boundary outlines the hydrologic system within which all living things are inextricably linked by the common waterway. The amount of storm runoff generated from a watershed depends on the watershed area, land uses, soil types, and depression losses on the surfaces. The movement of storm runoff through a watershed is characterized by the waterway's length, slope, floodplains, and vegetation condition.

3.1 Watershed

Watershed is also called *basin* or *catchment*. A watershed is defined by its boundaries that can be delineated from the topographic maps (as exemplified in Figure 3.2) and then verified by field inspections. The accuracy of the topographic map depends on the details of elevation contours and up-to-date developments within the watershed. *Field inspection* is always important to discover the latest natural or man-made changes to the waterways and to verify the existing conditions of the major drainage structures. For instance, the watershed depicted in Figure 3.3 exhibits several important drainage features, including

1. Natural *depression areas* such as wetlands, which delay the surface runoff movement
2. Man-made *storage facilities* such as detention basins, which reduce runoff flow rates
3. *Roads and highways*, which change the watershed boundary
4. Bridges and *crossing culverts*, which create runoff diversions
5. Any environmental changes including forest fires, landslides, etc.

3.1.1 Watershed area

The amount of runoff generated from a watershed is directly proportional to the watershed area upstream of the design point. Watershed area is the most important parameter to select a proper hydrologic method for flood flow predictions. Small watersheds can be analyzed by a linear method, whereas large watersheds are better modeled with nonlinear approaches. When conducting a hydrologic analysis, the watershed is often divided into smaller subareas, according to the drainage network, locations of hydraulic structures, and hydrologic homogeneity in land uses, vegetation, and soil types.

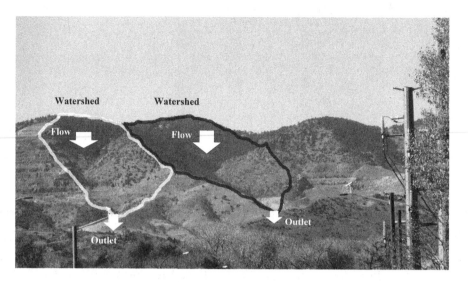

Figure 3.1 Definition of watershed.

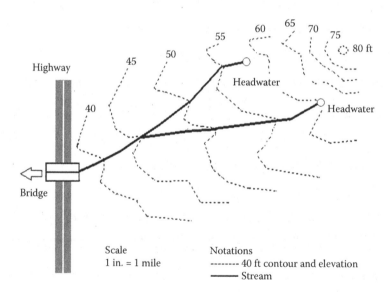

Figure 3.2 Watershed defined from topographic map.

3.1.2 Length of waterway

A natural waterway system is formed with many lateral branches. For the purpose of surface runoff predictions, the longest waterway from the most upstream boundary to the design point is selected as the representative length of the watershed. The length of a waterway has a direct impact on the flood wave travel time through the watershed. As shown in Figure 3.4, the longest waterway can further be divided into several reaches, including the most upstream reach for *overland flow*, the middle reach for *shallow swale*

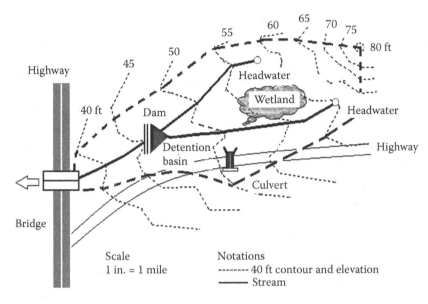

Figure 3.3 Watershed development verified by field inspection.

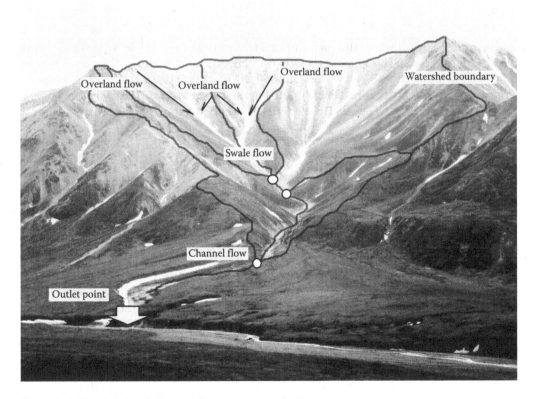

Figure 3.4 Overland and channel flows in watershed.

flow, and the downstream reach for *channel flow*. An overland flow is a two-dimensional sheet flow. A swale flow is a concentrated flow without a well-defined cross-section. A channel flow is a flow in a well-defined waterway and can be mathematically modeled by Manning's formula. The *time of concentration* is the period of time required for water to travel from the most upstream boundary to the watershed outlet. In practice, the time of concentration is the accumulated flow times through the overland, swale, and channel that reaches along the selected waterway.

On a topographic map, the reach of channel flow is often marked with a blue line. The most upstream point of a blue line or the beginning of the channel flow is termed headwater. From the headwater to the most upstream boundary of the watershed is the reach for overland and swale flows. Overland flows are shallow and slow. From field studies, the maximum length for an overland flow is approximately 300 ft in an urban area or 500 ft in a rural area. Often, street curbs and gutters are the indicators of urbanization. In an urban area, after the first 300 ft, overland flows are intercepted by street gutters and/or roadside ditches. In a rural area, after the first 500 ft, the overland flows are transformed into swale flows because the flows would have become concentrated and erosive. Swale flows will then drain into the downstream channel reach.

3.1.3 Watershed slope

Waterway slope is an important factor in determining the water flow velocity. The steeper the waterway is, the faster the water flow will be. A waterway is often divided into several segments. The slope for each segment is calculated as

$$S_i = \frac{H_i}{L_i} \tag{3.1}$$

in which S_i = slope for the ith reach in [L], H_i = vertical drop in elevation in [L], and L_i = length of reach in [L]. Equation 3.1 shall be applied to all reaches. The representative watershed slope is the weighted slope among all reaches. For instance, the *Colorado Urban Hydrograph Procedure* (CUHP) stated in the *Urban Stormwater Drainage Design Criteria Manual* (USWDCM, 2010) suggests that the watershed slope be calculated as

$$S_o = \left[\frac{1}{L} \sum_{i=1}^{i=n} \left(L_i S_i^{0.24} \right) \right]^{4.17} \tag{3.2}$$

in which S_o = waterway slope in percent, n = number of reaches and L = total length of waterway in [L].

3.1.4 Hydrologic types of soils

An infiltration rate reflects the ability of the soil medium to absorb water. This parameter is usually given in inch per hour or millimeter per hour. The Natural Resources Conservation Service (NRCSSCS, 1964) has developed a set of Soil Conservation Service (SCS) classifications on soils. In general, all soils are categorized into four hydrologic types—based on their infiltrating nature as follows:

Type A soil—soils having high infiltration rates between 0.30 and 0.45 in./h if thoroughly wetted and consisting mainly of moderately deep, well to excessively drained sand and gravel

Type B soil—soils having moderate infiltration rates between 0.15 and 0.30 in./h if thoroughly wetted and consisting mainly of moderately deep to deep, moderately well to well-drained soils with moderately fine to moderately coarse textures such as loamy soils

Type C soil—soils having slow infiltration rates between 0.05 and 0.15 in./h if thoroughly wetted and consisting mainly of soils with a layer that impedes the downward movement of water or soils with moderately fine to fine textures

Type D soil—soils having very slow infiltration rates between 0.01 and 0.05 in./h if thoroughly wetted and consisting mainly of clay soils with a swelling potential, soils with a permanent high water table, soils with a clay pan or clay layer at or near the surface, and shallow soils over nearly impervious materials

Infiltration rates are described by a decay function changing from a high rate at the beginning of the event when the soil is dry to a low rate when the soil becomes saturated. When the watershed has several types of soils, the representative infiltration rate can be determined as the area-weighted value.

3.1.5 Detention storage

A large storage volume can significantly reduce and delay the peak runoff flows. It is important to pay attention to the *depression storage capacity* in the watershed. For instance, significant storage volumes can exist along the river floodplains, natural lakes, man-made ponds, natural depression areas, and flood-control facilities. It is advisable that a significant storage area in a watershed be treated as an outlet. The storage effect on the incoming runoff must be evaluated by the flood flow routing process. The basic principle of flood routing is to solve the continuity principle with consideration to the change in the storage volume.

3.1.6 Land uses

Based on the surface textures, a watershed is divided into *impervious* and *pervious* areas. Soil infiltration losses occur only on the pervious area. The impervious area is subject to almost no infiltration loss. The runoff volume generated from an urban catchment is highly sensitive to the percentage of impervious area. The more the impervious area is, the faster the runoff flow will be. The more concentrated the surface runoff, the higher the runoff flow rates and volumes. Increase of impervious areas is an inevitable trend in an urbanized area. Many flood prediction methods directly correlate the urbanization impact on storm runoff with the percentage of impervious area in the watershed.

3.1.7 Development of watershed

Development of a watershed is a continuous process from its *historic* to *future condition*. The future condition of a watershed is often defined by the regional land use plans or city's zoning plans. To understand the current drainage problems in the watershed, the *existing condition* shall be studied. In order to mitigate a potential flooding problem, various *scenarios* shall be developed. The final selection on the remedial solution shall be based on the evaluations of their effectiveness and economics. The impact of a specific development project introduced to a watershed can be quantified by the comparison between the *predevelopment condition* without the project and the *postdevelopment* condition with the project. As a rule of thumb, a drainage design is for the purpose of not only collecting storm water but also mitigating the negative impacts due to the increased storm runoff.

Urbanization of a watershed often involves man-made drainage facilities. A location chosen for placing a drainage structure is called a *design point*. At a design point in the watershed, the design runoff discharge or volume must be known. For instance, at the site of a crossing bridge, the design discharge in the creek is a preknowledge for sizing the bridge's opening dimension. When designing a storm sewer system, the locations of manholes are design points because the sewers immediately downstream must be sized for the preknown design discharges. The concept of design point is important to the watershed analysis because the locations of design points dictate how the watershed is divided into subareas and how the flows shall be integrated along the drainage network.

3.2 Hydrologic losses

Not all raindrops can fall on the ground owing to vegetal interceptions and evaporation losses. Raindrops that reach the ground may become surface runoff after the soil losses. To model the rainfall–runoff process, it is assumed that the rainfall depth has to overcome the hydrologic losses before the overland flow occurs. The major hydrologic losses considered in the rainfall and runoff modeling techniques include *vegetal interception, depression storage, and soil infiltration.*

3.2.1 Interception losses

Interception is the portion of the precipitation that is retained by leaves and stems of vegetation, or other obstruction to prevent raindrops from falling on the ground. *Interception loss* is also called *initial loss* expressed in millimeter or inch per watershed. It was found that interception loss is proportional to the total precipitation in an event as (Horton, 1933)

$$I_a = 0.04 + 0.18P \tag{3.3}$$

in which I_a = interception loss in in. and P = precipitation in inches in Equation 3.3. In an urban area, the interception loss is relatively a small fraction of the precipitation. In practice, this amount is subtracted from the earliest precipitation amount at the beginning of the event.

3.2.2 Infiltration losses

Infiltration is the process by which surface runoff seeps into the soil throughout an event. A seepage flow may move laterally as an interflow to streams and lakes or vertically as a percolation flow to groundwater aquifers. Infiltration varies mainly according to soil texture and water content. Considering a column of soil sample, the soil porosity, θ_s, is defined as

$$\theta_s = \text{void volume / column volume ranging from 0.25 for loam to 0.45 for gravel.} \tag{3.4}$$

The water content, θ, is defined as

$$\theta = \text{water volume / void volume ranging from zero to } \theta_s. \tag{3.5}$$

Infiltration rate, $f(t)$, is the mass transfer rate in millimeter per hour or inch per hour from the water layer into the soil layer through the land surface. *Infiltration depth, $F(t)$,* is the cumulative water depth infiltrating into the soil column over a period of time, t.

At the beginning of a storm event (as shown in Figure 3.5), the initial infiltration rate, f_0, is high because the soil is not yet saturated, and then gradually decays to its final rate, f_c, when the soil is saturated.

There are several mathematical models derived to estimate infiltration rates and are briefly introduced in the following sections.

Green and Ampt

Among many empirical formulas, the *Green and Ampt formula* was developed based on the diffusion theory to describe the movement of infiltrating water through the soil column (Green and Ampt, 1911). The continuity principle for infiltrating water in the soils is derived as

$$\frac{\partial \theta}{\partial t} + \frac{\partial f}{\partial z} = 0 \tag{3.6}$$

in which f = infiltration rate in [L/T] equal to the vertical infiltrating velocity, t = time in [T], and z = vertical distance in soil medium in [L].

As illustrated in Figure 3.6, having the subsurface soil beneath the ponding area become saturated, the movement of the infiltrating water through the soil medium is dictated by the soil hydraulic conductivity and hydraulic head that includes the vertical

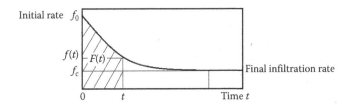

Figure 3.5 Soil infiltration rate.

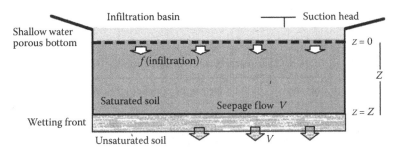

Figure 3.6 Wetting front and suction head.

distance, z, and the suction head, ψ in [L], due to soil capillary effects. A suction head is often expressed as a negative pressure in terms of water depth. According to the Darcy's law, the vertical velocity of the wetting front is

$$v = f = K_s i = K_s \left(\frac{z + \psi}{z} \right) = K_s \left(1 + \frac{\psi}{z} \right) \tag{3.7}$$

in which i = hydraulic gradient in [L/L] and K_s = hydraulic conductivity in [L/T]. Table 3.1 lists the recommended hydraulic conductivities for various soils.

After a period of time, the infiltration depth is estimated by the change of moisture content in the soil column as

$$F(t) = (\theta(t) - \theta_0) z \tag{3.8}$$

in which $F(t)$ = infiltration depth in [L] at time t, θ_0 = initial water content, $\theta(t)$ = water content at time t, and z = depth of wetting front in [L]. Rearranging Equation 3.8 yields

$$z = \frac{F(t)}{[\theta(t) - \theta_0]} = \frac{F(t)}{\Delta\theta} \tag{3.9}$$

Substituting Equation 3.9 into 3.7 yields

$$f(t) = K_s \left(1 + \frac{\psi \Delta\theta}{F(t)} \right) \tag{3.10}$$

in which $f(t)$ = infiltration rate in [L/T] at time t. From the initial water content to the saturated condition, the total vertical infiltration depth, F_s in [L], is

$$F_s = (\theta_s - \theta_0) z \tag{3.11}$$

in which θ_s = saturated water content. The final, saturated infiltration rate is

$$f_c = K_s \left(1 + \frac{\psi(\theta_s - \theta_0)}{F_s} \right) \tag{3.12}$$

The infiltration rate is varied with respect to the change of moisture content in the soil column. Equations 3.10 and 3.12 are used to determine the infiltration rate varied from the initial to the final rate.

Table 3.1 Saturated soil hydraulic conductivity

Material	Hydraulic conductivity (cm/s)
Gravel	10^{-2} to 10^{0}
Sand	10^{-3} to 10^{-1}
Silt/loam	10^{-5} to 10^{-3}
Clay	10^{-7} to 10^{-5}

Table 3.2 Recommended soil infiltration rates

Soil type	Initial rate (in./h)	Final rate (in./h)	Decay coefficient (1/h)
A	4.50	0.60	6.48
B	4.00	0.55	6.48
C	3.00	0.50	6.48
D	3.00	0.50	6.48

Horton's formula

Horton's formula is empirical. It describes the infiltration rate as an exponential decay curve as shown below:

$$f(t) = f_c + (f_0 - f_c)e^{-kt} \tag{3.13}$$

in which $f(t)$ = infiltration rate (mm/h) at time t, f_c = final rate (mm/h), f_0 = initial rate (mm/h), and k = decay constant (1/h). The design values for the parameters in Equation 3.13 are recommended in Table 3.2 (USDCM, 2010).

Infiltration amount, $F(t)$ in [L], is the area under the curve of infiltration rate over the period of elapsed time, which can be integrated as

$$F(t) = \int_{t=0}^{t} f(t)\,dt = f_c t + \frac{(f_0 - f_c)}{k}\left(1 - e^{-kt}\right) \tag{3.14}$$

When time is long enough, Equation 3.14 is reduced to

$$F(t) = f_c t + \frac{f_0 - f_c}{k} \ (t \Rightarrow \text{sufficiently long}) \tag{3.15}$$

In practice, at an elapsed time, t, Horton's formula is used to estimate the infiltration rate, $f(t)$, in Equation 3.13 and the infiltration amount, $F(t)$, in Equation 3.14. The Green and Ampt model provides the changes in the soil moisture using Equation 3.10 and the vertical movement of the wetting front in Equation 3.9.

EXAMPLE 3.1

As shown in Table 3.3, the infiltration of a soil is described by Horton's formula, with f_0 = 3.0 in./h, f_c = 0.60 in./h, and decay coefficient = 5.5/h. The soil has a porosity of 0.35, initial water content of 0.10, suction head of 3.5 in., and hydraulic conductivity of 0.39 in./h. Determine the increases of the soil moisture content and the wetting front movement during the first hour.

EXAMPLE 3.2

Continue with Example 3.1. Determine how deep the wetting front will be at $t = 120$ min.

Solution: At $t = 120$ min, $F(t) = 1.64$ in. According to Equation 3.9, the depth of saturation, z, is

$$z = \frac{F_s}{\theta_s - \theta_0} = \frac{1.64}{0.35 - 0.10} = 6.56 \text{ in.}$$

Table 3.3 Soil infiltration and movement of wetting front

Time (min)	f(t) (in./h)	F(t) (in.)	Δθ (in.)	z (in.)
	Equation 3.13	Equation 3.14	Equation 3.10	Equation 3.9
0.00	3.00	0.00	0.00	0.00
5.00	2.12	0.21	0.27	0.79
10.00	1.56	0.36	0.31	1.17
20.00	0.98	0.57	0.25	2.30
30.00	0.75	0.71	0.19	3.76
40.00	0.66	0.83	0.16	5.03
50.00	0.62	0.93	0.16	5.82
60.00	0.61	1.03	0.17	6.21
120.00	0.60	1.64	0.25	6.56

For this case, it takes 2 h to fill up the pores in the soil column. After $t > 120$ min, the infiltration rate on the land surface is reduced to the final infiltration rate. The saturated soil column acts like a pipe flow to pass the water flow at the final infiltration rate.

3.2.3 Depression losses

Rainfall excess yields overland flows that are first spread out to fill up depressed areas before the concentrated swale flows can be formed. As illustrated in Figure 3.7, *surface storage capacity* includes the following two portions: (1) *depression storage* in puddles and potholes and (2) *surface detention* under the water surface profile of the overland flow. Depression storage volumes will be part of the hydrologic losses, whereas the surface detention will be gradually released to become the recession hydrograph.

Depression storage accounts for the amount of water trapped in small puddles without running off. In practice, depression loss is the lump sum of all pothole volumes and is then expressed as an average depth over the entire watershed area in inches per watershed. Table 3.4 provides the recommended depression storage capacities for various land uses.

Values in Table 3.4 represent the *maximum depression storage volume*, D_m, available in a pothole. During an event, the *actual depression volume*, D, depends on the precipitation depth because the volume in the pothole is accumulated with respect to time. After it reaches the maximum value, overland flows will overtop the depressed area.

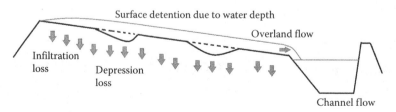

Figure 3.7 Surface storage.

Table 3.4 Recommended depression storage

Land cover	Range (in.)	Design value (in.)
Large paved area	0.05–0.15	0.10
Flat roofs	0.10–0.30	0.10
Sloped roofs	0.05–0.10	0.05
Lawn grass	0.20–0.50	0.03
Wooded area	0.20–0.60	0.40
Open fields	0.20–0.60	0.40
Sandy area		0.02
Loams		0.15
Clay		0.10

Table 3.5 Rainfall excess derived from hydrologic losses

Time, t (min)	Incremental precipitation, $\Delta P(t)$ (in.)	Initial loss, $I_a(t)$ (in.)	(2)–(3) (in.)	Incremental infiltration, $\Delta F(t)$ (in.)	(4)–(5) (in.)	Depression loss, $D(t)$ (in.)	Rainfall excess (7)–(8) (in.)
(1)	(2)	(3)	(4)	(5)	(6)	(7)	(8)
0.00	0.00	0.00	0.00				0.00
5.00	0.15	0.15	0.00				0.00
10.00	0.30	0.10	0.20	0.21	0.00[a]		0.00
15.00	0.45		0.45	0.15	0.30	0.30	0.00
20.00	0.46		0.46	0.11	0.35	0.05	0.30
25.00	0.08		0.08	0.09	0.00[a]		0.00
30.00	0.05		0.05	0.08	0.00[a]		0.00
Total	1.62	0.25	1.06	0.98	0.65	0.35	0.30

[a] Indicates that a zero replaces the negative value.

EXAMPLE 3.3

A rainfall event is observed with a total precipitation depth of 1.62 in. The rainfall distribution is described in Table 3.5. Considering the interception loss of 0.25 in., infiltration loss given by Example 3.1 and depression loss of 0.35 in. derive the amount of rainfall excess.

Solution: As illustrated in Table 3.5, it takes the first two 5-min rainfall blocks to compensate the initial loss. Furthermore, it is assumed that for each 5 min, the soils cannot infiltrate more than the available rainfall depth. For instance, at $t = 10$ min, the actual infiltrating depth is limited to the rainfall depth of 0.20 in., even though the soil is capable of infiltrating 0.21 in. At $t = 15$ min, the available rainfall depth is 0.45 in. that produces a potential runoff depth of 0.30 in. For this case, the potential runoff depth has to fill up surface depression first. As a result, at $t = 15$ min, the amount of 0.30 in. will be trapped in potholes. Because the maximum capacity of depression is 0.35 in., the additional amount of 0.05 in. applies to the next rainfall block at $t = 20$ min. Having subtracted all losses, the rainfall excess for this case is 0.3 in. with a duration of 5 min. In other words, this event is equivalent to a 5-min rainfall excess of 0.3 in.

3.3 Runoff hydrograph

Stream flows are recorded at a gage station continuously in time. Prior to a storm event, the stream gage registers the *base flow* in the river. During the storm event, the response of the watershed is registered as the *runoff hydrograph* that shows the variation of flow rates with respect to time. A *single-event hydrograph* is often used to design hydraulic structures, and the *long-term hydrographs* provide the design information for water resources planning. Unlike small catchments, stormwater characteristics in a large watershed are complicated because of the differences in elevations, slopes, soils, and vegetation covers. In that case, the delay of surface runoff becomes so significant that the interferences with subsurface and groundwater flows shall also be considered. In practice, the stream gage network in a large watershed is hardly adequate to provide complete rainfall–runoff information for the entire drainage area. Therefore, a procedure, such as *synthetic hydrograph,* needs to be developed so that the flood discharges at ungaged sites can be related to the observations at nearby gaged sites. In general, such a method requires a large amount of data and extensive efforts in calibrations.

Before it rains, the waterway carried a base flow that came from the local groundwater table. An observed runoff hydrograph consists of both the base flow and the direct flow, which are generated from the storm event. The two tangent lines (as illustrated in Figure 3.8) are used to separate the base flow from an observed runoff hydrograph. The *direct runoff volume (DRV)* is then calculated as

$$V = \int_{t=0}^{t=T_b} Q(t)\,\mathrm{d}t = \Delta t \left[\sum_{t=0}^{t=T_b} Q(t) \right] \quad \text{where } 0 \le t \le T_b \tag{3.16}$$

in which V = DRV in $[L^3]$, $Q(t)$ = direct runoff rate in $[L^3/T]$ at time t, T_s = ponding time when runoff starts, T_d = rainfall duration when rain ceases, T_c = time of concentration when peaking hydrograph starts, T_p = time to peak runoff, T_b = base time when runoff is depleted, and Δt = time interval. Equation 3.16 indicates the total DRV is equal to the sum of runoff ordinates multiplied by the time interval.

As illustrated in Figure 3.9, a *direct runoff hydrograph* (DRH) can be represented by the seven points, including

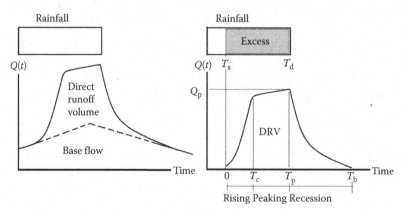

Figure 3.8 Runoff hydrograph.

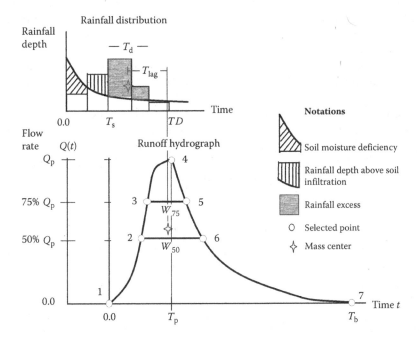

Figure 3.9 Direct runoff hydrograph.

1. The starting point at $(Q, t) = (0,0)$
2. The peak runoff rate, Q_p, at the time to peak, T_p
3. Two points to define the time-width, W_{50}, at 50% of peak runoff, Q_{50}
4. Two points to define the time-width, W_{75}, at 75% of peak runoff
5. The final point to define the base time, T_b

These seven points on the DRH are directly related to the following important time parameters on the DRH, including

1. The *ponding time*, T_s, which is defined from $t = 0$ to the beginning of runoff flow
2. The *duration of rainfall excess*, T_d, which is the period time to produce runoff flows
3. The *time to peak*, T_p, which is the period of time from the beginning to the peak flow
4. The *lag time*, T-lag, which is the time difference between the centers of rainfall excess and DRH

A database can be built based on the above information representing the observed DRH's recorded from the gaged watersheds. Regression analyses will generate the empirical formulas to reproduce the seven points on the storm hydrograph predicted for hydrologic designs and forecasting. For instance, the *CUHP* is the typical synthetic hydrograph prediction method using the regression formulas (CUHP, 2005).

3.4 Unitgraph and S-curve

The *unitgraph* is defined as a *direct runoff hydrograph* produced by *1 in. or 1 cm* of rainfall excess uniformly distributed over the entire watershed with a *specified duration*

(Sherman, 1932). Because the duration of the 1-in. excess rainfall can vary, a watershed has unlimited unitgraphs that are identified by the duration. For instance, a 10-min unitgraph means that the watershed is subject to 1 in. of excess rainfall over 10 min. As a result, its rainfall intensity is equal to 6.0 in./h. Similarly, a 30-min unitgraph is produced under a rainfall intensity of 2.0 in./h. As a result, the peak flow in the 10-min unitgraph is higher than that in the 30-min unitgraph. The major assumptions in the concept of unitgraph include

1. The linear relationship between the DRV and the runoff rates in the DRH
2. The constant lag time between the centers of the rainfall excess and the DRH
3. The applicability of linear superimposition among DRHs

A unitgraph for a watershed can be derived from (1) *observed hydrograph* or (2) *S-curve method*.

3.4.1 Unitgraph derived from observed hydrograph

A data set for deriving a unitgraph includes rainfall hyetographs, runoff hydrographs, base flows, and hydrologic losses. For each set of data, the duration and the amount of rainfall excess have to be derived from balancing the rainfall and runoff volumes. For convenience, runoff and rainfall volumes in the unitgraph method are expressed in inches or cm per watershed area. The *volume ratio* between the unitgraph and the observed direct runoff hydrograph is

$$K_v = \frac{1}{DRV} \tag{3.17}$$

where K_v = volume ratio and DRV = direct runoff volume (DRV) under the observed DRH in [L] per watershed. Based on the assumption of the linearity between runoff rate and runoff volume, the unitgraph can be derived from an observed DRH by multiplying DRH's runoff ordinates by the volume ratio, K_v. Under the assumption that the lag time is independent of runoff volume, this conversion process does not change the time scale on the observed DRH.

For mathematical convenience, a unitgraph can be converted into its *mass curve*, which is the plot of the cumulative runoff ordinates under a unitgraph. A mass curve appears like the shape of letter "S"; therefore, it is termed the *S-curve*.

EXAMPLE 3.4

Based on the volume balance analysis, it is concluded that the storm event has a rainfall excess of 0.30 in. over a duration of 5 min for the observed DRH listed in Table 3.6. Derive the unitgraph and S-curve from the observed DRH given in Table 3.6.

Solution: As stated earlier, the DRV = 0.3 in. for a duration of 5 min for this case. Therefore, the volume ratio is calculated as

$$K_v = \frac{1}{0.30} = 3.33$$

The 5-min unitgraph is derived by multiplying the ratio of 3.33 to the runoff ordinates under the observed DRH given in Table 3.6. The S-curve is then derived by accumulating the runoff

Table 3.6 Unitgraph and S-curve

Time (min)	DRH (cfs)	5-min unitgraph (cfs)	5-min S-curve (cfs)
0.00	0.00	0.00	0.00
5.00	5.00	16.50	16.50
10.00	24.00	79.20	95.70
15.00	80.00	264.00	359.70
20.00	72.00	237.60	597.30
25.00	60.00	198.00	795.30
30.00	48.00	158.40	953.70
35.00	35.00	115.50	1069.20
40.00	20.00	66.00	1135.20
45.00	10.00	33.00	1168.20
50.00	5.00	16.50	1184.70
55.00	3.00	9.90	1194.60
60.00	1.00	3.30	1197.90

DRH, direct runoff hydrograph.

ordinates for each time step. The total of runoff ordinates is amounted to 1197.90 cfs. According to Equation 3.16, the total runoff volume under this 5-min unitgraph is

$$V = \Delta t \left[\sum_{t=T_s}^{t=T_b} Q(t) \right] = (5 \times 60) \times 1197.9 = 359,370 \text{ ft}^3 \text{ under the 5-min unitgraph.}$$

Based on the definition of unitgraph, the watershed area for this case has to satisfy the following equation:

$$V = 1.0 \text{ in.} \times \text{watershed area} = 359,370 \text{ ft}^3$$

So, the watershed area = 4,312,440 ft² = 99 acres.

EXAMPLE 3.5

Use the unitgraph derived in Example 3.4 to determine the storm hydrograph for the three 5-min rainfall excess blocks as given in Table 3.7.

Solution: Column 2 in Table 3.7 lists three 5-min net rainfall depths. Column 3 is the runoff ordinates of the 5-min unitgraph. The first rainfall block of 0.2 in. produces its hydrograph as listed in column 4. Runoff rates in column 4 are generated by column 3 times 0.20 in. and then shifted by 5 min. The second hydrograph is generated by column 3 times 0.5 and then shifted by 10 min. The third hydrograph is generated by column 3 times 0.15 and then shifted by 15 min. The predicted storm hydrograph is equal to adding the three hydrographs together, according to the time step. For this case, the peak flow rate is found to be 198 cfs, which occurs at $t = 30$ min. This peak flow rate is composed of 39.6 cfs from the first hydrograph, 118.8 cfs from the second hydrograph, and 39.6 cfs from the third hydrograph.

Table 3.7 Storm hydrograph predicted by unitgraph

Time (min)	Rainfall excess (in.)	5-min unitgraph (cfs)	DRH-1 (cfs)	DRH-2 (cfs)	DRH-3 (cfs)	Storm DRH (cfs)
(1)	(2)	(3)	(4)	(5)	(6)	(7)
0.00	0:00	0.00				0.00
5.00	0.20	16.50	0.00			0.00
10.00	0.50	79.20	3.30	0.00		3.30
15.00	0.15	264.00	15.84	8.25	0.00	24.09
20.00		237.60	52.80	39.60	2.48	94.88
25.00		198.00	47.52	132.00	11.88	191.40
30.00		158.40	39.60	118.80	39.60	198.00
35.00		115.50	31.68	99.00	35.64	166.32
40.00		66.00	23.10	79.20	29.70	132.00
45.00		33.00	13.20	57.75	23.76	94.71
50.00		16.50	6.60	33.00	17.33	56.93
55.00		9.90	3.30	16.50	9.90	29.70
60.00		3.30	1.98	8.25	4.95	15.18
65.00			0.66	4.95	2.48	8.09
70.00				1.65	1.49	3.14
75.00					0.50	0.50
80.00						0.00

DRH, direct runoff hydrograph.

3.4.2 Unitgraph derived from S-curve

Unitgraphs with different durations can be derived from the S-curve. For instance, how to derive the 15-min unitgraph from the 5-min S-curve? First, we shall list two 5-min S-curves with a time shift of 15 min. The ordinates on the 15-min unitgraph are equal to the differences between the two 5-min S-curves divided by the ratio of 15/5, which is determined as

$$R = \frac{D_{\text{New}}}{D_{\text{Base}}} \tag{3.18}$$

in which R = ratio of duration, D_{New} = duration of new unitgraph to be derived, and D_{Base} = duration of the known S-curve.

EXAMPLE 3.6

Derive a 15-min unitgraph from the 5-min S-curve in Example 3.5.

Solution: Table 3.8 lists two 5-min S-curves with a time shift of 15 min. The difference between these two S-curves must be divided by the ratio as

$$R = \frac{15}{5} = 3.0$$

Table 3.8 15-min unitgraph derived from 5-min S-curve

5-min S-curve (cfs)	Shifted S-curve by 15 min (cfs)	15-min unitgraph (cfs)
(1)	(2)	[(1)–(2)]/3
0.00		0.00
16.50		5.50
95.70		31.90
359.70	0.00	119.90
597.30	16.50	193.60
795.30	95.70	233.20
953.70	359.70	198.00
1069.20	597.30	157.30
1135.20	795.30	113.30
1168.20	953.70	71.50
1184.70	1069.20	38.50
1194.60	1135.20	19.80
1197.90	1168.20	9.90
	1184.70	0.00
	1194.60	0.00
	1197.90	0.00

Column 3 in Table 3.8 is the 15-min unitgraph derived for this case. The sum of the runoff ordinates in column 3 is verified to be 1197.9 cfs or equivalent to 1 in. of water on the watershed of 99 acres. So, column 3 is the 15-min unitgraph.

3.5 Agricultural synthetic unitgraph

The *US NRCS* suggests that the runoff hydrograph from a rural watershed be approximated by a triangular shape (Figure 3.10). The NRCS *SCS unitgraph* is an empirical approach developed from forest mountain areas. As illustrated in Figure 3.10, the base time of the SCS triangular unit hydrograph is estimated as

$$T_b = \frac{5}{3} T_r \tag{3.19}$$

The volume of the SCS triangular hydrograph is

$$V = \frac{Q_p}{2} (T_r + T_b). \tag{3.20}$$

For a predetermined time to peak, T_r, in hours, the SCS unitgraph is defined to predict the peak flow rate in cfs based on 1.0 in. of net rainfall on a tributary area of 1.0 mile^2. Aided with Equations 3.19, 20, Equation 3.20 is transformed as

$$Q_p = \frac{2V}{(T_r + T_b)} = \frac{0.75V}{T_r} = 0.75 \times \left(\frac{1}{12} \frac{5280^2}{3600}\right) \frac{A}{T_r} = \frac{484A}{T_r} \tag{3.21}$$

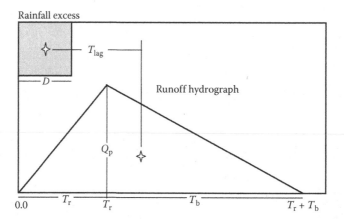

Figure 3.10 SCS triangular unitgraph.

where Q_p = peak flow rate in cfs, A = tributary area in square miles, and T_r = time to peak in hours. The lag time is defined as the time span from the center of the net rainfall excess to the time to peak. The empirical formula for such a lag time is

$$T_{lag} = \frac{L^{0.8}(S+1)^{0.7}}{1900\sqrt{S_o}} \text{ (only suitable for rural areas)} \tag{3.22}$$

where T_{lag} = lag time in hours, S_o = waterway slope in percent (%), L = waterway length in feet, and S = maximum soil retention volume, which is defined by the Curve Number (CN) as

$$S = \frac{1000}{CN} - 10 \tag{3.23}$$

CN in Table 3.9 is a special index system, which was developed to describe the soil infiltration loss used in the SCS unitgraph method. CN varies between 30 and 100. The more impervious area in the watershed, the higher the CN. According to Figure 3.10, the time to peak is calculated as

$$T_r = \frac{D}{2} + T_{lag} \tag{3.24}$$

The CNs in Table 3.9 were developed from agricultural watersheds, and they do not adequately represent the effect of impervious surfaces in urban areas. Impervious surfaces are more hydraulically efficient and often result in a shorter time to peak and a higher runoff volume. It is necessary to modify the SCS lag time for urban catchments as

$$T_{lag} = 0.6 \, T_c \text{ (for urbanized areas)} \tag{3.25}$$

in which T_c = time of concentration which is the travel time for water to flow through the waterway.

Table 3.9 SCS curve number for various land uses

Cover description		Curve numbers for hydrologic soil group			
Cover type and hydrologic condition	Average percent impervious area	A	B	C	D
Fully developed urban areas (vegetation established)					
Open space (lawns, parks, golf courses, cemeteries, etc.)					
Poor condition (grass cover <50%)		68	79	86	89
Fair condition (grass cover 50%–75%)		49	69	79	84
Good condition (grass cover >75%)		39	61	74	80
Impervious areas					
Paved parking lots, roofs, driveways, etc. (excluding right-of-way)		98	98	98	98
Streets and roads					
Paved; curbs and storm sewers (excluding right-of-way)		98	98	98	98
Paved; open ditches (including right-of-way)		83	89	92	93
Gravel (including right-of-way)		76	85	89	91
Dirt (including right-of-way)		72	82	87	89
Western desert urban areas					
Natural desert landscaping (pervious areas only)		63	77	85	88
Artificial desert landscaping (impervious weed barrier, desert shrub with 1- to 2- in. sand or gravel mulch and basin borders)		96	96	96	96
Urban districts					
Commercial and business	85	89	92	94	95
Industrial	72	81	88	91	93
Residential districts by average lot size					
1/8 acre or less (town houses)	65	77	85	90	92
1/4 acre	38	61	75	83	87
1/3 acre	30	57	72	81	86
1/2 acre	25	54	70	80	85
1 acre	20	51	68	79	84
2 acres	12	46	65	77	82

In the case that the watershed is urbanized with streets, gutters, and improved waterway, the kinematic wave method is a better approach.

EXAMPLE 3.7

Derive the SCS triangular unitgraph for the rural watershed with the following parameters: $D = 0.3\,h$, $A = 0.38$ mile^2, $L = 1.28$ mile, $L_c = 0.52$ mile, $CN = 85$ for south-west desert urban areas, and $S_o = 0.0102\,ft/ft$.

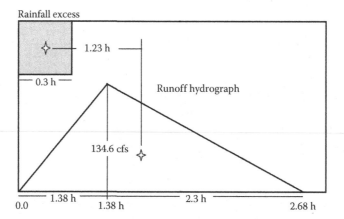

Figure 3.11 SCS triangular unitgraph for example watershed.

Solution:

$$S = \frac{1000}{85} - 10 = 1.765$$

$$T_{lag} = \frac{(1.28 \times 5280)^{0.8}(1.765+1)^{0.7}}{1900\sqrt{1.02}} = 1.23h$$

$$T_r = \frac{D}{2} + T_{lag} = \frac{0.3}{2} + 1.23 = 1.38h$$

$$T_b = \frac{5}{3}T_r = 2.3h$$

$$Q_p = \frac{484 \times 0.38}{1.38} = 134.6 \text{ cfs}$$

The hydrograph parameters for this case are plotted in Figure 3.11.

EXAMPLE 3.8

The synthetic unitgraph method was developed for large agricultural watersheds. Considering $D = 1.0h$, $A = 100$ mile^2, $L = 18$ miles, CN = 78 for south-west desert areas, and $S_o = 1.9\%$, determine the unitgraph.

Solution:

$S = 2.82$ max soil retention in in.
T-lag = 9.37 lag time in h
$T_r = 9.87$ time to peak flow in h
$T_b = 16.28$ time after peak in h
$Q_p = 4904.39$ peak flow rate in cfs
Rain volume = 5333.33 total unit volume in acre-ft
Runoff volume = 5300.00 volume before peak in acre-ft
Error in volume = 0.63%

3.6 Urban synthetic unitgraph

The CUHP was developed to predict storm runoff generated from urban areas. The method was calibrated with the rainfall–runoff data collected from several selected urban watersheds in the Metro Denver area, CO. The CUHP is recommended for hydrologic planning in the State of Colorado, USA. The CUHP applies Snyder's synthetic unit hydrograph procedures to determine the unit hydrograph for urbanized catchments. The empirical formulas are summarized as follows:

$$C_p = PC_t A^{0.15} \tag{3.26}$$

$$t_p = C_t \left(\frac{LL_c}{\sqrt{S_o}} \right)^{0.48} \tag{3.27}$$

$$q_p = 640 \frac{C_p}{t_p} \left(\text{cfs/mile}^2 \right) \tag{3.28}$$

$$T_p = 60 t_p + 0.5 t_u \ (\text{h}) \tag{3.29}$$

$$Q_p = q_p A \, (\text{cfs}) \tag{3.30}$$

$$t_u = \max\left(\frac{1}{3} t_p \times 60, 5 \right) (\text{min}) \tag{3.31}$$

in which A = drainage area (square mile), L = waterway length (mile), L_a = waterway length to the centroid of watershed (mile), S_o = watershed slope (ft/ft), C_t = coefficient for time to peak (Figure 3.12), P = coefficient for peak flow (Figure 3.13), t_u = rainfall duration (min), t_p = time to peak flow (h) from 1 mile², q_p = peak flow (cfs/mile²) from 1 mile², T_p = time to peak (min) on unitgraph, and Q_p = peak flow rate (cfs) on unitgraph.

Although Equation 3.31 is recommended for determining the rainfall duration, in practice, the selection is one of 5, 10, or 15 min. For convenience, the computational time increment is kept the same as the rainfall duration. Construction of a synthetic unitgraph requires the time widths at 50% and 75% of the peak flow (Figure 3.14). Empirical equations for these time widths are

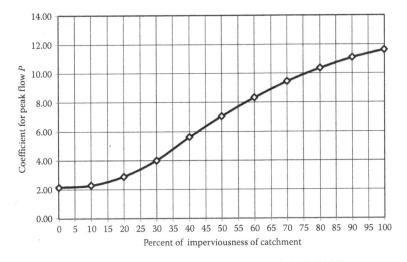

Figure 3.12 Coefficient, P, of peak flow for CUHP method (CUHP, 2005).

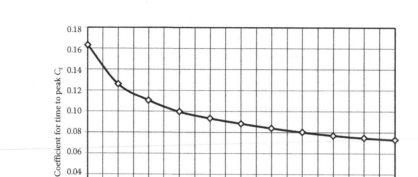

Figure 3.13 Coefficient, C_t, for time to peak in CUHP method (CUHP, 2005).

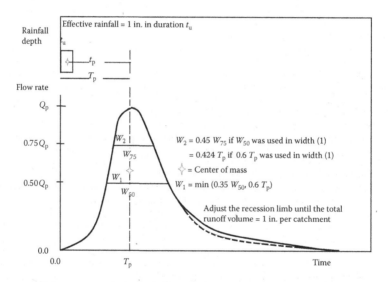

Figure 3.14 Shape of Colorado urban unitgraph (CUHP, 2005).

$$W_{50} = \frac{500}{q_p} \, (\text{h}) \tag{3.32}$$

$$W_{75} = \frac{260}{q_p} \, (\text{h}) \tag{3.33}$$

in which W_{50} = time width at 50% of peak discharge (h) and W_{75} = time width at 75% of peak discharge (h). Figure 3.14 illustrates how a unitgraph by the CUHP is defined by six points and the recession tail adjusted to satisfy the volume of 1 in. per watershed. The applicable range of CUHP is for urban watersheds between 5 and 3000 acres, and the waterway slope shall not be steeper than 6%.

EXAMPLE 3.9

Derive the 5-min unitgraph by the CUHP for the watershed with the following parameters: $A = 0.38$ mile^2, $L = 1.28$ mile, $L_c = 0.52$ mile, $I_a = 44\%$, and $S_o = 0.0102$ ft/ft.

Solutions:

Step 1: $C_t = 0.091$ based on 44% imperviousness from Figure 3.13.

Step 2: Because the watershed is larger than 90 acres, the time to peak shall be calculated by

$$t_p = 0.091 \left(\frac{1.28 \times 0.52}{\sqrt{0.0102}} \right)^{0.48} = 0.225\,h$$

Step 3: $t_u = \max\left(\frac{1}{3} t_p \times 60, 5 \right) = (4.5, 5.0) = 5.0\,min$

Step 4: $P = 6.2$ based on 44% imperviousness from Figure 3.12.

$$C_p = PC_t A^{0.15} = 6.2 \times 0.091 \times 0.38^{0.15} = 0.49$$

Step 5: $q_p = 640 \times \dfrac{0.49}{0.225} = 1394\,cfs/mile^2$

Step 6: $Q_p = 1394 \times 0.38 = 530\,cfs$

Step 7: $T_p = 60t_p + 0.5t_u = 60 \times 2.25 + 0.5 \times 5.0 = 16.0\,min$

Step 8: $W_{50} = 0.369\,h = 21.0\,min$ (7.4 min ahead of Q_p)
$W_{75} = 0.186\,h = 11.2\,min$ (5.0 min ahead of Q_p)

Step 9: The unit volume from the watershed $= 1/12 \times 0.38 \times 645 = 20.3$ acre-ft used to adjust the recession hydrograph for water volume balance.

Step 10: The unitgraph is constructed with the six points and a recession tail adjusted for 1-in. volume (Figure 3.15).

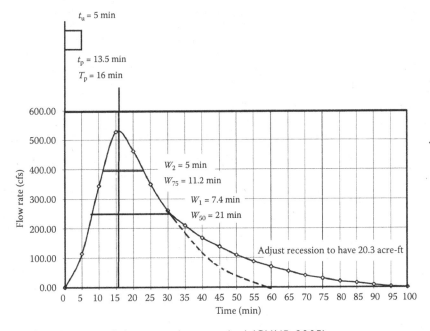

Figure 3.15 CUHP unitgraph for example watershed (CUHP, 2005).

EXAMPLE 3.10

The watershed in Example 3.9 has soil infiltration parameters as $f_0 = 3.0$ in./h, $f_c = 0.60$ in./h, and $k = 5.50$ l/h. The interception loss is 0.10 in. and depression loss is 0.10 in. The 10-year precipitation for the watershed is 1.61 in. Derive the 10-year storm hydrograph using the Denver's 10-year rainfall distribution.

1. Calculation of the 10-year rainfall excess (Table 3.10)
2. Calculation of the 10-year storm hydrograph (Table 3.11)

The individual and total hydrographs are presented in Figure 3.16.

Table 3.10 Calculation of 10-year rainfall excess for example watershed

Time, t (min)	$\Delta p(t)/PI$ (%)	Incremental precipitation, $\Delta P(t)$ (in.)	Initial loss, $\Delta I_a(t)$ (in.)	(2)−(3) (in.)	Incremental infiltration, $\Delta F(t)$ (in.)	(4)−(5) (in.)	Depression loss, $\Delta D(t)$ (in.)	Rainfall excess, (7)−(8) (in.)
(1)		(2)	(3)	(4)	(5)	(6)	(7)	(8)
0.00	0.00	0.00	0.00	0.00				0.00
5.00	2.00	0.03	0.03	0.00				0.00
10.00	3.70	0.06	0.06	0.00		0.00[a]		0.00
15.00	8.20	0.13	0.01	0.12	0.21	0.00[a]	0.00[a]	0.00
20.00	15.00	0.24		0.24	0.15	0.09	0.09	0.00
25.00	25.00	0.40		0.40	0.11	0.29	0.01	0.28
30.00	12.00	0.19		0.19	0.09	0.10	0.00	0.10
35.00	5.60	0.09		0.09	0.08	0.01	0.00	0.01
40.00	4.30	0.07		0.07	0.07	0.00[a]		0.00[a]
45.00	3.80	0.06		0.06	0.06	0.00[a]		0.00[a]
50.00	3.20	0.05		0.05	0.06	0.00[a]		0.00[a]
55.00	3.20	0.05		0.05	0.05	0.00[a]		0.00[a]
60.00	3.20	0.05		0.05	0.05	0.00[a]		0.00[a]
65.00	3.20	0.05		0.05	0.05	0.00[a]		0.00[a]
70.00	3.20	0.05		0.05	0.05	0.00[a]		0.00[a]
75.00	3.20	0.05		0.05	0.05	0.00[a]		0.00[a]
80.00	2.50	0.04		0.04	0.05	0.00[a]		0.00[a]
85.00	1.90	0.03		0.03	0.05	0.00[a]		0.00[a]
90.00	1.90	0.03		0.03	0.05	0.00[a]		0.00[a]
95.00	1.90	0.03		0.03	0.05	0.00[a]		0.00[a]
100.00	1.90	0.03		0.03	0.05	0.00[a]		0.00[a]
105.00	1.90	0.03		0.03	0.05	0.00[a]		0.00[a]
110.00	1.90	0.03		0.03	0.05	0.00[a]		0.00[a]
115.00	1.70	0.03		0.03	0.05	0.00[a]		0.00[a]
120.00	1.30	0.02		0.02	0.05	0.00[a]		0.00[a]
total	115.70	1.86	0.10	1.76	0.05	0.50	0.10	0.40

[a] 0.00 means that the rainfall excess is not enough to produce runoff.

Table 3.11 Calculation of 10-year storm hydrograph for example watershed

Time (min)	Rainfall excess (in.)	Unitgraph (cfs)	DRH-1 (cfs)	DRH-2 (cfs)	DRH-3 (cfs)	Storm DRH (cfs)
0.00	0.00	0.00				0.00
5.00	0.28	115.00	0.00			0.00
10.00	0.10	345.00	32.20	0.00		32.20
15.00	0.01	528.00	96.60	11.50	0.00	108.10
20.00		463.00	147.84	34.50	1.15	183.49
25.00		350.00	129.64	52.80	3.45	185.89
30.00		260.00	98.00	46.30	5.28	149.58
35.00		210.00	72.80	35.00	4.63	112.43
40.00		168.00	58.80	26.00	3.50	88.30
45.00		138.00	47.04	21.00	2.60	70.64
50.00		110.00	38.64	16.80	2.10	57.54
55.00		88.00	30.80	13.80	1.68	46.28
60.00		70.00	24.64	11.00	1.38	37.02
65.00		55.00	19.60	8.80	1.10	29.50
70.00		40.00	15.40	7.00	0.88	23.28
75.00		30.00	11.20	5.50	0.70	17.40
80.00		20.00	8.40	4.00	0.55	12.95
85.00		15.00	5.60	3.00	0.40	9.00
90.00		8.00	4.20	2.00	0.30	6.50
95.00		2.00	2.24	1.50	0.20	3.94
100.00		0.00	0.56	0.80	0.15	1.51
105.00			0.00	0.20	0.08	0.28
110.00				0.00	0.02	0.02
115.00					0.00	0.00

DRH, direct runoff hydrograph.

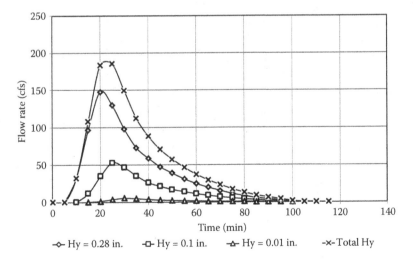

Figure 3.16 Convolution of unitgraphs for example watershed.

3.7 Information sources

Characteristics of flood flows, such as magnitude of peak discharges, frequency of occurrence, and flood volumes, are major considerations in flood predictions. Extensive stream flow data collected from large perennial streams can generally provide flood information necessary for the structure designs, and flood data collected from the same or similar regions can form a database for further regression analyses. The main sources of runoff information are

1.   US Geologic Survey, Water Resources Division, District Office for the information of peak discharge data, watershed characteristics, an index of urbanization
2.   US Geologic Survey, Topographic Division, District Office for topographic maps and land-use maps
3.   Department of Natural Resources Conservation Service for land-use data, soil data, and watershed characteristics
4.   Department of Commerce, Bureau of the Census for population data and urban growth

3.8 Homework

Q3.1 Conduct a watershed analysis for the design point at the highway in Figure Q3.1.

1.   Estimate the drainage area to the location of the bridge.
2.   Identify the representative waterway.
3.   Plot the variation of elevation along waterway length.
4.   Identify the lengths for overland, swale, and channel flows.
5.   Determine the weighted waterway slope.

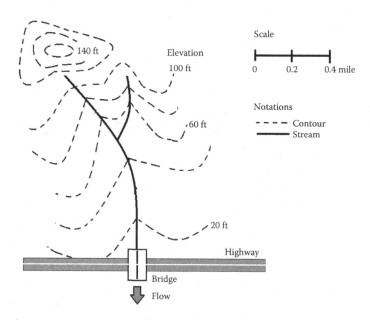

Figure Q3.1 Example watershed.

Q3.2 Continue *Q3.1*. Consider the lag time coefficient, $C_t = 1.8$, storage coefficient, $C_p = 0.6$, and design rainfall duration of 10 min. Construct the *Snyder Unit Hydrograph* for the watershed in *Q3.1*.

Q3.3 Continue *Q3.1*. Consider the SCS Curve Number, CN = 55 and design rainfall duration of 10 min. Construct the *SCS Unit Hydrograph* for the watershed in *Q3.1*.

Q3.4 Continue *Q3.1*. The depression loss for the watershed is 0.35 in. The soil infiltration is described with $f_0 = 3.0$ in./h, $f_c = 0.5$ in./h, and $k = 0.108$/min. For a given rainfall distribution below, fill in the calculation of net rainfall depth.

Time, t (min)	Precipitation, $\Delta p(t)$ (in.)	Infiltration, f(t) (in./h)	Infiltration, $\Delta F(t)$ (in.)	Runoff depth (in.)	Incremental depression (in.)	Cumulative depression (in.)	Net rain depth (in.)
(1)	(2)	(3)	(4)	(5) = (2)−(4)	(6)	(7)	(8) = (5)−(6)
0	0.000	3.000	0.000	0.000	0.000	0.000	
10	0.110	1.349	0.362	0.000	0.000	0.000	
20	0.220	0.788	0.178	0.042	0.042	0.042	
30	0.330						
40	0.250						0.062
50	0.120						0.033
60	0.050						
70	0.030						
80	0.010						
Total	1.120		1.089	0.445	0.350		

Q3.5 Continue *Q3.1*. Apply the net rainfall depth derived in *Q3.4* to the Snyder Unit Hydrograph derived in *Q3.2* to predict the storm hydrograph.

Bibliography

American Society of Civil Engineers. (1957). *Hydrology Handbook*, Manuals of Engineering Practice, No. 28, ASCE, New York.

Chow, V.T. (1957). *Hydrologic Determination of Waterway Areas for the Design of Drainage Structures in Small Drainage Basins*, Engineering Experiment Station Bulletin 462, University of Illinois at Champaign, Urbana, IL.

Chow, V.T., Maidment, D.R., and Mays, L.W. (1988). *Applied Hydrology*, McGraw Hill, Inc., New York.

CUHP. (2005). *Colorado Urban Unitgraph Procedure*, Chapter of Runoff, Urban Storm Water Design Criteria Manual, UDFCD, Denver, CO.

Dooge, J.C.I. (1959). "A General Theory of the Unit Hydrograph," *Journal of Geophysical Research*, Vol. 64, No. 2, pp. 241–256.

Eagleson, P.S. (1962). "Unit Hydrographs for Characteristics Sewered Areas," *Journal of the Hydraulics Division*, Vol. 88, No. HY2, pp. 1–25.

Green, W.H., and Ampt, G.A. (1911). "Studies of Soil Physics, I: The Flow of Air and Water Through Soils," *Journal of Agriculture Science*, Vol. 4, No. 1, pp. 1–24.

Guo, J.C.Y. (1988). "Colorado Unit Hydrograph Procedures—Its Synthetic Unit Hydrograph Characteristics," Proceedings of ASCE International Conference on Hydraulic Engineering held at Colorado Springs, CO.

Henderson, F.M., and Wooding, R.A. (1964). "Overland Flow and Groundwater Flow from Steady Rainfall of Finite Duration," *Journal of Geophysical Research*, Vol. 69, No. 8, pp. 1531–1539.

HMS. (2015). Hydrologic Analytic Model—*Flood Hydrograph Package Manual*, updated 1987, U.S. Army Corps of Engineers, Davis, CA.

Horton, R.E. (1933). "The Role of Infiltration in the Hydrologic Cycle," *Transactions American Geophysical Union*, Vol. 14, pp. 446–460.

Kohler, M.A., and Linsley, R.K. Jr. (1951). "Predicting the Runoff from Storm Rainfall," U.S. Weather Bureau, Research Paper 34.

Kuichling, E. (1889). "The Relation Between the Rainfall and the Discharge of Sewers in Populous Districts," *Transactions of the American Society of Civil Engineers*, Vol. 20, pp. 1–56.

Nash, J.E. 1958. "The Form of the Instantaneous Unit Hydrograph," General Assembly of Toronto, International Association Scientific Hydrology (Gentbrugge) Pub. 42, Compt. Rend. 3, pp. 114–118.

NRCS SCS. (1964). *SCS National Engineering Handbook*, "Section 4: Hydrology," updated 1972, U.S. Department of Agriculture, Washington, DC.

NRCS SCS. (1986). *Urban Hydrology for Small Watersheds*, 2nd ed., Technical Release No. 55 (NTIS PB87-101580), U.S. Department of Agriculture, Washington, DC.

Overton, D.E., and Meadows, M.E. (1976). *Stormwater Modeling*, Academic Press, New York.

Sherman, L.K. (1932). "Stream Flow from Rainfall by Unit-Graph Method," *Engineering News-Record*, Vol. 108, pp. 501–505.

Snyder, F.F. (1938). "Synthetic Unitgraphs," *Transactions American Geophysical Union*, Vol. 19, pp. 447–454.

Snyder, W.M. (1955). "Hydrograph Analysis by the Method of Least Squares," *ASCE, Journal of Hydraulics Division*, Vol. 81, pp. 1–25.

Tauxe, G.W. (1978). "S-Hydrographs and Change of Unit Hydrograph Duration," ASCE Technical Notes, *Journal of Hydraulics Division*, Vol. 104, No. HY3, pp. 439–444.

Taylor, A.B., and Schwartz, H.E. (1952). "Unit Hydrograph Lag and Peak Flow Related to Basin Characteristics," *Transactions American Geophysical Union*, Vol. 33, pp. 235–246.

USWDCM. (2010). *Urban Stormwater Design Criteria Manual*, Urban Drainage and Flood Control District, Denver, CO.

Viessman, W., Lewis, G.L., and Knapp, J.W. (1989). *Introduction to Hydrology*, 3rd ed., Harper & Row, New York.

Chapter 4

Hydrologic frequency analysis

The magnitude and time of occurrence of a hydrologic variable are random. When a set of hydrologic data is arranged in a decreasing or increasing order of magnitude, they can be analyzed by statistical methods. The reliability of a theoretical distribution derived from the data closely depends on the true representation of the sample data. In practice, it is rare to have a record that is long enough to cover the entire period of time of occurrence. As a result, the approximation of the observed patterns to the population distribution depends on the quality of sample data and the length of the record. Generally speaking, a 15- to 30-year record can produce a fairly reliable approximation for practical purposes.

Hydrologic variables such as rainfall and runoff are highly localized in their characteristics. Although many statistical models have been developed, none of them can be universally successful. Among them, frequency analysis is considered as a good means to predict the magnitude and recurrence interval of a hydrologic variable when the record of data adequately represents its population. Frequency analysis includes sample data processing, selection of the best-fitted probability distribution, and predictions by the sample statistics. Although the *Water Resources Council* has recommended the *log-Pearson type III distribution* for hydrologic frequency analyses, it remains a common practice to find the best-fitted probabilistic model to describe the distribution of the hydrologic variable involved in the design (Bulletin 17, 1982).

This chapter provides the definition of return period and a general formulation for normal, log-normal, Gumbel, exponential, Pearson, and log-Pearson distributions. This chapter also presents examples to illustrate database structure, calculations of sample statistics, determination of data skewness, predictions, and confidence intervals.

4.1 Basics of probability

Outcome of a hydrologic variable, q, is random. It is essential to know that the probability for $q =$ a specified magnitude is zero. We are interested in the probability for q to occur within a range. To study the outcome of a random event, q_i, a sample shall be collected from a large number of observations, N. Next, the range of the outcome is divided into a number of intervals with a sufficiently small increment of ΔQ. If there are M times of occurrence within the interval from $Q - 0.5\Delta Q$ to $Q + 0.5\Delta Q$, the relative probability for this interval is defined as

$$P((Q - \Delta Q) \le q < (Q + \Delta Q)) = \frac{M}{N} \tag{4.1}$$

A *histogram* is a plot of Equation 4.1 as a step function for all intervals used in the study. A *density function* is a mathematical expression that describes the histogram as a

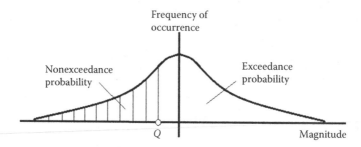

Figure 4.1 Probability density function.

continuous curve over the range of all intervals. The area under a density curve must be equal to unity. Integration of a density function yields the *accumulative probability curve* that begins with zero and ends with one. The probability of occurrence between any two limits is defined as the difference of the accumulative probabilities at the beginning and at the end of such an interval.

As shown in Figure 4.1, a *nonexceedance probability* is defined as the area under the density curve with a magnitude less than the specified magnitude, Q. Of course, the residual area is termed *exceedance probability* that is the chance for the event to be greater than or equal to the specified magnitude.

EXAMPLE 4.1

Out of 40 rainfall events, there are eight events with a rainfall depth, d, greater than 0.15 in. The exceedance probability for the depth of 0.15 in. is

$$P(d \geq 0.15) = \frac{8}{40} = 0.20.$$

The nonexceedance probability for the depth of 0.15 in. is

$$P(d < 0.15) = 1 - 0.20 = 0.80.$$

4.2 Hydrologic database

Flood frequency analysis is a statistical technique applied to the long-term record collected at a gage station under *a steady hydrologic condition* in a watershed. At a gage station, the hydrologic data are recorded continuously in time. It is important to make sure that the data record has not been significantly affected by the development of watershed activities. Urbanization is manifested by channelization, levees, reservoirs, highways, and streets. The watershed history and flood records must be carefully examined to ensure that no major hydrologic changes have occurred during the period of records. Care shall be taken when the watershed changes are incremental, which might not significantly alter the flow regimes from year to year, but will become noticeable by the accumulative effect after several years. Only records that represent relatively steady watershed hydrologic conditions should be used for frequency analysis. Predictions from such a database are only applicable to the similar hydrologic condition. Flood flow predictions for several stations under a similar hydrologic condition can then form a database for regional regression analyses.

A *mixed flood flow record* is formed by different types of events, such as snowmelt runoff mixed with rainfall storm runoff. A careful data screening is required when coping with a mixed data record. For instance, we have to separate runoff data by seasons or by the nature of events. An *incomplete flood flow record* is resulted from a temporary malfunction of the recording instruments when peak flow rates were too high or too low. Missing data in an incomplete record can be derived from the neighboring stations using their annual mean flood flow rates as weighting factors.

Reliability of a frequency analysis depends on the length of the record. A record of 10 years is barely acceptable for hydrologic predictions because most of the hydraulic designs are governed by 50- or 100-year events. It is important to know how to interpret the sample with confidence to extrapolate the derived relationship into extreme events. For this purpose, there are two types of databases developed from the *complete data series* (CDS) for the hydrologic frequency analysis: *annual maximum series* (AMS) and annual exceedance *series* (AES).

An AMS uses a period of 365 days as a hydrologic year. The representative flood for each period is the largest one observed within 365 days. As a result, an N-year record will include only N events, one from each year that is then used in the analysis. It often happens that the second highest event in a wet year exceeds the largest floods observed in some dry years. However, by definition, an AMS considers only the highest magnitude and ignores the rest.

An AES accepts hydrologic data as a continuous record. All observed events are ranked in a descending order in magnitude. An AES consists of the top N events from an N-year record. In doing so, there is a chance to include more than one event from a wet year and none from a dry year. In comparison, an AES is a more conservative approach than an AMS.

A CDS includes all observations. Obviously, the number of events in a CDS exceeds the years of the record. The ranked CDS offers a database for both frequency and duration analyses.

EXAMPLE 4.2

Table 4.1 presents a record of runoff peak discharges observed at a gage station from 1961 to 1975. The AMS and AES for this database are analyzed and summarized in Table 4.1.

Solution: The AMS consists of the highest event from each year. The second largest flood flow, 441 cfs, in 1974 was excluded from the AMS. However, it is noticed that the flow rate of 441 cfs was higher than many flood events observed in other years. The AES consists of the top 15 observed peak discharges from 1961 to 1975. The three events, 468, 543, and 441, observed in 1974 are all included in the AES. As expected, the AES has a higher mean value than that of the AMS. Therefore, an AES will produce more conservative (higher) predictions than its corresponding AMS.

Both an AMS and AES serve as a sample from which the statistical parameters of the population can be estimated. Predictions of certain events can then be modeled by the selected probability distribution. The validity and applicability of a database depend directly on the characteristics of the sample data used to estimate the model parameters. As a rule of thumb, the sample data used for parameter estimations must be representative of the situation in which the model is going to be used. In other words, the selection of sample data is dictated by the design criteria of the hydraulic structure under design. An AES may be considered if the flood damages are caused by their repetition such as traffic

Table 4.1 Databases for AMS and AES

Year	High	Flow (cfs)	Events	AMS Data	Ranked	Data	AES
1961	390			390	1	596	596
1962	374			374	2	591	591
1963	342			342	3	557	579
1964	507			507	4	549	557
1965	579	406	596	596	5	543	549
1966	416			416	6	533	543
1967	533			533	7	524	533
1968	505			505	8	507	524
1969	549	454		549	9	505	507
1970	414	410		414	10	505	505
1971	434	524		524	11	416	505
1972	505	415	406	505	12	414	497
1973	428	557	407	557	13	390	468
1974	468	543	441	543	14	374	454
1975	591	497		591	15	342	441

AMS, annual maximum series; AES, annual exceedance series.

interruptions by flooded culverts in highway drainage. In other cases where the design is controlled by the most critical condition such as spillway design, the AMS may be used.

4.3 Sample statistics

The variable, $q(i)$, can be represented by its mean and the departure of the variable from the mean. Such a departure can be expressed by a fraction of the standard deviation. Mean, standard deviation, and skewness of the distribution of variable, $q(i)$, are computed as

$$\overline{Q} = \frac{1}{n} \sum_{i=1}^{i=n} q(i) \tag{4.2}$$

$$S = \sqrt{\frac{1}{(n-1)} \sum_{i=1}^{i=n} \left(q(i) - \overline{Q}\right)^2} \tag{4.3}$$

$$g = \frac{1}{S^3} \frac{n}{(n-1)(n-2)} \sum_{i=1}^{i=n} \left(q(i) - \overline{Q}\right)^3 \tag{4.4}$$

in which $\overline{Q}$ = mean of variable, $q(i)$ = ith observation, n = number of observations, S = standard deviation, and g = skewness coefficient. Often, logarithmic values can narrow the differences between the predicted and observed values in the analysis. Using logarithmic values, the Equations 4.2 through 4.4 are converted to

$$\overline{Q}_{\log} = \frac{1}{n} \sum_{i=1}^{i=n} \log q(i) \tag{4.5}$$

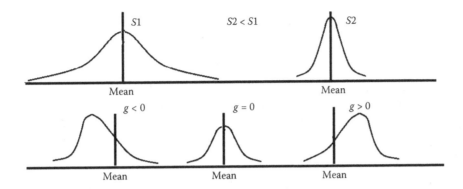

Figure 4.2 Statistic parameters for various distributions.

$$S_{\log} = \sqrt{\frac{1}{(n-1)}\sum_{i=1}^{i=n}\left[\log q(i) - \bar{Q}_{\log}\right]^2} \tag{4.6}$$

$$g_{\log} = \frac{1}{S_{\log}^3}\frac{n}{(n-1)(n-2)}\sum_{i=1}^{i=n}\left[\log q(i) - \bar{Q}_{\log}\right]^3 \tag{4.7}$$

in which $\bar{Q}_{\log}$ = mean of logarithmic values of $q(i)$, $S_{\log}$ = standard deviation of logarithmic values of $q(i)$, and $g_{\log}$ = skewness coefficient of logarithmic values of $q(i)$. As illustrated in Figure 4.2, a standard deviation represents the spread of $q(i)$. A higher standard deviation implies wider spread and vice versa. A skewness coefficient indicates the shape of the distribution. When $g = 0$, it is a symmetric distribution; $g > 0$ means more events greater than the mean, whereas $g < 0$ means more events smaller than the mean.

4.4 Plotting position

The primary purpose of hydrologic frequency analysis is to determine the recurrence intervals and the magnitudes of extreme events used for hydraulic structure designs. The recurrence interval of a magnitude, Q, is defined as the time interval for the random event to exceed or equal Q. For instance, the highest peak discharge over a 50-year record is equaled once in the period of 50 years; therefore, this highest value is considered to be close to the magnitude of the 50-year flood. Similarly, the second highest discharge has been equaled and exceeded twice in 50 years. Such a magnitude is termed *a 25-year flood* because, on the average, this value was equaled or exceeded once every 25 years. Many studies were conducted to define the relationship between rank and recurrence interval, because hydrologic data analysis begins with how to plot the observed data on a probabilistic graphic paper. The one that fits the data best shall be chosen for predictions. Plotting a data set on a probabilistic scale requires the determinations of relative plotting positions among the data points. The plotting position of a given variable, q, is defined by its *nonexceedance probability*, $P(q < Q)$. For instance, the exceedance probability of a 10-year flood is 1/10 (once every 10 years on an average), and its nonexceedance probability is 9/10.

Table 4.2 Empirical constants for plotting formulas

Empirical formula	a	b	$P(q < Q)$
California (1923)	0.00	0.00	$1 - m/n$
Hazen (1914)	−0.50	0.00	$1 - (m - 0.5)/n$
Weibull (1939)	0.00	1.00	$1 - m/(n + 1)$
Beard (1943)	−0.30	0.40	$1 - (m - 0.3)/(n + 0.4)$
Gringorten (1963)	0.44	0.12	$1 - (m + 0.44)/(n + 0.12)$
Cummane (1978)	0.40	0.20	$1 - (m + 0.40)/(n + 0.20)$

Graphical papers are made with the vertical axis to be the magnitude of a hydrologic event and the horizontal axis to be the nonexceedance probability. During the data analysis, the N-year data series shall be ranked in a decreasing order in magnitude. The plotting position for the *m*th event is determined by the total number of observations, *n*, and its rank, *m*. The empirical formula for determining the nonexceedance probability is

$$P(q < Q) = 1 - \frac{m + a}{n + b} \tag{4.8}$$

in which $P(q < Q)$ = nonexceedance probability, n = length of record in years, m = rank of an event in a decreasing order in magnitude, and a and b = empirical constants. The value of a is 0.00 for the uniform distribution, $a = 0.375$ for the normal distribution, and $a = 0.44$ for the Gumbel distribution. Table 4.2 shows the recommended values for the variables a and b. Selection of variables a and b is a matter of the local hydrology characteristics. In general, the California formula is recommended for an AES analysis, and the Weibull formula is recommended for an AMS analysis.

The fundamental definition in the statistical hydrology is the relationship between *return period* and *exceedance probability*. They are defined as

$$P(q \geq Q) = 1 - P(q < Q) \tag{4.9}$$

$$T_r = \frac{1}{P(q \geq Q)} \tag{4.10}$$

in which $P(q \leq Q)$ = exceedance probability and T_r = return period in years. The return period is defined as the average recurrence interval. However, it does not mean that an exceedance always occurs once every T_r years, but it means that the average time span between two adjacent exceedances is T_r years. Regardless of whether the return period is referring to an event greater than a value or to an event less than a value, the return period is related to a probability of an exceedance.

EXAMPLE 4.3

For comparison, both California formula and Hazen formula are applied to the AMS developed in Example 4.1. It is noted that a significant difference exists in dealing with the highest event (as shown in Table 4.3). For the highest flood magnitude, the Hazen formula assigns a return period of 30 years, but the California formula assigns a return period of 15 years. The difference between these two formulas diminishes as *m* increases. This implies that the Hazen formula is more suitable when dealing with outliers.

Table 4.3 Plotting positions by California and Hazen formulas

AMS			California formula		Hazen formula	
Peak, Q (cfs)	m	Q (cfs)	m/n	n/m	(m − 0.5)/n	n/(m − 0.5)
			P(Q ≥ q)	T_r (year)	P(Q ≥ q)	T_r (year)
390	1	596	0.07	15.00	0.03	30.00
374	2	591	0.13	7.50	0.10	10.00
342	3	557	0.20	5.00	0.17	6.00
507	4	549	0.27	3.75	0.23	4.29
596	5	543	0.33	3.00	0.30	3.33
416	6	533	0.40	2.50	0.37	2.73
533	7	524	0.47	2.14	0.43	2.31
505	8	507	0.53	1.88	0.50	2.00
549	9	505	0.60	1.67	0.57	1.76
414	10	505	0.67	1.50	0.63	1.58
524	11	416	0.73	1.36	0.70	1.43
505	12	414	0.80	1.25	0.77	1.30
557	13	390	0.87	1.15	0.83	1.20
543	14	374	0.93	1.07	0.90	1.11
591	15	342	1.00	1.00	0.97	1.03

4.5 Probability distributions

The general form for probability models is that the departure of a variable, $Q(T_r)$, from its mean can be expressed by the standard deviation, S, and frequency factor, $Z(T_r)$, as shown below (Chow et al., 1988):

$$Q(T_r) = \overline{Q} + Z(T_r)S \qquad (4.11)$$

Similarly, for logarithmic values, Equation 4.11, is converted to

$$\log Q(T_r) = \overline{Q}_{\log} + Z(T_r)S_{\log} \qquad (4.12)$$

The value of frequency factor in Equations 4.11 and 4.12 depends on the underlying probability distribution and the return periods. Theoretical formulas have been developed for the distributions discussed in the following sections.

Gumbel distribution

The frequency factor for the Gumbel distribution is defined as

$$Z_g(T_r) = -\frac{\sqrt{6}}{\pi}\left\{0.5772 + \ln\left[\ln\frac{T_r}{(T_r-1)}\right]\right\} \qquad (4.13)$$

in which $Z_g(T_r)$ = Gumbel frequency factor for return period T_r and π = 3.1416. According to Equation 4.11, the return period for the average magnitude is the one that has $Z_g = 0.0$. Setting Equation 4.13 equal to zero, the return period for the average magnitude is $T_r = 2.33$ years for the Gumbel distribution.

Exponential distribution

The frequency factor for exponential distribution is defined as

$$Z_e(T_r) = \frac{\sqrt{6}}{\pi}(\ln T_r - 0.5772) \tag{4.14}$$

in which Z_e = exponential frequency factor for return period T_r. Setting Equation 4.14 equal to zero, the return period for the average magnitude is 1.78 year for the exponential distribution.

Normal distribution

The normal distribution is symmetric with skewness = 0.0. The frequency factors for normal distribution are not analytically integratable, but they can be closely approximated by

$$B = \sqrt{\ln\left(\frac{1}{p^2}\right)} \tag{4.15}$$

in which the variable p is the exceedance probability. Having the variable B known by Equation 4.15, the value of z for the normal distribution is (Abramowitz and Stegun, 1965)

$$z = B - \frac{2.515517 + 0.802853B + 0.010328B^2}{1 + 1.432788B + 0.189269B^2 + 0.001308B^3} \tag{4.16}$$

Equation 4.16 is also valid to the log-normal distribution when logarithmic values are used. Equation 4.16 indicates that the return period for the average magnitude in Equation 4.11 is 2 years under the normal distribution.

Pearson type III distribution

Pearson type III distributions are a set of family curves. Each curve is identified by mean, standard deviation, and skewness coefficient. The frequency factors for a Pearson or log-Pearson type III distribution depend on the return period and the skewness coefficient determined by Equations 4.10 and 4.7. When the skewness coefficient is zero, the Pearson type III distribution is reduced to the normal distribution. The frequency factors for a symmetrical distribution are described by z in Equation 4.16. When the skewness coefficient is between 9.0 and −9.0 and the exceedance probability is from 0.0001 to 0.9999, the frequency factors of the Pearson type III distribution can be computed using the value of z in Equation 4.16 as (Harter, 1971; Kite, 1977)

$$Z_p = \frac{2}{g}\left\{\left[(z-k)k+1\right]^3 - 1\right\} \tag{4.17}$$

in which Z_p = frequency factor for the Pearson type III distribution, and k is defined as

$$k = \frac{g}{6} \text{ if real values are used and} \tag{4.18}$$

$$k = \frac{g_{log}}{6} \text{ if logarithmic values are used} \tag{4.19}$$

4.6 Probability graphic papers

Equation 4.11 indicates that the magnitude of the variable is linearly related to its frequency factor, Z. When plotting a sample on a probabilistic graph paper, the degree of linearity among (Q, Z) pairs can serve as a criterion to judge how well the probabilistic model can represent the sample. For convenience, graphic papers for various probability distributions have been produced using Equation 4.11. Example 4.4 illustrates how to prepare the Gumbel graphic paper.

EXAMPLE 4.4

A Gumbel graphic paper is prepared to plot the magnitude of the variable on the y-axis and the Gumbel frequency factors, Z_g, on the x-axis. As shown in Table 4.4, the values of Z_g for various return periods can be generated using Equation 4.13:

As illustrated in Figure 4.3, using the spacing defined by the frequency factor, Gumbel graphic papers can be produced to have return periods on the horizontal axis and event magnitudes on the vertical axis. For instance, the plotting position of the return period of 100 years is located at $Z_g = 4.14$ on the horizontal axis, and the plotting position for the return period of 50 years is located at $Z_g = 2.59$. For convenience, the horizontal axis is then marked with return period. It appears nonlinear when return periods are used on the horizontal axis. In fact, the linearity is preserved, or hidden, by the underlying frequency factors.

Table 4.4 Gumbel frequency factors

T_r	Z_g	$P(q \geq Q)$	$P(q < Q)$
2.00	−0.164	0.50	0.50
5.00	0.719	0.20	0.80
10.00	1.304	0.10	0.90
25.00	2.044	0.04	0.96
50.00	2.592	0.02	0.98
100.00	3.137	0.01	0.99

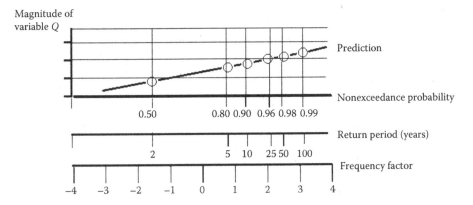

Figure 4.3 Scales used in Gumbel graphic paper.

EXAMPLE 4.5

The AES of 1-h rainfall depth has a mean, $\overline{Pl}$, of 0.95 in. and a standard deviation, Sl, of 0.45 in. Predict the 50-year, 1-h precipitation by Gumbel and normal distributions.

$$P(Pl \le P_{50}) = 1 - \frac{1}{T_r} = 0.98 \; \text{(exceedance probability)}$$

$$P(Pl > P_{50}) = \frac{1}{T_r} = 0.02 \; \text{(nonexceedance probability)}$$

Using the Gumbel distribution, we have

$$Z_g(50) = -\frac{\sqrt{6}}{\pi}\left\{0.5772 + \ln\left[\ln\frac{50}{(50-1)}\right]\right\} = 2.592$$

$$P1_{50} = \overline{P1} + Z_g(50)\,S1 = 0.95 + 2.592 \times 0.45 = 2.12 \text{ in.}$$

Using the normal distribution, the exceedance probability for a 50-year event is 0.02. With $p = 0.02$, Equation 4.15 yields

$$B = 2.7971$$

Substituting $B = 2.7971$ into Equation 4.16 yields $z = 2.054$. Then, the 50-year, 1-h precipitation depth predicted by the normal distribution is

$$Pl_{50} = 0.95 + 2.054 \times 0.45 = 1.87 \text{ in.}$$

4.7 Model predictions and best-fitted line

According to Equation 4.11, there are two parameters, mean and standard deviation, to be determined by the sample. As shown in Figure 4.4, these two parameters can be

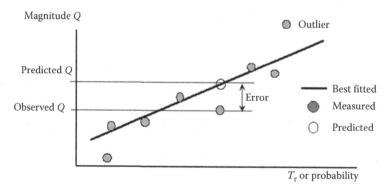

Figure 4.4 Best-fitted line on statistical graphic paper.

derived using the least square method to minimize the sum of error squared between the predicted, q_i^p, and the observed, q_i^o, for the ith test as

$$E = \sum_{i=1}^{i=n} \left(q_i^p - q_i^o \right)^2 = \sum_{i=1}^{i=n} \left(\bar{Q}' + K_i S' - q_i^o \right)^2 \tag{4.20}$$

in which E = sum of squared error, $\bar{Q}'$ = best-fitted value for mean, and S' = best-fitted value for standard deviation for the sample.

Equation 4.20 has two variables: $\bar{Q}'$ and S'. Minimization of the squared error can be achieved by setting the first derivatives of Equation 4.20 to be zero. Thus, we have

$$\frac{\partial E}{\partial \bar{Q}'} = 0 \tag{4.21}$$

and

$$\frac{\partial E}{\partial S'} = 0. \tag{4.22}$$

Solutions are

$$S' = \frac{n \sum_{i=1}^{i=n} (q_i K_i) - \sum_{i=1}^{i=n} K_i \sum_{i=1}^{i=n} q_i}{n \sum_{i=1}^{i=n} K_i^2 - \left(\sum_{i=1}^{i=n} K_i \right)^2} \tag{4.23}$$

and

$$Q' = \frac{1}{n} \sum_{i=1}^{i=n} q_i - \frac{S'}{n} \sum_{i=1}^{i=n} K_i \tag{4.24}$$

in which q_i = ith observed event and K_i = frequency factor for ith event. Replacing S with S' and Q with Q' in Equation 4.11, the linear equation of the best-fitted line is achieved as follows:

$$Q(T_r) = Q' + K(T_r)S' \tag{4.25}$$

When the sample is not large enough, Equation 4.25 may give better representation of the population. Otherwise, Equations 4.11 and 4.25 shall closely agree to each other. Of course, Equations 4.21 through 4.25 with the same procedure can be applied to logarithmic values.

4.8 Selection of probability model

The selection of the best-fitted model is a trial-and-error procedure. Figure 4.5 illustrates the steps. Decision making begins with the type of database, AMS or AES, and then the selection of plotting formula. Having the sample statistics calculated by Equations 4.2

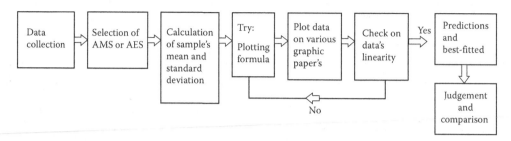

Figure 4.5 Procedures for model selection.

through 4.4, or Equations 4.5 through 4.7, the log-Pearson type III distribution is mostly recommended because of the nature of skewness in hydrologic data. Of course, the Gumbel, exponential, and log-normal distributions are also good probability models to provide comparisons.

Because the selection of the best-fitted probabilistic model is a trial-and-error process, the major effort is how to determine the plotting positions to display the data points on the selected probabilistic graphic paper. The best-fitted probabilistic model is the one that distributes the majority of the data points on a linear line, except a few outliers. This is an iterative process until the best model is found.

EXAMPLE 4.6

Apply Gumbel, exponential, normal, and Pearson type III distributions to the AMS in Example 4.2.

Aided by Equations 4.2, 4.3, and 4.4, the mean, standard deviation, and skewness coefficient are found to be 489.87 cfs, 81.32 cfs, and −0.53 for the AMS in Example 4.2. Let $a = 0$ and $b = 1$ in Equation 4.8. The return periods and exceedance probabilities are calculated as shown in Table 4.5. Peak flow rates predicted by various probability models are also listed in Table 4.5 for comparison.

4.9 Confidence limits

It is impractical to develop a sample as large as the population. As a result, the length of record affects the accuracy of the predictions. For instance, any 30-year continuous record out of a 50-year record may constitute a sample set. Therefore, at least 20 sets of 30-year sample can be derived from a 50-year continuous record. Each set of sample produces a set of estimates for population's mean and standard deviation. Therefore, for this case, we will have 30 estimates for the pairs of (*mean, standard deviation*).

The *central limit theorem* states that for a population with its finite mean, μ, and standard deviation, σ, the distribution of the sample's mean (such as the set of 30 data points as mentioned previously) will themselves be distributed as a *normal distribution* with a mean equal to μ and a standard deviation, σ_m, equal to

$$\sigma_m = \frac{\sigma}{\sqrt{n}} \qquad (4.26)$$

Table 4.5 Predictions by various probability models

Flow rate, Q (cfs)	Rank, m	T_r (years)	$P(Q > q)$	Gumbel, Z_g Equation 4.13	Gumbel, Q_g Equation 4.11	Expon, Z_e Equation 4.14	Expon, Q_e Equation 4.11	B Equation 4.15	Normal, Z Equation 4.16	Normal, Q_n Equation 4.11	Pearson, Z_p Equation 4.17	Pearson, Q_p Equation 4.11
596.00	1.00	16.00	0.94	1.69	627.00	1.71	629.10	2.35	1.53	614.60	1.40	603.80
591.00	2.00	8.00	0.88	1.12	580.90	1.17	585.10	2.04	1.15	583.40	1.11	580.00
557.00	3.00	5.33	0.81	0.78	552.90	0.86	559.40	1.83	0.89	562.00	0.89	562.50
549.00	4.00	4.00	0.75	0.52	532.30	0.63	541.20	1.67	0.67	544.70	0.71	547.80
544.00	5.00	3.20	0.69	0.32	515.50	0.46	527.00	1.53	0.49	529.60	0.55	534.40
534.00	6.00	2.67	0.63	0.14	501.10	0.32	515.50	1.40	0.32	515.70	0.39	521.80
524.00	7.00	2.29	0.56	-0.02	488.30	0.20	505.70	1.29	0.16	502.60	0.24	509.40
507.00	8.00	2.00	0.50	-0.16	476.50	0.09	497.20	1.18	0.00	489.90	0.09	497.00
505.00	9.00	1.78	0.44	-0.30	465.30	0.00	489.80	1.07	-0.16	477.10	-0.07	484.30
505.00	10.00	1.60	0.38	-0.44	454.50	-0.08	483.10	0.97	-0.32	464.10	-0.23	470.90
416.00	11.00	1.45	0.31	-0.57	443.70	-0.16	477.00	0.87	-0.48	450.40	-0.41	456.50
414.00	12.00	1.33	0.25	-0.71	432.60	-0.23	471.50	0.76	-0.67	435.60	-0.61	440.40
390.00	13.00	1.23	0.19	-0.85	420.60	-0.29	466.40	0.64	-0.87	418.90	-0.84	421.60
374.00	14.00	1.14	0.13	-1.02	406.90	-0.35	461.70	0.52	-1.12	398.70	-1.13	398.00
342.00	15.00	1.07	0.06	-1.25	388.60	-0.40	457.40	0.36	-1.46	370.90	-1.55	363.90

in which σ_m = standard deviation for the distribution of sample's mean, σ = standard deviation of population, and n = number of observations in the sample sets. *Confidence limits* are expressed as the exceedance probabilities and mark the reliability band above and below the frequency curve. As shown in Figure 4.6, the 5% (upper limit) and 95% (lower limit) confidence limits give a confidence level of 90%.

In general, the value of mean, μ, is determined by the mean of the sample tests, but the exact value of σ can only be approximated by the standard deviation, S, of the sample tests. US Water resources Council (Bulletin 17, 1982) suggested that these two confidence limits are determined as

$$Q_u = Q + SZ_u \tag{4.27}$$

$$Q_d = Q - SZ_d \tag{4.28}$$

$$Z_u = \frac{1}{c}\left(Z + \sqrt{Z^2 - cd}\right) \tag{4.29}$$

$$Z_d = \frac{1}{c}\left(Z - \sqrt{Z^2 - cd}\right) \tag{4.30}$$

$$c = 1 - \frac{z_*^2}{2(n-1)} \tag{4.31}$$

$$d = Z^2 - \frac{z_*^2}{n} \tag{4.32}$$

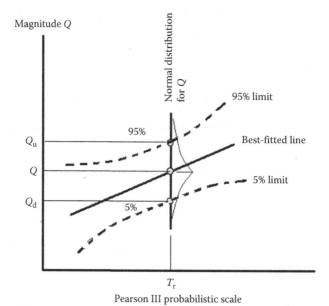

Figure 4.6 Confidence limits for the best-fitted line.

in which Q_u = upper limit, Q_d = lower limit, Z = frequency factor of the underlying probability distribution for the specified event such as the 100-year magnitude by the Gumbel distribution, and z_* = the frequency factor of the normal distribution for the limits selected. For instance, to determine the confidence limits between 5% and 95% requires $z_* = 1.645$, which is the frequency factor of the *normal distribution* with an exceedance probability of 5% or a nonexceedance probability of 95%.

EXAMPLE 4.7

Considering the 5% and 95% confidence limits for an AMS with $n = 20$, determine the values of Z_u and Z_d for the 100-year event described by the Gumbel distribution.

   The 5% and 95% confidence limits require $z_* = 1.645$ derived from the normal distribution. For the 100-year event, the frequency factor of the Gumbel distribution is $Z_g(100) = 3.137$ by Equation 4.13. Substituting z_* and Z into Equations 4.31 and 4.32 yields $c = 0.9288$ and $d = 9.705$. Equations 4.29 and 4.30 provide $Z_u = 4.356$ and $Z_d = 2.394$. The corresponding magnitudes for the limits can be further determined by Equations 4.27 and 4.28 when the mean and standard deviation are known. The same procedure can be applied to 2-, 5-, 10-, and 50-year events. Figure 4.6 is constructed with the 5% and 95% confidence limits.

EXAMPLE 4.8

Calculate the 95% and 5% confidence limits for the Gumbel distribution using the AMS in Example 4.7.

$z_* = 1.645$ for the exceedance probability of 5%.

$n = 15$

$c = 0.90 \, (\text{Equation 4.31})$

$d = Z_g^2 - 0.18 \, (\text{Equation 4.32})$

The prediction of peak flows is listed in Table 4.6.

Table 4.6 Example for confidence limits

Return period, T_r (year)	$P(Q < q)$	Gumbel freq. factor, Z_g	Prediction, Q_g (cfs)	d	Z_u	Z_d	Upper limit, Q_u (cfs)	Lower limit, Q_L (cfs)
2.00	0.50	−0.16	476.50	−0.15	0.27	−0.63	511.70	438.40
5.00	0.80	0.72	548.40	0.34	1.31	0.29	596.20	513.10
10.00	0.90	1.31	596.00	1.52	2.08	0.81	658.80	555.80
25.00	0.96	2.04	656.10	4.00	3.10	1.43	741.60	606.10
50.00	0.98	2.59	700.70	6.54	3.87	1.87	804.40	642.10
100.00	0.99	3.14	744.90	9.66	4.64	2.30	867.20	677.20
500.00	1.00	4.40	847.20	19.13	6.44	3.29	1013.70	757.20

4.10 Modification on sample skewness

Statistics of a sample are only estimates for the population. The error and bias in the skewness estimate decreases as the length of the record increases. Although hydrologic data lengths are often finite in length, the distribution of a random variable can be generated by Monte Carlo experiments, according to the parameters of the distribution. In 1982, the Interagency Advisory Committee on Water Data (Bulletin 17, 1982) recommended that the estimate of skewness coefficient derived from a finite record at a station can be modified by the generalized skewness coefficient, G, *from* (Figure 4.7) developed for the United States.

The weighting formula is

$$g_m = Kg + (1-K)G \tag{4.33}$$

in which g_m = modified skewness coefficient for the station, g = station skewness determined by Equation 4.7 for real values or Equation 4.10 for logarithmic values, G = generalized skewness coefficient at the station (Figure 4.7), and K = weighting factor to be determined by the variance of skewness coefficient.

$$K = \frac{V_G}{V_G + V_g} \tag{4.34}$$

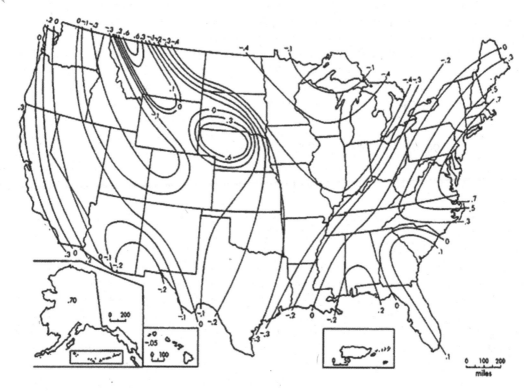

Figure 4.7 Generalized skewness coefficients, G, for the United States.

in which V_g = variance of g and V_G = variance of G. The value of variance is determined by the following empirical formulas:

$$V_G = 0.3025 \text{ (constant for Figure 4.7 for the United States)} \qquad (4.35)$$

and

$$V_g = 10^{\left[A - B\log\left(\frac{n}{10}\right)\right]} \qquad (4.36)$$

in which A and B are empirical values determined by

$$A = -0.33 + 0.08|g| \quad \text{if } |g| \leq 0.90 \qquad (4.37)$$

$$A = -0.52 + 0.30|g| \quad \text{if } |g| > 0.90 \qquad (4.38)$$

and

$$B = 0.94 - 0.26|g| \leq 1.50 \quad \text{if } |g| \leq 1.50 \qquad (4.39)$$

$$B = 0.55 \quad \text{if } |g| > 1.50. \qquad (4.40)$$

EXAMPLE 4.9

Calculate the weighted skewness coefficient for the AMS in Example 4.2.

Station skewness coefficients	$g = 0.5300$	
	$A = 0.2876$	Equation 4.37
	$B = 0.8022$	Equation 4.40
Station variance of g	$V_g = 0.3725$	Equation 4.36
	$G = -0.400$	Figure 4.7, at Chicago
General variance of G	$V_G = 0.3025$	Equation 4.34
Weighting factor	$K = 0.4481$	Equation 4.34
Modified skewness coefficient	$g_m = -0.4583$	Equation 4.33

With the modified skewness coefficient, the Pearson III distribution shall be tested again.

4.11 Tests of high and low outliers

Outliers are data points that depart significantly from the trend of the remaining data. Outliers can substantially affect the sample statistics. The sign of high outliers in the database is when the station's skewness coefficient is greater than +0.4, and of low outliers is when the station's skewness coefficient is smaller than −0.4. When logarithmic values are used for frequency analysis, the Water Resources Council (Bulletin 17, 1982) suggests that outliers be detected by the following:

$$Q_H = Q_{\log} + Z_0 S_{\log} \qquad (4.41)$$

$$Q_L = Q_{\log} - Z_0 S_{\log} \qquad (4.42)$$

Table 4.7 Outlier frequency factors based for log-Pearson type III distribution

Record length (year)	Frequency factor	Record length (year)	Frequency factor	Record length (year)	Frequency factor
10	2.036	45	2.727	80	2.940
15	2.247	50	2.768	85	2.961
20	2.395	55	2.804	90	2.981
25	2.486	60	2.837	95	3.000
30	2.563	65	2.866	100	3.017
35	2.628	70	2.893	105	3.033
40	2.682	75	2.917	110	3.049

Source: Bulletin 17. *Guidelines for Determining Flood Flow Frequency*, Interagency Advisory Committee on Water Data, Bulletin #17B of the Hydrology Subcommittee, OWDC, US Geological Survey, Reston, VA, 1982, 1983.

in which Q_H = value to detect high outliers, Q_L = value to detect low outliers, and Z_0 = recommended outlier frequency factor (see Table 4.7).

Any data whose magnitude is greater than Q_H is considered a high outlier. Similarly, any magnitude less than Q_L is a low outlier. Having detected outliers, it is necessary to compare the outlier with historic information at the station before deleting outliers from the AMS or AES. If the outlier is justified as historic flood data, it can be excluded from the database; otherwise, it shall be included in the frequency analysis.

4.12 Adjustments for zeros

The hydrologic study for the stream in an arid or semi-arid climate often comes across with years of no flow. Existence of zeros in a database causes a mathematical problem when using logarithmic values for the frequency analysis. Of course, we can add a small value to all zeros to overcome the mathematical problem without compromising the accuracy of the analysis. However, the total probability theorem offers a sound approach to cope with zeros in a database. The total probability consists of two components: one is under the condition $q = 0$, and the other is under the condition $q \neq 0$. As a result, the total probability is

$$P(q \geq Q) = P(q \geq Q / q = 0)P(q = 0) + P(q \geq Q / q \neq 0)P(q \neq 0). \tag{4.43}$$

Because $q = 0$ observed in the database cannot be coexisting with $q \geq Q$, it is concluded that

$$P(q \geq Q / q = 0) = 0. \tag{4.44}$$

As a result, Equation 4.43 becomes

$$P(q \geq Q) = P(q \geq Q / q \neq 0) \times P(q \neq 0). \tag{4.45}$$

Equation 4.45 is the total probability to make an adjustment to a record with zeros.

EXAMPLE 4.10

A 30-year runoff record has 3 years of zero flow. The Weibull formula is used to determine the plotting positions. With a record of 30 years, $n = 30$, the nonexceedance probability for the highest observed flow rate (i.e., $m = 1$) is

$$P(q \geq Q) = \frac{m}{n+1} = \frac{1}{31}$$

For this case, the number of nonzero years is

$$n = 30 - 3 = 27$$

The nonexceedance probability under the condition of nonzero flow for the highest flow rate is

$$P(q \geq Q / q \neq 0) = \frac{m}{n+1} = \frac{1}{28}$$

The probability for flow greater than zero is

$$P(q \neq 0) = \frac{30 - 3}{30} = \frac{9}{10}$$

According to Equation 4.45, the joint probability for being nonzero and also greater than the index, Q, is calculated as

$$P(q \geq Q) = \frac{1}{28} \times \frac{9}{10} = \frac{9}{280}$$

4.13 Mixed population

Reliability of flow-frequency relationship depends on the homogeneity of hydrologic data. Often, seasonal influence results in a nonuniform population. Table 4.8 summarizes several conditions in which a mixed population can be created, including snowmelt runoff mixed with rainfall runoff flows, hurricane events mixed with thunderstorm events, etc. As shown in Figure 4.8, a mixed population exhibits two distinct groups—low flows and high flows—because these two groups were caused by different reasons at different seasons.

As illustrated in Figure 4.8, the two groups of data offer two separate frequency curves that can be combined together to represent the mixed population. For a given magnitude, q, the conditional probability for a mixed population is estimated as

$$P_m = P_1 + P_2 - P_1 P_2 \tag{4.46}$$

where P_m = combined probability, P_1 = probability determined with the high-flow group, and P_2 = probability determined with the low-flow group.

Besides the mixing of two hydrologic data sets, the operation of a drainage facility may lead to a conditional chance for a subsequent consequence. For instance, a high tailwater

Table 4.8 Hydrologic data of mixed population

Type of study	High-flow group	Low-flow group
Cold region	Runoff in summer months	Snowmelt runoff in winter months
Coastal region	Hurricane runoff	Thunder storm runoff
Mountain region	Runoff at foothills (low elevation)	Runoff at mountains (high elevation)
Urban area	Before stormwater detention	After stormwater detention

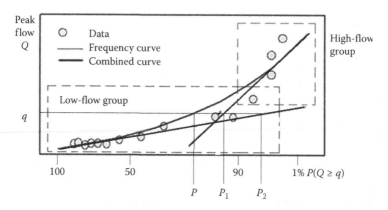

Figure 4.8 Mixed populations with distinct low flow and high flow.

in the downstream receiving water body will lead to damage to the upstream properties. This is a case of conditional probability as

$$P_c = P_3 P_4 \qquad (4.47)$$

where P_c = conditional probability, P_3 = probability of downstream outcome, and P_4 = probability of upstream hydrologic outcome.

EXAMPLE 4.11

The City of Aspen, CO, is located in the Rocky Mountain region at an elevation of 8000 ft. The snow season in the city is from September through April every year. The stream gage at the city is operated all year round. Therefore, runoff flows observed from May through August were induced by rain storms, and other flood events were related to snowmelt runoff. The analysis of the AES of peak flows from the years of 1983–2008 is presented in Figure 4.9. Obviously, the data set is divided into two groups. All high flows were recorded in summers, and the low-flow

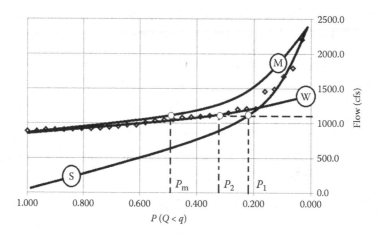

Figure 4.9 Combined probabilities.

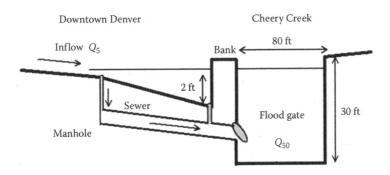

Figure 4.10 Conditional probability.

group was recorded in winters. The curve "S" in Figure 4.9 is the best-fitted model for peak flows observed in summers, whereas the curve "W" represents the model for the winter peak flows. The combined probability, $P(Q < 1100 \text{ cfs})$, is determined by $P_1 = 0.22$ from the S-Curve, and $P_2 = 0.32$ from the W-Curve. Aided by Equation 4.46, the combined probability $P(Q < 1100 \text{ cfs}) = 0.47$. Repeating this example, Curves S and W are merged into Curve M.

EXAMPLE 4.12

The lower downtown is a low land that drains into ABC Creek. The exit of the sewer line from the downtown area is protected with a flood gage from the high tailwater in ABC Creek. As shown in Figure 4.10, when the 50-year event occurs in ABC Creek, the flood gate will be shut down. If the sewer line does not drain, the lower downtown area will be inundated to 2 ft of water under a 5-year event. Determine the probability to have a 2-ft flood water in the downtown area.

Solution: The conditional probability for an inundation of 2 ft is calculated with $P_3 = 1/50$ and $P_4 = 1/5$ as

$$P_c = 1/5 \times 1/50 = 1/250.$$

4.14 Regional analysis

The hydrologic data collected from a hydrologic region, such as nearby gages along a river, can form a database for regional analyses. A regional analysis involves the selection of variables and determination of the relationship between the independent and dependent variables. All independent variables chosen for a regression analysis must be quantifiable. For instance, in a study of flood flow predictions, the peak runoff discharge is the dependent variable that can be correlated with watershed area, slope, index rainfall depth, and soil moisture condition as

$$Q = b_1 A^{b_2} S^{b_3} P^{b_4} P_5^{b_5} \tag{4.48}$$

in which Q = peak discharge, A = basin area, S = basin slope, P = index rainfall depth, P_5 = last 5-day cumulative precipitation depth as an index representing the soil antecedent moisture condition (AMC), and $b_1, b_2, \ldots,$ and b_5 are the parameters in regression analysis.

Equation 4.48 may be simplified to

$$Q = a_1 A^{a_2} P^{a_3} \text{ for the average antecedent soil condition.} \tag{4.49}$$

Further simplification can be

$$Q = c_1 A^{c_2} \text{ for a specified return period.} \tag{4.50}$$

To derive the parameters in the model, Equation 4.48 can be linearized by taking logarithms as

$$\log Q = \log b_1 + b_2 \log A + b_3 \log S + b_4 \log P + b_5 \log P_5 \tag{4.51}$$

Equation 4.51 is solved using the least square technique between the observed and predicted peak flows. Before conducting the regression analysis, it is important to screen the database and identify the range of applicability of the model. The reliability of the model can be evaluated by the root-mean-square between the predicted and observed values. However, it is important to know that an accidental correlation frequently occurs when the number of observations is small. Therefore, the physical relationships among variables shall also serve as the basis to examine the conclusion derived from the mathematical process. For example, the power for the variable of watershed slope in Equation 4.48 shall be compatible to Manning's formula recommended for open-channel hydraulics, and the power for the variable of watershed area shall not exceed one. A high correlation coefficient does not justify a negative power applied to the watershed slope. During the regression analysis, the proper limits for the variables shall be identified based on physical relationships. In the case that the derived power for a particular variable is close to zero, it implies that such a variable is not correlated well. Therefore, removal of such a variable shall be considered.

EXAMPLE 4.13

Table 4.9 is the summary of peak discharges derived from the gages along Fountain Creek in the City of Colorado Springs, CO. The regional analysis is performed using Equation 4.50. The logarithmic transform of Equation 4.50 is

$$\log Q = \log C_1 + C_2 \log A. \tag{4.52}$$

Based on the database in Table 4.9, we have

$$Q_{10} = 928.66 \, A^{0.41} \, \left(r^2 = 0.87 \right) \tag{4.53}$$

$$Q_{50} = 1221.37 \, A^{0.95} \, \left(r^2 = 0.95 \right) \tag{4.54}$$

$$Q_{100} = 1403.22 \, A^{0.58} \, \left(r^2 = 0.91 \right) \tag{4.55}$$

in which A = watershed area in square mile, Q_{10} = 10-year peak discharge in cfs, Q_{50} = 50-year peak discharge in cfs, Q_{100} = 100-year peak discharge in cfs, and r^2 = correlation coefficient.

Table 4.9 Regional analysis for Fountain Creek Basin in Colorado Springs, CO

Watershed and gage	Area (mile²)	Q_{10} (cfs)	Q_{50} (cfs)	Q_{100} (cfs)
Fountain Creek Downstream of Monument Creek	358.00	9225.00	28,511.00	42,206.00
Monument Creek near Fountain Creek	238.00	11,513.00	23,509.00	32,014.00
Fountain Creek Upstream of Monument Creek	121.00	4405.00	14,017.00	20,502.00
Cheyenne Creek near Fountain Creek	22.80	5603.00	10,612.00	13,307.00
Cottonwood Creek near Monument Creek	18.10	3102.00	5005.00	6403.00
Bear Creek near Monument Creek	10.71	3003.00	5001.00	6415.00
Pine Creek near Monument Creek	9.97	2301.00	5505.00	7605.00
Douglas Creek (North) near Monument Creek	6.21	2041.00	3581.00	4561.00
Fishers Canyon Creek near Fountain Creek	5.31	1465.00	2647.00	3084.00
Spring Run near I-25	3.69	961.00	1795.00	1238.00
Douglas Creek (south) near Monument Creek	3.49	1691.00	1908.00	3681.00
Mesa Basin near Monument Creek	2.21	1261.00	1881.00	2251.00
Rockrimmon Basin near Monument Creek	1.89	1511.00	2031.00	2481.00
C_1 in Equation 4.50		928.66	1221.37	1403.22
C_2 in Equation 4.50		0.41	0.54	0.58
r^2 for Equation 4.50		0.87	0.95	0.91

4.15 Flow-duration curve

There are two basic questions about hydrologic risk in engineering designs: (1) the chance of having the next event exceeding a specified magnitude and (2) the percentage of time for having the event exceeding a specified magnitude. For example, the *flow-frequency curve* provides the basis to estimate the chance of having the next event greater than or equal to the 100-year peak flow, while the *flow-duration curve* provides another basis to answer the question as to how many months in a year to have the flow greater than or equal to a specified amount. The flow-duration curve is a plot of the percentage of time that the flow in a stream is to equal or exceed a specified value. The array of data used in the study of flow-duration curve is the average values for a specified period, such as daily, weekly, or monthly discharge. Rank these average discharges from the largest value to the smallest value, involving a total of n values. The exceedance probability is calculated as

$$T_\% = \frac{m}{n} \times 100\% \tag{4.56}$$

where $T_\%$ = percentage of time in which the event, q, equals or exceeds the specified magnitude, Q, n = number of events in the record, and m = rank in a decreasing order in magnitude.

EXAMPLE 4.14

Table 4.10 presents the monthly average flow rates in a river. The ranked flow rates and their exceedance percentages of time are plotted in Figure 4.11.

Table 4.10 Flow duration analysis for monthly average flow ($n = 12$)

Month ($n = 12$)	Jan.	Feb.	Mar.	Apr.	May	Jun.	Jul.	Aug.	Sep.	Oct.	Nov.	Dec.
Flow (cfs)	100	150	200	350	550	700	1200	1500	900	450	280	125
m	1	2	3	4	5	6	7	8	9	10	11	12
Ranked flow	1500	1200	900	700	550	450	350	280	200	150	125	100
$T_\% = m/n$	8.33	16.67	25.00	33.33	41.67	50.00	58.33	66.67	75.00	83.33	91.67	100.00

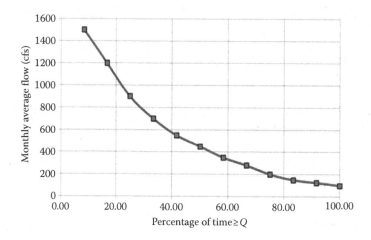

Figure 4.11 Flow duration curve for monthly average flows.

Figure 4.11 indicates that this river carries a flow rate ≥100 cfs for 100% of time. If we design a run-of-river power plant, we may expect a firm yield produced with 100 cfs. Similarly, this river is expected to provide a flow rate ≥450 cfs for 50% of the time.

The area under a flow-duration curve represents the average flow, which shall be close to the median monthly flow with $T_\% = 50\%$.

$$Q_a = \sum_{m=1}^{m=n-1} \frac{[Q(m)+Q(m+1)]}{2} \frac{[T_\%(m+1)-T_\%(m)]}{100} \tag{4.57}$$

in which Q_a = average flow in [L^3/T], $Q(m)$ = average monthly flow in [L^3/T], $T_\%(m)$ = percentage of time to exceed $Q(m)$, and m = rank from the highest to lowest. For the case, the average monthly flow rate is computed as

$$Q_a = \sum_{m=1}^{m=n-1} \frac{(1500+1200)}{2} \frac{(16.67-8.33)}{100} + \cdots + \frac{(125+100)}{2} \frac{(100-91.67)}{100} = 475.5\,\text{cfs} \tag{4.58}$$

A flow-duration curve provides important information to predict the benefits and costs for the operation of power plant, barge navigation, crop irrigation, etc. As illustrated in Figure 4.12, a flow duration curve is divided into five regions from high flow to low flow. The shape of a flow duration curve in its high-flow and low-flow regions is particularly significant in evaluating the

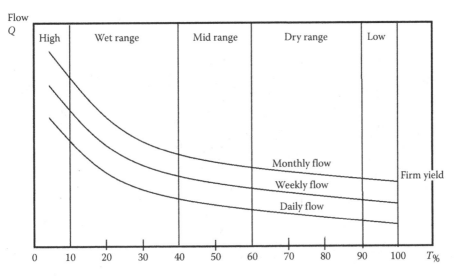

Figure 4.12 Sensitivity of flow duration curve.

pattern of river flows. The high-flow region indicates the type of flood regime the river is likely to have, while the low-flow region characterizes the ability of the river to sustain low flows during dry seasons. A very steep high-flow region implies that rain-induced floods dominate the peak flows in the river, and a very flat high-flow region would result from snowmelt floods that last for several months at high flow rates. In the low-flow region, an intermittent stream would exhibit the periods of no flow. A flat low-flow curve implies sufficient base flows sustained from the groundwater table.

4.16 Closing

The chapter follows the procedure outlined in Bulletin 17B that has been recommended as an official method for hydrologic frequency analyses using the log-Pearson III distribution. This procedure has been coded into computer software packages, including the *Peak FQ* computer model supported by the US Geologic Survey and the *HEC SSP 2.0* computer model for statistical analyses supported by the US Army Corps of Engineers. In fact, the package of HEC SSP includes all the functions in the Peak FQ model as a complete package for hydrologic frequency and duration curve analyses. Details can be found elsewhere as listed in the References.

4.17 Homework

Q4.1 Figure Q4.1 presents the complete record for 5-min rainfall depths, *P5*, recorded at a gage station from 1980 to 1990.

1. Construct the AMS.

Year	1980	1981	1982	1983	1984	1985	1986	1987	1988	1989	1990
Depth (in.)											

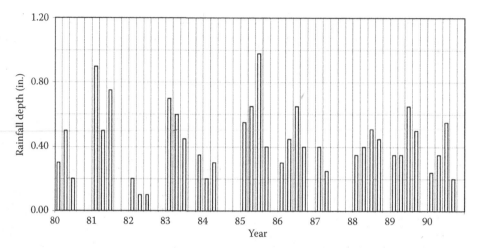

Figure Q4.1 Complete rainfall data record for AMS study.

2.   Rank the values in the table.
3.   Determine the return period for the magnitude of 0.7 in.
4.   Determine the exceedance probability, $P(P5 \geq 0.7 \text{ in.})$.

$Q4.2$ The annual peak runoff rates in Tenmile Creek at Frisco, CO, had been recorded from 1967 to 1982. Perform the following tasks:

1.   Calculate the mean, standard deviation, and skewness coefficient for the sample.
2.   Check if outliers exist.
3.   Adjust the sample skewness by the generalized skewness.
4.   Predict 2-, 10-, and 100-year peak flows by Gumbel distribution.
5.   Produce a graphical paper using the Gumbel scale.
6.   Determine the best-fitted line on the Gumbel probabilistic scale.
7.   Calculate the 5% and 95% confidence limits for Gumbel distribution.

Observed (Year)	Flow		Ranked	Data		
	Peak runoff, Q (cfs)	Rank, m	Ranked runoff, q (cfs)	$P(Q \geq q)$	Return period, T_r (year)	Predicted runoff, Q (cfs)
1967.00	606.00	1.00	1080.00			
1968.00	793.00	2.00	1060.00			
1969.00	606.00	3.00	943.00			
1970.00	943.00	4.00	864.00			
1971.00	864.00	5.00	830.00			
1972.00	781.00	6.00	799.00			
1973.00	788.00	7.00	793.00			
1974.00	830.00	8.00	788.00			
1975.00	1060.00	9.00	781.00			
1976.00	638.00	10.00	638.00			
1977.00	364.00	11.00	636.00			

(Continued)

(Continued)

Observed, (Year)	Flow		Ranked	Data		
	Peak runoff, Q (cfs)	Rank, m	Ranked runoff, q (cfs)	$P(Q \geq q)$	Return period, T_r (year)	Predicted runoff, Q (cfs)
1978.00	1080.00	12.00	636.00			
1980.00	799.00	13.00	606.00			
1981.00	636.00	14.00	606.00			
1982.00	636.00	15.00	364.00			

Q4.3 The duration curve of storage volume for a reservoir is presented in Figure Q4.3. Perform the following tasks:

1. Identify the firm yield.
2. Find the percentage of time to have a storage volume between 400 and 500 acre-ft.
3. Determine the average storage volume.

Q4.4 As shown in Figure Q4.4, the watershed hydrologic study produces two sets of flow predictions, including (a) the snowmelt flow-frequency curve derived from the stream gage data based on May–June flows, and (b) the extreme rainfall-induced flood flow-frequency

Flow rate, q (cfs)	Exceeding probability snow runoff, $P_s(Q > q)$	Exceeding probability rain runoff, $P_r(Q > q)$	Combined exceeding probability, $P_c = P_s + P_r - P_s P_r$	Combined return period, $T_r = 1/P_c$	Combined Gumbel frequency factor
500.00	0.5205	0.0795	0.5586	1.8	−0.287
800.00	0.1986	0.0508	0.2393	4.3	0.586
1000.00	0.1023	0.0386			
1300.00		0.0337		19	1.825
1900.00	0.0094				
2100.00		0.0205		45	2.509
2300.00	0.0018			60	2.736

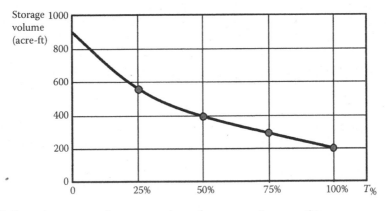

Figure Q4.3 Duration curve of storage volume for reservoir operation.

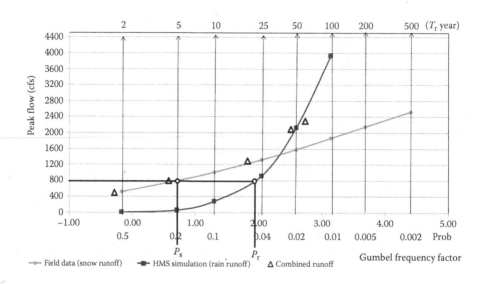

Figure Q4.4 Combined probability for mixed flow population.

curve derived from the numerical simulations for July–September storm events. Follow the example of $P(Q > 800$ cfs$)$ to construct the combined flow-frequency curve.

Bibliography

Abramowtiz, M., and Stegun, I.A. (1965). *Handbook of Mathematical Functions*, Dover, New York.

Beard, L.R. (1962). *Statistical Methods in Hydrology*, U.S. Army Corps of Engineers, Sacramento, CA.

Bulletin 17. (1982, 1983). *Guidelines for Determining Flood Flow Frequency*, U.S. Interagency Advisory Committee on Water Data, Bulletin #17B of the Hydrology Subcommittee, OWDC, US Geological Survey, Reston, VA.

Chow, V.T., Maidment, D.R., and Mays, L.M. (1988). *Applied Hydrology*, McGraw-Hill Company, New York.

Gringorten, I.I. (1963). "A Plotting Rule for Extreme Probability Paper," *Journal of Geophysical Research*, Vol. 68, No. 3, pp. 813–814.

Gumbel, E.J. (1954). "The Statistical Theory of Theory of Droughts," *Proceedings of the American Society of Civil Engineering*, Vol. 80, pp. 1–19.

Guo, J.C.Y. (1986). "Software for Hydrologic Frequency Analysis," the Fourth National Conference on the Application of Microcomputer to Civil Engineering, Orlando, FL, pp. 308–310.

Han, C.T. (1977). *Statistical Method in Hydrology*, The Iowa State University Press, Ames.

Harter, H.L. (1969). "A New Table of Percentage Points of the Pearson Type III Distribution," *Technometrics*, Vol. 11, No. 1, pp. 177–187.

Harter, H.L. (1971). "More Percentage Points of the Pearson Distribution," *Technometrics*, Vol. 13, No. 1, pp. 203–204.

HEC SSP. (2005). *Hydrologic Statistical Package*, Hydrologic Engineering Center, US Army Corps of Engineers, Davis, CA.

Kite, G.W. (1977). *Frequency and Risk Analysis in Hydrology*, Water Resources Publications, Littleton, CO.

Peak FQ Program. (2005). User's Manual for Computer Program PEAK FQ, office of surface water, USGS. http://water.usgs.gov/osw/bulletin17b/bulletin_17B.html.

Wallis, J.R., Matalas, N.C., and Slack, J.R. (1974). "Just a Moment," *Water Resources Research*, Vol. 10, No. 2, 211–219.

Chapter 5

Rational method

The rainfall and runoff process through a watershed depends on the nature of the drainage network. According to definition (AGU, 2015), a *small watershed* is one that has a direct and quick response to rainfall. In other words, the surface storage and depression characteristics do not significantly delay the runoff generation through a small watershed. The actual size of a small watershed may be up to 150 acres (60 ha), depending on the surface storage capacity. As a rule of thumb, the size of small watersheds is approximately 100–150 acres. In an urban area, a watershed is often divided into small catchments by streets, highways, and drainage channels. Urban catchments are often classified as "small watershed" because their tributary areas are small and highly impervious.

In this chapter, the concept of system is applied to a small watershed. It leads to the development of the *rational hydrograph method* (RHM) using the *time of concentration* as the *system memory* and the *runoff coefficient* as the *system parameter*. The RHM is a method developed to predict complete runoff hydrographs under a continuous non-uniform hyetograph. When the rainfall distribution is uniform, the RHM produces a triangular or trapezoidal storm hydrograph with its peak discharge consistent with the conventional rational method.

5.1 Rational method

The rational method is a simplified kinematic wave approach for peak flow estimation. The major variables in the rational method are rainfall intensity, watershed tributary area, and runoff coefficient. The rational method states that (Kuichling, 1889)

$$Q = K\, C\, I\, A \tag{5.1}$$

in which Q = flow rate of runoff in cfs or cms, C = runoff coefficient for design event, I = average intensity of rainfall over the watershed in in./h or mm/h, and A = tributary area in acre or hectare. The value of $K = 1$ if Q in cfs, I in in./h, and A in acre or $K = 1/360$ if Q in cm, I in mm/h, and A in hectare. The *time of concentration* and *runoff coefficient* are the two major parameters that describe the drainage characteristics of a small watershed. The *runoff coefficient*, C, represents the percentage of rainfall excess. The time of concentration defines the contributing rainfall amount to the peak runoff.

5.2 Intensity–duration–frequency formula

To predict a design event, the local rainfall statistics shall be used. A small watershed can be represented by the point rainfall statistics. The US Weather Bureau has published a set of rainfall statistics for the United States such as *HYDRO 35* for the eastern states and *Technical Paper 40* for the US Continent (Hershfield, 1961). More detailed information is also available in the *Precipitation Frequency Atlas 2* for 11 western states. These publications document the empirical procedures to derive rainfall depths for a specified return period and duration. For instance, Table 5.1 is an example of the rainfall depth–duration (P–D) relationship for the City of Denver, CO. The 1-h precipitation values are derived from the Volume 3 of *Rainfall Atlas 2* for the State of Colorado. Rainfall depths with its duration shorter than 60 min can be linearly related to the 1-h precipitation depth by a set of constant ratios (Table 5.1).

The relationship shown in Table 5.1 is the so-called precipitation–duration–frequency (PDF) that can be further converted into the intensity–duration–frequency (IDF) relationship. For instance, the 5-year 10-min precipitation is 0.45 in. (Table 5.1). The corresponding intensity is $0.45 \times 60/10 = 2.70$ in./h. Repeat this process to convert Table 5.1 into Table 5.2.

An IDF curve has a decay nature with respect to rainfall duration. It can be described by a hyperbolic function as

$$I = \frac{C_1 P_*}{(C_2 + T_d)^{C_3}} \tag{5.2}$$

in which I = intensity in in./h or mm/h, P_* = index rainfall depth in in. or mm, T_d = rainfall duration in minutes, and C_1, C_2, and C_3 are constants. Table 5.3 is an example for IDF

Table 5.1 Precipitation–duration–frequency (PDF) for Denver, CO

Return period (years)	5 min (in.)	10 min (in.)	15 min (in.)	30 min (in.)	60 min (in.)
2.00	0.28	0.43	0.54	0.75	0.95
5.00	0.39	0.45	0.61	1.07	1.35
10.00	0.46	0.72	0.91	1.26	1.60
50.00	0.65	1.01	1.28	1.78	2.25
100.00	0.75	1.17	1.48	2.06	2.60

Table 5.2 Intensity–duration–frequency (IDF) for Denver area

Return period (years)	5 min (in./h)	10 min (in./h)	15 min (in./h)	30 min (in./h)	60 min (in./h)
2.00	3.36	2.58	2.16	1.50	0.95
5.00	4.68	2.70	2.44	2.14	1.35
10.00	5.52	4.32	3.64	2.52	1.60
50.00	7.80	6.06	5.12	3.56	2.25
100.00	9.00	7.02	5.92	4.12	2.60

Table 5.3 Coefficients for 10-year intensity–duration formula for cities in the United States

Location	C_1P_{10}	C_2	C_3
Chicago	60.90	9.56	0.81
Denver	50.80	10.50	0.84
Houston	98.30	9.30	0.80
Los Angles	10.90	1.15	0.51
Miami	79.90	7.24	0.73
New York	51.40	7.85	0.75
Atlanta	64.10	8.16	0.76
St. Louis	61.00	8.96	0.78
Cleveland	47.60	8.86	0.79
Santa Fe	32.20	8.54	0.76

Note: P_{10} = 1-h 10-year precipitation depth.

Table 5.4 Index rainfall depth for intensity–duration–frequency formula for Denver, CO

Return period (year)	2	5	10	50	100
P_1 (in.)	0.95	1.35	1.61	2.21	2.65

coefficients in Equation 5.1 developed for the 10-year rainfall event for several metropolitan cities in the United States (Chen, 1976). The index rainfall depth was set to be P_{10}, which is the 10-year rainfall depths.

For the purpose of rainfall–runoff designs, the rational method requires IDF information. The design rainfall duration is assumed to be the time of concentration, T_c, of the watershed. It implies that the entire watershed is the tributary area to the generation of runoff flows. For instance, the IDF for the metro Denver area has three constants—$C_1 = 28.5$, $C_2 = 10.0$, and $C_3 = 0.789$—when using 1-h precipitation depths as the index.

$$I = \frac{28.5P_1}{(10+T_c)^{0.789}} \text{ for Denver, CO only} \tag{5.3}$$

Equation 5.3 represents the *average rainfall intensity* over the *rainfall duration*. In fact, the rainfall time distribution is hardly uniform. Therefore, the applicability of the rational method is limited to small, homogenous urban catchments. Table 5.4 summarizes the values of 1-h precipitation depths for Denver area.

5.3 Volume-based runoff coefficient

By definition, the *flow-based runoff coefficient* in the rational method is determined as

$$C = \frac{Q}{IA} \tag{5.4}$$

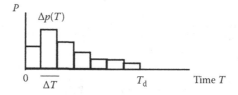

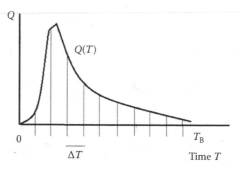

Figure 5.1 Rainfall and runoff volumes.

Equation 5.4 implies that the value of C varies with respect to flow. Referring to Figure 5.1, a *volume-based runoff coefficient* is the volume ratio of runoff hydrograph to rainfall hyetograph:

$$C = \frac{V_F}{V_R} = \frac{\sum\limits_{T=0}^{T=T_B} Q(T)\Delta T}{A \sum\limits_{T=0}^{T=T_d} \Delta P(T)} \tag{5.5}$$

in which V_F = runoff volume in $[L^3]$ under the hydrograph, V_R = rainfall volume in $[L^3]$ under the hyetograph, ΔT = incremental time step on runoff hydrograph such as 5 min, $Q(T)$ = runoff flow in $[L^3/T]$ at time T, $\Delta P(T)$ = incremental rainfall depth in $[L]$ at time T, A = tributary area in $[L^2]$, T_d = rainfall duration, and T_B = based time of runoff hydrograph.

In theory, both Equations 5.4 and 5.5 should yield identical runoff coefficients. In practice, the difference between the observed rainfall hyetograph and the design rainfall IDF curve derived from the rainfall statistics results in a gap between Equations 5.4 and 5.5. Generally, the volume-based runoff coefficient represents the average value for the entire event (Guo and Urbonas, 2014).

5.3.1 Drainage pattern

As illustrated in Figure 5.2, the conventional drainage design is a *two-flow system* that separates the impervious areas from pervious areas. As a result, the storm runoff can be

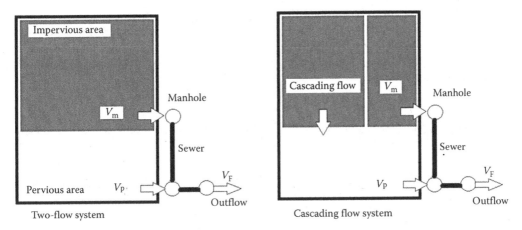

Figure 5.2 Two-flow and cascading drainage systems.

quickly and efficiently collected from the impervious area. In the recent years, under the concept of low-impact development (LID), the *cascading flow system* is recommended to spread stormwater from the upper impervious area onto the lower pervious area for more infiltrating benefits. As expected, two sets of runoff coefficients shall be separately derived for each of these two-flow systems.

5.3.2 Two-flow (distributed) system

Essentially, a two-flow system comprises two independent flow paths to drain runoff flows generated from the impervious and pervious areas, respectively. All individual impervious areas are connected together to collect and deliver stormwater directly into manholes in the street. Pervious areas are also linked through swales to pass stormwater to the streets. Under a rainfall event, the total rainfall volume on the watershed is

$$V_R = PA \tag{5.6}$$

where V_R = event rainfall volume in $[L^3]$ on watershed, P = event rainfall depth in $[L]$, and A = watershed area in $[L^2]$. The runoff volumes produced from the pervious and impervious areas are separately calculated as

$$V_m = (P - D_{vi})I_a A \tag{5.7}$$

$$V_p = m(P - D_{vp} - F)(1 - I_a)A \quad m = 1 \text{ if } V_p > 0; \text{ otherwise } m = 0 \tag{5.8}$$

$$V_F = V_m + V_p \tag{5.9}$$

where V_m = runoff volume in $[L^3]$ from impervious area, V_p = runoff volume in $[L^3]$ from pervious area, V_F = total runoff volume in $[L^3]$, D_{vi} = depression loss in $[L]$ on impervious area such as 0.1 in., I_a = impervious area ratio, $0 \leq I_a \leq 1$, D_{vp} = depression loss in $[L]$

on pervious area such as 0.4 in., F = infiltration amount in [L], and m = 1 if $V_p > 0$ or 0 if $V_p \leq 0$. The variable, m, is to warrant that V_p is numerically positive.

Infiltration loss depends on the nature of soils and the time of operation. Considering a period of 1 h, F = 1.8 in. for type A soils, F = 1.0 in. for type B soils, and F = 0.88 in. for type C and D soils (Guo and MacKenzie, 2014). By definition, the volume-based runoff coefficient is calculated as

$$C = \frac{V_F}{V_R} = n\left[\left(1-\frac{D_{vi}}{P}\right)I_a + m\left(1-\frac{D_{vp}}{P}-\frac{F}{P}\right)(1-I_a)\right] \quad n = 1 \text{ if } C > 0; \text{ otherwise } n = 0 \quad (5.10)$$

where C = volume-based runoff coefficient and n = variable to warrant $C \geq 0$. Equation 5.10 is the sum of two flows and is mostly dominated by the runoff volume, V_m, from the impervious areas. The runoff coefficient in Equation 5.10 is always greater than zero as long as $P > D_{vi}$.

EXAMPLE 5.1

Consider D_{vi} = 0.10 in., D_{vp} = 0.40 in., and F = 0.88 in. over 1 h for type C/D soils. A set of runoff coefficients are produced (as summarized in Table 5.5) using the 1-h rainfall depths (Table 5.4).

Using Table 5.5 as a template, runoff coefficients for type B soils can be produced with F = 1.0 in. (Table 5.6). Similarly, Table 5.7 is prepared for type A soils with F = 1.8 in.

Table 5.5 Runoff coefficients for type C/D soils (1 in. = 25.4 mm)

Soil type (C/D)			D_{vp} = 0.40 in.		
Infiltration F = 0.88 in.			D_{vi} = 0.10 in.		
Variable			Rainfall depth		
Return period	2 year	5 year	10 year	50 year	100 year
P (in.)	0.95 *	1.35	1.60	2.20	2.60
D_{vi}/P	0.11	0.07	0.06	0.05	0.04
D_{vp}/P	0.42	0.30	0.25	0.18	0.15
F/P	0.93	0.65	0.55	0.40	0.34
Imp I_a			Runoff coefficient		
0.05	0.04	0.10	0.24	0.45	0.53
0.10	0.09	0.14	0.27	0.47	0.55
0.20	0.18	0.23	0.35	0.53	0.60
0.30	0.27	0.31	0.42	0.58	0.64
0.40	0.36	0.40	0.50	0.63	0.69
0.50	0.45	0.49	0.57	0.69	0.73
0.60	0.54	0.58	0.64	0.74	0.78
0.70	0.63	0.66	0.72	0.79	0.83
0.80	0.72	0.75	0.79	0.85	0.87
0.90	0.81	0.84	0.86	0.90	0.92
0.99	0.89	0.92	0.93	0.95	0.96

Table 5.6 Runoff coefficients for type B soils

Soil B land use	Imp I_a	2 year	5 year	10 year	50 year	100 year
Lawns, sandy soil	0.02	0.02	0.02	0.14	0.38	0.47
Parks/cemeteries	0.05	0.04	0.05	0.17	0.39	0.49
Playground	0.10	0.09	0.09	0.21	0.42	0.51
Railroad yard area	0.20	0.18	0.19	0.29	0.48	0.56
Gravel streets	0.30	0.27	0.28	0.37	0.54	0.61
Low-density residential	0.40	0.36	0.37	0.45	0.60	0.66
Schools	0.50	0.45	0.46	0.53	0.66	0.71
High-density apartment	0.60	0.54	0.56	0.61	0.72	0.76
Business area	0.70	0.63	0.65	0.69	0.78	0.81
Light industrial	0.80	0.72	0.74	0.78	0.84	0.86
Commercial area	0.90	0.81	0.83	0.86	0.90	0.91
Roof, pavements	0.99	0.89	0.92	0.93	0.95	0.96

Table 5.7 Runoff coefficients for type A soils

Soil A land use	Imp I_a	2 year	5 year	10 year	50 year	100 year
Lawns, sandy soil	0.02	0.02	0.02	0.02	0.02	0.17
Parks/cemeteries	0.05	0.04	0.05	0.05	0.05	0.19
Playground	0.10	0.09	0.09	0.09	0.10	0.23
Railroad yard area	0.20	0.18	0.19	0.19	0.19	0.32
Gravel streets	0.30	0.27	0.28	0.28	0.29	0.40
Low-density residential	0.40	0.36	0.37	0.38	0.38	0.48
Schools	0.50	0.45	0.46	0.47	0.48	0.56
High-density apartment	0.60	0.54	0.56	0.56	0.57	0.64
Business area	0.70	0.63	0.65	0.66	0.67	0.72
Light industrial	0.80	0.72	0.74	0.75	0.76	0.80
Commercial area	0.90	0.81	0.83	0.84	0.86	0.88
Roof, pavements	0.99	0.89	0.92	0.93	0.95	0.95

5.3.3 Cascading flow (lumped) system

A *lumped system* represents a cascading flow process that drains stormwater generated from the upstream impervious areas onto the downstream pervious areas. The intercepted runoff volume is directly added to the lower pervious area for more infiltration benefits. Owing to the fact that the entire impervious area cannot be drained onto the receiving pervious area, a flow interception ratio is introduced to the calculation of runoff volume. The flow interception ratio is similar to the concept of routing percentage used in SWMM 2005 (Rossman, 2005). Under a cascading flow, the runoff volume is computed as (Guo and MacKenzie, 2014)

$$V_p = m[r(P - D_{vi})I_a A + (P - D_{vp} - F)(1 - I_a)A] \qquad (5.11)$$

where r = flow interception ratio ranging $0 \leq r \leq 1$. When $r = 1$, Equation 5.11 represents a complete flow interception, and when $r = 0$, Equation 5.11 is reduced to Equation 5.8 as a two-flow system. When $0 \leq r \leq 1$, the bypassing runoff volume, V_m, from the impervious area is directly released to the street as

$$V_m = (1-r)(P - D_{vi})I_a A \tag{5.12}$$

The resultant runoff coefficient, including the intercepted flow, is calculated as

$$C = \frac{V_F}{V_R} = n\left\{(1-r)\left(1 - \frac{D_{vi}}{P}\right)I_a + m\left[r\left(1 - \frac{D_{vi}}{P}\right)I_a + \left(1 - \frac{D_{vi}}{P} - \frac{F}{P}\right)(1 - I_a)\right]\right\} \tag{5.13}$$

Setting $r = 0$, Equation 5.13 is reduced to Equation 5.10. Equation 5.13 can result in $C = 0$ if the watershed is under a low development condition. On the other hand, Equation 5.13 is converged to Equation 5.10 when the watershed is so highly urbanized that the lower pervious area is overwhelmed with a large amount of runoff flow from the impervious area. In practice, the downstream pervious area needs to be comparable to the upstream impervious area in order to produce a significant infiltration benefit. In general, the cascading effect is more pronounced for frequent and small events and becomes diminished for extreme events. Figure 5.3 is an example for type C/D soils under a 2-year event. In general, the cascading infiltration benefit is limited to the range of $I_a < 45\%$.

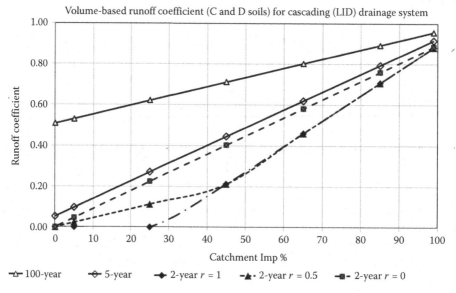

Figure 5.3 Reduction on 2-year runoff coefficients with 50% and 100% LID interception.

5.3.4 Weighted runoff coefficient

For development of mixed land uses, a watershed can be divided into various subareas based on different land uses. Using the area-based method, the weighted runoff coefficient for the entire watershed is calculated as

$$I_a = \frac{A_m}{A} = \frac{A - A_p}{A} \tag{5.14}$$

$$C = \frac{\sum_{i=1}^{i=n} C_i A_i}{A} \tag{5.15}$$

in which A_m = impervious area in $[L^2]$, A_p = pervious area in $[L^2]$, A = total area in $[L^2]$, C = weighted runoff coefficient for watershed, C_i = runoff coefficient for ith subarea, A_i = subarea in $[L^2]$, and n = number of subareas.

5.4 Time of concentration

By definition, the time of concentration of the watershed is the travel time required for stormwater to travel from the most upstream point along the waterway to the outlet. To estimate the time of concentration, it is recommended that the longest waterway be selected to represent the watershed. A waterway often begins with overland flows for a short distance, and then it becomes a gully or swale flow due to the concentration of flows. Further downstream, a waterway is formed by well-defined cross-sections through reaches. The time of concentration along a waterway is the cumulative flow times through the reaches:

$$T_c = T_o + \sum_{j=1}^{j=N} (T_f)_j \tag{5.16}$$

in which T_c = time of concentration in minutes, T_o = overland flow time in minutes, T_f = flow time through the gutter in minutes, j = jth reach, and N = number of reaches. Numerically, the flow time through each reach needs to be determined based on the topography and hydraulic roughness. Owing to the fact that the development of a watershed is a continuous process through multiple stages, it is recommended that the time of concentration be first estimated under the watershed's existing condition and then compared with that under the future condition, and whichever is shorter shall be selected for drainage designs.

5.4.1 Time of concentration for existing condition

An overland flow is a two-dimensional sheet flow. Overland flows occur over the areas upstream of concentrated flows. The maximum length of overland flows in urban areas is approximately 300 ft (90 m) before the overland flow is intercepted by a street gutter or inlet. For rural areas, a maximum length of 500 ft (150 m) is recommended for overland flows. Among many empirical formulas, the *airport formula* is recommended for urban drainage designs. The *airport formula* using English units states (UDFCD, 2010)

$$T_o = \frac{K_{T_o}(1.1 - C_5)\sqrt{L_o}}{S_o^{0.33}} \text{ for overland flow where } L_o \leq L_* \tag{5.17}$$

where T_o = overland flow in minutes, K_{T_o} = 0.395 for ft-s units or 0.715 for m-s units, L_o = overland flow length in [L], C_5 = runoff coefficient for a 5-year event, S_o = overland flow slope in [L/L], and L_* = maximum allowable distance in [L] such as 300 ft (90 m) for urban areas or 500 ft (150 m) for rural areas. Note that C_5 is recommended for Equation 5.17, whereas the design runoff coefficient, C, shall be used in Equation 5.1.

After the flow becomes concentrated, the *US Natural Resources Conservation Service* (NRCS) recommends that *the upland method* be used to estimate the flow time through a swale as (NRCS, 2013)

$$T_f = \frac{L_f}{60V_f} \text{ for shallow water flows} \tag{5.18}$$

$$V_f = K_f\sqrt{S_f} \tag{5.19}$$

where T_f = flow time in [min], L_f = flow length in [L], V_f = flow velocity in [L/T], S_f = flow line slope in [L/L], and K_f = conveyance coefficient in [L/T] (NRCS, 2013; McCuen, 1982). *The NRCS's Soil Conservation Service (SCS) upland method* classifies the surface linings in shallow swales into six categories. Their conveyance coefficients are developed for various roughness surfaces (as shown in Table 5.8).

The SCS upland method in Equation 5.19 was recommended for estimating flow velocities in shallow swales. For a well-defined stream or channel, Manning's formula shall be applied to estimate the flow velocity. However, when the design information is not readily available, types 5 and 6 conveyance coefficients listed in Table 5.8 are also recommended to estimate the shallow water flow velocities in streams, channels, and street gutters.

It is important to understand that the SCS upland method was developed to estimate the average flow time through the entire flow length along the waterway. During a storm event, the upstream reach of a waterway carries shallow sheet flows, and the discharge in a waterway increases downstream. Equation 5.19 estimates a *length-averaged velocity* through the entire waterway under an unsteady flow condition. In comparison, Manning's formula provides a *cross-sectional average velocity* for a steady flow. Therefore,

Table 5.8 Conveyance coefficients, K_f, for upland method

Type ID	Type of linings	K_f (ft/s)	K_f (m/s)
1	Forest or heavy meadow	1.5	0.46
2	Tillage or woodland	5.0	1.53
3	Short grain pasture	7.0	2.13
4	Bare soil	10.0	3.05
5	Grass swale	15.0	4.57
6	Paved gutter shallow flow	20.0	6.10

these two flow velocity equations are not comparable. Equation 5.19 gives much slower flow velocities than Manning's formula. Aided with Equations 5.17 and 5.18, the time of concentration, T_{C1} *in minutes*, for the existing condition is

$$T_{C1} = T_o + T_f \, (\text{min}) \tag{5.20}$$

5.4.2 Time of concentration for future condition

In an urban area, waterways are often equipped with drop structures for grade controls on the stream bed and/or check dams for flow diversions. As a result, Equations 5.17 and 5.18 do not reflect the future hydraulic condition along the waterway. To be conservative, a regional formula shall be developed for the *future time of concentration* under the postdevelopment condition. For instance, the Cities of Denver and Las Vegas (UDFCD, 2010; CCRFCD Manual, 1999) recommend the regional time of concentration be computed based on the future watershed's imperviousness ratio as (Guo and MacKenzie, 2014)

$$T_{C2} = T_* + \frac{L}{60V_*} \, (\text{min}) \tag{5.21}$$

$$V_* = kK_* \sqrt{S_a} \tag{5.22}$$

$$T_* = 18 - 15I_a \left(3 \leq T_* \leq 18 \, \text{min}, \; 0 \leq I_a \leq 1 \right) \tag{5.23}$$

$$K_* = 24I_a + 12 \left(12 \leq K_* \leq 36, \; 0 \leq I_a \leq 1 \right) \tag{5.24}$$

where T_{C2} = regional time of concentration in minutes, L = total length of waterway in [L], including all reaches along waterway, T_* = initial overland flow time in minutes, V_* = postdevelopment concentrated flow velocity in [L/T], K_* = conveyance factor in [L/T], $k = 1$ for foot-second units or 0.305 for meter-second units, and S_a = average slope along waterway. An *initial time* represents the overland flow time through the upland areas. Equation 5.23 implies that the initial time for urban overland flows is 18 min on a pervious surface with $I_a = 0$, and then it is reduced to 3 min for impervious surface under $I_a = 1.0$. Equation 5.23 reveals the fact that the higher the imperviousness in watershed, the shorter the length for overland flow. The *conveyance parameter* of K_* is for shallow, concentrated flows. Equation 5.24 reveals that K_* varies from 12 to 36 ft/s (3.6–11 m/s), depending on the watershed's imperviousness ratio. For instance, on a slope of 1%, Equation 5.24 sets the limits for the length-averaged flow velocity between 1.2 and 3.6 ft/s with an average velocity of 2.0 ft/s. It agrees well with the recommended $K_* = 20$ ft/s given in Table 5.8 for paved surface.

In practice, the design time of concentration is the smaller one between the *computed one representing the predevelopment condition* and the *regional times of concentration representing the postdevelopment condition*:

$$T_c = \min \left(T_{C1}, \, T_{C2} \right) \tag{5.25}$$

in which T_c = time of concentration for design in minutes.

5.4.3 Empirical formulas for time of concentration

There are two distinct approaches developed to estimate the time of concentration. One is *the velocity-based method* by which the time of concentration is defined as the flood wave travel time determined by the flow velocity and waterway parameters. Typical examples for the velocity-based method are Kirpich's formula (Kirpich, 1940) and the upland method recommended by the NRCS. The other approach is *the lag–time method* by which the time of concentration is defined as the time difference between the mass center of rainfall excess and the inflection point on the recession limb of runoff hydrograph. A typical example in this category is the NRCS–method. There are many empirical formulas developed for estimating the time of concentration. These empirical formulas indicate that the time of concentration is a function of waterway length, slope, hydraulic roughness, and rainfall amount. Examples are

$$T_c = 0.0078 \left(\frac{L}{\sqrt{S_o}} \right)^{0.77} \text{(Kirpich's formula in 1940)} \tag{5.26}$$

$$T_c = 0.00316 \frac{L^{0.8}}{\sqrt{S_o}} \left(\frac{1000.0}{\text{CN}} - 9.0 \right)^{0.7} \text{(SCS lag–time method in 1972)} \tag{5.27}$$

$$T_c = \frac{L}{180} + 10 \text{ (Colorado Unit Hydrograph Procedure in 1985)} \tag{5.28}$$

$$T_c = 0.93 N^{0.6} \frac{L^{0.6}}{\sqrt{S_o}} I^{-0.4} \text{ (Kinematic wave by Wooding in 1965)} \tag{5.29}$$

in which T_c = time of concentration in minutes, N = surface roughness such as 0.025 for bare soil and 0.015 for paved surface, CN = SCS curve number for the NRCS method, L = waterway length in feet, S_o = overland flow slope in ft/ft, and I = average rainfall intensity in in./h.

None of the abovementioned widely used formulas could provide either the true or reproducible values of the time of concentration. They are empirical for estimations only (McCuen et al., 1984). In fact, the time of concentration varies with respect to the antecedent soil condition and the distribution of rainfall. As a part of hydrograph convolution process, the time of concentration cannot be directly measured by the time difference between the hyetograph and the hydrograph (Singh and Cruise, 1992). However, it can be indirectly derived by minimizing the least square errors between the predicted and observed hydrographs using runoff coefficient and time of concentration as the system parameters (Guo, 2001a). Based on 44 observed events, the regression equation for estimating the time of concentration was derived as (Guo, 2001b)

$$T_c = M \left(\frac{L}{\sqrt{S_o}} \right)^{0.66} \left(M = 0.054 \text{ for metric unit or } 0.025 \text{ for English units} \right) \tag{5.30}$$

$$V = \frac{1}{N_v} L^{0.66} S_o^{0.33} \left(N_v = 3.48 \text{ for metric units or } 1.50 \text{ for English units} \right) \tag{5.31}$$

in which T_c = time of concentration in minutes, L = waterway length in meters or feet, V = length-averaged flow velocity in mps or fps, and N_v = conversion factor for units.

Equation 5.31 represents the length-averaged velocity over the entire waterway. Such an average value represents a spatial and temporal unsteady flow process. As expected, Equation 5.31 gives smaller velocities than those from Manning's formula. For instance, Equation 5.31 results in a length-averaged flow velocity of 3.25 fps along a 3000-ft waterway on a slope of 5.0%, while the cross-sectional average flow velocity is estimated to be 7.0–9.0 fps by Manning's formula with a roughness coefficient of 0.030. Applying the SCS upland method to the same example, the predicted flow velocity is 3.35 ft/s in a grass waterway. Therefore, Equation 5.31 is similar to and numerically equivalent to the SCS upland method (NRCS, 2013).

EXAMPLE 5.2

A subdivision shown in Figure 5.4 has an area of 350 ft × 500 ft. The local soil is type B. The land uses consist of residential and commercial areas. Determine the 5- and 100-year runoff coefficients for the entire area.

For this case, the residential area is $350 × 350/43,560 = 2.81$ acres, and the commercial area is $150 × 350/43,560 = 1.21$ acre. The runoff coefficients for 5 and 100 years are tabulated in Table 5.9.

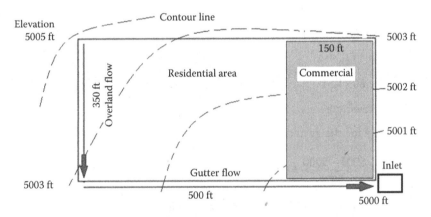

Figure 5.4 Example for mixed land uses.

Table 5.9 Area-weighting method for runoff coefficients

Variable	Land Use Residential	Use Commercial	Weighted C_5
For 5-year event			
Area (acres)	2.810	1.210	
C_5 (5-year)	0.37	0.83	0.51
For design event			
Area (acres)	2.810	1.210	
C-design	0.66	0.91	0.74

EXAMPLE 5.3

Estimate the 100-year peak flow for the subdivision in Figure 5.4.

1. Calculation of runoff coefficient

    From Table 5.5, the weighted $C_5 = 0.51$ and $C_{100} = 0.74$.

2. Calculation of T_{C1} under the existing condition:
    a. Overland flow time

        The slope for the overland flow is

        $$S_o = \frac{5005 - 5003}{350} = 0.0057$$

        The overland flow time is calculated up to 300 ft as

        $$T_o = \frac{0.393(1.1 - 0.54)\sqrt{300}}{0.0057^{0.33}} = 20.9 \, \text{min}.$$

    b. Swale flow time from 300 to 350 ft as

        Let $K = 20$ and $S_o = 0.0057$. The swale flow velocity is

        $$V_2 = 20 \times \sqrt{0.0057} = 1.51 \, \text{fps}.$$

        The swale flow time is

        $$T_2 = \frac{50}{60 \times 1.51} = 0.55 \, \text{min}.$$

    c. Gutter flow time

        The slope for the gutter flow is

        $$S_3 = \frac{5003 - 5000}{500} = 0.006$$

        Let $K = 20$ and $S_3 = 0.006$. The gutter flow velocity is

        $$V_3 = 20 \times \sqrt{0.0060} = 1.55 \, \text{fps}.$$

        The gutter flow time is

        $$T_3 = \frac{500}{60 \times 1.55} = 5.38 \, \text{min}.$$

    d. Time of concentration,
        T_{C1}, under the existing condition is calculated as

        $$T_{C1} = 20.9 + 0.55 + 5.38 = 26.83 \, \text{min}$$

3. Calculation of T_{C2} under the postdevelopment condition:

    With $I_a = 0.55$, $T_* = 18 - 15 I_a = 9.75 \, \text{min}$. and $K_* = 24 \, I_a + 12 = 25.2 \, \text{fps}$

$$S_a = \frac{5005 - 5000}{(350 + 500)} = 0.006$$

$$T_{C2} = 9.75 + \frac{(350 + 500)}{60 \times 25.2\sqrt{0.006}} = 17.1\,\text{min}$$

4. Prediction of peak flow is calculated as

$$T_c = \min(T_{C1}, T_{C2}) = 17.1\,\text{min}$$

$$I = \frac{28.5 \times 2.62}{(10 + 17.1)^{0.789}} = 5.50\,\text{in./h}$$

$$Q = 0.74 \times 5.50 \times \frac{(350 \times 500)}{43,560} = 15.70\,\text{cfs}$$

EXAMPLE 5.4

Estimate the 5-year peak runoff for the street inlet in Figure 5.5 in the City of Denver. The overland flow time is predicted with $C_5 = 0.92$ (see Table 5.6) as

$$T_o = \frac{0.395(1.1 - 0.92)\sqrt{40}}{0.02^{0.33}} = 1.64\ \text{min}$$

The gutter flow time is predicted with $K_f = 20$ as

$$T_s = 400/(60 \times 20.0\sqrt{0.01}) = 3.33\,\text{min}$$

$$T_{C1} = 1.64 + 3.33 = 4.97\,\text{min}$$

The future T_{C2} is calculated with $I_a = 1.0$, $T_* = 3.0\,\text{min}$, $K_* = 36\,\text{fps}$ and

$$S_a = \frac{S_o L_o + S_f L_f}{L_o + L_f} = \frac{0.02 \times 40 + 0.01 \times 400}{(40 + 400)} = 0.011$$

$$T_{C2} = 3.00 + \frac{(40 + 400)}{60 \times 36\sqrt{0.011}} = 4.95\,\text{min}$$

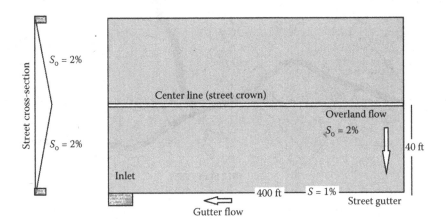

Figure 5.5 Catchment for street drainage.

Use $T_c = 5\,min$ for this case. Therefore, the 5-min rainfall intensity applies to this case. The 5 year 1-h precipitation in Denver is 1.35 in. (see Table 5.4). The design rainfall intensity and peak discharge are

$$I = \frac{28.5 \times 1.35}{(10+5)^{0.789}} = 4.55\,in./h$$

$$Q_p = 0.92 \times 4.55 \times \left(\frac{40 \times 400}{43,560}\right) = 1.54\,cfs$$

EXAMPLE 5.5

A residential subdivision has a tributary area of 33 acres and imperviousness ratio of 0.55. The soil type in this watershed is type C/D soils. As shown in Figure 5.6, the waterway is marked from point 1 to point 2 as overland flow, from point 2 to point 3 as street gutter flow, and from point 3 to point 4 as a grass swale flow. The lengths and slopes for the flow segments are summarized in Table 5.10.

Predict the 100-year peak flow using the 100-year IDF formula,

$$I\,(in./h) = \frac{74.5}{(10+T_c)^{0.789}} \text{ in which } T_c = \text{time of concentration in minutes}$$

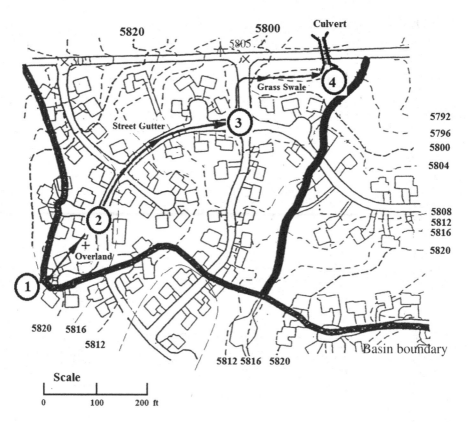

Figure 5.6 Example watershed for calculating time of concentration.

Table 5.10 Flow lengths and slopes for example watershed

Reach	Length (ft)	Type	Slope (%)
1–2	150.00	Overland flow	2.0
2–3	800.00	Gutter flow ($K_f = 20.0$)	1.12
3–4	250.00	Grass flow ($K_f = 15.0$)	4.8

With $I_a = 0.55$ for soil C/D, the 5-year runoff coefficient is $C_5 = 0.55$ (from Table 5.5), which is used to calculate the overland flow time, and the 100-year runoff coefficient is $C_{100} = 0.75$ (from Table 5.5), which is used to calculate the 100-year peak flow.

The times of concentration are determined as shown in Table 5.11. The postdevelopment condition offers a faster flow. As a result, for this case, the time of concentration is determined to be 16.42 min for the design condition, and the 100-year peak flow is determined to be 138.52 cfs.

5.5 Peak flow prediction with multiple tributary areas

Hydrologic homogeneity is one of the basic assumptions for small watershed hydrology. If the land uses or soil types in a watershed vary from one area to another, such a watershed shall be divided into basins. Each basin has its outlet point. All outlets shall be connected together by waterways such as street gutters, sewers, or roadside ditches. To model stormwater movements, all outlets and waterways are expressed as nodes and links. The physical layout of a watershed is then converted into a node-link system. The drainage study of multiple basins starts from the most upstream subbasin and then accumulates the flow time through the drainage network. At the nth node on the waterway, the accumulated contributing area is

$$(A_e)_n = C_n A_n + \sum_{i=1}^{i=n-1} C_i A_i \tag{5.32}$$

The accumulated flow time through the waterway system is

$$(T_c)_n = (T_c)_{n-1} + \frac{L_n}{60 V_n} \tag{5.33}$$

in which A_e = effective contributing area in $[L^2]$, T_c = time of concentration accumulated through the system in minutes through the waterway system, L = waterway length in $[L]$, V = average flow velocity in $[L/T]$, $i = i$th node upstream, and $n = n$th node.

At a design point, the flow times are calculated and accumulated along the incoming waterways. The longest one is selected as the design rainfall duration to predict the peak flow. Having known the contributing area and flow time, the peak discharge is predicted by the rational method.

EXAMPLE 5.6

Three basins in the City of New York are designed to drain into a detention system (Figure 5.7). The watershed parameters for these three basins are summarized in Table 5.12. Determine the 10-year peak discharge at point B.

Table 5.11 Example of flow prediction using rational method

5-year runoff coefficient	Overland flow			Street gutter flow				Swale flow				Existing, T_c (min)
C_5	Slope (%)	Length (ft)	Time (min)	Slope (%)	Length (ft)	SCS K (fbs)	Time (min)	Slope (%)	Length (ft)	SCS K (fps)	Time (min)	
	S_o	L_o	T_o	S_2	L_2	K_2	T_2	S_3	L_3	K_3	T_3	T_{C1}
0.55	2.00	150.00	9.68	1.11	800.00	20.00	6.33	4.80	250.00	15.00	1.27	17.27

Drainage area (acre)	Imp ratio	Design runoff coefficient (C-design)	Total flow length (ft)	Average slope (%)	Initial time (min)	Convey factor (fps)	Flow velocity (fps)	Flow time (min)	Future T_c (min)	Design storm duration (min)	Rainfall intensity 100-year (in/h)	Peak flow (cfs)
A	I_a		L	S_a	T_*	K_*	V_*	T_f	T_{C2}	T_d	I	Q_p
33.00	0.55	0.75	1200.00	1.42	9.75	25.20	3.00	6.67	16.42	16.42	5.60	138.52

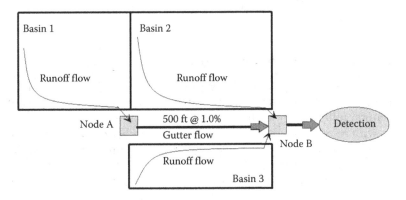

Figure 5.7 Layout for multiple basins.

Table 5.12 Basin parameters

Basin ID	Area (acres)	C	T_c (min)
1.00	2.00	0.55	15.00
2.00	5.00	0.65	22.00
3.00	1.50	0.81	12.00

Figure 5.7 indicates that there are three flow paths to reach point B. Their flow times are as follows:

1. From basin 2: $T_2 = 22$ min

2. From basin 3: $T_3 = 12$ min

3. From basin 1: The flow time shall be counted for the time of concentration of basin 1, and the flow time from point A to point B through the street. According to the SCS upland method, the conveyance parameter for the paved gutter flow is 20.0 on a slope of 0.01. The flow time from basin 1 to point B is the sum of the time of concentration of basin 1 and the flow time through the 500-ft gutter as

$$T_1 = 15 + \frac{500}{60 \times 20 \times \sqrt{0.01}} = 19.17 \, min$$

At point B, the design rainfall duration $T_d = \max (T_1, T_2, T_3) = 22$ min. From Table 5.3, the 10-year design rainfall intensity for New York is

$$I = \frac{51.40}{(7.85 + 22)^{0.75}} = 4.02 \, in./h$$

According to Equation 5.32, the accumulated effective area at point B is

$$A_e = 0.81 \times 1.50 \times 0.55 \times 2.0 \times 0.65 \times 5.0 = 5.65 \, acres.$$

The 10-year peak discharge is

$$Q = I\,A_e = 22.40\text{ cfs}$$

5.6 Concept of rational hydrograph

Stormwater movement is a continuous accumulation of overland flows through space and time. As the flow moves downstream, the waterway collects lateral inflows from both banks. The lateral inflows vary with respect to the rainfall intensity, and the rainfall intensity varies with respect to time and to space as well. To better explain this complicated process, let us consider an ideal watershed (as shown in Figure 5.8); this ideal watershed is divided into four identical basins that drain into the collector channel. Let us assume that (1) the time of concentration, T_c, for each basin is 5 min, and (2) the flow time, T_f, through each segment of the collector channel is also 5 min.

As illustrated in Figure 5.8, the uniform rainfall starts at t = 5 min and ceases at t = 30 min. Runoff flows are accumulated through space and time toward the watershed's outlet. For instance, the flow, q1, was generated at t = 5 min from basin 1, and it takes 15 min for q1 to reach the outlet. The flow, q2, was generated at t = 10 min from basin 2, and it takes 10 min for q2 to reach the outlet. The flow, q3, was generated at t = 15 min from basin 3, and it takes 5 min for q3 to reach the outlet. As a result, the peak flow at t = 20 min is the sum of q1 from basin 1, q2 from basin 2, q3 from basin 3, and q4 from basin 4. At t = 20 min, the entire watershed has become the tributary to the runoff flow at the outlet. Therefore, the time of concentration for this watershed is 20 min. The peak flow continues until the rain ceases at $t = T_d$. As shown in Figure 5.8, the rainfall–runoff accumulation process is divided into three segments: (1) *Before* T_c, *the runoff flow is rising*, (2) *between* T_c *and* T_d, *the runoff flow is peaking*, and (3) *after* T_d, *the runoff flow is receding.*

By repeating the runoff accumulation (as shown in Figure 5.8), we can reach the conclusion that such an ideal condition will produce triangular and trapezoidal hydrographs,

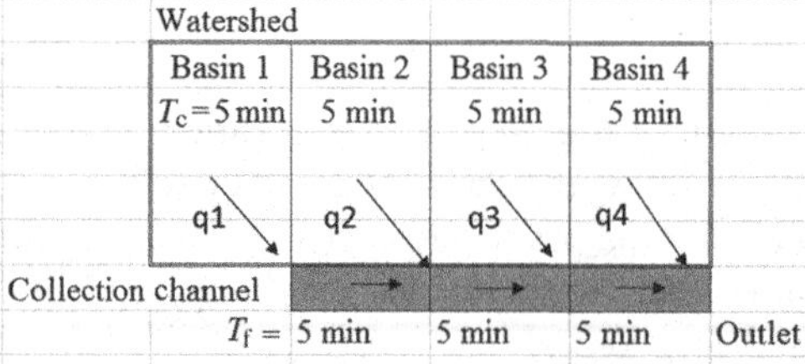

Time (min)	Rainfall	Basin 1	Basin 2	Basin 3	Basin 4	Flow at Outlet	Comments	
0	0	0	0	0	0	0	Dry bed	
5	i	q1	q2	q3	q4	q4		
10	i	q1	q2	q3	q4	q3+q4	T_c = 20 min	Rising
15	i	q1	q2	q3	q4	q2+q3+q4		
20	i	q1	q2	q3	q4	q1+q2+q3+q4	Peak occurs at T_c	
25	i	q1	q2	q3	q4	q1+q2+q3+q4		Peaking
30	i	q1	q2	q3	q4	q1+q2+q3+q4	Rain stops at T_d	
35	0	0	0	0	0	q1+q2+q3		
40	0	0	0	0	0	q1+q2	T_c = 20 min	Recession
45	0	0	0	0	0	q1		
50	0	0	0	0	0	0	Dry bed	
55	0	0	0	0	0	0		

Figure 5.8 Generation of runoff hydrograph.

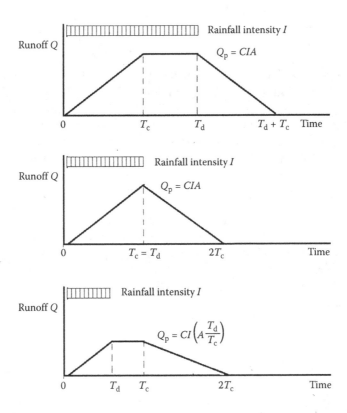

Figure 5.9 Trapezoidal runoff hydrograph using rational method.

depending on whether the rainfall duration is longer or shorter than the time of concentration at the outlet. As illustrated in Figure 5.9, under a long event, the runoff hydrograph reaches its peak at $t = T_c$ and then remains peaking till $t = T_d$. It takes a period of T_c for flow to complete the rising hydrograph. Moreover, it takes a period of T_c to deplete the recession hydrograph. Under a short rainfall event, the entire watershed does not have a chance to become the tributary to the runoff flow. As a result, the rational method is revised to

$$Q = KCIA\left(\frac{T_d}{T_c}\right) \quad \text{for } T_d \leq T_c \tag{5.34}$$

The ratio of T_d/T_c defines the ratio of tributary area to total watershed area when the watershed is under a short rainfall event.

5.7 Rational hydrograph method

The ideal condition to produce triangular or trapezoidal hydrographs does not exist in engineering practice. Therefore, we shall expand the rational method from the ideal case under a uniform rainfall distribution to the real case under a nonuniform hyetograph.

Runoff generation from a watershed is a response to the loading of rainfall amount on the watershed. Hydrologic systems are continuous in time and also causal because the output cannot precede its corresponding inputs over a length of time in the past. Such a time period is termed the *system memory* over which the historical input affects the present system behavior (Chow et al., 1988; Singh, 1982). When introducing the rational method to stormwater drainage designs, Kuichling (1889) stated that the peak rate of runoff at a design point is a direct function of the tributary area and the tributary rainfall amount over the past up to the time of concentration of the watershed. As illustrated in Figure 5.10, a hyetograph consists of a series of rain blocks. The loading of rain blocks onto a watershed is similar to the weight of a train onto a bridge. No matter how long the train is, the loading on the bridge depends on the bridge length.

Applying the same analogy to the rainfall loading on the watershed, the waterway length is converted to the time of concentration as the system memory, T_c, which defines the rainfall amount as the input in the past for producing a flow rate as the output at the present. The flow rate, $Q(T)$ at time T, on the hydrograph (Figure 5.10) depends on the contributing rainfall amount from $(T - T_c)$ to T. Both can be computed as (Guo and Urbonas, 2014)

$$I(T) = \frac{1}{T_c} \sum_{t=T-T_c}^{t=T} \Delta P(t) \text{ where } T_c \leq T \leq T_d \tag{5.35}$$

$$Q(T) = KCAI(T) \text{ where } T_c \leq T \leq T_d \tag{5.36}$$

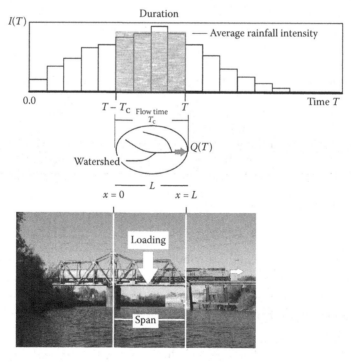

Figure 5.10 Illustration of rational method with a given rainfall event.

in which $I(T)$ = moving average rainfall intensity at time T for a period of T_c prior to T, $Q(T)$ = runoff rate at time T, t = time variable, T_d = event duration, A = tributary area, and $\Delta P(t)$ = incremental rainfall depth at time t. Aided with Equations 5.35 and 5.36, the rational method is expanded into the rainfall–runoff convolution process to convert a nonuniform rainfall distribution into the corresponding runoff hydrograph (Guo, 2001a,b).

Equations 5.35 and 5.36 are applicable to the peaking hydrograph from T_c to T_d. Before the time of concentration, the rising hydrograph represents the runoff flow from the tributary area, which is on an increasing rate through each time step until the entire watershed becomes tributary at $T = T_c$. Under the assumption of linear increasing rate, the tributary area is approximated as (Guo, 2000)

$$A_e = A \frac{T}{T_c} \quad \text{where } T \le T_c \tag{5.37}$$

in which A_e = tributary area to runoff flow at time T. Aided by Equation 5.37, the corresponding runoff flow on the rising hydrograph is estimated as

$$Q(T) = KCA_e I(T) = KCA \frac{T}{T_c} I(T) \quad \text{for } T_c \ge T \ge 0 \tag{5.38}$$

$$I(T) = \frac{1}{T} \sum_{t=0}^{t=T} P(t) \quad \text{for } T_c \ge T \ge 0 \tag{5.39}$$

The recession hydrograph begins as soon as the rain ceases. Runoff flows on the recession hydrograph is a decay curve in nature. For a small watershed, a linear approximation is developed as

$$Q(T) = Q(T_d)\left(1 - \frac{T - T_d}{T_c}\right) \quad \text{for } T_d \le T \le (T_d + T_c) \tag{5.40}$$

$$Q(T_d) = KCAI(T_d) \tag{5.41}$$

in which $Q(T_d)$ = runoff flow at T_d determined by Equation 5.36 and $I(T_d)$ = average rainfall from $(T_d - T_c)$ to T_d determined by Equation 5.35.

EXAMPLE 5.7

The rainfall–runoff event was recorded on November 6, 1977 at the USGS Gage Station 06714300 located at the Concourse D Storm Drain in Stapleton Airport, Denver, CO (USGS Open File 82-873). The watershed area is 96.75 acres with an imperviousness of 38%. The waterway has a length of 2530 ft on a slope of 0.012 ft/ft. The total precipitation for this event was 0.25 in. with duration of 65 min. The observed peak runoff rate was 12.0 cfs.

The time of concentration of this watershed is estimated to be 20 min, and the runoff coefficient is approximately 0.32. Therefore, the runoff flows were predicted by Equations 5.38 and 5.39 before 20.0 min, Equations 5.35 and 5.36 after 20.0 min, and Equations 5.40 and 5.41 after

65 min. Table 5.13 presents a comparison between the observed and predicted hydrographs by RHM. As shown in Figure 5.11, the predicted hydrograph by the RHM reflects the temporal variations on the hyetograph and results in good agreement with the observed.

In this case, the peaking rainfall blocks changed from nonuniform to uniform, so were the runoff flows. The recession began when the rainfall ceased.

Table 5.13 Predicted hydrograph for Stapleton Airport Watershed in Denver, CO

Time (min)	Incremental precipitation (in.)	Observed hydrograph (cfs)	Rational (in./h)	Hydrograph (cfs)	Remarks
			Moving average intensity	Predicted runoff rate	
0.00	0.00	0.00	0.00	0.00	Rising
5.00	0.03	0.30	0.36	2.79	
10.00	0.03	4.20	0.36	5.57	
15.00	0.04	9.80	0.40	9.29	
20.00	0.03	**12.00**	0.39	**12.07**	Peaking
25.00	0.03	12.00	0.39	12.07	
30.00	0.02	10.00	0.36	11.15	
35.00	0.01	7.90	0.27	8.36	
40.00	0.01	6.10	0.21	6.50	
45.00	0.01	5.00	0.15	4.64	
50.00	0.01	4.60	0.12	3.72	
55.00	0.01	4.40	0.12	3.72	
60.00	0.01	4.60	0.12	3.72	
65.00	0.01	2.60	0.12	3.72	$Q(T_d) = 3.72$
70.00	0.00	1.60	0.09	2.97	Recession
75.00	0.00	1.10	0.06	2.23	
80.00	0.00	0.80	0.00	1.49	
85.00	0.00	0.70	0.00	0.74	

Note: A = 96.75 acres, T_c = 20 min, C = 0.32 for 38% imperviousness.

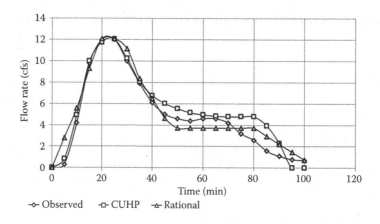

Figure 5.11 Case study for Stapleton Airport Watershed, Denver, CO.

5.8 Applicability limit

The basic assumptions for the applications of the rational method are summarized as follows:

1. Runoff flow rate in the rational method is linearly varied with rainfall depth. Over a single storm event, the maximum rainfall amount over a period of time of concentration is the contributing rainfall depth to the peak runoff flow rate. Statistically, it means that the 100-year rainfall depth produces the 100-year peak flow, and so on and so forth.
2. The hydrologic losses in the watershed are homogenous and uniform. The runoff coefficient varies with respect to the type of soils, imperviousness ratio, and rainfall frequency. A runoff coefficient used in design represents the average soil antecedent moisture condition.
3. A time of concentration is assumed to be equivalent to the time of equilibrium when the entire watershed becomes the tributary area to the peak flow. Under the condition of composite soils and land uses, the area-weighted method is recommended to derive the average hydrologic parameters.
4. This method does not involve any hydrograph routing; as a result, it is not applicable to watersheds with a significant depression area or storage capacity such as ponds and lakes.
5. This method tends to slightly overestimate the combined peak flow at a design point where several upstream flows come together, because the accumulation of time of concentrations is not adequate to compensate the flow attenuation through the hydrograph routing.

The applicable limit of the RHM was examined by the SCS unitgraph method in the HEC-1 Flood Prediction Package using a series of hypothetical square watersheds ranging from 0.01 to 1.0 mile^2. The watershed slope was assumed to be 0.01 ft/ft for all test watersheds, and the runoff coefficient was assigned to be 0.75, equivalent to SCS Curve Number of 85. The SCS 6-h rainfall distribution curve was adopted as the design rainfall distribution with a total precipitation of 2.77 in. Figure 5.12 presents the comparison of the predicted peak flows by HEC-1 and RHM models. It can be seen that the predicted peak runoff rates are very similar until the watershed area exceeds 150 acres.

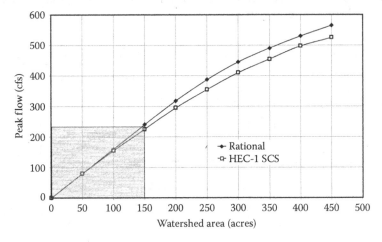

Figure 5.12 Applicability limit for rational hydrograph method.

As a linear model, the major assumptions in the RHM are that the surface storage effect in the watershed is negligible, and the present runoff flow is linearly related to the cumulative rainfall depth within a period of the time of concentration. In general, the RHM tends to overestimate the rising hydrograph. After the entire watershed becomes tributary to runoff, the RHM does fairly reflect the temporal changes in the rainfall distribution. When applying a uniform design rainfall distribution to a small watershed, the RHM produces a triangular hydrograph when the rainfall duration is equal to the time of concentration, or a trapezoidal hydrograph when the rainfall duration is longer or shorter than the time of concentration. Hydrographs predicted by the RHM are comparable with sophisticated models such as CUHP, SWMM, and HEC-1 models. Considering the variations of natural depression in watersheds, it is suggested that the RHM be applicable up to 150 acres.

5.9 Homework

Q5.1 Find the runoff coefficient for an area in which 30% of the area is developed into a business district and another 70% of the area is developed into industrial parks.

Q5.2 A 10-acre lots is to be developed with 2 acres of business areas, 7 acres of residential area, and 1 acre of open space. Find the area-weighted runoff coefficient.

Q5.3 Figure Q5.3 presents a street drainage project located in Denver, CO. The 10-year rainfall IDF in in./h is calculated as

$$I = \frac{28.5 \times 1.61}{(10 + T_c)^{0.789}} \text{ in which } T_c = \text{time of concentration in minutes.}$$

From point A to point B is an overland flow on a ground slope of 0.02. From point B to point C is a street gutter flow on a ground slope of 0.015. From point C to point D is another gutter flow on a ground slope of 0.01.

1. Determine the 10-year peak discharge at point C.
2. Determine the 10-year peak discharge at point D.

Q5.4 A residential area is divided into four smaller areas. Their tributary areas, runoff coefficients, and times of concentration are summarized in Figure Q5.4. The overland

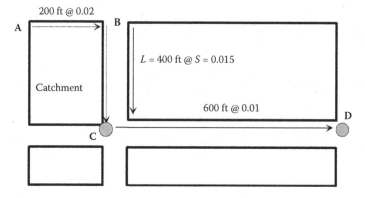

Figure Q5.3 Example of street drainage.

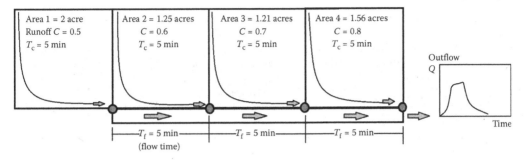

Figure Q5.4 Layout of residential development.

Table Q5.4 Prediction of runoff hydrograph

Subarea ID	1.00	2.00	3.00	4.00
Sub area, A (acres)	2.00	1.25	1.21	1.56
Runoff coefficient, C	0.5	0.6	0.7	0.8
CA (acres)	1.00	0.75	0.85	1.25

flow from each subarea is collected by the channel as shown in Figure Q5.4. The time of concentration, T_c, for each subarea is 5 min, and the flow time, T_f, for each channel segment is also 5 min. The observed rainfall intensity distribution is given in Table Q5.4.

Use the rational method to predict the runoff hydrograph at the outlet.

Solution: Calculation of runoff hydrograph

Time (min)	Rainfall intensity (in./h)	Area1 CA I	Area2 CA 0.75	Area3 CA 0.85	Area4 CA 1.25	Sum of flow at outlet	Discharge at outlet	
0	0	0	0	0	0	0	0.0	
5	1.0	1.0	0.8	0.9	1.3	1.3	1.3	
10	2.0	2.0	1.5	1.7	2.5	2.5 + 0.9	3.4	
15	4.0	4.0	3.0	3.4	5.0	5.0 + 1.7 + 0.8	7.5	
20	5.0	5.0	3.8	4.3	6.3	6.3 + 3.4 + 1.5 + 1.0	12.2	T_c = 20 min
25	2.0	2.0	1.5	1.7	2.5	2.5 + 4.3 + 3.0 + 2.0	11.8	
30	1.0	1.0	0.8	0.9	1.3	1.3 + 1.7 + 3.8 + 4.0	10.7	
35	0.5	0.5	0.4	0.4	0.6	0.6 + 0.9 + 1.5 + 5.0	8.0	
40	0.1	0.1	0.1	0.1	0.1	0.1 + 0.4 + 0.8 + 2.0	3.3	T_d = 40 min
45	0.0	0.0	0.0	0.0	0.0	0.1 + 0.4 + 1.0	1.5	
50	0.0	0.0	0.0	0.0	0.0	0.1 + 0.5	0.6	
55	0.0	0.0	0.0	0.0	0.0	0.1	0.1	
60	0.0	0.0	0.0	0.0	0.0	0	0.0	
65	0.0	0.0	0.0	0.0	0.0	0	0.0	

Bibliography

AGU Committee. (2015). American Geophysical Union, *Committee on Runoff*. http://sites.agu.org/.

CCRFCD Manual. (1999). Clark County Regional Flood Control District, *Design Manual*, http://gustfront.ccrfcd.org/pdf_arch1/hcddm/Current%20Manual%20by%20Section/Section%20100.pdf.

Chen, C.L. (1976). "Urban Storm Runoff Inlet Hydrograph Study," Volume 4, Synthetic Storms for Design of Urban Highway Drainage Facilities, Report, Utah Water Research Laboratory, Utah State University, Logan.

Chow, V.T., Maidment, D.R., and Mays, L.M., (1988). *Applied Hydrology*, McGraw-Hill Company, New York.

CUHP. (2010). "Colorado Urban Hydrograph Procedure," *Urban Stormwater Drainage Manual*, Urban Drainage and Flood Control District, Denver, CO.

Guo, J.C.Y. (1998). "Overland Flow on a Pervious Surface," *International Journal of Water*, Vol. 23, No. 2, pp. 91–95.

Guo, J.C.Y. (2000). "A Semi Virtual Watershed Model by Neural Networks," *Journal of Computer-Aided Civil and Infrastructure Engineering*, Vol. 15, pp. 439–444.

Guo, J.C.Y. (2001a). "Rational Hydrograph Method for Small Urban Watersheds," *Journal of Hydrologic Engineering*, Vol. 6, No. 4, pp. 352–357.

Guo, J.C.Y. (2001b). "Storm Hydrographs from Small Urban Catchment," *Journal of International Water*, Vol. 25, No. 3, pp. 481–487.

Guo, J.C.Y. (2003). Response to Discussion on "Rational Hydrograph Method for Small Urban Catchments," *Journal of Hydrologic Engineering*, Vol. 2, No. 1.

Guo, J.C.Y., and MacKenzie, K. (2014). "Modeling Consistency for Small to Large Watershed Studies," *Journal of Hydrologic Engineering*, Vol. 19, No. 8, 04014009-1 to -7.

Guo, J.C.Y., and Urbonas, B. (2014). "Volume-Based Runoff Coefficient," *Journal of Irrigation and Drainage Engineering*, Vol. 140, No. 2, 04013013-1 to -5.

HEC-1. (2005). *Hydrologic Package for Watershed Analysis*, U.S. Army Corps of Engineers, Hydrologic Engineering Center, Davis, CA.

Hershfield, D.M. (1961). "Rainfall Frequency Atlas of the United States for Durations from 30 Minutes to 24 Hours and Return Periods from 1 to 100 Years," Technical Paper No. 40, U.S. Department of Commerce, Weather Bureau, Washington, DC.

HMS. (2010). *Hydrologic Modeling Simulation Package*, U.S. Army Corps of Engineers, Hydrologic Engineering Center, Davis, CA.

Kirpich, Z.P. (1940). "Time of Concentration for Small Agricultural Watersheds," *Civil Engineering*, Vol. 10, No. 6, p. 362.

Kuichling, E. (1889). "The Relation Between Rainfall and the Discharge of Sewers in Populous Districts," *Transactions of the American Society of Civil Engineers*, Vol. 20, pp. 1–56.

McCuen, R. (1982). *A Guide to Hydrologic Analysis Using SCS Methods*, Prentice-Hall, Inc., Englewood Cliffs, NJ.

McCuen, R.H., Wong, S.L., and Rawls, W.J. (1984). "Estimating Urban Time of Concentration," *Journal of Hydraulic Engineering*, Vol. 110, No. 7, pp. 887–904.

McPherson, M.B. (1968). *Urban Water Resources Research*, ASCS Task Committee, published by ASCE Urban Hydrology Research Council, New York.

Morgali, J.R. (1970). "Laminar and Turbulent Overland Flow Hydrographs," *Journal of Hydraulic Engineering*, Vol. 96, No. HY2, pp. 441–360.

NRCS. (2013). SCS Upland Method. http://www.nrcs.usda.gov/wps/portal/nrcs/site/national/home/.

Rossman, L. (2005). EPA SWMM5. http://www2.epa.gov/water-research/storm-water-management-model-swmm.

Schaake, J.C. (1965). "Synthesis of the Inlet Hydrograph," John Hopkins Storm Drainage Research Project, Technical Report No. 3.

Singh, V.P. (1982). *Hydrologic Systems: Rainfall-Runoff Modeling*, Volume 1, Prentice-Hall, Inc., Englewood Cliffs, NJ.

Singh, V.P., and Cruise, J.F. (1992). "Analysis of the Rational Formula Using a System Approach," *Catchment Runoff and Rational Formula*, edited by B.C. Yen, Water Resources Publication, Littleton, CO, pp. 39–51.

Soil Conservation Service (SCS) or Natural Resource Conservation Service (NRCS). (1976). "Rainfall-Runoff for Small Watersheds," Technical Release No. 55, U.S. Government Printing Office, Washington, DC.

UDFCD. (2010). "Rainfall and Runoff," Volume 1, *Urban Storm Water Design Criteria Manual* (USWDCM), Urban Drainage and Flood Control District, Denver, CO.

USGS Open File Report 82-873. (1983). "Rainfall-Runoff Data from Small Watersheds in Colorado, October 1977 through September 1980," USGS, Lakewood, CO.

Wooding, R.A. (1965). "A Hydraulic Model for a Catchment-Stream Problem," *Journal of Hydrology*, Vol. 3, pp. 254–267.

Yen, B.C., and Chow, V.T. (1974). *Experimental Investigation of Watershed Surface Runoff*, Hydraulic Engineering Series, No. 29, Department of Civil Engineering, University of Illinois at Urbana–Champaign.

Yu, Y.S., and McNown, R.K. (1965). "Runoff from Impervious Surfaces," *Journal of Hydraulic Research*, Vol. 2, No. 1, pp. 3–24.

Chapter 6

Watershed modeling

The development process of a watershed is dynamic. Runoff records collected in the past reflect only the historic situation in the watershed, thus little use for the future flood predictions. As long as the watershed continues changing, any hydrologic analyses on an inconsistent database will not produce reliable conclusions. Therefore, numerical modeling is an alternative to quantify the impact of the development on the hydrologic condition in the watershed. Predictions of rainfall and runoff from a watershed rely on numerical modeling techniques that apply the concept of system to define the relationships among the rainfall as the input, the watershed as the throughput, and the outflow as the output. Empirical formulas shall be developed from observed rainfall and runoff data to establish a reliable, consistent input–output relationship. Although hydrologic analyses rely on the risk-based approach, it is important to distinguish between taking a risk and living with uncertainty. *Risk* indicates those conditions in which there exists a chance for a certain event to occur. For instance, on an average, the risk for the 100-year event to occur in a year is a chance of 1/100. *Uncertainty* is the condition whereby a lack of information prohibits knowing something for certain. For instance, there is an uncertainty about the soil infiltration rate that may be varied between 0.5 and 1.0 in./h.

Before the watershed is ready for development, a numerical computer model shall be formulated to simulate stormwater movements under the existing and future watershed condition. Various design alternatives shall be evaluated for decision makings at the level of *watershed master drainage (hydrologic) planning*, and then the final selection shall be adopted into the *hydraulic designs of the outfall structures,* including sewer trunk lines, regional channels, and major detention facilities. In general, *hydrologic planning* is a study to determine the design flows at the major design points based on the local climate and watershed information. Later, *hydraulic designs* are performed to size the drainage facilities to convey and to store flood flows. The goal of a master drainage planning is to set forth a regional consistent level of protection from potential floods. For engineering applications, *hydraulic analyses* are often performed using a deterministic approach for a preselected risk level. For instance, at the level of hydrologic planning, the magnitude of the 100-year rainfall depth has to be selected within its 5% and 95% confidence limits. Next, the hydraulic design follows the standards and criteria to determine the floodplain's width and water depth using the principles of energy and momentum. Often, the error tolerance in hydraulic designs is much less than that used in hydrologic analyses.

Over the last century, the major research efforts in the flood predictions yielded many useful computer models by which the physical laws developed for surface hydrology have been incorporated into the numerical algorithms. The physical process between rainfall and runoff is simulated by a series of numerical processes. The drainage features in a watershed are described by a set of parameters used in the modeling techniques. Each parameter may vary in a range to reflect the changes in the watershed. The impacts of the watershed development at various stages are quantified by these hydrologic parameters. These numerical simulation techniques have extended hydrologic modeling from the analyses of historical data to the predictions of flood flows under various development conditions.

6.1 Watershed numerical model

Stormwater modeling begins with the watershed topographic maps from which the boundaries of the tributary area, the length of the waterway, and the locations of major drainage structures can be identified. The main purposes of watershed numerical modeling are

1. To predict the magnitudes, frequencies, and distributions of flood flow,
2. To understand the existing flood problems,
3. To evaluate design alternatives for decision making,
4. To conduct scenario studies for future planning, and
5. To assess the impacts on the environment.

For the various purposes, watershed stormwater models are generally classified into (1) *event-based* and (2) *continuous simulation* models. Event-based models are developed to predict the storm runoff due to a single storm event. Examples are the *hydrologic model system (HMS)* published by the Hydrologic Engineering Center (HEC), Army Corps of Engineers since 2000, and the *Technical Release 20 (TR 20)* supported by the Natural Resources Conservation Service. Event-based models are often employed to predict rainfall and runoff for a short period of time. Long-term continuous models are developed using a long-term rainfall record to predict the statistics for the surface and subsurface runoff flows. For instance, the Stormwater Management Model (EPA SWMM, 2005) is supported by the US Environmental Protection Agency (EPA) for long-term rainfall-runoff simulations for a watershed with or without low-impact development (LID) facilities.

Watershed modeling for drainage designs is subject to public hearings and the regulatory agency's approval. As a common practice, the regulatory agency recommends the proper hydrologic methods and models for the project site. It is important to observe the local drainage and design criteria. All drainage studies are affected by the upstream and downstream developments. Therefore, the understanding of the published *regional master drainage plan* is the key factor for conducting an on-site drainage study. Any and all on-site drainage designs must comply with the published regional master drainage plan as a joint effort to reduce the overall flood damage. Any deviation from the published master drainage plan requires negotiations among the property owners and approval by the local community and regulatory agencies.

6.2 Watershed drainage network

An urban stormwater drainage system consists of the following components: (1) watershed areas to generate storm runoff, (2) waterways to collect runoff, (3) detention facilities to store runoff, (4) diversion facilities to transfer runoff out of the watershed, and (5) inflow facilities to receive the runoff flows pumped into the watershed. Each component needs to be prescribed by a set of parameters used in the model.

From the topographic map, the tributary area to a *design point* is delineated based on the elevation contours. A large area is then divided into smaller subbasins, depending on drainage network, hydrologic homogeneity, soil distribution, and land uses. The shape of a subbasin shall have a proper aspect ratio. As a rule of thumb, the width-to-length ratio shall not exceed four. In general, the input data for a subbasin include surface area, waterway length and slope, shape factor, watershed development indicator, and hydrologic losses. The shape of a subbasin can be described by the *waterway length* from the outlet to the subbasin centroid, the *flow time* through the subarea such as time of concentration, or the *time of lag* between the centers of the rainfall and runoff distributions. A watershed development indicator may be the *imperviousness percentage, runoff coefficient,* or *curve number.* Soil infiltration parameters include depression and infiltration losses under a specified soil antecedent moisture condition.

The storm hydrograph produced from a subbasin, such as B1–B5 (shown in Figure 6.1), shall be predicted by a proper hydrologic method. In general, a *unitgraph method* is more applicable to irregular, large watersheds, whereas the *kinematic wave (KW) method* is more applicable to small, urbanized watersheds. Design parameters used in a hydrologic method shall be chosen based on the recommendations in the local design criteria manual or calibrated by the local data. During the numerical process, a watershed is often converted into a *node–link representation.* As illustrated in Figure 6.1, a node represents a subbasin outlet, whereas a link represents the waterway. The runoff flow generated from a subbasin is placed at the subbasin outlet points 2, 4, and 6 in Figure 6.1. The runoff enters the downstream waterway through links 25, 45, 57, and 67. Applying an appropriate routing method, the hydrograph is transferred through a link and then placed at the downstream end of the link. At a confluence node 5, all hydrographs are combined, according to the time sequence.

Case A in Figure 6.1 will have a difficulty to print out the inflow hydrograph to the detention pond at node 9, because the hydrograph at node 9 represents the after-detention hydrograph. As revised in case B, a *dummy node, node 7,* and a *dummy link, link 79,* are added into the node–link system. Node 7 offers the combined hydrograph. Link 79 provides the inflow hydrograph to the pond. A dummy unit does not have a length. The *Direct Routing Method* is used to transfer the hydrograph through a dummy link without any attenuation.

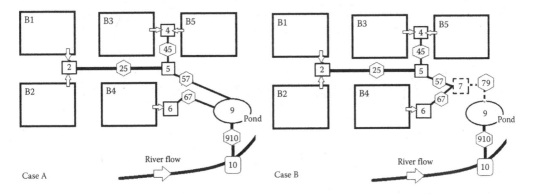

Figure 6.1 Node–link system.

6.3 Kinematic overland flow

Prediction techniques of stormwater hydrograph are generally classified into two major categories: (1) *unit hydrograph (UG) method* and (2) *KW approach.* A common practice in the UG method is to use the length between the outlet and the watershed's centroid to represent the watershed shape factor. The KW procedure employs the unit-width approach that converts a natural, irregular watershed into its equivalent rectangular KW plane using the preselected plane width (Rossman, 2010; Guo, 2006). The overland flow is then modeled by the KW theory using the rectangular *x-y* coordinates. Parking lots (shown in Figure 6.2) are the best example of KW rectangular plane.

For modeling convenience, the sloping plane in Figure 6.3 is further divided into the left-impervious plane and the right-pervious plane, according to the impervious area ratio, I_a, in the watershed. The two flows from the left and right planes will be merged at the watershed outlet. The total flow, Q, collected through the central channel is the sum of the unit-width flows, q, from the left and right planes. The *unit-width KW model* for overland flow consists of the continuity and simplified momentum principles:

$$\frac{\partial y}{\partial t} + \frac{\partial q}{\partial x} = I_e \tag{6.1}$$

and

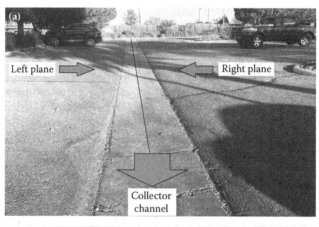

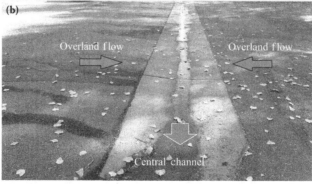

Figure 6.2 Kinematic wave rectangular planes. (a) Two KW planes and central channel and (b) inlet as design point.

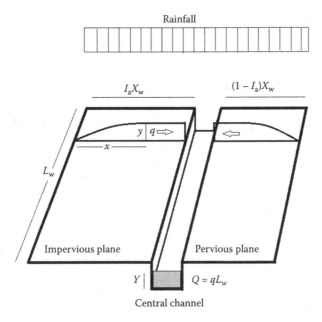

Figure 6.3 Unit-width approach used on kinematic wave rectangular plane.

$$S_f = S_o \tag{6.2}$$

in which y = overland flow depth in [L], q = flow rate in [L^2/T], I_e = excess rainfall intensity in [L/T], which is equal to the design rainfall intensity because of no hydrologic loss on an impervious surface, S_f = friction loss, and S_o = ground slope.

6.3.1 Conversion of watershed into rectangular plane

The KW procedure requires the conversion of the real watershed into its virtual rectangle on the KW plane. Figure 6.4 illustrates the major parameters between the real and virtual systems. According to the principle of continuity, the total tributary area must satisfy the following:

$$A = X_w L_w \tag{6.3}$$

in which A = watershed area in [L^2], X_w = length of overland flow on KW plane in [L], and L_w = width of KW plane in [L]. The fall over the waterway is the elevation difference from the high point on the upstream boundary to the outlet. Between these two systems, the potential energy in terms of the vertical fall along the waterway must be preserved:

$$S_o L = S_w (X_w + L_w) \tag{6.4}$$

in which S_o = longitudinal slope along the waterway through watershed, S_w = slope on KW plane, and L = length of waterway in [L].

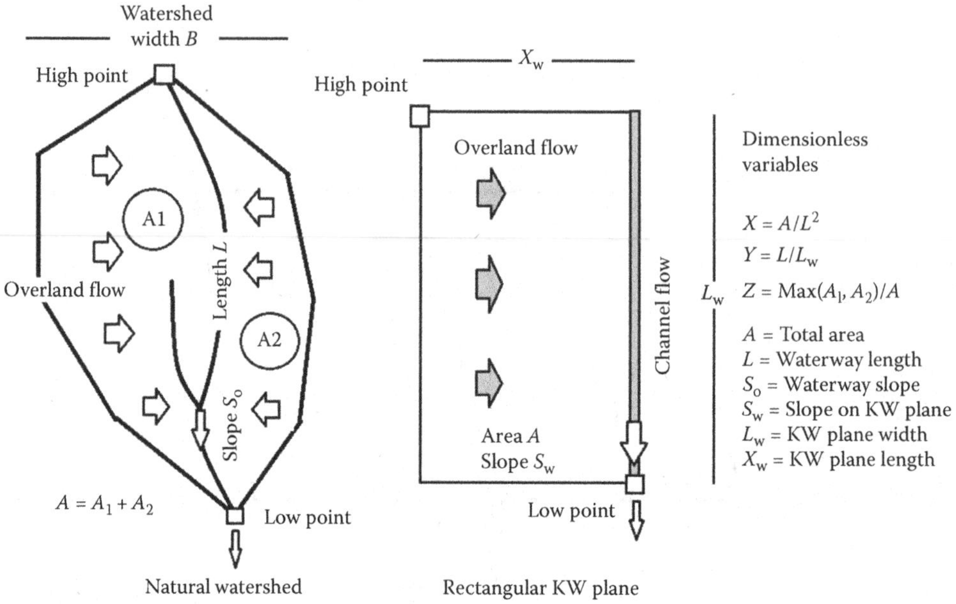

Figure 6.4 Natural watershed and kinematic wave rectangular wave plane.

Using the waterway length, L, to normalize the parameters, Equations 6.3 and 6.4 are converted into

$$\frac{A}{L^2} = \frac{X_w}{L} \frac{L_w}{L} \tag{6.5}$$

$$\frac{S_o}{S_w} = \frac{X_w}{L} + \frac{L_w}{L} \tag{6.6}$$

Equation 6.5 implies that the watershed shape must be preserved between these two systems. Watershed shape factor represents how the overland flows are collected into the waterway. Referring to Figure 6.5, the shape factors for the real watershed and virtual KW plane are defined as

$$X = \frac{A}{L^2} \cong \frac{B}{L} (\leq K) \tag{6.7}$$

$$Y = \frac{L_w}{L} \tag{6.8}$$

in which X = watershed shape factor, B = average width of watershed, Y = KW shape factor for the KW plane, and K = upper limit of shape factor. In practice, it is advisable that a large watershed be divided into smaller subareas, and each subarea should have a shape factor not exceeding the limit $K \leq 4$ (UDFCD, 2005); otherwise, the peak runoff

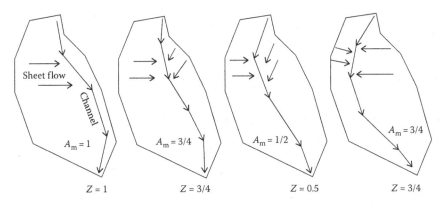

Figure 6.5 Area skewness.

may be overestimated because the subarea is too wide in shape. Aided by Equations 6.5 and 6.8, Equation 6.6 becomes

$$\frac{S_o}{S_w} = \frac{X}{Y} + Y(X \le K) \tag{6.9}$$

The relationship between X and Y was derived using the parabolic equation as (Guo and Urbonas, 2009)

$$Y = (1.5 - Z)\left[\frac{2}{1-2K}X^2 - \frac{4K}{1-2K}X\right] \tag{6.10}$$

$$Z = \frac{A_m}{A} = \frac{\max(A_1, A_2)}{A} \tag{6.11}$$

in which Z = area skewness coefficient between 0.5 and 1.0 and A_m = larger one between A_1 and A_2 that are the two subareas divided by the waterway (Figure 6.4). As illustrated in Figure 6.5, $Z = 0.5$ for a symmetric watershed and $Z = 1.0$ for a side channel along the watershed boundary.

It is noted that Equation 6.10 is reduced to $Y = 2$ for a square watershed with a central channel and $Y = 1$ for a square watershed with a side channel. As indicated in Equation 6.10, the relationship between X and Y depends on the application limit of the watershed shape factor, X. For instance, the width-to-length ratio should not exceed 4. Substituting $K = 4$ into Equation 6.10 yields

$$Y = (1.5 - Z)(2.286X - 0.286X^2) \text{ for all watersheds with } K = 4. \tag{6.12}$$

Table 6.1 tabulates special cases of Equation 6.12. In practice, the shape factor, X, is first determined from the natural watershed. Next, Equation 6.10 or 6.12 gives the equivalent KW shape factor, Y, which defines the width of the KW plane.

Table 6.1 Summary of KW shape factors for conversion to KW rectangular plane

Condition	K = specified variable	K = 4
General formula	$\dfrac{L_w}{L} = (1.5 - Z)\left[\dfrac{2}{1-2K}\left(\dfrac{A}{L^2}\right)^2 - \dfrac{4K}{1-2K}\dfrac{A}{L^2}\right]$	$\dfrac{L_w}{L} = (1.5 - Z)\left[2.286\dfrac{A}{L^2} - 0.286\left(\dfrac{A}{L^2}\right)^2\right]$
Central channel (Z = 0.5)	$\dfrac{L_w}{L} = \dfrac{2}{1-2K}\left(\dfrac{A}{L^2}\right)^2 - \dfrac{4K}{1-2K}\dfrac{A}{L^2}$	$\dfrac{L_w}{L} = 2.286\dfrac{A}{L^2} - 0.286\left(\dfrac{A}{L^2}\right)^2$
Rectangle (Z = 0.5 and $A/L^2 = B/L$)	$\dfrac{L_w}{L} = \dfrac{2}{1-2K}\left(\dfrac{B}{L}\right)^2 - \dfrac{4K}{1-2K}\dfrac{B}{L}$	$\dfrac{L_w}{L} = 2.286\dfrac{B}{L} - 0.286\left(\dfrac{B}{L}\right)^2$
Square (Z = 0.5 and B/L = 1)	$\dfrac{L_w}{L} = 2$	$\dfrac{L_w}{L} = 2$
Side channel (Z = 1.0)	$\dfrac{L_w}{L} = \dfrac{1}{1-2K}\left(\dfrac{A}{L^2}\right)^2 - \dfrac{2K}{1-2K}\dfrac{A}{L^2}$	$\dfrac{L_w}{L} = 1.143\dfrac{A}{L^2} - 0.143\left(\dfrac{A}{L^2}\right)^2$
Rectangle (Z = 1.0 and $A/L^2 = B/L$)	$\dfrac{L_w}{L} = \dfrac{1}{1-2K}\left(\dfrac{B}{L}\right)^2 - \dfrac{2K}{1-2K}\dfrac{B}{L}$	$\dfrac{L_w}{L} = 1.143\dfrac{B}{L} - 0.143\left(\dfrac{B}{L}\right)^2$
Square (Z = 0.5 and B/L = 1)	$\dfrac{L_w}{L} = 1$	$\dfrac{L_w}{L} = 1$
At $A/L^2 = 0$	$\dfrac{L_w}{L} = 0$	$\dfrac{L_w}{L} = 0$
At $A/L^2 = K$		$\dfrac{L_w}{L} \approx 2.286(1.5 - Z)$

KW, kinematic wave.

EXAMPLE 6.1

Determine the KW plane width for the watershed illustrated in Figure 6.6 with $L = 2323\,\text{ft}$, $A = 67.9$ acres (1 acre = $43{,}560\,\text{ft}^2$), impervious percent $I_a = 60\%$, and $S_o = 2\%$.

Solution:

$$X = \frac{A}{L^2} = \frac{67.9 \times 43,560}{2323^2} = 0.55$$

$Z = 0.6$ (estimated from Figure 6.6)

$$Y = (1.5 - Z)(2.286X - 0.286X^2) = (1.5 - 0.6)(2.286 \times 0.55 - 0.286 \times 0.55^2) = 1.17$$

$$Y = \frac{L_w}{L} = \frac{L_w}{2323} = 1.17. \text{ So we have}: L_w = 2709\,\text{ft}$$

$$X_w = \frac{A}{L_w} = \frac{67.9 \times 43,560}{2709} = 1092\,\text{ft}$$

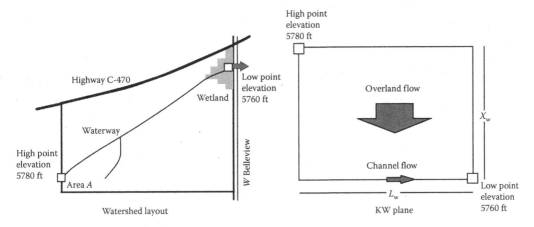

Figure 6.6 Watershed to be converted into kinematic wave plane.

$$\frac{S_o}{S_w} = \frac{X}{Y} + Y = \frac{0.55}{1.17} + 1.17 = 1.64. \text{ So, we have } S_w = 0.012\,\text{ft/ft}$$

It is critically important to understand that the overland flow length, X_w, is derived for the virtual KW plane, which is not a real or rough surface. This KW flow length is resulted from a conformal mapping approach that projects the actual flow motion onto a virtual surface as a "shadow" motion. As a result, the overland flow length, X_w, on the KW plane is not subject to the maximum allowable overland flow length of 300–500 ft as recommended for real watersheds.

EXAMPLE 6.2

Conduct the sensitivity study of KW shape factor on a set of square watersheds illustrated in Figure 6.7 with $A = 50$ m $\times$ 50 m and $S_o = 2\%$.

Solution: Table 6.2 summarizes the detailed calculations for watershed shape conversions. Although these five cases in Figure 6.7 appear similar, none of their KW shape factors is duplicated. As expected, only case 2 has a ratio of $L_w/L = 2.0$, which is a special case. The rest have the ratio of $L_w/L < 2.0$. This test reveals that Equation 6.12 is adequately sensitive to the difference in watershed shape, X, and can produce the various KW shape factors accordingly.

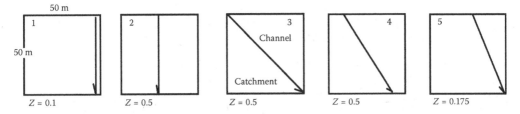

Figure 6.7 Family of equivalent square watersheds.

Table 6.2 Comparison of parameters for KW plane

Area ID	A (m²)	L (m)	Z = Am/A	X = A/L²	Y = Lw/L	Lw (m)	So/Sw	Sw (%)
S1	2500	50.00	1.00	1.00	1.00	50.00	2.00	1.000
S2	2500	50.00	0.50	1.00	2.00	100.00	2.50	0.800
S3	2500	70.70	0.50	0.50	1.07	75.78	1.54	1.300
S4	2500	55.90	0.50	0.80	1.65	92.00	2.13	0.938
S5	2500	55.90	0.75	0.80	1.23	69.00	1.88	1.062

KW, kinematic wave.

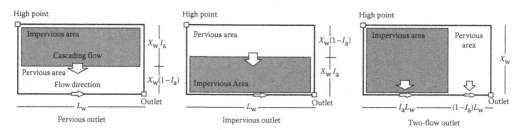

Figure 6.8 Cascading flow versus conventional two-flow system.

In the numerical computation, the net rainfall depth is set to be the difference between the hyetograph and infiltration decayed curve. The KW governing equation applies the difference $(i - f)$ to calculate the net rainfall depth during the event and $(0 - f)$ after the rain ceases. It implies that the KW method is path dependent—the longer the path, the more the loss, and the lesser the peak. For instance, when dividing a large watershed into several smaller subareas, a narrower planar width will increase the flow length and decrease the peak flow. On the contrary, a wider planar width will decrease the flow length and increase the peak flow. Care has to be taken when selecting the planar width for the KW computations. The conventional approach in stormwater modeling is to separate the impervious area from the pervious area. As a result, there are two flow paths developed for an on-site drainage system. On the contrary, the latest concept of LID promotes cascading flows from the impervious area onto the pervious area. As illustrated in Figure 6.8, the length of X_w and L_w can be further divided into two segments, according to the area imperviousness percentage (I_a) in the watershed.

6.3.2 KW flow on impervious plane

On an impervious surface, infiltration and depression losses are negligible. Overland flows are generated from a unit-width area. The rating curve for an overland flow is expressed as

$$q = \alpha y^m \tag{6.13}$$

in which α and m are constants. Empirical formulas are developed from Equation 6.13 as

$$\alpha = \frac{k_n}{N}\sqrt{S_0} \quad \text{and} \quad m = \frac{5}{3} \quad \text{for Manning's formula} \tag{6.14}$$

$$\alpha = C_c\sqrt{S_0} \quad \text{and} \quad m = \frac{3}{2} \quad \text{for Chezy's formula} \tag{6.15}$$

in which N = Manning's Roughness, k_n = 1.0 for meter-second units or 1.486 for foot-second units, and C_c = Chezy's conductivity coefficient. The value of m varies between 3.0 for laminar flow and 3/2 for fully turbulent flow. Taking the first derivative of Equation 6.13 with respect to x yields

$$\frac{\partial q}{\partial x} = \alpha m y^{m-1}\frac{\partial y}{\partial x} \tag{6.16}$$

Substituting Equation 6.16 into Equation 6.1 yields

$$\frac{\partial y}{\partial t} + \alpha m y^{m-1}\frac{\partial y}{\partial x} \approx \frac{\partial y}{\partial t} + u\frac{\partial y}{\partial x} = I_e \tag{6.17}$$

where u = flow velocity in [L/T]. Equation 6.17 is the total derivative of flow depth. Solutions for Equation 6.17 are composed of two characteristic curves:

$$\frac{dy}{dt} = I_e \tag{6.18}$$

$$\frac{dx}{dt} = u = \alpha m y^{m-1} \tag{6.19}$$

The initial condition for the overland flow in Figure 6.3 is a dry bed everywhere:

$$y(t,x) = y(0,x) = 0.0 \tag{6.20}$$

The upstream boundary does not have any inflow. As a result, the boundary condition is

$$y(t,x) = y(t,0) = 0.0 \tag{6.21}$$

Aided by the initial and boundary conditions, integrating Equation 6.18 yields

$$y = I_e t \tag{6.22}$$

Integrating Equation 6.19 yields

$$x = \alpha I_e^{m-1} t^m \tag{6.23}$$

Substituting Equation 6.22 into Equation 6.23 yields the water surface profile (x, y):

$$x = \alpha \frac{y^m}{I_e} \tag{6.24}$$

When the KW reaches the outlet at $x = L$, the flow time, t, is termed *the time of equilibrium* of the watershed, T_E. Therefore, Equation 6.23 is converted into the equation for time of equilibrium:

$$T_E = \left(\frac{L}{\alpha I_e^{m-1}} \right)^{\frac{1}{m}} \tag{6.25}$$

When Manning's formula is used, $m = 5/3$, the time of equilibrium in Equation 6.25 becomes

$$T_E = \left(\frac{NL}{k_n \sqrt{S_0} I_e^{0.67}} \right)^{0.60} \tag{6.26}$$

The equilibrium flow depth, Y_e, and discharge, q_e, at the outlet are (Wooding, 1965)

$$Y_e = I_e T_E \tag{6.27}$$

$$q_e = \alpha Y_e^m \tag{6.28}$$

The time of equilibrium is similar to the time of concentration, except that the rainfall excess must be uniform in time and space. After the time of equilibrium, the entire unit-width area becomes tributary to the flow at the outlet. In comparison, a long rainfall event that has a duration longer than the time of equilibrium is more critical to the design. Therefore, further discussions of overland hydrograph are based on the assumption of a long event. To apply Equations 6.22, 6.23, and 6.26 to the unit-width area, the overland runoff hydrograph can be predicted, as shown in Figure 6.9. An overland runoff hydrograph consists of three segments: *the rising limb* before the time of equilibrium, *the peaking portion* between the time of equilibrium and the end of rainfall, and *the recession* after the rain ceases.

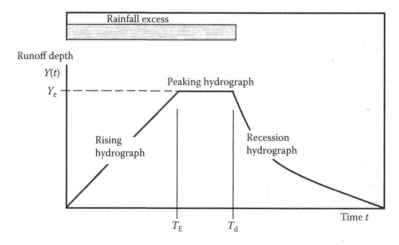

Figure 6.9 Runoff hydrograph.

1. *Rising hydrograph* $0 \leq t \leq T_E$. Equation 6.22 is converted to

$$Y(t) = I_e t \tag{6.29}$$

in which $Y(t)$ = runoff depth in [L] at time t.

2. *Peaking hydrograph* $T_E \leq t \leq T_d$. Equation 6.29 is applied to the peak depth:

$$Y(t) = I_e T_E \tag{6.30}$$

in which T_d = rainfall duration in [T].

3. Recession $t \geq T_d$

After the rain ceases, the equilibrium water profile begins to recede. As shown in Figure 6.10, the water depth y at the station x has to travel through the distance of $(L - x)$ to reach the outlet. Aided by Equation 6.19, the recession wave movement is described as

$$\frac{\mathrm{d}x}{\mathrm{d}t} = \frac{L - x}{t - T_d} = u \tag{6.31}$$

Substituting Equations 6.24 for x and Equation 6.19 for u into Equation 6.31 yields

$$L = \alpha \frac{y^m}{I_e} + \alpha m y^{m-1}(t - T_d) \tag{6.32}$$

For a specified t, $t \geq T_d$, the outlet depth, $Y(t) = y$ in Equation 6.32 can be solved iteratively.

The previous derivation also implies that the rising hydrograph can be related to the equilibrium variables as

$$Y(t) = Y_e \frac{t}{T_E} \tag{6.33}$$

$$q(t) = q_e \left(\frac{t}{T_E}\right)^m \tag{6.34}$$

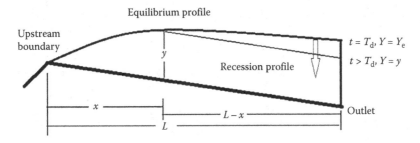

Figure 6.10 Recession of equilibrium water profile for overland flow.

EXAMPLE 6.3

Predict the overland flow hydrograph for $I_e = 6.0$ in./h applied to the impervious surface in the watershed in Example 6.1 with $A = 67.9$ acres, $L_w = 2709$ ft, $S_o = 0.012$, $N = 0.014$, and $T_d = 60$ min.

$$\alpha = \frac{k}{N}\sqrt{S_o} = \frac{1.486}{0.014}\sqrt{0.012} = 11.63 \text{ and } m = 5/3 = 1.67$$

$$I_e = 6 \text{ in./h} = 0.000138 \text{ ft/s}$$

$$L = X_w = A/L_w = (67.9 \times 43,560)/2709 = 1091.8 \text{ ft as the length of the unit-width}$$
$$\text{overland flow}$$

$$T_E = \left(\frac{NL}{k_n\sqrt{S_o}I_e^{0.67}}\right)^{0.60} = \left(\frac{0.014 \times 1091.8}{1.486 \times \sqrt{0.012} \times 0.000138^{0.67}}\right)^{0.60} = 535.5 \text{ s}$$

The rising hydrograph is calculated as

$$Y(t) = I_e t = 0.000138 \, t \text{ for } t < 535.5 \text{ s}$$

The peaking hydrograph remains constant:

$$Y_e = I_e T_E = 0.000138 \times 535.5 = 0.0744 \text{ ft}$$

$$q_e = \alpha Y_e^m = 11.63 \times 0.0744^{1.67} = 0.152 \text{ cfs/ft}$$

Table 6.3 Overland flow hydrograph from impervious surface

t (s)	Guessed Y (ft)	Calculated Y (ft)	$q(t)$ (cfs/ft)
0		0.0000	0.0000
360		0.0500	0.0781
536		0.0744	0.1516
1080		0.0744	0.1516
1440		0.0744	0.1516
1800		0.0744	0.1516
2160		0.0744	0.1516
2520		0.0744	0.1516
2880		0.0744	0.1516
3240		0.0744	0.1516
3600		0.0744	0.1516
3720	0.0590	0.0590	0.1030
3840	0.0464	0.0464	0.0689
3960	0.0364	0.0364	0.0461
4080	0.0300	0.0300	0.0333
4200	0.0200	0.0200	0.0169
4320	0.0200	0.0200	0.0169
4440	0.0100	0.0100	0.0053

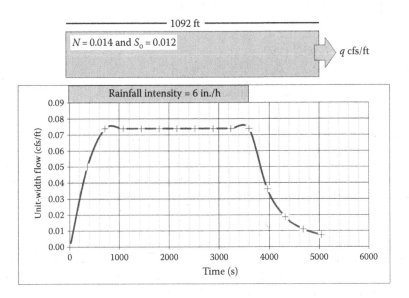

Figure 6.11 Predicted kinematic wave overland flow from impervious area.

The recession hydrograph, $Y(t) = y$, that is determined by an iterative procedure with guessing the water depth for a given time, t, in Equation 6.32 is

$$1091.8 = 10.85\frac{y^{1.67}}{0.000138} + 11.63 \times 1.67 \times y^{m-1} \times (t - 3600) \quad \text{for } t > 3600\,\text{s}$$

The rating curve in Equation 6.13 is used to convert flow depths to flow rates. The predicted overland runoff hydrograph is summarized in Table 6.3 and is plotted in Figure 6.11.

6.3.3 Overland flow on pervious surface

On a pervious surface, the rainfall excess is subject to infiltration losses, which can be described as

$$I_e = I - f_t \tag{6.35}$$

in which f_t = infiltration rate in [L/T] at time t described by Horton's formula in Chapter 3. With a decay infiltration rate, the rainfall distribution becomes nonuniform. The KW travel time through a watershed under such a nonuniform rainfall excess is no longer a constant, but it varies between the *time of concentration* (the longest) and *the time of equilibrium* (the shortest). Considering the decay nature of soil infiltration, at the beginning of an event, the higher infiltration rates result in less runoff depths and longer travel times. As time goes on, the infiltration rate gradually reaches its final constant rate, and the KW travel time is also gradually reduced to the time of equilibrium. Guo (1998) suggested that $m = 2$ for a wide overland flow. As a result, Equation 6.28 becomes

$$q = \alpha y^2 \tag{6.36}$$

Substituting Equation 6.35 into Equation 6.18 yields

$$\frac{dy}{dt} = I - \left[f_c - (f_0 - f_c) e^{-kt} \right] \qquad (6.37)$$

Substituting Equation 6.36 into Equation 6.19 yields

$$\frac{dx}{dt} = 2\alpha y \qquad (6.38)$$

The overland runoff hydrograph generated from a pervious surface under a long rainfall can be divided into three distinct portions: *Rising, Peaking,* and *Recession.* Details are discussed as follows:

1. **Rising portion**
    During an event, runoff occurs until the rainfall amount exceeds hydrologic losses. The ponding time, T_S, is the period of time with no runoff at the beginning of the rainfall event because $I \leq f_0$. Of course, $T_S = 0$ if $I \geq f_0$. As shown in Figure 6.12, the rising hydrograph starts from the ponding time, T_S, and ends at the time of concentration $(T_S + T_C)$. The time of concentration is the period of time for water to travel through the waterway to reach the outlet. Before $(T_S + T_C)$, the flow is on a rising curve, and after $(T_S + T_C)$, the flow becomes peaking because the entire watershed has become contributing to the outflow. The rising hydrograph is represented by integrating Equation 6.37 from T_S to t as

$$Y(t) = (I - f_c)(t - T_S) + \frac{f_0 - f_c}{k} \left(e^{-kt} - e^{-kT_S} \right) \quad \text{for} \quad T_S \leq t \leq (T_S + T_C) \qquad (6.39)$$

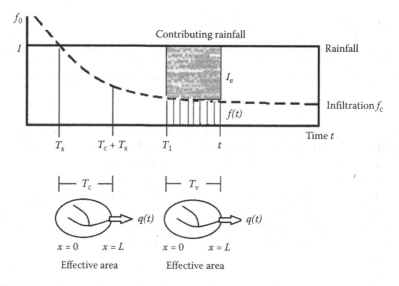

Figure 6.12 Kinematic wave integration domains for time and space.

Correspondingly, the travel distance for the KW is the integration of Equation 6.38 from $x = 0$ to x:

$$\frac{x}{2\alpha} = \frac{(I - f_c)}{2}(t - T_S)^2 + \frac{(f_0 - f_c)}{k^2}\left\{[1 - k(t - T_S)]e^{-kTS} - e^{-kt}\right\} \quad \text{for } 0 \le x \le L \quad (6.40)$$

Equations 6.39 and 6.40 provide direct solutions for flow depth and discharge at time $t \le (T_S + T_C)$.

2. **Peaking portion**

Referred to Figure 6.12, the peaking portion starts at $(T_S + T_C)$ till T_d when the rain ceases. According to the definition of time of concentration, substituting $t = T_S + T_C$ and $x = L$ into Equations 6.39 and 6.40 yields

$$Y_c = (I - f_c)T_C + \frac{f_0 - f_c}{k}e^{-kTS}\left(e^{-kTC} - 1\right) \text{ at } t = T_S + T_C \quad (6.41)$$

$$\frac{L}{2\alpha} = \frac{(I - f_c)}{2}T_C^2 + \frac{(f_0 - f_c)}{k^2}e^{-kTS}\left(1 - kT_C - e^{-kTC}\right) \quad (6.42)$$

in which Y_c = water depth in [L] at $t = (T_C + T_S)$ and T_C = time of concentration, which can be iteratively solved using Equation 6.42. With the known T_C, the water depth $Y(t) = Y_c$ can be determined by Equation 6.41.

The KW speed on a pervious surface is proportional to the rainfall excess. As time goes on, the infiltration rate decays. The more the rainfall excess, the faster the KW. After $t > (T_S + T_C)$, the contributing rainfall to the water depth in Figure 6.12 is the rain amount from T_1 to t. The period between T_1 and t is termed the *time of travel*, T_v, which is similar to T_C, but shorter than T_C. During each period of time of travel, the integration domains are from $x = 0$ to $x = L$ for distance and $t = T_1$ to $t = t$ for time. Integrating Equations 6.37 and 6.38 yields

$$Y(t) = (I - f_c)T_v + \frac{f_0 - f_c}{k}e^{-k(t - T_v)}\left(e^{-kT_v} - 1\right) \quad (T_S + T_C) \le t \le T_d \quad (6.43)$$

$$\frac{L}{2\alpha} = \frac{(I - f_c)}{2}T_v^2 + \frac{(f_0 - f_c)}{k^2}e^{-k(t - T_v)}\left(1 - kT_v - e^{-kT_v}\right) \quad (6.44)$$

$$T_v = T_1 - t \quad (6.45)$$

Equation 6.44 has only one unknown T_v that can be iteratively solved. The flow depth during the peaking period is calculated by Equation 6.43. The peaking hydrograph ends at T_d when the rain ceases. The maximum flow depth is calculated at $t = T_d$ in Equation 6.43 as

$$Y_m = (I - f_c)T_v + \frac{f_0 - f_c}{k}e^{-k(T_d - T_v)}\left(e^{-kT_v} - 1\right) \quad (6.46)$$

When time t is long enough for the infiltration rate to reach its final constant rate, Equations 6.43 and 6.44 are reduced to

$$Y_e = (I - f_c)T_E = I_e T_E \quad (6.47)$$

$$\frac{L}{2\alpha} = \frac{I - f_c}{2} T_E^2 \tag{6.48}$$

The time of equilibrium can be directly solved using Equation 6.48, and then the equilibrium depth is provided by Equation 6.47. After $T_S + T_C$, each flow depth at the outlet is determined with its own time of travel, T_v, which is the rainfall duration contributing to the rainfall depth. T_v varies between T_C and T_E. Or, T_C is the longest T_v and T_E is the shortest T_v.

3. **Recession portion**

The recession hydrograph starts at $t = T_d$. During the recession, the flow depth at the outlet is predicted by integrating Equation 6.37 with $I = 0$ as

$$\int_{Y_m}^{y} dy = -\int_{T_d}^{t} f(t) dt \tag{6.49}$$

Substituting Horton's formula into Equation 6.49 yields

$$y = Y_m - f_c(t - T_d) + \frac{f_0 - f_c}{k}\left(e^{-kt} - e^{-kT_d}\right) \quad \text{for } t > T_d \tag{6.50}$$

For a given time $t > T_d$, Equation 6.50 provides the flow depth at the outlet point.

EXAMPLE 6.4

Apply $I_e = 6.0$ in./h to the pervious area in the watershed in Example 6.3 with $A = 67.9$ acres, $L_w = 2709$ ft, $S_o = 0.012$, $N = 0.030$, $f_0 = 3.0$ in./h, $f_c = 0.5$ in./h, $k = 0.0018$ l/s, and $T_d = 60$ min.

$$\alpha = \frac{k_n}{N}\sqrt{S_o} = \frac{1.486}{0.03}\sqrt{0.012} = 5.43 \text{ and } m = 2$$

$I_e = 6$ in./h $= 0.000138$ ft/s, $f_c = 0.5$ in./h $= 0.000012$ ft/s, and $f_0 = 0.000069$ ft/s. Because $I_e > f_0$, so $T_S = 0$

$$L = X_w = A/L_w = (67.9 \times 43,560)/2709 = 1091.8 \text{ ft.}$$

First, we have to determine the time of concentration for this case. Substituting the design variables into Equation 6.42 yields

$$\frac{1091.8}{2 \times 5.43} = \frac{(0.000138 - 0.000012)}{2}T_c^2$$
$$+ \frac{(0.000069 - 0.000012)}{0.0018^2}e^{-0.0018 \times 0.0}\left(1 - 0.0018 \times T_c - e^{-0.0018T_c}\right)$$

By trial and error, $T_C = 1429.4$ s.

1. **Rising hydrograph before $t < T_C$**

Apply Equation 6.39 to the given time, t, to determine the flow depth:

$$Y(t) = (0.000138 - 0.000012)(t - 0) + \frac{0.000069 - 0.000012}{0.0018}\left(e^{-0.0018t} - e^{-0.0018 \times 0.0}\right)$$

2. Peaking hydrograph $T_C \leq t \leq T_d$

Apply Equation 6.44 to the given time t to determine the time of travel:

$$\frac{1091.8}{2 \times 5.43} = \frac{(0.000138 - 0.000012)}{2} T_v^2$$
$$+ \frac{(0.000069 - 0.000012)}{0.0018^2} e^{-0.0018(t-T_v)} \left(1 - 0.0018 T_v - e^{-0.0018 T_v}\right)$$

Having known T_v, use Equation 6.43 to find the flow depth:

$$Y(t) = (0.000138 - 0.000012) T_v + \frac{0.000069 - 0.000012}{0.0018} e^{-0.0018(t-T_v)} \left(e^{-0.0018 T_v} - 1\right)$$

The maximum flow depth and flow rate, Y_m, for this case is determined using Equation 6.46 as

$$Y_m = (0.000138 - 0.000012) T_v + \frac{0.000069 - 0.000012}{0.018} e^{-0.0018(3600-T_v)} \left(e^{-kT_v} - 1\right)$$

By trial and error, $T_v = 1259.4$ s, $Y_m = 0.16$ m, and $q_m = 0.139$ cfs/ft.

3. Recession hydrograph $t > T_d$

For a given $t > T_d$, the flow depth is determined using Equation 6.50 as

$$y = 0.16 - 0.000012(t - 3600) + \frac{0.000069 - 0.000012}{0.0018} \left(e^{-0.0018t} - e^{-0.0018 \times 3600}\right)$$

The details of calculations are summarized in Table 6.4 and plotted in Figure 6.13. It is noticed that after T_C, the peak flow depth continues increasing because of the decrease in the infiltration

Table 6.4 Overland flow hydrograph from pervious surface

	t (s)	T_v (s)	Y(t) (ft)	q(t) (cfs/ft)
t = 0	0		0.0000	0.000
	320		0.0267	0.004
	600		0.0552	0.017
	900		0.0888	0.043
	1200		0.1243	0.084
$T = T_S + T_C$	1429	1429.00	0.1522	0.126
	1800	1325.50	0.1563	0.133
	2400	1277.70	0.1588	0.137
	3000	1263.90	0.1596	0.138
t = T_d	3600	1259.40	0.1599	0.139
	4800		0.1460	0.116
	6000		0.1321	0.095
	7200		0.1182	0.076
	8400		0.1043	0.059
	9600		0.0904	0.044
	10800		0.0765	0.032
	12000		0.0626	0.021
	13200		0.0488	0.013

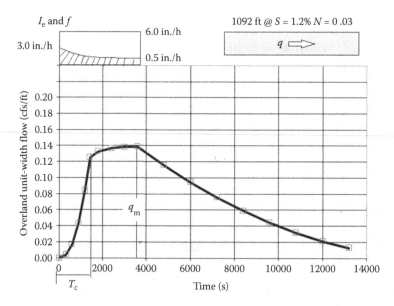

Figure 6.13 Overland flow hydrographs from pervious surface.

loss. The maximum depth occurs at $t = T_d$. The time of travel varies between the time of concentration and the time of equilibrium. In comparison with Example 6.3, the peak flow from the pervious surface is 0.139 versus 0.152 cfs/ft from the impervious surface. The time to peak for the overland flow from the impervious surface is much faster than that on the pervious surface.

EXAMPLE 6.5

Apply $I_e = 6.0$ in./h to the watershed in Example 6.1. The watershed has $A = 67.9$ acres, $I_a = 60\%$, and $L_w = 2709$ ft. As shown in Figure 6.14, the watershed is converted into a rectangular sloping plane, and then the plane is divided into an impervious area of 60% the total and another pervious area of 40% the total. Assuming that the drainage system is composed of two independent flows, determine the overland flow hydrograph from this watershed.

Examples 6.3 and 6.4 offer the unit-width overland flows. The total runoff flow at the outlet is the sum of

$$Q(t) = q_{imp}(t)I_aL_w + q_{perv}(t)(1 - I_a)L_w \tag{6.51}$$

in which $Q(t)$ = total flow in $[L/T^3]$, $q(t)_{imp}$ = unit-width flow in $[L^2/T]$ from impervious surface, and $q(t)_{perv}$ = unit-width flow in $[L^2/T]$ from pervious surface. For instance, at $t = 3600$ s, the total flow is accumulated as

$$Q = 0.152 \times 0.6 \times 2709 + 0.139 \times (1 - 0.6) \times 2709 = 390.53 \text{ cfs}$$

This approach ignores the routing effect through the collector channel because the storage volume in the collector channel is negligible.

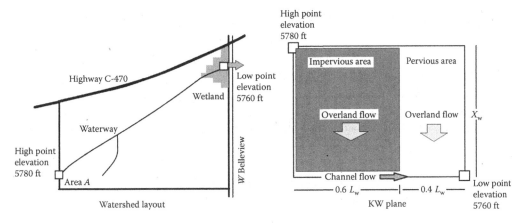

Figure 6.14 Overland hydrographs from watershed.

6.4 Rational method: A special case of KW flow

Taking the first directive of Equation 6.13 with respect to t yields

$$\frac{\partial q}{\partial t} = \alpha m y^{m-1} \frac{\partial y}{\partial t} \tag{6.52}$$

Substituting Equation 6.52 into Equation 6.1 yields

$$\frac{\partial q}{\partial t} + \alpha m y^{m-1} \frac{\partial q}{\partial x} = \alpha m y^{m-1} I_e \tag{6.53}$$

Equation 6.53 is the total derivative of variable q with respect to t:

$$\frac{dq}{dt} = \alpha m y^{m-1} I_e \approx \overline{I}_e \alpha m y^{m-1} \tag{6.54}$$

Equation 6.54 can be simplified using the average rainfall intensity, $\overline{I}_e$, when dealing with a nonuniform event. As illustrated in Figure 6.15, the rainfall amount contributing to the runoff rate, $q(T)$, is defined by the drainage area in terms of its waterway length from $x = 0$ to $x = L$. This waterway length is then converted to its equivalent flow time, i.e., the time of concentration, T_C, as the ratio of waterway length to average flow velocity, V (Figure 6.15). As a result, Equations 6.19 and 6.54 are integrated as (Guo, 1998)

$$L = \int_{T-T_C}^{T} \alpha m y^{m-1} \, dt \tag{6.55}$$

$$q(T) = \overline{I}_e \int_{T-T_C}^{T} \alpha m y^{m-1} \, dt = \overline{I}_e L \tag{6.56}$$

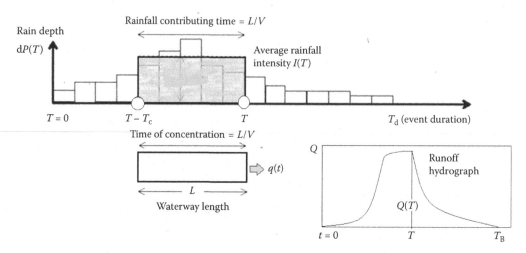

Figure 6.15 Rational method as kinematic wave.

To expand the runoff per unit-width into the entire watershed area yields

$$Q(T) = L_w \times q(T) = L_w \times (LI_e) = A \times I_e(T) \tag{6.57}$$

$$I_e(T) = CI(T) \tag{6.58}$$

in which L_w = watershed width similar to the width of the KW plane, L = watershed length, A = catchment area, C = runoff coefficient, $I(T)$ = average rainfall intensity, $I_e(T)$ = net rainfall intensity, and $Q(T)$ = runoff from the watershed at time T. A runoff coefficient represents the ratio of rainfall volume that can be converted into runoff volume. Of course, $(1 - C)$ represents the ratio of hydrologic losses. More discussions of runoff coefficient can be found in Chapter 5.

During the peaking time, the peak runoff is linearly related to the average peak rainfall intensity as

$$Q_p = CI_pA \tag{6.59}$$

in which Q_p = peak runoff in $[L^3/T]$ and I_P = average peak rainfall intensity in $[L/T]$. Equation 6.59 is the rational method that is applicable to watersheds smaller than 100 acres (40 ha).

EXAMPLE 6.6

Apply $I_e = 6.0$ in./h to the watershed in Example 6.1. The watershed area is $A = 67.9$ acres. According to the KW method, the peak flow is predicted to be 390.53 cfs. Determine the runoff coefficient.

Solution: Consider Equation 6.59 for this case. Care has to be taken when converting units to feet-second as

$$390.53 = C \times \left(6.0 \times \frac{1}{12} \times \frac{1}{3600}\right) \times (67.9 \times 43{,}560), \quad \text{So } C = 0.95$$

In this case, the time to peak is $t = 60$ min when the rain ceases. The soil infiltration rate has already reached its final rate of 0.5 in./h on the pervious area.

6.5 Conveyance waterway

In an urban area, overland flows are collected by the street gutters and then conveyed by sewers, roadside ditches, and channels. As illustrated in Figure 6.16, these conveyance facilities are modeled as follows:

* Street with an underground sewer
* Street with a roadside ditch
* Channel with a low flow pipe
* Floodplain section with a low flow channel

A conveyance facility is described by its hydraulic parameters, cross-sectional geometry, and bankfull depth. When a roadside ditch (shown in Figure 6.17) is depicted by a single cross-section, the KW approach may assume that the flow capacity can continue increasing with the assumed vertical walls above the bankfull depth, or the computer model will only release the flow up to the bankfull capacity. The former produces an unrealistic high release, whereas the latter results in a flat and prolonged low flow. Neither can represent the true flow until the dynamic wave approach is applied to more a detailed cross-section.

6.6 Detention basin

A detention basin (Figure 6.18) is a hydraulic structure designed for flood control. It stores the excess stormwater during the peak time and then gradually releases the stored volume at a rate not exceeding the downstream drainage capacity. Numerically, a detention system is modeled as a node with a large storage volume. According to the basin geometry and outlet work, the hydraulic performance of a detention system shall be described by the *storage-outflow curve* that is formed with pairs of *storage volume* versus *outflow rate*. For simplicity, the depletion process in a detention basin can be approximated by orifice and weir hydraulics when the outlet system is not affected by tailwater.

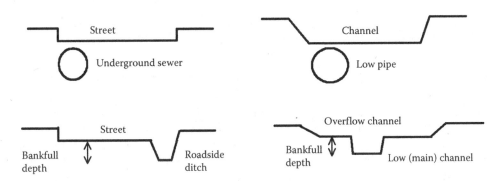

Figure 6.16 Conveyance facilities in urban area.

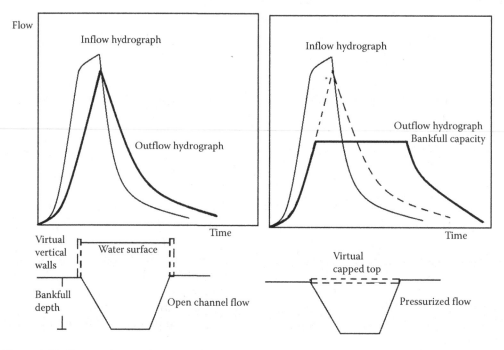

Figure 6.17 Composite sections in urban area.

Figure 6.18 Stormwater detention basins. (a) Stormwater detention in garden and (b) stormwater detention in fence wall.

In practice, roadway embankments and culvert entrances impose storage effects to flow attenuation. It is important to understand that these minor storage volumes are not reliably maintained for flood control purposes. During a major event, these minor storage volumes may become nullified after the embankment is washed away. Therefore, it is suggested that random storage volumes in the watershed be excluded from the *regional flood mitigation plan* but shall be included in the computer model when investigating the existing flooding problems such as in forensic studies.

6.7 Diversion facilities

Diversion of storm runoff occurs when storm runoff is transferred across the physical boundary of a watershed. In an urban area, a runoff diversion is often caused by street intersections or drainage structures. To model a stormwater diversion, the relationship of inflow versus diverted flow is required. Such a relationship can be developed from the configuration of the drainage structure. Three common cases are discussed in the following sections.

6.7.1 Culvert across a highway embankment

At the place where a highway intercepts a natural stream (Figure 6.19), a culvert or bridge is built to sustain the continuity of runoff. When the inflow to the culvert is greater than the capacity of the culvert, the upstream pool begins to be filled up. At the bankfull level, the flow overtops the bank to spill into the roadside ditch parallel to the highway. Without knowing the downstream condition, the culvert hydraulics can be estimated by its inlet control capacity with the headwater depth equal to the flow depth in the approaching channel. The diverted flow is the difference between the inflow to the culvert and the release through the culvert.

The watershed may involve irrigation canals. A man-made canal is often built with a flat bottom that allows the water to flow in both directions. During the major storm event, a canal may be modeled as a flowing, full linear reservoir that does not result in any runoff diversion at all.

6.7.2 Flow across street intersection

A street intersection can induce flow diversion. For instance, Figure 6.20 depicts a case in which the incoming stormwater can continue flowing into the downstream street section or curves into the alley. How to split the stormwater on the street depends on the street cross-sectional geometry. Under the major event, the water depth on a street may be 6 in. above the street crown (centerline). As a result, the street flow can be modeled as a wide channel flow. Manning's equation for wide channel states

$$Q = \frac{k_n}{N} Y^{\frac{5}{3}} W \sqrt{S} \tag{6.60}$$

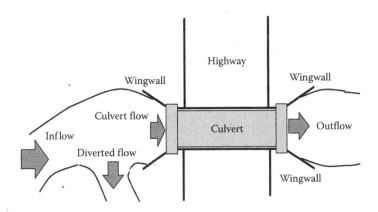

Figure 6.19 Runoff diversions by culvert.

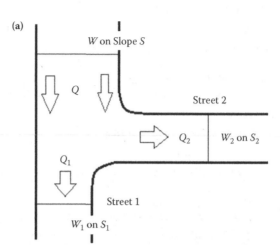

Figure 6.20 Split flow at street intersection. (a) Parameters at street intersection and (b) flow at street intersection.

in which Q = street flow in [T^3/T], Y = flow depth in [T], W = street width in [T], and S = street slope.

At the intersection, the water flow will be split between the street and the alley. Considering the intersection is inundated and acts like a pool, the flow depth, Y, at the intersection is applied to both street flows. As a result, the hydraulic capacity ratio, R, between these two streets downstream of the intersection is

$$R = \frac{Q_1}{Q_2} = \frac{W_1\sqrt{S_1}}{W_2\sqrt{S_2}} \tag{6.61}$$

$$Q_1 = RQ_2 \tag{6.62}$$

$$Q = (1+R)Q_2 \tag{6.63}$$

in which Q = inflow to street intersection in [L^3/T], Q_1 = flow in [L^3/T] remained on street 1, Q_2 = diverted flow in [L^3/T], W = width of flow in [L], and S = street slope in [L/L]. The subscript 1 represents the variables of street 1, and the subscript 2 represents

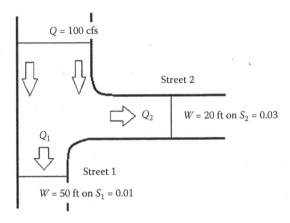

Figure 6.21 Example for flow diversion at street intersection.

Table 6.5 Example runoff diversion at street intersection

Inflow, Q (cfs)	100	200	300	400	500
Diverted, Q (cfs)	41	82	123	164	205

the variables of street 2 (Figure 6.20). For a given Q as the inflow, the diverted flow to street 2 (the alley) is

$$Q_2 = \frac{1}{1+R}Q \tag{6.64}$$

EXAMPLE 6.7

A discharge of 100 cfs is carried by a street of 50 ft wide on a slope of 2.0% (Figure 6.21). At the intersection, the flow is diverted into an alley of 20 ft wide on a slope of 3.0%. Construct the table of flow diversion at the intersection.

$$R = \frac{50\sqrt{0.01}}{20\sqrt{0.03}} = 1.44$$

$$Q_2 = \frac{1}{1+1.44}Q = 41\,\text{cfs} \quad \text{when } Q = 100\,\text{cfs}$$

For this case, the inflows are split according to Table 6.5.

6.7.3 Flow intercepted by inlet and sewer

Storm sewer systems are usually designed to carry a minor event such as a 5-year peak discharge. When modeling the 100-year event, the flow on the street (as shown in Figure 6.22) is the flow difference between the 100- and 5-year events.

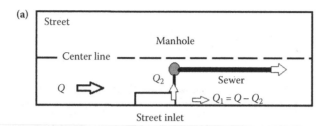

Figure 6.22 Street inlet for flow diversion. (a) Parameters in street flow interception and (b) interception inlet.

The flow interception along a street gutter depends on the inlet and sewer capacities, depending on whichever is less. The table of runoff diversion shall cover the rising hydrograph of the major event with a maximum diverted flow not exceeding the design capacity of the sewer system

EXAMPLE 6.8

A 36-in. sewer is designed to carry a 5-year peak discharge of 40 cfs. The 100-year peak flow on the street is 350 cfs. Determine the table of flow rates intercepted by the sewer line from the street gutter during the 100-year storm event.

Solution: It is assumed that the sewer line can intercept the street flow up to 40 cfs. Table 6.6 shows the diverted flow versus the flow on the street.

Table 6.6 Flow interception at street inlet

Q-100 year (inflow rate), cfs	0	40	41	350
Q-2 year (diverted flow), cfs	0	40	40	40

6.8 Flowchart of numerical process

Stormwater modeling studies are classified into (1) *simulation of an observed event* and (2) *prediction of a design event*. Simulations of an observed event evaluate the accuracy of the model and also calibrate the model parameters. Predictions of design events evaluate the consistency of the model and also calibrate the relative magnitudes among design flows. Both accuracy and consistency are important to a deterministic model. Figure 6.23 presents a flowchart for a watershed modeling study. It begins with the selection of simulation or prediction. Under a simulation, the observed rainfall and runoff must be provided for comparison purposes. Otherwise, the design rainfall distribution must be used for prediction purposes.

Having all storm hydrographs computed from the subbasins, the channel routing process will begin from the most upstream nodes. By following the drainage network, each subbasin hydrograph will be routed through the outfall link to the downstream node. At a node, all the incoming hydrographs are added up, according to the time sequence. At a diversion node, the diverted flow is arithmetically subtracted from the incoming hydrograph, and the remaining flow is routed through the downstream link. At an inflow node, an additional inflow hydrograph can be introduced into the hydrograph composition process. At a pump station node, the pumped-out flow is determined by the pump characteristic curve and then subtracted from the inflow hydrograph. At a storage node, the reservoir routing is applied to balance the volumes among storage change, inflow, and outflow. The stored water volume will be released through the downstream channel toward the outlet of the watershed.

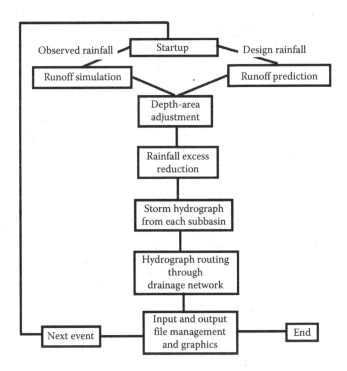

Figure 6.23 Numerical process of watershed modeling.

6.9 Small urban watershed modeling

A small watershed will produce runoff as a quick response to rainfall. Urban watersheds are equipped with efficient drainage systems and tend to produce flash floods as a small watershed. *The historic hydrologic condition* of a watershed provides valuable information about the formations of the natural drainage ways and sediment transport processes. Developments in urban areas change the natural drainage patterns from sheet flows to concentrated flows. Through several developing stages, the watershed is expected to become fully developed, according to the land use plan. The major *outfall systems* identified in the master drainage plan shall be sized for the *fully developed future condition*. The historic and future conditions represent two extreme limits during the watershed development. The existing watershed condition serves as the basis for flood insurance purposes. Between the development stages, the impact study is to identify the differences of runoff flows between the *predevelopment* and *postdevelopment conditions*.

EXAMPLE 6.9

The watershed in Figure 6.24 is located at the southeast corner of West Belleview Avenue and Interstate 470 in the City of Denver, CO. The watershed has a drainage area of $0.29\,\text{mile}^2$ (187.0 acres) and is divided into four subbasins (Figure 6.25).

Both subbasins 7 and 8 are to be developed into residential areas. Subbasin 6 is to be developed for high-density condominium. Subbasin 10 remains as open space for flood detention and wetland. The outfall structure for the detention basin is a 24-in. pipe under West Belleview Avenue. Using Equation 6.12, the subbasin hydrologic parameters are converted into KW sloping planes (as listed in Table 6.7). Table 6.8 provides standard design variables for hydrologic losses for pervious and impervious surfaces. The existing imperviousness increases from 5% to 15% and the future imperviousness will increase from 50% to 75% as shown in Table 6.7. The runoff flow will be collected by streets and roads as shown in Figure 6.25. The purpose of the watershed study is to determine the developed 100-year flood flows and to size the regional detention basin at the watershed outlet.

For this case, the flood flows are modeled for the pre- and postconditions as follows:

1. **Predevelopment condition**
   The goal of the hydrologic study is to detect the flooding problems in the watershed and serves as the basis to quantify the impacts after the watershed becomes developed.

Figure 6.24 Watershed under master drainage plan study. (a) Subarea 10 with wetland and (b) subareas 7 and 8 on left and 6 on right.

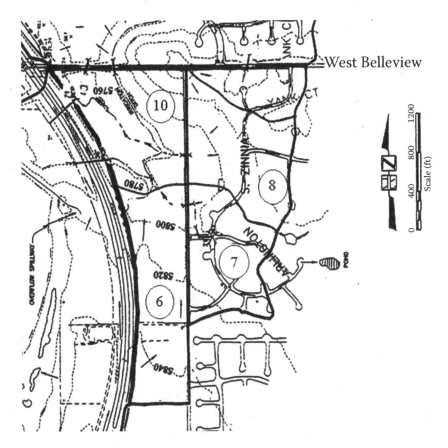

Figure 6.25 Watershed for case I.

Table 6.7 Subbasin hydrologic parameters used in case I study

Subarea	A (acre)	L (mile)	S_o (ft/ft)	$X = A/L^2$	$Z = A_m/A$	$Y = L_w/L$	L_w (ft)	S_w (ft/ft)	Existing (Imp%)	Future (Imp%)
6	67.84	0.45	0.03	0.52	0.50	1.12	2656.98	0.019	5	75
7	21.76	0.3	0.02	0.38	0.50	0.82	1303.29	0.016	15	65
8	25.60	0.25	0.04	0.64	0.50	1.35	1776.58	0.022	10	65
10	67.84	0.44	0.03	0.55	0.60	1.05	2437.75	0.019	5	50

Table 6.8 Subbasin soil loss information used in case I study

Basin no.	Depression loss		Soil loss		Coefficient
	Impervious (in.)	Pervious (in.)	f_0 (in./h)	k (I/s)	f_c (in./h)
6	0.10	0.40	3.0	0.0018	0.50
7	0.10	0.40	3.0	0.0018	0.50
8	0.10	0.40	3.0	0.0018	0.50
10	0.10	0.40	3.0	0.0018	0.50

The computer model, EPA SWMM, is used to mimic the physical layout of the existing drainage network. For instance, a steep waterway reveals the erosion potential. It is necessary to add drops or check dams to the waterway to slow down the flow movement. The existing 100-year peak discharge at the watershed's outlet is determined to be 95 cfs. It is necessary to investigate if the downstream drainage capacity is adequate to pass this flow. If not, what is the allowable release flow rate? The allowable release rate is critically important when formulating the future flood mitigation plans.

2. **Postdeveloped condition**

   The alternatives for the development condition include (1) *conveyance model*, i.e., channels only, or (2) *conveyance and storage model*, i.e., a detention basin at the watershed's outlet. For this case, the 100-year peak flow under the postdevelopment condition will be 415.0 cfs. For this case, the downstream existing sewer line only allows no more than 80 cfs. Therefore, a detention basin is proposed to be installed in subbasin 10. This proposed detention basin will be shaped as a wetland area using the 10- and 100-year storage volumes. Details of detention basin design will be discussed in Chapter 13.

6.10   Large watershed modeling (>100 mile^2)

Flood flows generated from a watershed of several hundred square miles are often controlled by the area covered under the thunderstorm. In other words, only the watershed area covered by the storm cell can be the effective tributary area to produce flood flows. *Storm-centering test* is often conducted to model a large watershed in order to determine the worst but probable condition for conservative designs. For instance, to model a watershed of 500 mile^2, the tributary area outlined by the topographic contours can be far greater than the average size of summer storms. Recognizing that the areal average rainfall depth is inversely proportional to the tributary area, the concept of "*the larger watershed, the more runoff*" is no longer always true. Therefore, it is necessary to investigate the tradeoff between the contributing area and the point rainfall depth reduction.

The basic input information for storm-centering tests includes watershed hydrologic parameters, point rainfall depth, and depth-area reduction factor (DARF). The major task of the storm-centering test is to identify the critical storm coverage area by which the highest runoff flow or volume can be produced. Among many hydrologic parameters, the *watershed tributary area* and *rainfall depth* are the most decisive factors. Before conducting a storm-centering test, a large watershed shall be divided into several major subbasins based on the prevailing wind direction, orographic condition, waterway network, and regional flood mitigation facilities. Next, the subareas can be grouped into several storm coverage areas. For each possible coverage area, the proper DARF is applied to the predictions of design flows. Of course, the worst condition is recommended for design.

The *Rainfall Atlas* published by the *National Ocean and Atmospheric Administration* (NOAA) was derived from the measurements at point rain gages. A value obtained from an isopluvial map represents the statistical precipitation depth at that point. For the purpose of hydrologic designs, it is critically important that a point rainfall depth be converted to the average value for the area of study. The NOAA recommends a set of DARF that can be applied to the point value as

Table 6.9 Rainfall events observed in McCarran Airport, NV

Storm cover area (mile²)	Storm cover area (km²)	Observed rainfall events						Observed average DARF
		June 13, 1955	Oct 21, 1957	July 3, 1975	Aug 10, 1981	Aug 10, 1983	Oct 17, 2006	
0.1	0.3	1.00	1.00	1.00	1.00	1.00	1.00	1.000
1	2.6	0.92	0.90	0.95	0.96	0.97	0.95	0.942
10	26.4	0.75	0.79	0.87	0.81	0.88	0.87	0.828
50	132.0	0.69	0.76	0.75	0.70	0.82	0.75	0.745
100	263.9	0.51	0.67	0.66	0.52	0.65	0.67	0.613
200	527.9							
300	791.8							
400	1055.8							
500	1319.7			0.41	0.34		0.48	0.410

Source: Mark Group. *Flood Control Master Plan for Maopa Valley, Nevada*, Mark Group, Engineering and Geologists, Inc., Las Vegas, NV, 1988.

$$P_c = RP \tag{6.65}$$

in which P = point rainfall depth and R = DARF. A decay formula of DARF with respect to watershed area can be derived as

$$R = R_c + (1 - R_c)e^{-kA} \tag{6.66}$$

in which R_c = final constant when watershed is greater than $300\,mile^2$, A = contributing area in square miles, and k = decay coefficient. Details of Equation 6.66 can be found in Chapter 2. For this case, severe storms recorded in the Las Vegas area were collected and analyzed (Table 6.9) for deriving the local rainfall DARFs. The best-fitted values to Equation 6.66 are derived as

$$R = 0.41 + 0.59e^{-0.0138A} \text{ for the Las Vegas area} \tag{6.67}$$

The database used to derive Equation 6.67 had rainfall duration from 6 to 12 h (Guo, 2012, 2014).

6.10.1 Storm-centering test

A large watershed ($>100\,mile^2$) is not always completely covered by a single storm event. Therefore, the runoff generation from a large watershed is sensitive to where the center of storm is located and which portion of the watershed is under the storm—upper valley, middle valley, or lower valley. To be conservative, the critical location of storm center needs to be identified. The larger the storm-covering area, the higher the value of DARF. The tradeoff between tributary area and DARF provides a basis on how to choose the storm coverage in order to conservatively predict the design flood flows. Before the storm-centering test, a large watershed is divided into subbasins. As illustrated in Figure 6.26, several likely storm centers are developed according to the storm cell

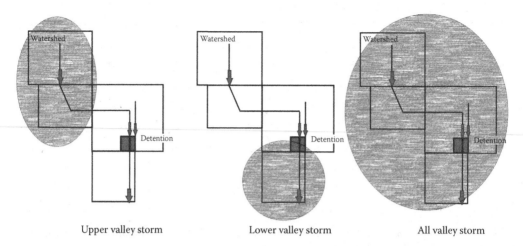

Upper valley storm   Lower valley storm   All valley storm

Figure 6.26 Illustrations of possible storm centers.

movement, waterway network, and orographic and topographic conditions. The case of a valley storm has the largest tributary area and the lowest DARF, whereas the case of a lower valley storm has the smallest tributary area with the highest DARF.

For each case of storm centering, the accumulated tributary area toward the design point is the sum as follows:

$$A = \sum_{i=1}^{i=M} A_i \tag{6.68}$$

in which A = total tributary area in $[L^2]$ for the selected case of storm centering, A_i = area of subbasin in $[L^2]$, i = ith subbasin, and M = number of subbasins contributing to design point. In essence, Equation 6.68 represents the size of the storm-covering area. As a result, the corresponding DARF can be computed using Equation 6.66 as

$$R = R_c + (1 - R_c)e^{-k\sum_{i=1}^{i=M} A_i} \tag{6.69}$$

In an urban area, flood flows are intercepted by either a conveyance facility like flood channels or a storage facility like a detention basin. In comparison, an urban channel has a negligible attenuation on peak flows compared to the detention basin. In this case, all urban channels are treated as a *direct release*, whereas all detention basins are converted into an *extended release*. For a direct release, the unit-area peak flow (discharge per tributary area) is defined as

$$q_i = \frac{Q_i}{A_i} \tag{6.70}$$

in which q_i = unit-area peak release in cfs/acre or cms/ha from subbasin, A_i and Q_i = peak discharge in cfs or cms generated from subbasin A_i. For an extended release, the detention basin temporarily stores the runoff volume and then gradually releases it over time. Based on the concept of unit-area peak release in Equation 6.70, an extended release, O_i, can be converted to its equivalent tributary area as

$$A_{E_i} = \frac{O_i}{Q_i} A_i \qquad (6.71)$$

in which A_{E_i} = equivalent tributary area for ith subbasin and O_i = extended peak discharge released from ith subbasin through its detention basin. In general, $O_i \leq Q_i$ because of the detention effect. For a special case, when $Q_i = O_i$, which has no detention effect at all, Equation 6.71 is reduced to a direct release. Substituting Equation 6.71 into Equation 6.68 yield

$$A_E = \sum_{i=1}^{i=M} \frac{O_i}{Q_i} A_i \qquad (6.72)$$

in which A_E = total equivalent tributary area for the selected storm center. In essence, Equation 6.72 represents the runoff-producing area. In practice, the design information required in Equation 6.72 is readily available from the *regional master drainage plan*. In this study, the maximal runoff volume at the design point is determined as

$$\text{Max } V = \text{Max}(P_A \cdot A_E) \qquad (6.73)$$

in which V = runoff volume in in.-mile^2. Aided with Equations 6.69 and 6.72, Equation 6.73 is expanded into

$$\text{Max } V = \text{Max}\left\{ P_o \times (R_c + (1 - R_c)e^{-k \sum\limits_{i=1}^{i=M} A_i} \times \left[\sum_{i=1}^{i=M} \frac{O_i}{Q_i} A_i \right] \right\} \qquad (6.74)$$

Equation 6.74 shall be tested for all possible storm coverage areas to identify the one that produces the maximal runoff volume. In practice, the operation of summation in Equation 6.74 can be easily processed through tabulations.

EXAMPLE 6.10

Storm-centering test is employed to maximize the design peak discharge at the site of the *Lower Detention Basin (LDB)* located on the Western Tributary Wash that drains into the Las Vegas Wash in the City of North Las Vegas, NV. As illustrated in Figure 6.27, there are two existing regional detention systems built upstream of the proposed LDB. They are *the Kyle Canyon Detention Basin (KCDB)* and the *Rancho Detention Basin (RDB)*. The total tributary area to the location of the proposed LDB is 92.0 mile^2 that is divided into four subareas: a tributary area of 58.0 mile^2 that drains into *KCDB*, another tributary area of 6.0 mile^2 that drains into RDB, and two more separate areas of 22.6 and 5.4 mile^2 that directly drain into the proposed *LDB*.

According to the Master Drainage Plan published for the Las Vegas Valley (Master Plan Update, 2000), the tributary areas and design peak inflows and outflows to the proposed and existing LDB, RDB, and KCDB are summarized in Table 6.10. Based on the major

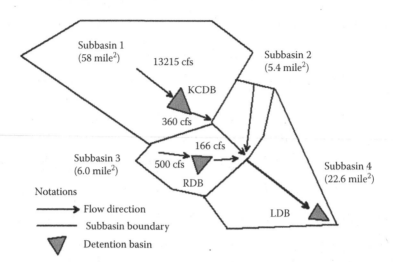

Figure 6.27 Detention basins and subareas in case study.

Table 6.10 Analysis of equivalent tributary area to LDB

Case	Tributary area, A_i (mile^2)	Peak inflow, Q_i (cfs)	Peak outflow, O_i (cfs)	Equivalent tributary area, A_{E_i} (mile^2)
(1) Subbasin 4 draining into LDB	22.6			22.6
(2) Subbasin 2 draining into LDB	5.4			5.4
(3) Subbasin 3 draining into RDB	6.0	500.0	166.0	2.0
(4) Subbasin 1 draining into KCDB	58.0	13,215.0	360.0	1.6

LDB, Lower Detention Basin; KCDB, Kyle Canyon Detention Basin; RDB, Rancho Detention Basin.

waterways through the watershed, four possible storm centers are investigated for this case as follows:

1. *Storm center 1*: covering subbasin 4 of 22.6 mile^2 to LDB
2. *Storm center 2*: covering subbasins 4 and 2 or 28.0 mile^2 to LDB
3. *Storm center 3*: covering subbasins 4, 2, and 3 or 34.0 mile^2 to LDB
4. *Storm center 4*: covering the entire tributary area of 92.0 mile^2 to LDB

Table 6.10 is the analysis of the equivalent tributary area from these four subareas. For instance, subbasins 4 and 2 are a direct release. As a result, their tributary areas are the equivalent areas. However, Subareas 3 and 1 are an extended release. Their equivalent area shall be weighted by the ratio of peak outflow to peak inflow. For instance, subbasin 1 has a tributary area of 58 mile^2. Aided by Equation 6.71, its equivalent area is computed as

$$A_{E1} = \frac{O_1}{Q_1} A_1 = \frac{360}{13215} \times 59 = 1.61 \ \text{mile}^2$$

The subscript 1 represents the parameter for subbasin 1, etc.

Table 6.10 presents the total cumulative runoff volumes for four cases. For example, case 4 has a total tributary area of 92 mile^2 under the design storm. Based on Equation 6.67, the value of DARF for case 4 is computed as

$$R = 0.41 + 0.59e^{-0.0138A} = 0.41 + 0.59e^{-0.0138 \times 92} = 0.56 \text{ for case 4}$$

Referring to Table 6.10, the area-averaged rainfall depth is reduced from its point depth of 3.0 to 1.71 in., which shall be applied to the equivalent area of 30.57 mile^2 that is accumulated as

$$A_E = \sum_{i=1}^{i=M} \frac{O_i}{Q_i} A_i = 22.6 + 5.4 + 6.0 \times \frac{166}{500} + 58.0 \times \frac{360}{13215} = 31.57 \text{ mile}^2 \text{ for case 4}$$

The runoff volume for case 4 is the product of the area-averaged rainfall depth and its equivalent area as

$$V = 1.71 \times 31.57 = 53.17 \text{ in.-mile}^2 \text{ for case 4}$$

Repeat the same process for cases 1, 2, and 3 as shown in Table 6.11. The maximal runoff volume is derived from case 3, or the storm cell covering Subareas 4, 3, and 2 shall produce the highest peak flow at the proposed location for LDB.

For this case, detailed hydrologic models were derived for cases 1 to 4 using the HEC Hydrologic Modeling System (HECHMS) model (HECHMS, 2015). The entire 92 mile^2 watershed was divided into 35 small subareas. The 24-h Soil Conservation Service (SCS) Type II rainfall distribution and the SCS UG method were employed to predict the storm hydrographs from these 35 subbasins. These storm hydrographs were then routed through the flood channels and detention basins to predict the peak inflow discharge at the LDB site. The HECHMS model confirms that case 3 produces the highest peak discharge and storage volume at the proposed site. In 1996, the LDB was designed for a tributary area of 34.0 mile^2. The construction of the LDB was completed in 2000.

Table 6.11 Storm-centering tests for lower detention basin in Las Vegas, NV

Case ID	Storm cover area (mile^2)	Total tributary area, A (mile^2)	Accumulated equivalent area, A_E (mile^2)	6-h point rainfall depth, P_o (in.)	Las Vegas DARF value, R factor	Areal rainfall depth, P_A (in.)	Runoff volume, V (in.-mile^2)
1	Subbasin 4	22.60	22.60	3.00	0.84	2.53	57.08
2	Subbasins 4 + 2	28.00	28.00	3.00	0.80	2.41	67.58
3	Subbasins 4 + 3 + 2	34.00	29.99	3.00	0.77	2.31	69.42
4	All subbasins	92.00	31.57	3.00	0.56	1.68	53.17

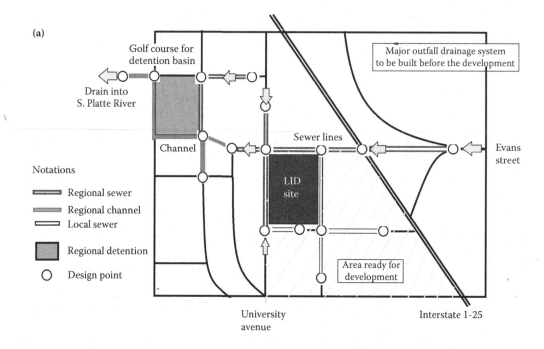

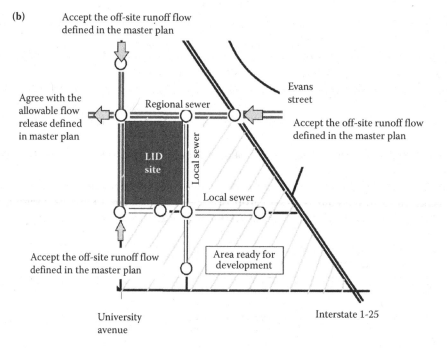

Figure 6.28 Large- and small-scale models. (a) Large-scale model for entire watershed and (b) small-scale model for site drainage.

6.11 Closing

In comparison, the current numerical modeling techniques are too simple to fully describe the dynamics of a storm cell. Hydrologic methods are sensitive to the size of the watershed. In practice, we build several layers of watershed stormwater models from a large-scale model for regional purposes to a small-scale model for site details. When conducting a regional watershed study (>200 mile^2), a large-scale model may use subareas ranging from 5 to 10 mile^2. The computer model, *HEC1 or HMS*, applies a set of DARFs to determine the design flows at the major design points. Next, we zoom in to work on small sites such as 1-mile^2 area using EPASWMM for drainage details. All off-site runoff flows to this 1-mile^2 study area have already been defined by the large-scale model. Using the boundary values, we link the small-scale and large-scale models together on the basis of consistency. The 1-mile^2 study area can numerically be treated either as one lumped area without details or as a multiple link-node model for details. The final check is to make sure that the outflow from this 1-mile^2 study area closely agrees with the design flow defined by the large-scale model. The procedure is illustrated in Figure 6.28. In doing so, the modeling skill warrants upstream and downstream consistency based on the regional master drainage plan.

The main purpose of stormwater modeling is to predict the future postdevelopment condition. There is not enough or even no field data available for model calibration. Therefore, a regional stormwater model shall be developed and examined on the basis of consistency, which preserves the basic relationship between the magnitudes of flood flows and the tributary areas. Of course, the ultimate goal is to improve a consistent model to become an accurate model when more field data become available.

6.12 Homework

Q6.1 Determine the KW plane for the given watershed in Figure Q6.1 with waterway length, $L = 1450$ ft, drainage area, $A = 22.5$ acres (1 acre = 43,560 ft^2), waterway slope, So = 0.02, and imperviousness of 60%.

1. Convert the watershed into a KW plane using a two-flow system in which the pervious and impervious areas are separated by the collector channel. Determine L_w, X_{w1}, and X_{w2}.
2. Layout the KW plane for a cascading flow system in which the impervious area drains into the downstream pervious area. Determine L_w, X_{w1}, and X_{w2}.

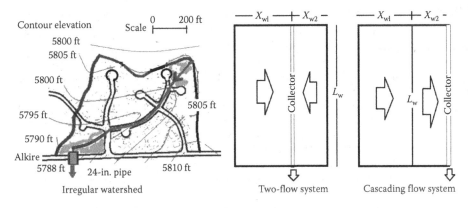

Figure Q6.1 Watershed for developing kinematic wave plane.

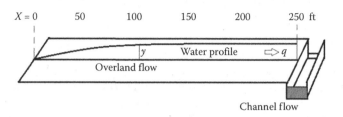

Figure Q6.2 Overland flow in a parking lot.

Q6.2 A concrete parking lot in Figure Q6.2 is under a uniform rainfall intensity of 10 in./h for a duration of 30 min. The overland flow has a slope of 2% for a length of 250 ft. The Manning's roughness coefficient is 0.05.

1. Determine the time of concentration at the outlet.
2. Construct the rising overland flow hydrograph in cfs/ft at the outlet.
3. Construct the peaking overland flow hydrograph in cfs/ft at the outlet.
4. Construct the recession hydrograph at the outlet.

Q6.3 A grass area is under a uniform rainfall intensity of 10 in./h for a duration of 30 min. The overland flow has a slope of 2% for a length of 250 ft. The Manning's roughness coefficient is 0.25. The infiltration loss is defined as

$f(t) = 1.5 + 2.5e^{-0.12t}$ in which $f(t)$ = infiltration rate in in./h, and
   t = elapsed time in minute

1. Determine the time of concentration for the outlet.
2. Construct the rising overland flow hydrograph in cfs/ft at the outlet.
3. Construct the peaking overland flow hydrograph in cfs/ft at the outlet.
4. Construct the recession hydrograph at the outlet.

Bibliography

Bedient, P.B., Huber, W.C., and Vieux, B.E. (2008). *Hydrology and Floodplain Analysis*, 4th ed., Prentice-Hall, Upper Saddle River, NJ.

Blackler, G., and Guo, J.C.Y. (2012). "Field Test of Paved Area Reduction Factors Using a Storm Water Management Model and Water Quality Test Site," *ASCE Journal of Irrigation and Drainage Engineering*, Vol. 17, No. 8.

Chow, V.T. (1964). *Handbook of Applied Hydrology*, McGraw-Hill Book Company, New York.

Eagleson, P.S. (1970). *Dynamic Hydrology*, McGraw Hill, New York, pp. 344–346.

EPA SWMM5. (2005). Stormwater Management Model published by US EPA. http://www2.epa. gov/water-research/storm-water-management-model-swmm.

Guo, J.C.Y. (1984). "Effects of Infiltration on Hydrograph," Proceedings of ASCE International Conference on Irrigation and Drainage Engineering, held at Flagstaff, Arizona.

Guo, J.C.Y. (1988). "Dynamics and Kinematics of Overland Flow," Proceedings of ASCE International Conference on Hydraulic Engineering held at Colorado Springs, CO.

Guo, J.C.Y. (1998). "Overland Flow on a Pervious Surface," *IWRA International Journal of Water*, Vol. 23, No. 2, 91–96.

Guo, J.C. Y. (2000). "Storm Hydrographs for Small Catchments," *International Journal of Water, AWRA*, Vol. 23, No. 2.

Guo, J.C.Y. (2001). "Rational Hydrograph Method for Small Urban Catchments," *ASCE Journal of Hydrologic Engineering*, Vol. 6, No. 4.

Guo, J.C.Y. (2006). "Dimensionless Kinematic Wave Unit Hydrograph for Storm Water Predictions," *ASCE Journal of Irrigation and Drainage Engineering*, Vol. 132, No. 4.

Guo, J.C.Y. (2012). "Storm Centering Approach for Flood Predictions from Large Watersheds," *ASCE Journal of Hydrologic Engineering*, Vol. 17, No. 9.

Guo, J.C.Y. (2014). Closure on "Storm Centering Approach for Flood Predictions from Large Watersheds," *ASCE Journal of Hydrologic Engineering*, Vol. 19, No. 1, pp. 272–274.

Guo, J.C.Y., and Urbanos, B. (2009). "Conversion of Natural Watershed to Kinematic Wave Cascading Plane," *Journal of Hydrologic Engineering*, Vol. 14, No. 8, pp. 839–846.

Guo, J.C.Y., Cheng, J., and Wright, L. (2012). "Field Test on Conversion of Natural Watershed into Kinematic Wave Rectangular Planes," *ASCE Journal of Hydrologic Engineering*, Vol. 17, No. 8.

HEC-1. (1993). *Introduction and Application of Kinematic Wave Routing Techniques Using HEC-1*, United States Army Corps of Engineers, Davis, CA.

HECHMS. (2015). *HEC Hydrologic Modeling System*, Volume 4, U.S. Army Corps of Engineers, Davis, CA.

Henderson, F.M., and Wooding, F.A. (1964). "Overland Flow and Groundwater Flow from Steady Rainfall of Finite Duration," *Journal of Geophysical Research*, Vol. 69, No. 8, pp. 1531–1539.

Izzard, C.F. (1946). "Hydraulics of Runoff from Developed Surfaces," Proceedings of the Twenty-Sixth Annual Meeting of the Highway Research Board held at Washington, DC. December 5–8, 1946, Highway Research Board Proceedings, pp. 129–146.

Kuichling, E. (1889). "The Relation Between the Rainfall and the Discharge of Sewers in Populous Districts," *Transactions of the American Society of Civil Engineers*, Vol. 20, pp. 1–56.

Langford, K.J., and Turner, A.K. (1973). "An Experimental Study of the Application of Kinematic Wave Theory to Overland Flow," *Journal of Hydrology*, Vol. 18, pp. 125–245.

Lighthill, M.J., and Hitham, G.B. (1955). "On Kinematic Waves, Part 1: Flood Movement in Long Rivers," *Proceedings of the Royal Society of London A*, Vol. 229, No. 1178, pp. 281–316.

Mark Group. (1988). *Flood Control Master Plan for Maopa Valley, Nevada*, Technical Report, prepared by the Mark Group, Engineering and Geologists, Inc., Las Vegas, NV.

Master Plan Update. (2000). Master Drainage Plan Update for Las Vega Valley, prepared by Clark County Regional Flood Control District, Las Vegas, NV.

Morris, E.M., and Woolhiser, D.A. (1980). "Unsteady One Dimensional Flow Over a Plane: Partial Equilibrium and Recession Hydrographs," *Water Resources Research*, Vol. 16, No. 2, pp. 335–360.

Overton, D.E., and Meadows, M.E. (1976). *Stormwater Modeling*, Academic Press, New York.

Rossman, L.A. (2010). *Storm Water Management Model, User's Manual, Version 5.0*, EPA/600/R-05/040, EPA, Cincinnatim, OH.

Singh, V.P., and Woolhiser, D.A. (1976). "A Nonlinear Kinematic Wave Model for Watershed Surface Runoff," *Journal of Hydrology*, Vol. 31, pp. 221–243.

TR20. (2005). Technical Release 20 published by the Naural Resource Conservation Service. http://www.nrcs.usda.gov/wps/portal/nrcs/detailfull/national/water/manage/?&cid=stelprdb1042924.

UDFCD. (2005). *Urban Stormwater Design Criteria Manual*, supported by the Urban Drainage and Flood Control District, Denver, CO.

Wooding, R.A. (1965). "A Hydraulic Mode for a Catchment-Stream Problem," *Journal of Hydrology*, Vol. 3, pp. 254–267.

Woolhiser, D.A., and Ligget, J.A. (1967). "Unsteady One Dimensional Flow Over a Plane: The Rising Hydrograph," *Water Resources Research*, Vol. 3, No. 3, pp. 753–771.

Yen, B.C., and Chow, V.T. (1974). *Experimental Investigation of Watershed Surface Runoff*, Hydraulic Engineer Series, No. 29, Department of Civil Engineering, University of Illinois at Urbana–Champaign.

Yu, Y.S., and McNown, R.K. (1965). "Runoff From Impervious Surfaces," *Journal of Hydraulic Research, IAHR*, Vol. 2, No. 1, pp. 3–24.

Chapter 7

Flood channel design

Flood water in an urban area must be quickly carried away by streets and channels. Design of flood channel takes into consideration both the channel hydraulic capacity and the future maintainability. In an urban area, the alignment of a flood channel is often selected subject to many constraints and has to compromise with the existing underground utilities. Therefore, it is necessary to investigate various alternatives for comparison. The final selection is often dictated by the cost-effectiveness and public safety (ASCE and WEF, 1992).

Open-channel flows are analyzed by the empirical formulas that calculate the cross-sectional average flow velocity and hydraulic parameters. *Manning's formula* is the most popular for determining the flow condition (Manning, 1891):

$$U = \frac{k}{N} R^{\frac{2}{3}} \sqrt{S_e} \tag{7.1}$$

$$Q = UA \tag{7.2}$$

$$R = \frac{A}{P} \tag{7.3}$$

$$D = \frac{A}{T} \tag{7.4}$$

$$F_r = \frac{U}{\sqrt{gD}} = \frac{\text{Flow velocity}}{\text{Wave celerity}} \tag{7.5}$$

$$V_w = \sqrt{gD} \tag{7.6}$$

in which U = cross-sectional flow velocity in [L/T], Q = design flow rate in [L^3/T], R = hydraulic radius in [L], P = wetted perimeter in [L], A = flow area in [L^2], D = hydraulic depth in [L], T = top width in [L], g = gravitational acceleration in [L/T^2], F_r = Froude number, S_e = energy slope in [L/L], k = 1.486 for using feet-second units and 1.0 for using meter-second units, and V_w = wave celerity in [L/T]. As illustrated in Figure 7.1, a trapezoidal channel has a bottom width, B, and two side slopes, Z_1 and Z_2. The wetted perimeter is the length measured from the right bank to the left bank under the water surface. The top width is the length measured from the right to left bank above the water surface.

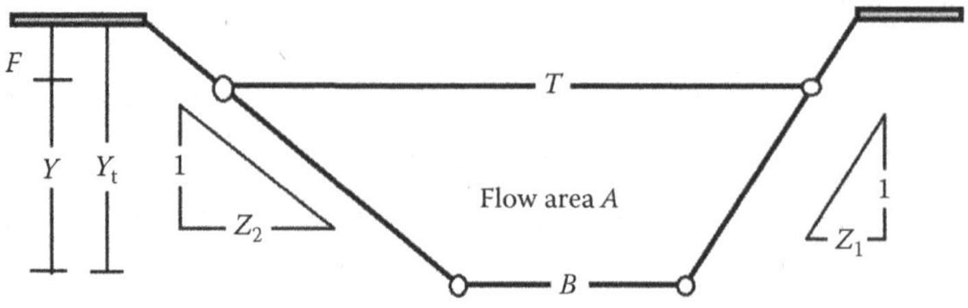

Figure 7.1 Trapezoidal channel cross-section.

In practice, the design discharge for a channel is often given, and the task is to determine the normal flow. For convenience, substituting Equations 7.1 and 7.3 into Equation 7.2 yields

$$Q = \frac{k}{N} P^{\frac{-2}{3}} A^{\frac{5}{3}} \sqrt{S_e}$$

(7.7)

Substituting Equations 7.2 and 7.4 into Equation 7.5 yields

$$F_r^2 = \frac{TQ^2}{gA^3}$$

(7.8)

Froude number is the ratio of flow velocity to wave celerity in shallow water. When the flow velocity is greater than its wave celerity or $F_r > 1$, the channel flow is *supercritical*; otherwise, the channel flow is *subcritical* ($F_r < 1$). In case that $F_r = 1$, the channel flow is *critical*. Aided with Equation 7.8, the critical flow depth is solved as follows:

$$F_r^2 = \frac{T_c Q^2}{gA_c^3} = 1.0$$

(7.9)

where the subscript, c, represents the variables associated with the critical flow. There are six variables in Manning's formula: U, Y, B, S, N, and side slope, Z. To design a channel, the engineer has to select five variables and then calculate the sixth variable. The slope in Manning's formula is referred to as the energy slope. When designing a new channel, not all information is available yet. As a result, a uniform flow depth under the *normal flow* condition is assumed. A normal flow can only be developed in a *prismatic channel* that has a uniform cross-section in a long and straight reach on a constant slope. For the given design discharge, the normal depth is determined by Equation 7.7 under the assumption that the energy slope is equal to the channel bottom slope:

$$S_e = S_o$$

(7.10)

where S_o = channel bottom (invert) slope in [L/L]. When working with an existing channel, the flow depth and channel cross-section are known. As a result, the energy slope associated with the design flow is determined by rearranging Equation 7.1 to yield

$$S_e = \frac{N^2 U^2}{k^2 R^{1.33}} \quad k^2 = 1 \text{ for meter-second or 2.21 for feet-second unit.}$$

(7.11)

On the contrary, when designing a new channel, its invert slope shall be selected with the evaluation of the existing topographic condition along the proposed channel alignment. Next, the channel cross-section is calculated based on the normal flow condition and then added with a freeboard for safety. The *height of freeboard* is the vertical distance above the water surface to the top of the banks. The freeboard section prevents water spills due to the superelevation in a curve reach or waves in a straight reach. For a trapezoidal channel, the excavated channel cross-section is defined as

$$Y_t = Y_n + F \tag{7.12}$$

$$T_n = B + (Z_1 + Z_2)Y_n \tag{7.13}$$

$$A_n = \frac{Y_n}{2}(B + T_n) \tag{7.14}$$

$$P_n = B + Y_n\left(\sqrt{1 + Z_1^2} + \sqrt{1 + Z_2^2}\right) \tag{7.15}$$

in which Y_t = channel excavated depth in [L], F = height of freeboard in [L], Y_n = normal depth in [L] determined in a straight reach, B = channel bottom width in [L], Z_1 = right side slope in [L/L], Z_2 = left side slope in [L/L], T_n = top width in [L], A_n = normal flow area in [L^2], and P_n = wetted perimeter in [L] for normal flow. In a curved reach, the superelevation effect is mostly sensitive to the flow velocity. Using the flow Froude number as a basis, two sets of empirical formulas are developed and recommended for freeboard design as follows:

$$F = 1.0 \text{ ft} + \frac{U_n^2}{2g} \text{ for subcritical flow} \tag{7.16}$$

$$F = \max\left[\frac{1}{6}\left(Y_n + \frac{U_n^2}{2g}\right), 3\text{ft}\right] \text{ for supercritical flow} \tag{7.17}$$

in which F = height of freeboard in [L], U_n = normal velocity in (ft/s) and Y_n = normal flow depth in (ft) for design discharge.

Equation 7.17 demands a higher freeboard than Equation 7.16 because it takes into consideration the potential hydraulic jumps in a high gradient channel.

EXAMPLE 7.1

A trapezoidal channel is designed to carry a discharge of 1000 cfs with $B = 10$ ft, $Z_1 = 4$, $Z_2 = 5$, $S_o = 0.04$ ft/ft, and $N = 0.03$. Determine the normal and critical flow conditions.

Solution: Set flow depth $Y = Y_n$. The cross-sectional parameters are

$$A_n = (10 + 4.5Y_n)$$

$$P_n = 10 + \sqrt{1 + 4^2}Y_n + \sqrt{1 + 5^2}Y_n$$

Substituting A_n and P_n into Equation 7.7 with the channel bottom slope $S_o = 0.04$ yields a range of normal flow conditions (Table 7.1). The relationship between flow depth and discharge is

Table 7.1 Rating curve for normal flow

Design information	Bottom width	$B = 10.00\,\text{ft}$
	Left side slope	$Z_1 = 4.00\,\text{ft/ft}$
	Right side slope	$Z_2 = 5.00\,\text{ft/ft}$
	Manning's N	$N = 0.03$
	Channel bottom slope	$S_0 = 0.0400\,\text{ft/ft}$

Flow depth, Y (ft)	Top width top (ft)	Flow area, A (ft^2)	Wetted P-meter, (ft)	Hydraulic radius, R (ft)	Flow velocity, U (fps)	Flow rate, Q (cfs)	Froude number, F_r
1.96	27.62	36.82	28.05	1.31	11.89	437.7	1.81
2.44	31.93	51.08	32.47	1.57	13.42	685.5	1.87
2.92	**36.24**	**67.40**	**36.89**	**1.83**	**14.84**	**1000.0**	**1.92**
3.39	40.55	85.78	41.30	2.08	16.17	1386.8	1.96
3.87	44.86	106.23	45.72	2.32	17.43	1851.5	2.00

termed *the rating curve* for the given channel cross-section. For the specified design flow of 1000 cfs, the normal flow condition is found to be $Y_n = 2.92\,\text{ft}$, $U_n = 12.84\,\text{ft/s}$, and $Fr = 1.92$. This is a supercritical flow.

The critical flow condition is defined by $F_r = 1$. Try $Y_c = 4.02\,\text{ft}$. Equation 7.9 is solved as

$$A_c = Y_c(10 + 4.5Y_c) = 112.74\,\text{ft}^2$$

$$T_c = 10 + 9Y_c = 46.14\,\text{ft}$$

Check if $F_r^2 = T_c Q^2 / g A_c^3 = 1.0$ to confirm that the critical depth is $Y_c = 4.02\,\text{ft}$.

In summary, it takes the following three steps to design a channel:

1. Evaluating the existing topographic condition to select the channel bottom slope.
2. Sizing the channel cross-section based on the normal flow condition.
3. Making adjustments on the height of freeboard to accommodate waves and jumps.

7.1 Grade control

Man-made waterways carry concentrated flows with high velocities that result in severe bank scours and channel bed erosion. It is imperative that a flood channel be protected by proper linings and grade controls across the channel bed where the energy dissipation is necessary. Grade control by drop structure is a common practice to design a channel on a steep slope. A drop structure creates a vertical fall within a short horizontal distance along the channel longitudinal alignment. In general, the channel will be stabilized after the channel bottom slope becomes milder upstream of the drop structure. Figure 7.2 presents examples of drops built across the waterway.

Drop structures can be built with gabions, sheet piles, grouted riprap, and/or concrete walls with footings. In urban areas, it is more desirable to use a series of low-head vertical drops. A high-head drop can be laid over a sloped chute (as shown in

Figure 7.2 Drop structure for grade control. (a) Drop built on natural waterway and (b) drop across grass channel.

Figure 7.3). The downstream plunging pool is sized to dissipate the energy associated with the impinging jumps.

Drop structures are susceptible to bank and bottom erosions. In many cases, additional riprap protections are needed around a drop structure. Failures associated with a drop may be caused by overtopping flows, seepage piping flows, degradation on the channel bed, and/or scours along embankments. A drop structure should be located in a relatively straight reach with at least 100 ft upstream or downstream from a curve reach. The foundation material must be designed to meet the required supporting strength to resist the potential sliding force and overturning moment. Design considerations for a drop structure include the following:

1. **The degree of protection**
   Sediment erosion control can be achieved by reducing flow velocity to permissible velocity or flow Froude number to be below 0.80. Safety criteria determine the type and height of the drop structure.
2. **The selection of building materials**
   Examinations on flow hydraulics, channel morphologic, and soil gradations on the channel bed provide basic guidelines on the selection of building materials. In general, gabions are not recommended for a permanent drop structure because of fast deterioration.
3. **The forces on the structure**
   Considerations include the range of flow rates, water depths and velocities, head differences, storage capacities upstream of the submerged dam, seepage and uplift forces, and channel flow regime, i.e., subcritical flow or supercritical flow.
4. **The width and height**
   The drop structure has to be wide enough to pass the design discharge. For a riprap drop, the unit discharge shall not exceed 35 cfs/ft. The height of a drop structure is governed by the degree of protection, available construction material, required structural stability, and cost. Riprap drops built across channel floors are most economical but not stable for a drop more than 4 ft because of their structural instability. Grouting is a way to improve the strength of a riprap drop structure, but it requires an underdrain system in order to relieve uplift force due to seepage. The drop height of a vertical drop is limited to 7 ft because of expensive construction costs on retaining walls and footings.

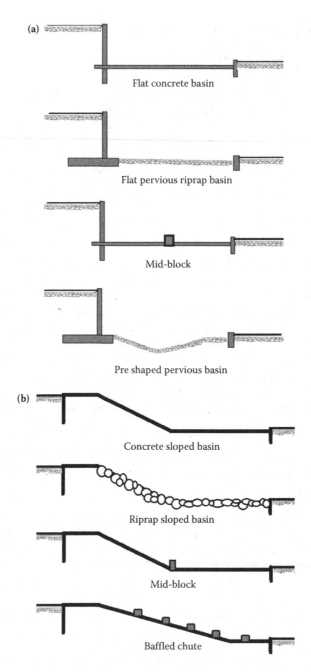

Figure 7.3 Vertical and sloping drops. (a) Vertical drops and (b) sloping drops.

7.1.1 Drop height

As illustrated in Figure 7.4, the top width of a drop structure has to be wide enough to pass the design discharge. The weir section on top of the drop structure shall be designed with multiple sections to pass low and high flows. For a given design discharge, the drop

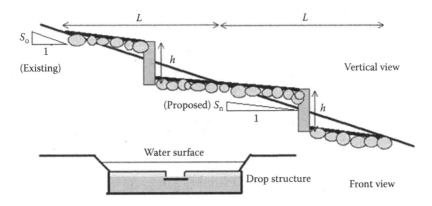

Figure 7.4 Illustration of drop height and reach length.

height is determined by the preselected channel slope determined by the permissible flow velocity. The height of a drop is calculated as (Guo, 2009)

$$h = (S_o - S_n)L \tag{7.18}$$

in which h = drop height in [L], S_n = proposed channel bed slope in [L/L], S_o = existing channel bottom slope in [L/L], and L = length of reach in [L].

The purpose of a drop structure is to reduce the flow velocity and to satisfy a set of design criteria on permissible flow Froude number. Therefore, it is convenient to relate Equation 7.11 to flow Froude number. Aided with Equation 7.5, the proposed slope is calculated as

$$S_n = \frac{N^2 U^2}{k^2 R^{4/3}} = \frac{N^2 g D}{k^2 R^{4/3}} F_r^2 \tag{7.19}$$

Design parameters for energy dissipation over a drop structure include channel lining roughness, flow Froude number, and jumps in the stilling basin. All these design parameters shall be collectively utilized to maximize the energy dissipation. Dimensional analysis is employed to provide a clue as to how to select the permissible Froude number (Guo, 2009). Using the normal depth as the characteristic length, the dimensionless form of Equation 7.18 is derived as

$$h_* = \left(S_o - N_*^2 F_r^2 \frac{D_*}{R_*^{4/3}} \right) L_* \tag{7.20}$$

The dimensionless variables in Equation 7.20 are defined as

$$h_* = \frac{h}{Y_n} \tag{7.21}$$

$$L_* = \frac{L}{Y_n} \tag{7.22}$$

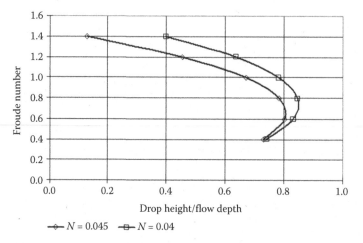

Figure 7.5 Energy dissipation over drop structure.

$$D_* = \frac{D}{Y_n} \tag{7.23}$$

$$R_* = \frac{R}{Y_n} \tag{7.24}$$

$$N_* = \left(\frac{N^2 g}{k^2 Y_n^{1/3}}\right)^{0.5} \tag{7.25}$$

The dimensionless variable, h_*, represents the energy dissipation defined as the ratio of drop height to normal flow depth. A drop structure is always protected with riprap blankets that have a Manning's roughness coefficient of 0.04–0.045. For a range of Froude number, Equation 7.20 is tested and plotted in Figure 7.5. It suggests that the maximized h_* be achieved with $F_r = 0.7$–0.8 (Guo, 2009).

To protect the channel bed from erosion, the normal flow velocity for the design discharge shall be reduced to 7.0 fps for cohesive soil bed or 5.0 fps for sandy soil bed. For both cases, the flow Froude number shall be reduced to 0.80 or smaller. To satisfy the above mentioned criteria, installing a drop structure is necessary.

EXAMPLE 7.2

The trapezoidal channel in Example 7.1 shall be designed with a permissible flow velocity not exceeding 5.0 ft/s. Determine the drop height for a reach of 100 ft. Let $U = 5.0$ ft/s

$$A = \frac{Q}{U} = \frac{1000.0}{5.0} = 20.0 \, \text{ft}^2$$

$$T = B + (Z_1 + Z_2)Y = 10 + (4+5)Y$$

$$A = \frac{Y}{2}(B+T) = \frac{Y}{2}[10 + (10 + 9Y)]$$

So, $Y = 5.65$ ft, $U = 5.0$ ft/s, $R = 3.22$ ft, $T = 60.83$ ft, and $F_r = 0.49$

$$S_n = \frac{N^2 U^2}{k^2 R^{4/3}} = \frac{0.03^2 \times 5.0^2}{1.468^2 \times 3.22^{4/3}} = 0.0021$$

$h = (S_o - S_n)L = (0.04 - 0.0021) \times 100 = 3.79$ ft and the top width $= 60.83$ ft for this case.

EXAMPLE 7.3

The trapezoidal channel in Example 7.1 shall be designed with a permissible Froude number not exceeding 0.7. Determine the drop height for a channel reach of 100 ft.

$$F_r^2 = \frac{TQ^2}{gA^3} = \frac{(10+9Y) \times 1000^2}{32.2 \times \left[\frac{Y}{2}[10 + (10+9Y)]\right]^3} = 0.7^2 \text{ So, } Y = 4.76 \text{ ft, } U = 6.69 \text{ ft/s and } R = 2.77 \text{ ft}$$

$$S_n = \frac{N^2 U^2}{k^2 R^{4/3}} = \frac{0.03^2 \times 6.69^2}{1.486^2 \times 2.77^{4/3}} = 0.0047$$

$h = (S_o - S_n)L = (0.04 - 0.0047) \times 100 = 3.53$ ft

In comparison, the drop height for this channel in Example 7.1 is dominated by the criterion of permissible flow velocity that demands a drop of 3.79 ft per every 100 ft of reach.

7.1.2 Plunging pool

A vertical drop structure produces a turbulent jet with a parabolic trajectory to impinge on the downstream channel floor. The geometry of this jet flow (shown in Figure 7.6) is described by the drop height and drop number that are defined as

$$q = \frac{Q}{T} \tag{7.26}$$

$$D_n = \frac{q^2}{gh^3} \tag{7.27}$$

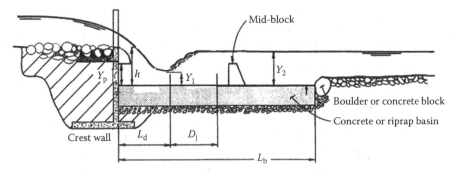

Figure 7.6 Plunging pool downstream of drop structure.

The drop-pool section is designed to have a rectangular cross-section with its width equal to the top width, T, of the drop structure.

$$\frac{L_d}{h} = 4.30 D_n^{0.27} \tag{7.28}$$

$$\frac{Y_p}{h} = 1.00 D_n^{0.22} \tag{7.29}$$

$$\frac{Y_1}{h} = 0.54 D_n^{0.425} \tag{7.30}$$

$$\frac{Y_2}{h} = 1.66 D_n^{0.27} \tag{7.31}$$

$$\frac{D_j}{Y_1} = 10(F_{r1} - 1) = 10\left(\frac{U_1}{\sqrt{gY_1}} - 1\right) \tag{7.32}$$

in which q = unit-width discharge in $[L^2/T]$, D_n = drop number, L_d = distance in $[L]$ from drop wall to depth Y_1, Y_p = pool depth in $[L]$ near drop wall, Y_1 = pool depth in $[L]$, U_1 = average flow at section 1 in $[L/T]$, Y_2 = tailwater depth in $[L]$, D_j = jump length in $[L]$, F_{r1} = Froude number based on flow depth Y_1 in $[L]$, and L_b = length of pool in $[L]$.

EXAMPLE 7.4

The trapezoidal channel in Example 7.2 shall be designed with Q = 1000 cfs, T = 60.83 ft, and h = 3.79 ft. Design the dimension of the plunging pool. The solution is summarized in Table 7.2.

7.1.3 Weir on top of drop structure

From the point of view of safety, a drop structure is always sized to withstand the extreme event. However, its daily operation is to pass frequent, small flows. As a result, the top weir should be designed with a low-flow notch. The low-flow rate is ~1–3% of the

Table 7.2 Design of drop pool

Channel reach distance	L = 100 ft
Drop height per reach	h = 3.79 ft
Unit flow rate per foot	$q = Q/T$ = 16.44 cfs/ft
Drop number	D_n = 0.15
Water depth in pool	Y_p = 2.51 ft
Water depth before jump	Y_1 = 0.92 ft
Water depth after jump	Y_2 = 3.80 ft
Location of jet impingement	L_d = 9.84 ft
Length of jump	D_j = 20.88 ft
Minimal total length of pool	$L_d + D_j$ = 30.71 ft

design flow such as the 100-year peak flow. Applying the weir flow formula to the low-flow notch for the low-flow rate yields

$$Q_L = \frac{2}{3}C_o\sqrt{2g}\,W_1 H_1^{\frac{3}{2}} = C_w W_1 H_1^{\frac{3}{2}} \tag{7.33}$$

where Q_L = low flow in [L^3/T], C_o = discharge coefficient such 0.6–0.7, C_w = weir co-efficient, g = gravitation acceleration in [L/T^2], W_1 = notch width in [L], and H_1 = low head depth in [L] on notch weir. In practice, the low head depth is preselected from the range of 1 to 2 ft. The only unknown W_1 in Equation 7.33 can be determined. Under the design flow, the high head depth on top of the drop structure applies to both the notch weir in the center and the two overflow weirs on both sides. The weir flow formula for the composite weir section under the high head depth is written as

$$Q_H = \frac{2}{3}C_o\sqrt{2g}\left[W_1(H_1+H_2)^{\frac{3}{2}} + (W_2-W_1)H_2^{\frac{3}{2}} \right] \tag{7.34}$$

where Q_H = design flow [L^3/T] under high head depth, W_2 = top width in [L] of the drop structure, and H_2 = water depth in [L] on top of overflow weir. In practice, W_2 in [L] is set to be the top width of the approaching channel section upstream of the drop structure. As a result, the only unknown H_2 in Equation 7.34 can be solved for the design flow (Figure 7.7).

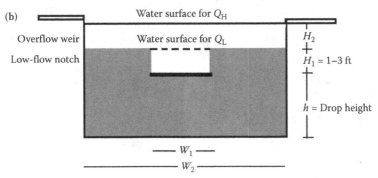

Figure 7.7 Weir section on top of drop structure. (a) Front view and (b) low-flow notch.

Table 7.3 Top weir cross-section

Weir flow coefficient	3.21	**Low flow**	**Weir**
Low flow	Q-Low	cfs	20
WS elevation	Enter H_1	ft	1.00
Weir width	W_1 (guessed)	ft	6.23
Diff. from Q-low	Check	cfs	0.00
		Overflow	**Weir**
Design flow	Q-high	Low weir	1000
WS elevation	H_2 (guessed)	ft	2.86
Weir width	W_2	ft	60.83
Diff. from Q-high	Check	cfs	0.00

EXAMPLE 7.5

The trapezoidal channel in Example 7.1 shall be designed with Q_L = 20 cfs, Q_H = 1000 cfs, H_1 = 1.0 ft, and C_o = 0.6 or C_w = 3.21 using feet-second unit. The upstream top width T = 60.83 ft. The dimensions of the low-flow and overflow weirs are determined as shown in Table 7.3.

7.2 Natural waterway

Hydraulic properties of a natural waterway are varied along the channel reach. The cross-sections as shown in Figure 7.8 in a natural waterway have been shaped with the long-term erosion by a wide range of flood flows. They are usually stable and do not have severe degradation or aggregation problems.

However, urbanization process significantly increases runoff rates and volumes. Therefore, it is important to assess the urbanization impacts on a natural waterway. Dealing with a natural waterway, the engineer must prepare the channel cross-sections and collect field data such as sediment samples, bank materials, etc. Study of the existing hydraulic conditions provides the understanding of the channel stability and serves as a basis to evaluate the impacts of urban developments on the natural waterway. The recommended criteria and considerations for natural waterway improvements are as follows:

Figure 7.8 Natural waterway with drops. (a) Mild channel with drops and (b) steep channel with continuous drops.

1. Natural waterways shall have adequate capacity to pass the major storm event.
2. Floodplains along a major drainage way must be defined and published in order to avoid inadvertent developments into the existing floodway and floodplain areas.
3. Roughness coefficients must be carefully determined and used for the analysis of water surface profiles.
4. The velocity and Froude number for the design discharge must be computed. If the flow Froude number is greater than unity, erosion protection must be considered.
5. Any grade-control structures such as drops and check dams installed on a natural waterway must be protected from scours.

7.3 Grass channel

Grass-lined channel in Figure 7.9 is the most desirable waterway in an urban area. Grass linings provide protection to the channel bed and banks from erosion. Table 7.4 provides the classification of vegetal hydraulic resistance based on the heights of grass stems.

Figure 7.9 Grass channel. (a) Reinforced grass channel and (b) grass channel.

Table 7.4 Retardance of various kinds of grass linings

Retardance	Cover	Condition	Height (in.)
A	Weeping love grass	Excellent stand	30–36
B	Kudzu	Dense growth and Uncut	12
	Bermuda grass	Good stand	24
	Weeping love grass	Good stand	12
	Alfalfa	Good stand	
	Mixed grass	Uncut and un-mowed	
C	Crabgrass	Fair stand	10–48
	Bermuda grass	Good stand, mowed	6
	Common Lespedeza	Uncut	11
	Mixed grass	Good Stand, uncut	6–8
	Centipede grass	Very dense	6
	Kentucky blue	Good stand	11
D	Bermuda grass	Good stand, cut	2.5
	Common Lespedeza	Excellent stand, uncut	4.5
	Buffalo grass	Good stand, uncut	3–6
	Mixed grass	Good stand, uncut	4–5
	Lespedeza	Good stand, cut	2
E	Bermuda grass	Good stand, cut	1.0–1.5

The hydraulic resistance of various grass linings to the flow has been classified into five categories:

1. Class A: very high vegetal retardance
2. Class B: high vegetal retardance
3. Class C: average vegetal retardance
4. Class D: low vegetal retardance
5. Class E: very low vegetal retardance

The roughness coefficient, Manning's N, of a grass channel has been found to be a function of the product, UR in $[L^2/T]$, in which U is the cross-sectional average flow velocity and R is the hydraulic radius. The design curves shown in Figure 7.10 developed by the US Natural Resources Conservation Service (NRCS) are often used to determine the value of Manning's N by an iterative process. As shown in Figure 7.10, it is noted that the limiting values of Manning's N exist for all types of grass linings. For instance, the Manning's N for type A grass converges to the value of 0.06. These limiting values (Table 7.5) may serve as an initial estimate of roughness coefficient when designing a grass-lined channel.

The design criteria for a grass-lined channel include the following:

1. The maximum permissible velocity for the major storm event in a grass channel is 7.0 fps for cohesive soil and 5.0 fps for sandy soil. A minimum of 2.0 fps should be maintained for the minor event.
2. The flow Froude number for the design discharge must be <0.8.
3. In places where the natural topography is steeper than desired, drop structures may be considered to control the channel grade.

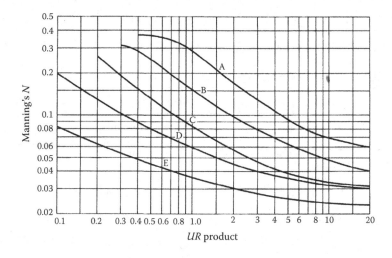

Figure 7.10 Variations of Manning's N in grass channel.

Table 7.5 Limiting Manning's N for grass linings

Type of grass linings	A	B	C	D	E
Limiting Manning's N	0.0600	0.0400	0.0330	0.0300	0.0240

4. The center line curvature shall have a radius twice the top width of the design flow but not <100 ft.
5. The flow depth shall be kept <5 ft for the major event.
6. The roughness coefficient must be determined by the empirical relationships using the product of UR for various vegetal retardance levels.

A grass channel is designed to have the maximum cross-sectional average flow velocity, U_m, not exceeding 7.0 fps for cohesive soil bed and 5.0 fps for sandy soil bed. The design process of a grass channel is iterative until the design criteria are satisfied. The Manning's roughness coefficient may start with the limiting value for the type of grass (as shown in Table 7.5). The hydraulic radius corresponding to the maximum permissible velocity is determined as

$$R = \left(\frac{NU_m}{kS_0^{\frac{1}{2}}} \right)^{\frac{3}{2}} \tag{7.35}$$

Next, we compute the product, RU_m, by which we can revise the value of Manning's N, according to Figure 7.10. Substituting the new N value into Equation 7.35 yields a new value for R. Repeat this process until the value of Manning's N is converged to Figure 7.10. The solution is achieved if the difference between the new and previous N values is <3.0%.

$$Error = \left| \frac{N_{new} - N_{old}}{N_{new}} \right| \leq 0.03 \tag{7.36}$$

in which N_{new} = Manning's N updated and N_{old} = Manning's N for the last iteration.
   The cross-sectional elements including flow depth, Y, channel width, B, and side slope, Z, are then calculated by solving the following two equations simultaneously:

$$A = \frac{Q}{U_m} \tag{7.37}$$

and

$$R = \frac{A}{P} \tag{7.38}$$

More than one solution may be achieved by solving Equations 7.37 and 7.38. Final selection depends on the design constraints and site conditions.

EXAMPLE 7.6

A triangular channel is to be lined by a 6-in. Bermuda grass on a slope of 0.8%. The maximum permissible velocity is 5.0 fps. Determine the channel cross-section to carry a design flow of 100.0 cfs.
   According to Table 7.4, a 6-in. Bermuda grass presents type C vegetal retardance. From Table 7.5, the limiting Manning's N for type C grass is 0.032.

Table 7.6 Iterative procedure for grass channel design

Trial no.	Manning's N used	Computed, R	Product $(R \cdot U_m)$	New Manning's, N
1.0	0.0320	1.32	6.58	0.0370
2.0	0.0370	1.64	8.18	0.0360
3.0	0.0360	1.58	7.90	0.0355

$$R = \left(\frac{NU_m}{kS_0^{\frac{1}{2}}}\right)^{\frac{3}{2}} = \left(\frac{0.032 \times 5}{1.486 \times \sqrt{0.008}}\right)^{\frac{3}{2}} = 1.32 \text{ ft}$$

So, the product, $UR = R \times U_m = 1.32 \times 5.0 = 6.58$.

From Figure 7.10, Manning's $N = 0.037$ for $UR = 6.58\,\text{ft}^2/\text{s}$. With the new value of Manning's N, the iterative procedure is summarized in Table 7.6.

Check the convergence:

$$\left|\frac{0.036 - 0.0355}{0.036}\right| = 0.014 < 0.03$$

The third iteration satisfies the convergence criteria. The triangular cross-sections have to satisfy the following:

$$A = ZY^2 = \frac{Q}{U_m} = \frac{100}{5} \text{ or } Y = \sqrt{\frac{20}{Z}}$$

$$R = \frac{A}{P} = \frac{20}{2Y\sqrt{1 + Z^2}} < 1.58$$

Solutions can be achieved by trial and error as listed in Table 7.7.

Considering the channel's top width and available right of way, $Z = 3.0$ is accepted. The normal flow condition includes $Y_n = 2.58\,\text{ft}$ and $U_n = 4.25\,\text{fps}$. The height of freeboard for this case is determined to be

$$F = 1.0 + \frac{U_n^2}{2g} = 1.0 + \frac{4.15^2}{2 \times 32.2} = 1.27 \text{ ft}$$

The channel excavation depth for the straight reach is

$$Y_t = 2.58 + 1.27 = 3.85 \text{ ft}$$

Table 7.7 Multiple solutions for grass channel design

Side slope, Z (ft/ft)	Flow depth, Y (ft)	Hydraulic, R (ft)	Flow velocity (fps)
2.00	3.16	1.42	4.90
3.00	2.58	1.23	4.25
4.00	1.24	1.09	3.88

Figure 7.11 Examples of riprap channel. (a) Riprap channel with drop structure and (b) riprap channel on steep slope.

7.4 Riprap channel

Riprap linings are flexible and generally less expensive than rigid linings. Riprap lining is a good choice if the channel is expected to experience frost heave and/or swelling of the underlying soils that can change the shape of the channel lining. Riprap blankets permit infiltration and exfiltration and can be vegetated to have a natural appearance. More natural behavior offers better habitat opportunities for local flora and fauna. In many cases, riprap linings are designed to provide only transitional protection against erosion before vegetation becomes the permanent lining in the channel. Vegetative channel lining is recognized as the best management practice for stormwater quality design in highway drainage systems.

Riprap lining (Figure 7.11) is suitable for a short but steep channel reach. The nature of high friction of rocks contributes to the effectiveness of energy dissipation. Especially where right of way for the drainage way is limited, riprap channel may be considered because steeper side slopes can be acceptable. There are two approaches recommended for designs of riprap lining. The shear stress-based method is recommended for reaches with a supercritical flow, whereas the stream power-based method is recommended for subcritical flows.

7.4.1 Stream power-based method

To protect the upstream reach of a drop structure, the stream power-based method was developed using the product of US_o as the basis to determine the riprap rock size, in which U is the cross-sectional average flow velocity and S_o is the channel bottom slope. Manning's roughness, N, is correlated to the intermediate riprap rock size, D_{50}, in feet as

$$N = 0.0395 D_{50}^{\frac{1}{6}}$$

(7.39)

As recommended (USWDCM, 2001), the resistance indicator, K, for a stable riprap channel is defined by

$$K = \frac{US_0^{0.17}}{(S_s - 1)^{0.66}}$$

(7.40)

Table 7.8 Commercial types of riprap rock

Resistance K value	Riprap rock type	Size of D_{50} (in.)
1.5–4.0	VL	6.0
4.0–5.0	L	9.0
5.0–5.8	M	12.0
5.8–7.1	H	18.0
7.1–8.2	VH	24.0

in which S_s = specific gravity of rock such as 2.5–2.65. Equation 7.40 is recommended when the flow Froude number is <0.80. Table 7.8 provides a basis to select a riprap blanket based on the resistance indicator, K, described by Equation 7.40.

It takes an iterative process to size the riprap lining. Assign a value for D_{50} in Equation 7.39 to calculate Manning's N. Use Manning's formula to determine the normal flow condition. Check if the product of US_o in Equation 7.40 satisfies the recommended range for the resistance indicator, K. Repeat this procedure until the proper D_{50} is found. The stream power-based method is good for reaches with a subcritical flow and if the ratio of water depth to D_{50} is 3 or higher.

7.4.2 Shear stress-based method

Friction force of water flowing in a channel is termed the tractive force that is expressed as the shear stress per unit area. The basis for stable riprap channel design is that the flow-induced shear stress should not exceed the permissible shear stress determined for the lining materials. Based on the logarithmic turbulent flow velocity distribution, Manning's N for riprap lining is directly related to the selected riprap rock size, D_{50}, as (HEC 15)

$$N = \frac{R^{1/6}}{3.82\left(2.25 + 5.23 \, \log(R/D_{50})\right)} \tag{7.41}$$

Using the normal flow as the basis, the flow shear stress applied to the wetted perimeter is equal to

$$\tau_o = \gamma R S_o \tag{7.42}$$

where τ_o = flow shear stress in N/m^2 or lb/ft^2 and γ = unit weight of water such as $9810 \, N/m^3$ or $62.4 \, lb/ft^3$. The permissible shear stress is set to be the critical shear stress that will induce the incipient motion to each individual riprap rock. The permissible shear stress is determined as

$$\tau_p = F(S_s - 1)\gamma D_{50} \tag{7.43}$$

where τ_p = permissible shear stress for the channel lining in N/m^2 (lb/ft^2). The value of F is determined by an interpolation between the two limits defined by the flow Reynolds number as

$$F = 0.15 \text{ if } Re > 2 \times 10^5 \tag{7.44}$$

$$F = 0.047 \text{ if } Re > 4 \times 10^4 \tag{7.45}$$

$$Re = \frac{UR}{v} \tag{7.46}$$

where Re = flow Reynolds number and v = kinematic viscosity of water. The basic design procedure is to make sure the computed flow shear stress is not exceeding the permissible shear stress determined for the selected riprap lining. To be conservative, a safety factor is introduced to the comparison as

$$S_f = \frac{\tau_p}{\tau_o} > 1.0 - 1.5 \tag{7.47}$$

The safety factor, S_f, provides a measure of uncertainty. A safety factor of 1.0 is acceptable for most cases. However, a higher safety factor up to 1.5 is recommended if there exists significant uncertainty regarding the design discharge and/or the consequences of riprap failure lead to huge damage.

EXAMPLE 7.7

A riprap channel is designed to carry a flow of 500.00 cfs. The channel has an invert slope of 0.03, bottom width of 3.00 ft, and side slope of 1V:2H. Design the riprap channel cross-section using the stream power method.

Solution: First, select a type of rock riprap. Let us begin with the VL type rock that has $D_{50} = 6.0$ in. and specific weight of 2.5.

$D_{50} = 6.0$ in. $= 0.5$ ft

$N = 0.0395\, D_{50}^{(1/6)} = 0.0352$

Next, the normal flow condition is determined by Manning's formula as: $Y = 3.52$ ft and $U = 11.80$ fps. Substituting the normal depth and velocity into the rock resistance indicator equation yields

$$K = \frac{US_0^{0.17}}{(S_s - 1)^{0.66}} = \frac{11.80 \times 0.012^{0.17}}{(2.5 - 1)^{0.66}} = 4.98$$

Based on Table 7.8, the *L type rock* should be used. The D_{50} for the *L type rock* is 9 in. Repeating the previous procedures, we have

$N = 0.04$, $Y = 4.10$ ft, $U = 10.89$ fps

Using the normal flow depth and velocity, the rock resistance indicator, K, is

$$K = \frac{US_0^{0.17}}{(S_s - 1)^{0.66}} = \frac{10.89 \times 0.012^{0.17}}{(2.5 - 1)^{0.66}} = 4.59$$

According to Table 7.8, the selected *type L rock* is recommended for this case. The channel cross-section is designed as shown in Figure 7.12.

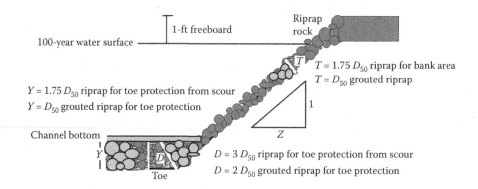

Figure 7.12 Riprap toe and bed in cross-section.

Table 7.9 Riprap design with a drop

Normal flow depth	Guess $Y = 6.07\,\text{ft}$
Top width	$T = 64.62\,\text{ft}$
Flow area	$A = 226.44\,\text{ft}^2$
Wetted perimeter	$P = 65.97\,\text{ft}$
Hydraulic radius	$R = 3.43\,\text{ft}$
Flow velocity	$V = 4.42\,\text{fps}$
Hydraulic depth	$D = 3.50\,\text{ft}$
Froude number	$F_r = 0.42$
Difference from design Q	Check on $dQ = 0.00\,\text{cfs}$
Resistance indicator	$K = 1.11$

EXAMPLE 7.8

The trapezoidal channel in Example 7.1 carries a discharge of 1000 cfs with $B = 10\,\text{ft}$, $Z_1 = 4$, and $Z_2 = 5$. The existing slope of 4% is reduced to 0.21% with the proposed drop structure. Determine the riprap rock blanket to protect the upstream of the drop structure.

Solution: Try type VL riprap rock with $D_{50} = 6$ in. and $S_s = 2.65$.

$$N = 0.0395(D_{50})^{\frac{1}{6}} = 0.0395(0.5)^{\frac{1}{6}} = 0.0352.$$

The solution for normal flow condition is summarized in Table 7.9.

For this case, $F_r = 0.42 < 1.0$. It is a subcritical flow. $K = 1.11$ implies that type VL riprap blanket is sufficient.

EXAMPLE 7.9

The trapezoidal channel in Example 7.1 carries a discharge of 1000 cfs with $B = 10\,\text{ft}$, $Z_1 = 4$, and $Z_2 = 5$. Determine the riprap blanket to protect the channel on a slope of 4%.

Solution: Try type L riprap rock with $D_{50} = 9$ in. and $S_s = 2.65$. Solution is presented in Table 7.10.

For this case, $F_r = 0.98$. $S_f > 1.50$ implies that type L riprap blanket is sufficient.

Table 7.10 Riprap design for steep channel

Enter roughness height	$D_{50} = 9.00$ in.
Normal flow depth	*Guess*, $Y = 4.06$ ft
Top width	$T = 46.58$ ft
Flow area	$A = 114.96$ ft^2
Wetted perimeter	$P = 47.48$ ft
Hydraulic radius	$R = 2.42$ ft
Manning's N	$N = 0.0618$
Flow velocity	$V = 8.70$ fps
Hydraulic depth	$D = 2.47$ ft
Froude number	$F_r = 0.98$
Difference from design Q	*Check* $\Delta Q = 0.00$ cfs
Flow shear stress	$\tau_o = 6.04$ lb/ft^2
Flow Reynolds number	$Re = 1.76E + 06$ lb/ft^2
Value of variable F	$F = 0.15$ lb/ft^2
Permissible shear stress	$\tau_p = 11.58$ lb/ft^2
Shear safety factor	$S_f = 1.92$ lb/ft^2
Required safety factor	$S_f\text{-req.} = 1.50$ lb/ft^2

7.5 Composite channel

In an urban area, the major waterway has to be designed to safely pass major flood flows and also frequent nuisance flows. As a result, the concept of multiple design events is critically important for urban channel design. Often, the wide floodplain areas shall be further developed into river parks and greenbelts for recreational uses. To accommodate the wide spectrum of storm events, the cross-section of a major waterway is composed of multiple layers, including the lower section to serve as the *main channel* to pass the frequent low flows and the overbank section reserved as the *overbank channel* to convey the major events such as 50- and 100-year storm runoff. A composite channel (as illustrated in Figure 7.13) is aesthetic and also provides significant detention storage volume during the major storm event. Examples of urban composite waterway are presented in Figure 7.14.

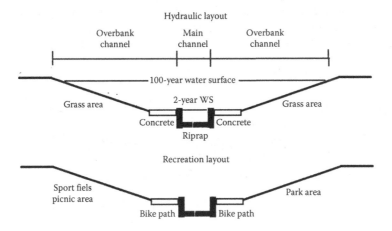

Figure 7.13 Cross-section of urban composite channel.

Figure 7.14 Cherry Creek in dry time and wet time, Denver, CO. (a) Cherry Creek, Denver, CO and (b) flood flow in Cherry Creek.

The concept of multiple land uses on the wide floodplains and multiple layers to convey all storm events is the most preferable approach to lay out an urban waterway when the easement and right of way are available.

The main channel of a composite cross-section is usually shallow, narrow, and hard bottomed with a side slope ranging as 1V:1–2H. The capacity of a main channel is sized to carry a low flow up to the 2-year storm event. The overbank areas are usually wide, flat, and grass-lined to serve as open space. In order to enhance recreational uses, the overbank areas are laid out on a very mild side slope such as 1V:7–10H. There are three major types of composite channels, depending on the capacity and functions of the main channel. They are (1) low-flow channel, (2) trickle-flow channel, and (3) wetland flow channel.

7.5.1 Low-flow channel

A low-flow channel (Figure 7.15) is often used along the major waterway in an urban area. When the overbank areas are used for recreational trails or bike path as shown in Figure 7.16, the main channel should have a capacity of passing up to the 2-year peak discharge without a freeboard. The bankfull depth of the main channel ranges from 3 to 5 ft. Both banks of the main channel shall be stabilized with riprap or concrete lining protection. Above the main channel are overflow bank areas that are lined with grass on a mild side slope.

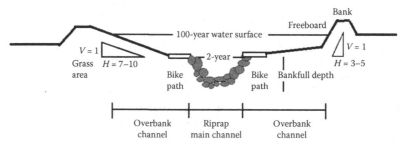

Figure 7.15 Low-flow channel section.

Figure 7.16 Examples of low-flow channel. (a) Low-flow channel with overbank areas and (b) low-flow channel with vertical banks.

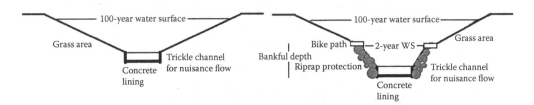

Figure 7.17 Trickle channel through composite channel.

7.5.2 Trickle channel

A trickle channel (Figure 7.17) is suitable for river parks and recreational areas on floodplains where return flows from irrigation and base flows from groundwater cause nuisance. A trickle channel is usually laid along the center of the floodplain with a shallow and narrow cross-section. The capacity of a trickle channel is about 1.0%–3.0% of the major flow. A composite channel through a river park may consist of three layers, including overbank area for major event, main channel for minor event, and trickle channel for base flows. As expected, a trickle channel is susceptible to bed erosion because it carries nuisance flows continually. It is advisable that the trickle channel be protected by concrete linings or riprap linings with excessive plant growth (Figure 7.18).

7.5.3 Wetland channel

Wetlands are created by the active interaction between surface storm runoff and groundwater. Natural wetlands are located at the low point on a wide floodplain where the water flow is spread into a vegetative area. A wetland area is expected to have a constant water depth of 4–6 in. Wherever the existing wetland areas are affected by surface drainage, a wetland channel (as shown in Figure 7.19) must be considered. A wetland channel has a similar layout as the low-flow channel, except that they provide habitats for aquatic, terrestrial, and avian wildlife. Channel erosion protections must not affect the interactions between groundwater and surface water. Wetland channel provides water quality enhancement as the flows move through the marshy vegetation. The abundant bottom vegetation traps

Figure 7.18 Examples of trickle channel. (a) Concrete trickle channel and (b) grass trickle channel.

Figure 7.19 Examples of wetland channel. (a) Constructed wetland channel and (b) natural wetland channel.

sediments and reduces flow velocity. Therefore, it is suggested that a wetland channel be widened to diffuse the flood water and deepened by 1 or 2 ft for excessive sediment deposit.

The hydraulic roughness of man-made grass-lined channels depends on the length and type of grass, as well as the water depth of flow. Table 7.11 is recommended for the determination of hydraulic roughness in composite grass-lined waterways and channels. It is noted that a flow depth exceeding 2 ft will begin to lay the grass down to form a smoother

Table 7.11 Manning's roughness for composite grass channel

Grass type	Grass length (in.)	0.7 ft < depth < 1.5 ft	Depth >3.0 ft
Bermuda	2	0.0350	0.0300
	4	0.0400	0.0300
Kentucky	2	0.0350	0.0300
	4	0.0400	0.0300
Any grass			
Good stand	12	0.0700	0.0350
	24	0.1000	0.0350
Fair stand	12	0.0600	0.0350
	24	0.0700	0.0350

bottom. Therefore, tall grass does not significantly increase the roughness coefficient to a major runoff flow.

7.5.4 Design criteria for composite channel

The following design criteria are recommended for composite channel design:

1. **Channel capacity**
   A trickle channel is sized to carry the base flow or up to 3% of the peak discharge of the major event. The capacity of a main channel shall be designed not exceeding the 2-year event. Often, 1/3 to 1/2 of the 2-year peak discharge is recommended. A wetland channel shall be capable of delivering the existing flow to sustain the habitats in the wide and deep sections.
2. **Flow depth**
   The bankfull depth for a main channel ranges from 1 to 3 ft. The maximum depth in the overbank areas shall not exceed 5–7 ft. A wetland channel adds 1–2 ft to the flow depth in the widened channel section as the storage volume for future sediment deposit.
3. **Flow velocity**
   The normal flow velocity for the major design discharge should not exceed 7.0 fps for erosion-resistant soil and 5.0 fps for easily eroded soil. Froude number for the major event in a composite grass channel should not exceed 0.8.
4. **Freeboard**
   No freeboard is needed for the main channel. A freeboard of 1 ft is recommended for the overbank areas if the flow is a subcritical flow. Superelevation and cross waves in the supercritical flow often require a higher freeboard.

7.5.5 Conveyance capacity in composite channel

Owing to different types of vegetation on overbank areas, a composite channel is divided into three flow areas: *main channel, left bank area, and right bank area*. The main channel is further divided into the *low-flow section*, which has a depth less than the bankfull depth, Y_m, and the *high-flow section*, which submerges the overbank areas. For a given depth, the flow capacity in a composite channel is analyzed under the assumption that the internal friction on the intersurface is negligible.

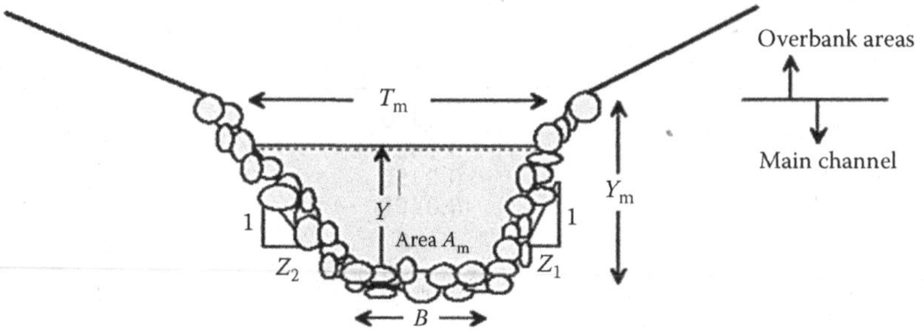

Figure 7.20 Low-flow channel capacity.

7.5.5.1 Flow in main channel (Y ≤ Y$_m$)

When the flow depth in Figure 7.20 is shallower than Y_m, the flow is confined within the main channel only. The capacity of the main channel at the depth Y is calculated as

$$A_m = [B + 0.5(Z_1 + Z_2)Y]Y \tag{7.48}$$

$$P_m = B + \left[\sqrt{1 + Z_1^2} + \sqrt{1 + Z_2^2}\right]Y \tag{7.49}$$

$$Q_m = \frac{K}{N_m} A_m^{\frac{5}{3}} P_m^{\frac{-2}{3}} \sqrt{S_0} \tag{7.50}$$

$$T_m = B + (Z_1 + Z_2)Y_m \tag{7.51}$$

$$F_{rm} = \left(\frac{Q_m^2 T_m}{g A_m^3}\right)^{0.5} \tag{7.52}$$

7.5.5.2 Flow in overbank areas (Y ≤ Y$_m$)

When the flow exceeds the bankfull capacity of the main channel, water is spilt into the overbank areas. Assuming that the internal frictions on the intersurfaces are negligible, the capacity of a composite channel with a depth $Y > Y_m$ (as illustrated in Figure 7.21) is separately estimated by the following steps:

1. **Flow in main channel**

$$A_m = [B + (Z_1 + Z_2)Y_m](Y - Y_m) + [B + 0.5 \times (Z_1 + Z_2)Y_m]Y_m \tag{7.53}$$

$$P_m = B + \left(\sqrt{1 + Z_1^2} + \sqrt{1 + Z_2^2}\right)Y_m \tag{7.54}$$

$$Q_m = \frac{k}{N_m} A_m^{\frac{5}{3}} P_m^{\frac{-2}{3}} \sqrt{S_0} \tag{7.55}$$

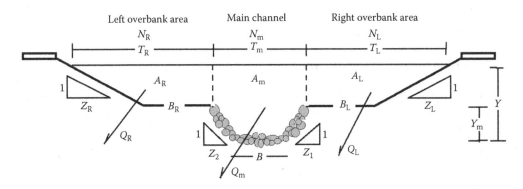

Figure 7.21 Hydraulic capacity of a composite channel.

$$T_m = B + (Z_1 + Z_2)Y_m \tag{7.56}$$

$$F_{rm} = \left(\frac{Q_m^2 T_m}{g A_m^3}\right)^{0.5} \tag{7.57}$$

2. **Flow in left overbank area**

$$A_L = B_L(Y - Y_m) + 0.5 Z_L (Y - Y_m)^2 \tag{7.58}$$

$$P_L = B_L + \sqrt{1 + Z_L^2}(Y - Y_m) \tag{7.59}$$

$$Q_m = \frac{k}{N_m} A_m^{\frac{5}{3}} P_m^{\frac{-2}{3}} \sqrt{S_0} \tag{7.60}$$

$$T_L = B_L + (Y - Y_m)Z_L \tag{7.61}$$

$$F_{rL} = \left(\frac{Q_L^2 T_L}{g A_L^3}\right)^{0.5} \tag{7.62}$$

3. **Flow in right overbank area**

$$A_R = B_R(Y - Y_m) + 0.5 Z_R (Y - Y_m)^2 \tag{7.63}$$

$$P_R = B_R + \sqrt{1 + Z_R^2}(Y - Y_m) \tag{7.64}$$

$$Q_R = \frac{k}{N_R} A_R^{\frac{5}{3}} P_R^{\frac{-2}{3}} \sqrt{S_0} \tag{7.65}$$

$$T_R = B_R + (Y - Y_m)Z_R \tag{7.66}$$

$$F_{rL}=\left(\frac{Q_L{}^2 T_L}{gA_L{}^3}\right)^{0.5} \tag{7.67}$$

As a result, the flow parameters for the entire composite channel section are summed up as

$$A = A_L + A_R + A_m \tag{7.68}$$

$$P = P_L + P_R + P_m \tag{7.69}$$

$$Q = Q_L + Q_R + Q_m \tag{7.70}$$

$$T = T_L + T_R + T_m \tag{7.71}$$

$$N = \frac{k}{Q} A^{\frac{5}{3}} P^{\frac{-2}{3}} \sqrt{S_0} \tag{7.72}$$

$$F_r = \left(\frac{Q^2 T}{gA^3}\right)^{0.5} \tag{7.73}$$

in which A = flow area in $[L^2]$, B = bottom width in $[L]$, N = Manning's roughness, T = top width in $[L]$, P = wetted perimeter in $[L]$, S_o = channel bottom slope in $[L/L]$, Z = channel side slope in $[L/L]$, Q = discharge in $[L^3/T]$, Y = flow depth in $[L]$, and F_r = Froude number. Subscript m represents variables related to the main channel, subscript L represents variables in the left overbank flow area, and subscript R is for the variables in the right overbank flow area.

The rating curve of a channel section describes the relationship between stage and conveyance capacity. Using Equations 7.48 through 7.73, a rating curve for a composite channel can be established.

EXAMPLE 7.10

A drainage channel in Figure 7.22 is laid on a slope of 2% and designed to pass 100 cfs in the main channel and 550 cfs by the entire section. The main channel is lined by riprap rocks with a side slope of IV:IH. The overbank areas are lined by 12-in. mixed grass with a side slope of IV:5H. Determine the bankfull and maximum water depths for this channel.

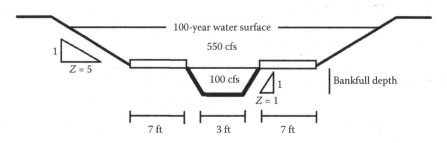

Figure 7.22 Hydraulic capacity in composite channel.

1. **Flow analysis for main channel section**
   First, apply the riprap design procedure to confirm that the VL type rock blanket with D50 of 6.0 in. is adequate for the main channel. The hydraulic roughness in this main channel is 0.0352. The discharge of 100 cfs has a depth of 2.48 ft. For convenience, the main channel for this case is set to have a bankfull depth of 2.50 ft.

2. **Flow analysis for the composite channel section**
   According to Table 7.11, the roughness coefficient is found to be 0.07 for a 12-in. mixed grass. The capacity of the composite channel at a specified depth can be determined using Equations 7.53 through 7.67. For example, the following is the calculation of flow rates at a depth of 4.5 ft:

   a. Main channel

   The design parameters for the main channel are as follows: $B = 3.0$ ft, $N_m = 0.035$, $Z_1 = Z_2 = 1$ ft/ft, $Y_m = 2.5$ ft, and $S_o = 0.02$ ft/ft. Aided with Equations 7.53 through 7.57, the flow parameters are determined as

   $$A_m = [3.0 + (1+1) \times 2.5] \times (4.5 - 2.5) + [3.0 + 0.5 \times (1+1) \times 2.5] \times 2.5 = 29.7 \, \text{ft}^2$$

   $$P_m = 3.0 + 2.5 \times [(1+1^2)^{0.5} + (1+1^2)^{0.5}] = 10.07 \, \text{ft}$$

   $$Q_m = \frac{1.486}{0.035} \times 29.75^{\frac{5}{3}} \times 10.07^{-\frac{2}{3}} \times \sqrt{0.02} = 366.65 \, \text{cfs}$$

   $$T_m = 3.0 + (1+1) \times 2.5 = 8.0 \, \text{ft}$$

   $$F_{rm} = \sqrt{\frac{366.65^2 \times 8}{32.2 \times 29.75^3}} = 1.126$$

   b. Left overbank flow

   The design parameters for the left overflow section are as follows: $B_L = 7.0$ ft, $N_L = 0.07$, $Z_L = 5.0$ ft/ft, and $S_o = 0.02$ ft/ft. Using Equations 7.58 through 7.62, we have

   $$A_L = 7.0 \times (4.5 - 2.5) + 0.5 \times 5.0 \times (4.5 - 2.5)^2 = 24.0 \, \text{ft}^2$$

   $$P_L = 7.0 + (4.5 - 2.5) \times \sqrt{1 + 5.0^2} = 17.20 \, \text{ft}$$

   $$Q_L = \frac{1.486}{0.070} \times 24^{\frac{5}{3}} \times 17.20^{-\frac{2}{3}} \times \sqrt{0.02} = 90.22 \, \text{cfs}$$

   $$T_L = 7.0 + (4.5 - 2.5) \times 5.0 = 17.0 \, \text{ft}$$

   $$F_{rL} = \sqrt{\frac{90.22^2 \times 17.0}{32.2 \times 24.0^3}} = 0.558$$

c. Right overbank flow

The right overflow section is composed of the following parameters: $B_R = 7.0\,\text{ft}$, $N_R = 0.07$, $Z_R = 5.0\,\text{ft/ft}$, and $S_o = 0.02\,\text{ft/ft}$. Using Equations 7.62 through 7.67, we have

$$A_R = 7.0 \times (4.5 - 2.5) + 0.5 \times 5.0 \times (4.5 - 2.5)^2 = 24.0\,\text{ft}^2$$

$$P_R = 7.0 + (4.5 - 2.5) \times \sqrt{1 + 5.0^2} = 17.20\,\text{ft}$$

$$Q_R = \frac{1.486}{0.070} \times 24^{\frac{5}{3}} \times 17.20^{-\frac{2}{3}} \times \sqrt{0.02} = 90.22\,\text{cfs}$$

$$T_R = 7.0 + (4.5 - 2.5) \times 5.0 = 17.0\,\text{ft}$$

$$F_{rR} = \sqrt{\frac{90.22^2 \times 17.0}{32.2 \times 24.0^3}} = 0.558$$

d. As a result, the flow parameters for the entire section are

$$A = A_L + A_R + A_m = 77.75\,\text{ft}^2$$

$$P = P_L + P_R + P_m = 44.47\,\text{ft}$$

$$Q = Q_L + Q_R + Q_m = 549.5\,\text{cfs} \left(\text{close to the design flow of } 550\,\text{fs}\right)$$

$$T = T_L + T_R + T_m = 42.0\,\text{ft}$$

$$N = \frac{1.486}{547.10} \times 77.75^{\frac{5}{3}} \times 44.47^{-\frac{2}{3}} \sqrt{0.02} = 0.043$$

$$F_r = \left(\frac{547.10^2 \times 42}{32.2 \times 77.75^3}\right)^{0.5} = 0.911$$

Figure 7.23 Public safety on and along urban floodplains. (a) Vertical and sloping composite drop and (b) trail between high banks without exit.

7.6 Closing

Floodplains along an urban channel provide open space for recreational uses, including parks, sports, bike paths, rafting, boating, etc. Public safety is always the major concern in the design of any and all hydraulic structures along or across an urban channel. It is critically import that riparian trails are not in a closed system between channel banks. Walkways to get out of the floodplains must be readily available for emergent evacuation. The riparian trail in Figure 7.23 was forced to shut down after loss of human life during an event of flash flood. The entire bike path system in this case needs a redesign to add more exits to the high banks. Signs of flash flood warming must be posted and operated timely to enhance public safety.

A grade-control structure (Figure 7.23) was built across the waterway for multiple purposes. It has a vertical drop for passing the water flow and also a sloping drop to act as a rafting passage. Drop structures must be safely designed to accommodate human's water recreational activities under the low- and high-flow conditions. Often, hydraulic structures were designed to achieve their hydraulic purposes, but they do need habitations to accommodate other needs after years in use.

7.7 Homework

Q7.1 A trapezoidal channel in Figure Q7.1 has a bottom width of 5 ft and side slopes of 1V:4H and 1V:3H. The design discharge is 1000 cfs with a Manning's coefficient of 0.033. The length of channel is 200 ft on a ground slope of 0.03 ft/ft. The design criteria for this case include that the flow velocity <7 fps and Froude Number <0.8. Perform the following tasks:

1. Estimate the height of the drop structure that is to be placed in the middle of this 200-ft reach.
2. Size the low-flow weir to pass 20 cfs on top of the proposed drop structure.

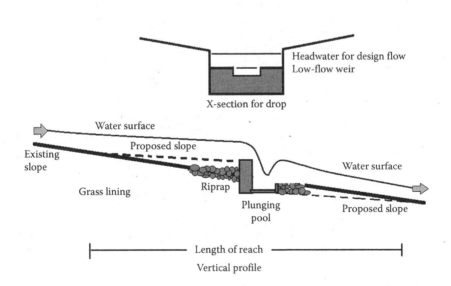

Figure Q7.1 Drop structure design.

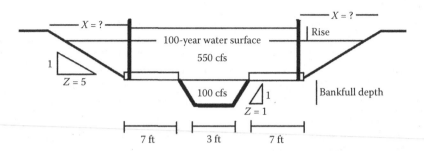

Figure Q7.2 Floodplain encroachment.

3. Determine the width and water depth on top of the drop.
4. Design the plunging pool downstream of the drop.
5. Suggest the riprap rock size, D_{50}, to protect upstream of the drop.
6. Determine the Manning's N for type C grass linings for the rest of the reach.

Q7.2 The drainage channel is laid on a slope of 2%. The 100-year peak discharge is 550 cfs. The main channel has Manning $N = 0.035$ with a side slope of 1V:1H. The overbank areas have Manning $N = 0.07$ with a side slope of 1V:5H. The bankfull depth is set to be 2.5 ft. Determine the encroachment distance, X, in Figure Q7.2, if the 100-year water depth can be raised by 0.5 ft.

Q7.3 Figure Q7.3 illustrates the top view of a flood channel. This channel starts with a straight reach and then turns into a curved reach with a radius of 250 ft. The outfall reach is straight and drains into a lake. This channel is to be lined by grass with drops.

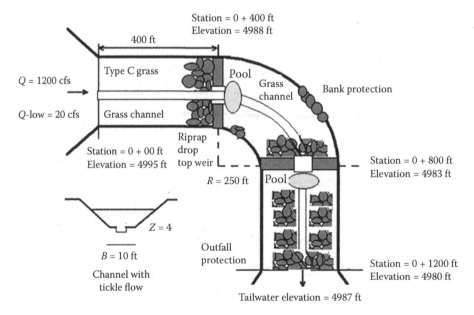

Figure Q7.3 Layout of flood channel.

The design discharge for a major storm event is 1200 cfs, and the trickle channel is designed to carry 20 cfs. The design parameters for this channel are as follows:

1. The channel cross-section shall be trapezoidal with $B = 10$ ft and $Z = 4.0$.
2. The channel will be lined by type C grass.
3. The permissible velocity in the channel is 7 fps with its Froude number <0.70.
4. The upstream of a drop needs to be protected by riprap blankets.
5. The downstream of a drop shall be equipped with a plunging pool.
6. Add a freeboard of 6-in. to take care of superelevation.

The elevations along the centerline of the channel reach are given as follows:

Section	Ground elevation (ft)	Reach	Length (ft)
1	4995	1	400
2	4988	2	400
3	4983	3	400
4	4980	4	In lake

Your task: Design drops at Station 0 + 400 ft and Station 0 + 800

1. Propose a new slope using a drop.
2. Determine the drop height for design flow $Q = 1200$ cfs.
3. Size the top weir to pass a low-flow Q-low $= 20$ cfs.
4. Feature the plunging pool downstream of the drop.
5. Choose a riprap blanket to protect the approach channel upstream of the drop.
6. Design the grass channel using type C grass.
7. Add a freeboard.
8. Discuss how to add the maintenance access ramp.

Design the riprap channel from Station 0 + 800 ft and Station 0 + 1200 using the shear stress method, and discuss the backwater effects from the lake.

Bibliography

ADOWR. (2000). *Design Guideline for Diversion Channels*, Flood Warning and Dam Safety Section, Arizona Department of Water Resources, Phoenix, AZ.

ASCE and WEF. (1992). "Design and Construction of Urban Stormwater Management Systems," American Society of Civil Engineers and Water Environment Federation, Reports of Engineering Practice No. 77 and WEF Manual of Practice FD-20, ASCE, New York.

Barnes, H.H. (1967). "Roughness Characteristics of Natural Channels," Water Supply Paper No. 1849, U.S. Department of the Interior, Geologic Survey, Washington, DC.

Chow, V.T. (1959). *Open Channel Hydraulics*, McGraw-Hill, New York.

City and County of Sacramento, California. (1992). *Hydrologic Standards*, prepared by Brown and Caldwell, Sacramento, CA.

Clark County. (1995). *Hydrologic Criteria and Drainage Design Manual*, Section 700 Open Channels, Clark County Regional Flood Control District, Clark County, Las Vegas, NV.

French, R.H. (1985). *Open Channel Hydraulics*, McGraw Hill, New York.

Guo, J.C.Y. (1999a). *Channel Design and Flow Analysis*, Water Resources Publication, Littleton, CO.

Guo, J.C.Y. (1999b). "Critical Flow Section in a Collector Channel," *ASCE Journal of Hydraulic Engineering*, Vol. 125, No. 4, doi: 10.1061/(ASCE)0733-9429.

Guo, J.C.Y. (2009). "Grade Control for Urban Channel Design," *Journal of Hydro-environment Research*, pp. 239–242, doi: 10.1016/j.jher.2009.01.001.

Little, W.C. and Murphey, J.B. (1982). "Model Study of Low Drop Grade Control Structures," *Journal of the Hydraulic Division*, Vol. 108, No. HY10, pp. 1132–1146.

Manning, R. (1891). "On the Flow of Water in Open Channel and Pipes," *Transactions, Institution of Civil Engineers of Ireland, Dublin*, Vol. 20, pp. 161–207.

US ACOE. (2008). *Hydraulic Design of Flood Control Channels*, Department of the Army, US Army Corps of Engineers, Washington, DC, 20314–1000, EM 1110-2-1601.

USWDCM. (2001). *Urban Storm Water Design Criteria Manual*, Volumes 1 and 2, Urban Drainage and Flood Control District, Denver, CO.

Chapter 8

High-gradient concrete channel

Channels with rigid linings such as concrete or shotcrete are considered highly durable and stable. In comparison with other channel linings, a concrete channel has the lowest Manning's roughness coefficient and the highest conveyance capacity to pass flood flows. Owing to the high initial cost, concrete channel is recommended only in places where grass lining is not possible. Generally, a concrete channel is placed on a high-gradient alignment where neither a drop structure nor a check dam is suitable to control the flow condition. As a rigid wall, the side slope in a concrete channel is not restricted by the soil stability and can be as steep as 1V:1H or vertical. Thus, concrete channels (Figure 8.1) are the solution for narrow right of way and for preventing potential scours and erosion on the channel bed.

8.1 Concrete channel

A smooth concrete surface has a Manning's N of 0.013–0.022. Water flow can run 20–30 ft/s in a concrete channel on a slope of 1%–3%. As expected, a concrete channel is likely designed to carry shallow supercritical flows. The design criteria for a concrete channel are essentially developed to control the water flow velocity for safety.

1. The side slope of a concrete channel can be as steep as vertical to 2V:1H.
2. The minimum thickness of concrete lining is 6 in.
3. Flow velocities in a concrete channel shall not exceed 18 ft/s.
4. Any potential for hydraulic jump shall be assessed.
5. The superelevation of the water surface at a bend must be included.
6. Manning's N depends on the surface finish as suggested in Table 8.1.

8.1.1 Weeping hole and underdrain system

The major advantage of concrete channel is to be able to pass the water flow at a high velocity. Care must be taken because a high velocity may cause a high static pressure when seepage flows are developed through the cracks and joints between concrete panels. A concrete channel shall be equipped with a subdrain system that has adequate weeping holes to release the static pressure built up in the saturated soil layers behind and/or beneath the concrete panels. After years in service, it is necessary to repair the deteriorated concrete linings and to identify the potential creeping motions of the soil layers beneath the concrete linings.

Figure 8.1 Concrete channel. (a) Matured vegetation along concrete channel and (b) concrete channel.

Table 8.1 Manning's roughness for concrete surface

Surface finish	Manning N
Trowel finish	0.0130
Float finish	0.0150
Unfinished	0.0170
Shotcrete, troweled, but not wavy	0.0180
Shotcrete, troweled, and wavy	0.020
Shotcrete, unfinished	0.0220

EXAMPLE 8.1

A concrete channel (Figure 8.2) carries the design flow at 18 ft/s with a depth of 3 ft. During the period of flood rising, the seepage flow through cracks and joints will saturate the soil layer underneath the concrete panels. The seepage flow is so slow such that its velocity head is negligible. As a result, the hydrostatic head in the saturated soil layer is estimated by the energy principle as

$$\frac{P_s}{\gamma} = \frac{U^2}{2g} + Y = \frac{18.0^2}{2 \times 32.2} + 3.0 = 8.0 \, \text{ft}$$

in which P_s = hydrostatic pressure in seepage flow in [L], γ = specific weight of water, 62.4 lb/ft^3, and g = gravitational acceleration in [L/T^2].

During the recession of flood flow, the water flow is emptied out of the channel much faster than the seepage flow through the saturated soil layers. For this case, after the channel dries out, the saturated pressure of 8 ft of water built up behind the concrete wall will impose a hydrostatic force of 2000 pound per linear foot on the bank wall. Such a high static force can cause a failure to the concrete linings (Figure 8.3). This example explains why the flow velocity in a concrete channel shall not exceed 18 ft/s, otherwise the concrete linings may become vulnerable to the saturated hydrostatic force built up through cracks and joints. To alleviate such potential damage, weeping holes and subdrains shall be installed to quickly remove the moisture out of the saturated soil layers. A concrete channel demands timely inspections and regular maintenance before and after a flood season, because weeping holes may become clogged after a severe event.

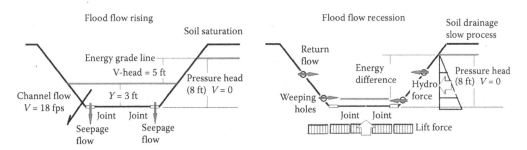

Figure 8.2 Illustration of hydrostatic force on concrete lining.

Figure 8.3 Failures in concrete channel. (a) Failure of concrete channel and (b) cracks on concrete panels.

8.2 Efficient channel sections

As indicated in Manning's formula, the hydraulic capacity in a rigid channel solely depends on the sectional elements and the longitudinal slope. For convenience, Manning's formula is further simplified to

$$Q = K\sqrt{S_o} \tag{8.1}$$

in which Q = design discharge in $[L^3/T]$, S_o = channel slope in $[L/L]$, and K = conveyance of channel section defined as

$$K = \frac{k_n}{N} A_f^{\frac{5}{3}} P^{\frac{-2}{3}} \tag{8.2}$$

where k_n = 1.486 for feet-second units or 1.0 for meter-second, N = Manning's roughness coefficient, A_f = flow area in $[L^2]$, and P = wetted perimeter in $[L]$. As shown in

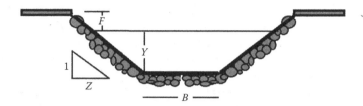

Figure 8.4 Channel excavated area.

Figure 8.4, the excavated area of a channel section includes both the flow area and the required freeboard height, F. For a trapezoidal channel, we have

$$A_t = (Y+F)B + Z(Y+F)^2 \qquad (8.3)$$

$$A_f = YB + ZY^2 \qquad (8.4)$$

in which A_t = excavated area in $[L^2]$, Y = flow depth in $[L]$, B = bottom width in $[L]$, Z = side slope in $[L/L]$, and F = height of freeboard in $[L]$. In practice, a freeboard of 1.0 ft is recommended for a major waterway (10- to 100-year event), whereas 0.5 ft is acceptable for a minor waterway (2- to 5-year event).

The optimal channel section is defined as the cross-section that provides a minimum excavation for the given discharge as follows:

Objective Function = Minimization of excavated area, A_t
subject to : Q = given design discharge.

Notice that the frictional resistance to water flow is proportional to the wetted perimeter. Therefore, the minimization of the excavated area is equivalent to the minimization of hydraulic resistance. As a result, the earlier objective will lead to *the most hydraulically efficient channel section*. According to the principle of duality, the above mentioned statement is identical to

Objective Function = Maximization of channel capacity, Q
subject to : A_t = specified constant.

As a result, it is concluded that the optimal channel section can be achieved by either maximizing the capacity for a specified channel sectional area or minimizing the excavated area for a specified capacity. A specified channel sectional area can be translated into a fixed cost, while a specified capacity implies a given design discharge. Both objective functions, discussed earlier, lead to the same conclusion (Guo and Hughes, 1984):

$$\frac{dA_t}{dY} = 0 \qquad (8.5)$$

and

$$\frac{dQ}{dY} = \frac{dK}{dY} = 0 \tag{8.6}$$

In general, the value of Z is predetermined by the soil stability such as the angle of repose. As a result, A_t in Equation 8.5 and K in Equation 8.6 are dependent on both the depth Y and bottom width B. Therefore, the total derivative of A_t is written as

$$\frac{dA_t}{dY} = \frac{\partial A_t}{\partial Y} + \frac{\partial A_t}{\partial B}\frac{dB}{dY} = 0 \tag{8.7}$$

Similarly, Equation 8.6 becomes

$$\frac{dK}{dY} = \frac{\partial K}{\partial Y} + \frac{\partial K}{\partial B}\frac{dB}{dY} = 0 \tag{8.8}$$

The optimal trapezoidal section is expressed by its depth-to-width ratio. Substituting Equation 8.3 into Equation 8.7 yields

$$\frac{dB}{dY} = -\frac{B}{Y+F} - 2Z \tag{8.9}$$

For convenience, let the following variables be defined as

$$m = \frac{F}{Y} \tag{8.10}$$

$$n = \frac{B}{Y} \tag{8.11}$$

$$Q' = \left(\frac{NQ}{k_n\sqrt{S_o}}\right)^3 \tag{8.12}$$

The variable m is the ratio of freeboard height to flow depth, and n is the ratio of channel bottom width to flow depth. Substituting Equations 8.10 through 8.12 into Equation 8.9 yields the optimal trapezoidal shape to be

$$n = \frac{1}{5m+2}\left[-(3km+2Zm-2k+3Z)+\sqrt{M}\right] \tag{8.13}$$

$$M = 9k^2m^2 - 16Z^2m^2 + 4k^2 + Z^2 + 32m^2kZ - 12k^2m$$
$$+ 38mkZ - 16Z^2m - 4kZ \tag{8.14}$$

$$k = \sqrt{1+Z^2} \tag{8.15}$$

$$A'_t = \frac{A_t}{Y^2} = n(1+m) + Z(1+m)^2 \tag{8.16}$$

$$A'_f = \frac{A_f}{Y^2} = n + 2 \tag{8.17}$$

$$P' = \frac{P}{Y} = n + 2k \tag{8.18}$$

$$Q'' = \frac{Q'}{Y^8} = (n + Z)^5 (n + 2k)^{-2} \tag{8.19}$$

Several special cases need to be further discussed.

8.2.1 Efficient trapezoidal channel without freeboard

Without a freeboard, we have $m = 0$. Equation 8.13 is reduced to

$$n = \frac{B}{Y} = 2(k - Z) \tag{8.20}$$

Substituting Equation 8.20 into Equations 8.16 through 8.19 yields

$$\frac{A_t}{Y^2} = n + Z \tag{8.21}$$

$$\frac{A_f}{Y^2} = 2k - Z \tag{8.22}$$

$$\frac{P}{Y} = 4k - 2Z \tag{8.23}$$

$$\frac{Q'}{Y^8} = (2k - Z)^5 (4k - 2Z)^{-2} \tag{8.24}$$

To minimize the wetted perimeter, we can differentiate Equation 8.23 with respect to Z and then set it equal to zero. The best side slope is found to be

$$Z = \frac{1}{Z_o} \tag{8.25}$$

in which $Z_o = \sqrt{3}$. Equation 8.25 implies that the most efficient channel section is a half hexagon. Substituting Equation 8.25 into Equations 8.21 through 8.24 yields

$$\frac{A_t}{Y^2} = Z_o \tag{8.26}$$

$$\frac{A_f}{Y^2} = Z_o \tag{8.27}$$

$$\frac{P}{Y} = 2Z_o \tag{8.28}$$

$$\frac{Q'}{Y^8} = \frac{Z_o^3}{4} \tag{8.29}$$

8.2.2 Efficient rectangular channel with freeboard

Because $Z = 0$ and $k = 1$, Equation 8.13 becomes

$$n = \frac{B}{Y} = \frac{2(2-3m)}{(5m+2)} \tag{8.30}$$

Substituting Equation 8.30 into Equations 8.16 through 8.19 yields

$$\frac{A_t}{Y^2} = \frac{2(2-3m)(1+m)}{(5m+2)} \tag{8.31}$$

$$\frac{A_f}{Y^2} = \frac{2(2-3m)}{(5m+2)} \tag{8.32}$$

$$\frac{P}{Y} = \frac{4(2+m)}{(5m+2)} \tag{8.33}$$

$$\frac{Q'}{Y^8} = \frac{2(2-3m)^5(2+m)^{-2}}{(5m+2)^3} \tag{8.34}$$

8.2.3 Efficient rectangular channel with no freeboard

For the case of a rectangular channel without a freeboard, we have $F = 0$, $Z = 0$, $m = 0$, and $k = 1$. Equation 8.30 becomes

$$n = \frac{B}{Y} = 2.0 \tag{8.35}$$

Substituting Equation 8.35 into Equations 8.31 through 8.34 yields

$$\frac{A_t}{Y^2} = 2.0 \tag{8.36}$$

$$\frac{A_f}{Y^2} = 2.0 \tag{8.37}$$

$$\frac{P}{Y} = 4.0 \tag{8.38}$$

$$\frac{Q}{Y^8} = 2.0 \tag{8.39}$$

Equations 8.36 through 8.39 agree with the conventional solution for the optimal rectangular channel without a freeboard (Chow, 1959).

EXAMPLE 8.2

A rectangular channel is designed to carry 100 cfs on a slope of 1%. The Manning's roughness is 0.03. Determine the optimal channel section with a freeboard of 0.5 ft.

Solution: The following iterative procedure is recommended to derive the solution for the optimal channel cross-section:

1. Guess the value of m. For instance, let m be 0.15.
2. Compute

$$Q' = \left(\frac{NQ}{k_n\sqrt{S_o}}\right)^3 = 8162.15 \text{ where } k = 0.1486.$$

3. Determine the ratio Q'/Y^8 by Equation 8.34 as

$$\frac{Q'}{Y^8} = \frac{2(2-3m)^5(2+m)^{-2}}{(5m+2)^3} = \frac{3.87}{20.80} = 0.186, \text{ or } Y = 3.80 \text{ ft}$$

4. Calculate the value of m:

$$m = \frac{F}{Y} = \frac{0.50}{3.80} = 0.132$$

Table 8.2 Optimal trapezoidal cross-sections

Design flow	$Q = 100.00$ cfs
Manning's roughness	$N = 0.0300$
Longitudinal slope	$S_o = 0.01000$ ft/ft
Freeboard to flow depth ratio	$m = F/Y = 0.135$
Dimensionless flow	$Q' = 8162$

Side slope, Z	Side parameter, $k = (1 + Z^2)^{0.5}$	M parameter	Width-to-depth ratio, $n = B/Y$	Total area ratio, A_t/Y^2	Flow area ratio, A_f/Y^2	Wetted P-meter ratio, P/Y	Flow rate-to-depth ratio, Q'/Y^8
0.00	1.00	2.54	1.19	1.35	1.19	3.19	0.24
1.00	1.41	6.07	0.54	1.90	1.54	3.37	0.76
2.00	2.24	14.60	0.32	2.94	2.32	4.79	2.90
3.00	3.16	28.69	0.22	4.11	3.22	6.54	8.06
4.00	4.12	48.38	0.17	5.34	4.17	8.41	17.73
5.00	5.10	73.69	0.13	6.60	5.13	10.33	33.41
6.00	6.08	104.62	0.11	7.86	6.11	12.28	56.58

Flow depth, Y (ft)	Bottom width, B (ft)	Free board, F (ft)	Wetted parameter, P (ft)	Flow area, A_f (ft²)	Excavated area, A_t (ft²)	Froude number, F_r	Specific energy, E_s (ft)
3.69	4.40	0.50	11.79	16.25	18.45	0.56	4.28
3.19	1.72	0.43	10.74	15.66	19.34	0.81	3.82
2.70	0.85	0.37	12.92	16.86	21.38	0.87	3.24
2.37	0.52	0.32	15.54	18.15	23.21	0.88	2.85
2.15	0.36	0.29	18.10	19.30	24.76	0.87	2.57
1.99	0.27	0.27	20.54	20.30	26.08	0.87	2.37
1.86	0.21	0.25	22.86	21.18	27.24	0.86	2.21

The computed value of m does not agree with the guessed value of 0.15.

5.   Let the new value of m be 0.135. Repeat steps 2 to 4 to find $Y = 3.69$ ft and $m = 0.134$, which is close enough to the guessed value. Substituting $m = 0.135$ into Equations 8.31 through 8.33 yields

$$\frac{A_t}{Y^2} = 1.354, \quad \text{so,} \quad A_t = 18.45 \text{ ft}^2$$

$$\frac{A_f}{Y^2} = 1.193, \quad \text{so,} \quad A_f = 16.25 \text{ ft}^2$$

$$\frac{P}{Y} = 3.194, \quad \text{so,} \quad P = 11.79 \text{ ft}$$

This case may be expanded into an optimal design for trapezoidal channels with various side slopes. Setting $m = 0.135$, the lengthy iterative procedure is summarized in Table 8.2.

As shown in Table 8.2, the case with $Z = 0$ provides the minimal excavated area, whereas another case with $Z = 1$ provides the minimal flow area. When Z increases, the channel bottom width decreases. The trend for the most efficient channel cross-section shifts toward the triangular shape with the least freeboard.

8.3   Waves in a high-gradient channel

Supercritical flows occur in steep channels. A small perturbation in a supercritical flow may be dampened out or amplified into waves, depending on the stability of the flow, lining roughness, and channel alignment.

1.   **Roll waves in a straight reach**

In a straight and steep reach, the flow is undertaking a significant acceleration. Consequently, the uniform flow as predicted by Manning's equation breaks into a train of traveling waves (as shown in Figure 8.5). These waves form a pulsating flow as a response to the temporary force balance among gravitational force, skin friction, and internal turbulent stress. Roll waves in pulsating flow exhibit similar characteristics to moving oblique jumps. They are in the transitional phase from a wall boundary layer flow to a turbulent flow (Chow, 1959).

Figure 8.5 Roll waves in concrete channel. (a) Supercritical flow and (b) roll waves.

After a pulsating flow is developed in a steep channel, roll waves begin to grow in wave height downstream. Eventually, these waves break down to form hydraulic bores or shock waves with rough tumbling heads and smooth tails.

2. **Superelevation and cross waves along a curve reach**

In addition to roll waves, the flow along a curve reach is also dominated by the centrifugal forces that are manifested by the superelevation along the outer bank as shown in Figure 8.6. The streamlines of the flow through a curve channel are not only curvilinear but also interwoven, resulting in significant spiral currents and cross waves. The complexity of the flow through a curve path causes difficulties in estimating water surface rises, energy losses, jumps, and discharge allocation at each section.

3. **Oblique jumps at a bend**

The forces from the bank walls can trigger instability to a supercritical flow. For instance, at a bend in Figure 8.7, the outer wall introduces continuous oblique jumps, whereas the inner wall produces expansion waves or continuous hydraulic drops. These two waves form cross waves on the water surface, bouncing back and forth between the bank walls.

It is clear that the uniform flow approach does not adequately address waves and energy dissipation in a supercritical flow with roll waves. The assumption of uniform flow always results in a wide and shallow channel on a high gradient. In fact, a steep channel

Figure 8.6 Superelevation in concrete curve channel. (a) Superelevation at bank and (b) superelevation in laboratory.

Figure 8.7 Oblique jumps at bend in concrete channel. (a) Cross waves in oscillatory flow and (b) oblique jumps.

must be equipped with a sufficient freeboard on top of uniform flow depth to prevent water spills due to roll waves, cross waves, oblique jumps, and superelevation.

8.4 Analysis of flow stability

The challenge in the design of a high-gradient channel is the uncertainty involving the energy interchange between the flow depth and the velocity head. The stability of a supercritical flow is sensitive to channel shape, lining roughness, curvature of alignment, and invert slope. Often, the flow velocity in a concrete channel can be as high as 20–30 ft/s or equivalent to a velocity head of 6–14 ft. One of the major concerns in the design of a high-gradient channel is to make sure that the selected channel cross-section can sufficiently accommodate the increase of flow depth due to roll waves. Using the flow Froude number as a criterion, open-channel flows are classified into two regimes: *subcritical* and *supercritical* flows. Furthermore, using the ratio of incremental flow velocity to incremental wave celerity as a criterion, supercritical open-channel flows are further divided into two regimes: *stable uniform* and *pulsating (unstable) flows* (Escoffier and Boyd, 1962). Roll waves in a pulsating flow are formed through a continuous growth of disturbances. The complex structure of roll waves consists of a series of bores separated by smooth variable water depth (Mayer, 1957). The heights of roll waves are random and temporal. Several studies applied normal probability distribution to describe laboratory and field data of roll waves with Froude number between 3.45 and 5.60 (Brock, 1969; Stonstreet, 1996). The channel cross-section has a decisive influence on the generation of roll waves. A proper selection of cross-sectional elements may eliminate roll waves (Thorsky and Haggman, 1970).

Studies of roll waves were performed primarily in connection with the mechanism of instability of uniform flow on a steep slope. Vedernikov number was developed to identify the existence of a pulsating flow in a high-gradient channel (Vedernikov, 1945, 1946). Vedernikov number is defined as

$$N_V = m_v \left(1 - R \frac{dP}{dA}\right) F \tag{8.40}$$

in which $m_v = 2/3$ when Manning's formula is used or 1/2 when Chezy's formula is used, R = hydraulic radius in [L], P = wetted perimeter I [L], F = Froude number, and A = flow area in [L^2]. To be a stable uniform flow, N_v shall be less than or equal to unity. To apply Equation 8.40 to a trapezoidal channel, the parameters are derived as

$$R = \frac{A}{P} = \frac{y(b+zy)}{b+2ky} \tag{8.41}$$

$$k = \sqrt{1+z^2} \tag{8.42}$$

$$\frac{dP}{dA} = \frac{2k}{b+2zy} \tag{8.43}$$

in which y = flow depth in [L], b = channel bottom width in [L], and z = channel side slope in [L/L]. Normalizing Equations 8.41 through 8.43 by channel bottom width and then substituting Equations 8.41, 8.42, and 8.43 into Equation 8.40 yields the limiting Froude number, F_L:

$$F_L = \frac{3}{2}\left[\frac{(1+2kY^*)(1+2zY^*)}{1+2zY^*+2kzY^{*2}}\right] \tag{8.44}$$

$$Y^* = \frac{y}{b} \tag{8.45}$$

Equation 8.44 is the definition of *limiting Froude number* for having a stable uniform supercritical flow in a trapezoidal channel. In other words, when the flow Froude number is greater than the limiting Froude number, the flow becomes unstable. When $z = 0$, Equation 8.44 is reduced to

$$F_L = \frac{3}{2}(2Y^*+1) \tag{8.46}$$

Figure 8.8 presents the family curves for limiting Froude numbers with $z = 0.0, 1.0, 2.0,$ and 3.0. These curves are converged at $F_r = 1.5$. It implies that supercritical uniform flows can be sustained in trapezoidal channels when the flow Froude number $F_r < 1.5$; otherwise, roll waves will be developed. Applying the limiting Froude number as a criterion, flows with Froude number >1.5 are further divided into two categories: (1) *stable uniform flows* when $F_r \le F_L$ and (2) *pulsating flows* when $F_r > F_L$. As a rule of thumb, a deep and narrow channel tends to carry stable flows, whereas a shallow and wide channel tends to carry pulsating flows.

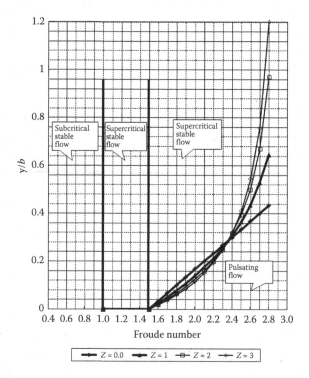

Figure 8.8 Flow regimes for open-channel flows.

Table 8.3 Flow stability analysis for steep concrete channel

Width, B (ft)	Depth, y (ft)	Flow velocity (ft/s)	Flow Froude number, F_r	Limiting Froude number, F_L	Comment
10	5.30	28.3	2.16	<3.09	Stable
12	4.45	28.1	2.35	<2.61	Stable
15	3.65	27.4	2.53	>2.23	Unstable
20	2.90	25.9	2.67	>1.94	Unstable

Care has to be taken when designing a steep channel. The channel cross-sectional geometry in terms of y/b ratio, channel slope, and lining roughness shall be selected to avoid roll waves or to add an adequate freeboard if the roll waves are developed. Otherwise, an additional freeboard section has to be provided.

EXAMPLE 8.3

A rectangular concrete channel is designed to carry 1500 cfs on a slope of 2.0%. Consider Manning's $N = 0.014$. Determine the maximum width-to-depth ratio, b/y, in order to have a stable flow.

1. To select a range of width, b, for instance, $b = 10, 12, 15, 20$ ft
2. To calculate the normal flow and flow Froude number, F_r, using Manning's formula for each width
3. To calculate the limiting Froude number by Equation 8.44
4. To identify the cases that satisfy $F_r < F_L$

Table 8.3 is the comparison between F_r and F_L for various channel widths. As expected, the narrower the channel, the more stable the flow. For this case, the channel width has to be narrower than 12.50 ft in order to have a stable flow.

When the channel width is >12 ft for this case, the flow becomes unstable and tends to generate roll waves. Therefore, an additional freeboard greater than or equal to the height of roll waves shall be considered. The height of roll waves can be determined using the model of moving hydraulic jump.

8.5 Roll waves

Development of roll waves is a continuous amplification of a small perturbation. For the purpose of channel design, this complicated process is approximated by the model of positive surges that has an advancing front with the profile similar to a moving hydraulic jump. When the height of the surge is small, the surge appears like an undular jump. When the jump height increases, the undulation will eventually be transformed into a surge with a sharp and steep front. As illustrated in Figure 8.9, the wave front is considered as a positive surge. The unsteady flow pattern can be converted into a steady flow by adding a negative wave speed to the entire flow field. As illustrated in Figure 8.9, we "freeze" the wave movement to produce a steady flow field. With a negative wave speed imposed to the flow field, the continuity principle becomes

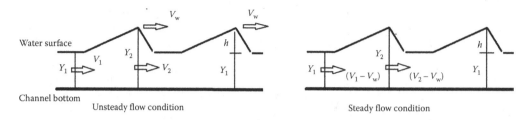

Figure 8.9 Moving jump.

$$(V_2 - V_w)A_2 = (V_1 - V_w)A_1 \tag{8.47}$$

Rearranging Equation 8.47 yields

$$V_2 = \frac{(V_1 - V_w)A_1 + V_w A_2}{A_2} \tag{8.48}$$

And the momentum principle between sections 1 and 2 is

$$(V_w - V_2)(V_2 - V_1) = \left(\bar{y}_2 - \frac{A_1}{A_2}\bar{y}_1\right)g \tag{8.49}$$

Solving Equations 8.47, 8.48, and 8.49 simultaneously yields (Chow, 1959)

$$V_w = V_1 + \sqrt{\frac{(A_2\bar{y}_2 - A_1\bar{y}_1)g}{A_1\left(1 - \dfrac{A_1}{A_2}\right)}} \tag{8.50}$$

Let the height of roll waves be the difference between flow depths:

$$h = y_2 - y_1 \tag{8.51}$$

$$C = V_w - V_2 \tag{8.52}$$

in which h = roll wave height in [L] and C = wave celerity in [L/T]. Considering that the roll waves near the center of the channel section are similar to that in a rectangular channel, the representative height of roll waves can be derived by simplifying Equations 8.48 and 8.49 with $z = 0$. After canceling the wave velocity in Equations 8.48 and 8.49, the height of roll waves is derived as (Guo, 1999a, 1999b)

$$h = \frac{C^2}{g}\left(\frac{2y_1}{y_1 + y_2}\right)\left(\frac{V_2}{C} - \frac{V_1}{C}\right) = \frac{C^2}{g}\left(\frac{2y_1}{y_1 + y_2}\right)(F_2 - F_1) \tag{8.53}$$

$$F_1 = \frac{V_1}{C} \tag{8.54}$$

$$F_2 = \frac{V_2}{C} \tag{8.55}$$

In practice, F_2 is set to be the flow Froude number for the design discharge, and F_1 is set to be the limiting Froude number, F_L, determined by Vedernikov number by Equation 8.44. Equation 8.53 indicates that the only condition for roll wave to exist is $F_2 > F_1$; otherwise, $h = 0$. When the height of roll waves is small compared with the depth of flow, i.e., $y_1 \approx y_2$, Equation 8.53 is reduced to

$$h = \frac{C^2}{g}\left(\frac{V_2}{C} - \frac{V_1}{C}\right) = \frac{(V_w - V_2)^2}{g}(F_2 - F_1) \tag{8.56}$$

Equation 8.56 agrees with the two-dimensional surge model (Chow, 1959) and can provide an estimate of the height of roll waves in a steep, straight concrete channel. To apply the abovementioned procedure to design a high-gradient channel, the flow condition, at first, shall be evaluated by Equation 8.44 or Figure 8.8. If the flow Froude number is greater than the limiting, a pulsating flow is expected. The height of roll waves shall then be determined by Equation 8.53 for the freeboard design.

EXAMPLE 8.4

A concrete channel is designed to carry a discharge of 5000 cfs on a slope of 3.0% with Manning's N of 0.014 and a side slope of 1V:2H. Estimate the height of roll waves when the bottom width is 15 ft.

1. **Stability analysis**

   Based on Manning's formula, the normal flow condition for 5000 cfs includes a flow depth, $Y_n = 4.92$ ft, flow area, $A_n = 122.27$ ft^2, flow velocity, $V_n = 40.89$ fps, and flow Froude number, $F_r = 3.84$. As a result, the ratio of normal flow depth to channel width is

$$Y^* = \frac{y}{b} = \frac{4.92}{15.0} = 0.3275$$

   The limiting Froude number by Equation 8.44 is computed with $z = 2$ and $k = \sqrt{5}$ as

$$F_L = \frac{3}{2}\frac{(1+2\sqrt{5}\times 0.3275)(1+2\times 2\times 0.3275)}{(1+2\times 2\times 0.3275 + 2\sqrt{5}\times 2\times 0.3725^2)} = 2.61$$

   For this case, the flow Froude number is greater than the limiting Froude number. As a result, this is a case of pulsating flow. Next, we shall determine the height of roll waves.

2. **Height of roll waves**

   For this case, the concrete channel can only sustain supercritical stable flows with a flow Froude number of up to 2.61. Referring to Figure 8.9, the section 1 before the jump is the location of the limiting flow. Therefore, the flow parameters in section 1 are determined to be

$$y_1 = 4.92 \text{ ft}, \ A_1 = 122.27 \text{ ft}^2, \ V_1 = 40.89 \text{ fps},$$
$$\bar{y}_1 = 0.5y_1 = 2.46 \text{ ft, and } F_1 = F = 2.61$$

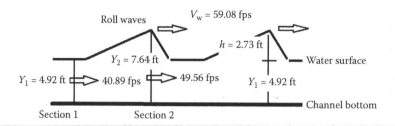

Figure 8.10 Roll waves in moving jump.

Assume that $h = 2.73$ ft. After a moving jump, the flow condition becomes

$$y_2 = 4.92 + h = 7.64 \text{ ft}, \ A_2 = 231.45 \text{ ft}^2$$

$$\bar{y}_2 = 0.5y_2 = 3.82 \text{ ft, and } F_2 = F_r = 3.82$$

Aided with Equation 8.50, the wave speed is computed as

$$V_w = 59.08 \text{ fps}$$

Aided by Equation 8.48, the flow velocity at section 2 is computed as

$$V_2 = 49.56 \text{ fps}$$

For this case, the wave celerity is the difference as

$$C = V_w - V_2 = 9.52 \text{ fps}$$

Substituting the above variables into Equation 8.56, the height of roll waves is calcu-lated to be 2.72 ft, which is close to the assumed value. Otherwise, this process can be repeated until the calculated wave height is converged. Having known the wave height, the freeboard for this channel shall be designed to accommodate the roll wave height of 2.73 ft. Figure 8.10 represents the flow condition through the moving jump. Of course, the engineer can also choose a deeper and narrower cross-section to avoid the occur-rence of roll waves.

EXAMPLE 8.5

A 9.8-ft wide rectangular concrete channel was used to illustrate how to use the wave fre-quency distribution to estimate the height of roll waves (French, 1985). The maximum height of roll waves in this channel was determined to be 1.8 ft when the discharge was 320 cfs on a 10% slope. Let us apply the moving hydraulic jump approach to this case. It takes the following steps:

1. **Check stability of design flow**
    According to Manning's formula, the design discharge for the example produces a uni-form flow depth of 1.06 ft with a flow Froude number of 5.26. Using Equation 8.44 or Figure 8.8, the limiting Froude number for this case is found to be 1.8. $F_r > F_L$ for this case; therefore, roll waves are expected.

2. **Calculation of roll wave height**
    - Guess the height of roll waves, h ($h = 1.51$ ft).
    - Solve for y_2 and V_2 by Equation 8.51 and Equation 8.48, respectively ($y_2 = 2.57$ ft, $V_2 = 37.7$ fps).
    - Solve for V_w by Equation 8.50 and C by Equation 8.52 ($V_w = 42.63$ fps).
    - Calculate the height of roll waves by Equation 8.53 ($h = 1.52$ ft).
    - Repeat the above steps until the calculated height is close to the guessed value.
    - This procedure leads to a prediction of roll wave height of 1.52 ft, which is comparable to the method of wave frequency distributions (French, 1985).

EXAMPLE 8.6

A sensitivity study is conducted on roll waves in a concrete channel with a discharge of 5000 cfs, Manning $N = 0.014$, and slope of 3.0%. Channel shapes considered are the combinations among bottom widths of 10, 15, and 20 ft, and side slopes of 0, 1, 2, and 3 ft/ft.

Solution: As shown in Table 8.4, the wider the channel, the higher the roll waves. Therefore, there exists a tradeoff between channel width and depth when roll wave is a concern. It is noticed that among all cases in Table 8.4, the minimal flow area ($15 \times 7.51 = 112.6$ ft^2) is achieved by the 15-ft wide rectangular channel that carries a stable supercritical flow with no roll waves.

8.6 Cross waves at a bend

Cross waves as shown in Figure 8.11 have been observed at a bend in a steep channel. At the bend, the outer wall turns inward to the direction of flow. An oblique jump will be triggered at the outer bend, and then the positive wave front with a higher water depth crosses the water surface toward the inner wall. The inner wall, which turns away from the direction of flow, will induce an oblique drop, which is an expansion wave with a shallower water depth. The expansion wave front crosses the water surface toward the outer wall. These two wave fronts are bounced back and forth between the two walls. Interference between these two wave fronts results in a disturbance pattern of cross waves until the wave heights are diffused downstream of the bend.

Cross waves at a bend can be analyzed using the unit-width approach under the following assumptions:

1. The channel is horizontal or the weight of water body is ignored.
2. The friction loss in the flow is negligible.

The semitheoretical analysis derived for the flat, frictionless channel provides good estimates of the flow patterns observed in a steep concrete channel (Ippen, 1951).

8.6.1 Oblique jump at outer bank

Oblique jump is similar to the shock wave in a high-speed air flow. When a supercritical flow is deflected inward to the center of the channel at a bend, an oblique jump is induced along the wave front, CD (Figure 8.12). The flow velocity is decomposed into two components: one is parallel to the wave front, and the other is normal to the wave front. The

Table 8.4 Sensitivity test on flow stability

Channel width, b (ft)	Side slope, z (ft/ft)	Depth width ratio, $Y* = y/b$	Limiting Froude number, $F_l = F_L$	Limiting flow depth, Y_l (ft)	Limiting flow velocity, V_l (fps)	Roll wave height, H (ft)	Wave speed, V_w (fps)	Design flow depth, Y_2 (ft)	Flow Froude number, $F_2 = F_r$	Design flow velocity, V_2 (fps)	Remarks
10	0	1.18	5.03	11.75	42.55	0.00	62.02	11.76	2.19	42.57	No wave
10	1	0.67	2.81	6.65	45.23	3.32	65.45	9.97	3.66	54.21	
10	2	0.56	2.82	5.64	41.74	2.7	60.27	8.34	3.83	50.28	
10	3	0.51	2.86	5.10	38.8	2.57	56.87	7.67	3.84	47.66	
15	0	0.50	3.00	7.51	44.49	0.00	60.05	7.51	2.86	44.52	No wave
15	1	0.37	2.51	5.53	44.14	3.95	65.03	9.48	3.73	54.81	
15	2	0.33	2.61	4.92	40.9	3.26	60.31	8.18	3.84	51.06	
15	3	0.3	2.71	4.57	38.23	2.72	56.28	7.29	3.83	47.47	
20	0	0.29	2.36	5.73	43.64	3.9	64.03	9.63	3.21	51.9	
20	1	0.24	2.26	4.75	42.62	4.22	63.67	8.97	3.76	54.15	
20	2	0.22	2.42	4.37	39.83	3.42	59.26	7.79	3.83	50.46	
20	3	0.21	2.54	4.12	37.46	2.84	55.5	6.96	3.82	47.04	
25	0	0.19	2.07	4.76	42.07	4.73	63.45	9.49	3.40	52.73	
25	1	0.17	2.09	4.18	40.98	4.27	61.86	8.45	3.78	52.85	
25	2	0.16	2.25	3.94	38.67	3.49	57.99	7.43	3.82	49.54	
25	3	0.15	2.39	3.77	36.62	2.92	54.6	6.69	3.81	46.44	

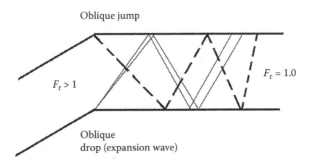

Figure 8.11 Cross waves at a bend.

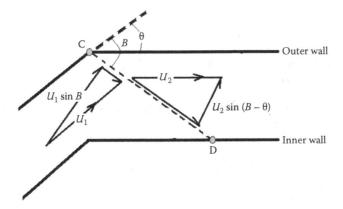

Figure 8.12 Oblique jump at outer bend in steep concrete channel.

normal velocity component must satisfy the continuity principle for the same flow rate crossing the wave front. As a result, we have

$$Y_1 U_1 \sin B = Y_2 U_2 \sin(B - \theta) \tag{8.57}$$

where Y = flow depth in [L], U = flow velocity in [L/T] normal to the wave front, θ = angle of wall deflection, and B = angle of wave front. The subscription of "1" means the variables associated with section 1, etc. Similarly, the momentum of the flow shall be balanced across the wave front as

$$\frac{Y_2}{Y_1} = \frac{\tan B}{\tan(B - \theta)} = \frac{1}{2}\left(\sqrt{1 + 8(F_1 \sin B)^2} - 1\right) \tag{8.58}$$

Based on the geometry of the band, the empirical formula was developed to determine the angle of wave front as (Ippen, 1949)

$$\tan \theta = \frac{\tan B\left(\sqrt{1 + 8(F_1 \sin B)^2} - 3\right)}{2(\tan B)^2 + \sqrt{1 + 8(F_1 \sin B)^2} - 1} \tag{8.59}$$

$$F_1 = \frac{U_1}{\sqrt{gY_1}} \tag{8.60}$$

The energy losses associated with this jump is estimated as

$$\Delta E = \frac{(Y_2 - Y_1)^3}{4Y_1Y_2} \tag{8.61}$$

in which F_1 = Froude number before the oblique jump and ΔE = energy losses in [L] through the oblique jump.

The wave front of an oblique jump will be continuously deflected between the bank walls until the flow Froude number becomes <1, i.e., subcritical flow.

EXAMPLE 8.7

A transition is built to convert the rectangular channel from 32 ft wide at the upstream station to 12 ft wide at the downstream station (Figure 8.13). The approach flow velocity at the upstream station is 20.0 fps, and the flow depth is 0.40 ft. Analyze the cross waves.

As illustrated in Figure 8.14, at point A, the incoming flow is deflected to have a flow direction parallel to wall AC. From the layout, the wall deflection angle is calculated as

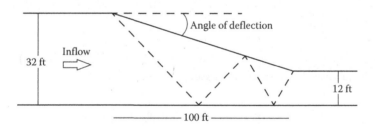

Figure 8.13 Top view of contraction section.

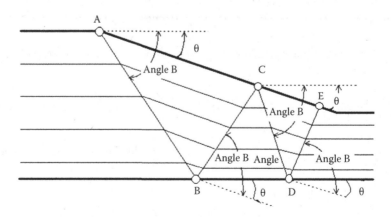

Figure 8.14 Flow zones with oblique jumps.

Table 8.5 Oblique jumps in cross waves

Wave front angle (°)	U_1 (fps)	Y_1 (ft)	F_{r1}	U_2 (fps)	Y_2 (ft)	F_{r2}	Comments
20.51	20.0	0.40	5.57	18.97	0.92	3.48	Jump again
27.1	18.97	0.92	3.48	17.55	1.67	2.40	Jump again
36.50	17.55	1.67	2.60	15.58	2.63	1.70	Jump again
53.86	15.58	2.63	1.70	12.45	2.94	1.10[a]	End

[a] It is close to one.

$$\theta = \tan^{-1}\frac{32-12}{100} = 11.5°$$

The approach flow condition is

$$U_1 = 20.0 \, \text{fps}, \; Y_1 = 0.40 \, \text{ft}$$

$$F_1 = \frac{20.0}{\sqrt{32.2\times 0.40}} = 5.57$$

$$\tan(11.5°) = \frac{\tan\left(\sqrt{1+8(1.76\sin B)^2}-3\right)}{2(\tan B)^2 + \left(\sqrt{1+8(1.76\sin B)^2}-1\right)}$$

By trial and error, the wave front angle, B, is found to be 20.51°.

$$Y_2 = 0.40\times\left(\sqrt{1+8(1.76\sin(20.51°)^2}-1\right) = 0.923 \, \text{ft}$$

$$U_2 = \frac{20.0\times 4.0\times\sin(20.51°)}{0.923\times\sin(20.51°-11.50°)} = 18.98 \, \text{ft}$$

$$F_2 = \frac{18.98}{\sqrt{32.2\times 0.923}} = 3.48 > 1.0$$

The above calculation is for zone ABC in Figure 8.14. Because the Froude number in zone ABC is still greater than one, the flow will jump again at point B. The wave front at point B is so deflected that the flow direction becomes parallel to wall BD. The flow condition in zone BCD can be calculated using the approach flow in zone ABC. For this case, a sequence of oblique jumps is triggered until the flow Froude number becomes less than one. Computations of these oblique jumps for this example are tabulated in Table 8.5 and plotted in Figure 8.15.

8.6.2 Oblique drop at inner bank

Along the inner bank at a bend, an expansion wave or oblique drop is induced. Because of the sudden expansion at point D in Figure 8.16, the flow depth continuously decreases through a fan-shaped zone, delineated by the wave angles of B_1 and B_2 in Figure 8.16.

$$B_1 = \sin^{-1}\frac{1}{F_{r1}} \tag{8.62}$$

$$B_2 = \sin^{-1}\frac{1}{F_{r2}} \tag{8.63}$$

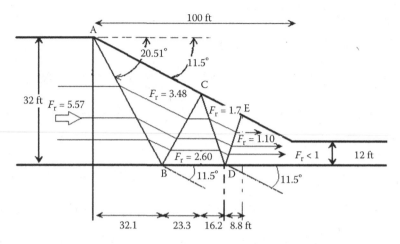

Figure 8.15 Oblique jumps predicted through channel transition.

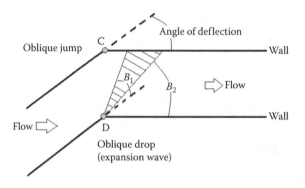

Figure 8.16 Expansion wave at inner bend.

$$\frac{Y_2}{Y_1} = \frac{F_{r1}+2}{F_{r2}+2} \tag{8.64}$$

$$\theta = \left(\sqrt{3}\tan^{-1}\frac{\sqrt{3}}{\sqrt{F_{r2}-1}} - \tan^{-1}\frac{1}{\sqrt{F_{r2}-1}}\right) + \left(\tan^{-1}\frac{1}{\sqrt{F_{r1}-1}} - \sqrt{3}\tan^{-1}\frac{\sqrt{3}}{\sqrt{F_{r1}-1}}\right) \tag{8.65}$$

in which θ = wall deflection angle, which shall be a negative value for expansion. The subscript 1 represents the variables before the expansion wave, and the subscript 2 represents the variables after the expansion wave.

After an oblique drop, the flow depth is decreased. The wave angle of each succeeding wavelet depends on the local Froude number, which changes through the continuous drops. Oblique drop is more a process to diffuse the energy through a series of infinitesimal steps or wavelets. The disturbance of expansion wave is either washed away or diffused into the turbulent flow within a short distance.

EXAMPLE 8.8

The transition section in Example 8.7 is reduced in length. As shown in Figure 8.17, the channel wall is deflected with an angle of 11.3° at point A for a length of 45 ft and then deflected back to the original direction at point C. As shown in Example 8.7, an oblique jump is triggered at point A. In this example, expansion waves begin to spread out at point C. Considering that the wall deflection angle, θ, is a value of negative 11.3° and Froude number of the incoming flow, Fr_1, is 2.60, Equation 8.65 becomes

$$-11.3° = \left(\sqrt{3} \tan^{-1} \frac{\sqrt{3}}{\sqrt{F_{r2}-1}} - \tan^{-1} \frac{1}{\sqrt{F_{r2}-1}} \right)$$
$$+ \left(\tan^{-1} \frac{1}{\sqrt{2.60^2-1}} - \sqrt{3} \tan^{-1} \frac{\sqrt{3}}{\sqrt{2.60^2-1}} \right)$$

By trial and error, the value of $F_{r2} = 3.75$.

$$\frac{Y_2}{Y_1} = \frac{F_{r2}+2}{F_{r1}+2} = \frac{2.60^2+2}{3.75^2+2} = 0.87$$

So, the flow depth after expansion is $Y_2 = 1.44$ ft. The boundaries of the expansion wave can be determined by Equations 8.62 and 8.63 as

$$B_1 = \sin^{-1} \left(\frac{1}{2.60} \right) = 22.62°$$

$$B_2 = \sin^{-1} \left(\frac{1}{3.75} \right) = 15.47°$$

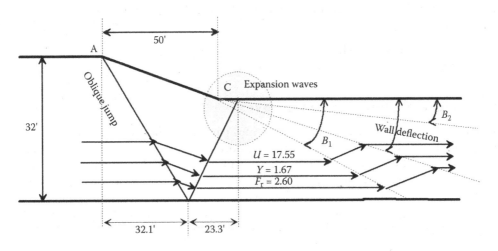

Figure 8.17 Expansion waves through transition.

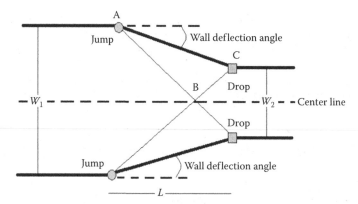

Figure 8.18 Recommended length for transition section.

In design, the distance from point A to point C shall be matched with the wave length so that the expansion waves can be canceled by the oblique jump waves. As a rule of thumb, the length of the transition section illustrated in Figure 8.18 is recommended as

$$L = \frac{W_1 - W_2}{2 \tan \theta} \tag{8.66}$$

in which L = length of the transition section in feet, W_1 = transition entrance width in feet, W_2 = transition exit width in feet, and θ = the wall deflection angle.

EXAMPLE 8.9

The wave length after the second oblique jumps in Figure 8.17 is 55.4 ft (32.1 + 23.3 = 55.4 ft). The expansion waves start at 50 ft downstream of point A. Determine the wall deflection angle in Figure 8.18, knowing that the length of transition is 50 ft, the entrance width is 32 ft, and the exit width is 12 ft. Using Equation 8.66 yields the angle of wall deflection to be

$$L = \frac{32 - 12}{2 \tan \theta} = 50; \text{ so, } \theta = 11.3°$$

8.7 Superelevation and cross waves

A prismatic channel produces flow streamlines parallel to the channel centerline in the direction of the alignment. In practice, it is too ideal to expect that the channel under design is long and straight. It is almost unavoidable that the channel alignment has curves and bends. Through a curve reach, the flow is no longer unidirectional. A significant lateral flow or *secondary current* can be developed. This transverse movement is to shift the flow volume from the inner bank to the outer bank. As a result, a spiral flow pattern is developed along a helical path in the longitudinal direction. To account for the superelevation around a curve reach, a freeboard section must be added to the normal depth. The superelevation is calculated as

$$\Delta H = K_s \frac{U^2 T}{g R_c} \tag{8.67}$$

in which ΔH = superelevation in [L], T = top width in [L] of the channel, R_c = centerline radius in [L] of curvature, and K_s = water rise constant as recommended in Table 8.6.

As a rule of thumb, the radius of the curvature must be greater than three times the channel top width. The water rise shall not produce a transverse water surface profile exceeding 10°. Usually, a curve channel is designed to have a simple circular curve as the central portion with or without transitions upstream and downstream. In general, a subcritical flow has a small amount of superelevation that does not require any transitions upstream and downstream of the curved reach. The height of the outside wall shall be increased by the amount of superelevation over the full length of the curvature. The height of the inner wall may remain the same as the normal flow condition predicted for the straight channel. However, the disturbance caused by the channel curvature to a supercritical flow will affect the flow not only through the length of the curvature but also many times the channel widths downstream. Therefore, it is recommended that for a supercritical flow, increasing the wall height by the amount of superelevation applies to both the curved reach and a considerable distance downstream.

Supercritical flow around a curve channel (Figure 8.19) is hydraulically complicated. In practice, a set of simplified equations has been developed for calculating the cross waves in a curved channel as (Ippen, 1951, 1949)

$$B = \sin^{-1} \frac{\sqrt{g y_1}}{U_1} \tag{8.68}$$

$$\theta = \tan^{-1} \frac{2b}{(2R_c + b)\tan B} \tag{8.69}$$

$$y_2 = \frac{\left[U_1 \sin\left(B + \frac{\theta}{2} \right) \right]^2}{g} \tag{8.70}$$

$$U_2 = \frac{Q}{A_2} \tag{8.71}$$

Table 8.6 Values for superelevation constant

Flow type	Froude number	Channel section	Type of curve	Value of K_s
Subcritical	$F < 0.85$	Rectangular	Simple circular	0.50
Subcritical	$F < 0.85$	Trapezoidal	Simple circular	0.50
Supercritical	$F > 1.13$	Rectangular	Simple circular	1.00
Supercritical	$F > 1.13$	Trapezoidal	Simple circular	1.00
Supercritical	$F > 1.13$	Rectangular	Spiral transition	1.00
Supercritical	$F > 1.13$	Trapezoidal	Spiral transition	1.00
Supercritical	$F > 1.13$	Rectangular	Spiral banked	1.00

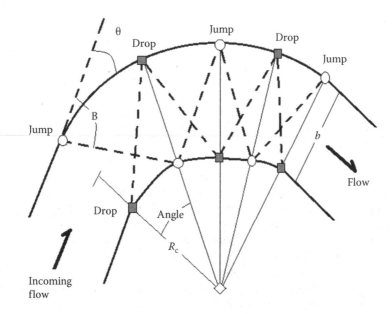

Figure 8.19 Illustration of cross waves along a curve channel.

in which B = wave front angle, U_1 = incoming velocity in [L/T], y_1 = incoming flow depth in [L], θ = central angle, b = width in [L] of rectangular channel, R_c = radius in [L] of channel centerline, U_2 = flow velocity in [L/T] after jump, and y_2 = flow depth in [L] after jump. The continuous deflections of oblique jumps form cross waves through a curve reach. Although Equations 8.68 through 8.71 are simplified solutions to estimate cross waves, they have been verified by flow measurements.

EXAMPLE 8.10

A rectangular concrete channel is laid on a slope of 2%. The bottom width for this channel is 10 ft. This channel goes through a curve with a radius of 300 ft along the centerline. Calculate the cross waves through this curve channel.

Solution: According to Manning's formula, the normal flow condition includes normal depth of 2.23 ft, normal flow velocity of 22.45 fps, and flow Froude number of 2.65. According to Table 8.6, $K_s = 1.0$ to calculate the superelevation as

$$\Delta H = K_s \frac{U^2 T}{g R_c} = 1.0 \times \frac{22.45^2 \times 10}{32.2 \times 300} = 0.52\,\text{ft}$$

Using the normal flow condition as the inflow to enter the curve channel, the first jump has a wave front angle of 0.387 rad aided with Equation 8.68 and a central angle of 0.157 rad using Equation 8.79. With these two angles, Equations 8.70 and 8.71 are used to predict the flow depth (=3.15 ft) and velocity (=15.86 fps) after the first jump. As indicated in Table 8.7, the above procedure is repeated until the flow Froude number is close to unity.

Table 8.7 Calculations of cross waves through curve channel

Upstream conditions			Wave front angle (rad)	Central angle (rad)	Downstream conditions			Jump length (ft)
Depth, y_1 (ft)	Velocity, V_1 (ft/s)	Froude number, F_1			Depth, y_2 (ft)	Velocity, V_2 (ft/s)	Froude number, F_2	
2.23	22.45	2.65	0.387	0.157	3.15	15.86	1.57	47.1
3.15	15.86	1.57	0.688	0.078	3.45	14.48	1.37	23.5
3.45	14.48	1.37	0.816	0.061	3.65	13.70	1.26	18.2

8.8 Channel depth

Manning's formula tends to give a shallow and wide cross-section for a supercritical flow in a high-gradient channel. Caution must be taken because the dynamic energy in terms of the flow velocity head is interchangeable with the flow depth. An obstruction in a supercritical flow may trigger continuous oblique jumps that gradually convert the flow dynamic energy to its flow depth. This complicated process results in water spills and disturbances on the water surface. The design procedure for a high-gradient channel is recommended as follows:

1. To select a channel cross-section for the design flow
2. To determinate the uniform flow depth, Y_n, and freeboard, F
3. To calculate the limiting Froude number
4. To modify the channel depth for the straight reach as

$$Y_t = \max(Y_n + F, \; Y_n + h) \tag{8.72}$$

in which Y_t = channel excavated depth in [L], Y_n = normal depth in [L], F = recommended freeboard in [L], according to the design criteria, and h = height of roll waves in [L].
5. To modify the channel depth at the bend as

$$Y_t = \max(Y_n + F, \; Y_n + \text{height of oblique jump}) \tag{8.73}$$

6. To modify the channel depth along the curve reach as

$$Y_t = \max(Y_n + F, \; Y_n + \max(\text{superelevation, height of cross waves})) \tag{8.74}$$

In addition, a supercritical flow is also sensitive to channel alignment, wall expansion, sectional contraction, and invert drop. Caution must be taken wherever a possibility exists to trigger a hydraulic jump. When the flow condition becomes too complicated, it is advisable that the channel depth on a high gradient be designed using the specific energy (normal flow depth plus its velocity head) as the flow depth.

8.9 Homework

Q8.1 A rectangular channel is designed to carry a flow of 1000 cfs. This channel is laid on a slope of 0.015 with Manning's roughness coefficient of 0.045. Determine the optimal channel section and normal flow condition for the following cases: (A) without a freeboard and (B) with a freeboard of 1.0 ft.

Solution:

Z	$k = (1 + Z^2)^{0.5}$	$n = B/Y$	A_t/Y^2	A_f/Y^2	P/Y	Q'/Y^8	Y (ft)	B (ft)	F (ft)
0.00	1.00	2.00	2.00	2.00	4.00	2.00	7.23	14.47	0.00
0.00	1.00	1.30	1.45	1.30	3.30	0.35	9.01	11.75	1.00

Q8.2 A trapezoidal channel is designed to carry a flow of 1000 cfs. This channel has a side slope of 1V:1H, Manning's roughness of 0.045, and channel slope of 0.015. Determine the optimal channel section and normal flow condition for the following cases: (A) without a freeboard and (B) with a freeboard of 1.0 ft.

Solution:

Z	$k = (1 + Z^2)^{0.5}$	$n = B/Y$	A_t/Y^2	A_f/Y^2	P/Y	Q'/Y^8	Y (ft)	B (ft)	F (ft)
1.00	1.41	0.83	1.83	1.83	3.66	1.53	7.48	6.20	0.00
1.00	1.41	0.56	1.89	1.56	3.39	0.80	8.11	4.52	1.00

Q8.3 A concrete trapezoidal channel is designed to carry a flow of 5000 cfs. This channel has a channel bottom width of 20 ft, side slope of 1V:1H, Manning's roughness of 0.014, and channel slope of 0.025. (1) Determine if the normal flow is unstable. (2) If it is unstable, estimate the height of roll waves. (3) Suggest the height of freeboard for this channel.

Solution: For this case, the normal flow depth is 5.00 ft.

Flow depth, Y (ft)	Channel width, B (ft)	Side slope, z (ft/ft)	Parameter, k	Ratio y/b, Y*	Limiting Froude number, F_l	Flow Froude number, F_2	Channel flow condition
5.00	20.0	1.00	1.4142	0.2500	2.29	3.45	Unstable

Try wave height, h (ft)	Flow condition before bore				
	Flow depth, y_i (ft)	Flow area, A_1 (ft^2)	Centroid depth, y_{c1} (ft)	Flow velocity, V_1 (ft/s)	Limiting Froude number, F_l
2.96	5.00	125.02	2.50	39.99	2.29

Flow condition with bore						Wave celerity, C (ft/s)	Calculated wave height (ft)
Flow depth, y_2 (ft)	Flow area, A_2 (ft^2)	Centroid depth, y_{c2} (ft)	Wave speed, V_w (ft/s)	Flow velocity, V_2 (ft/s)	Flow Froude number, F_2		
7.96	222.58	3.98	58.35	48.04	3.45	10.31	2.96

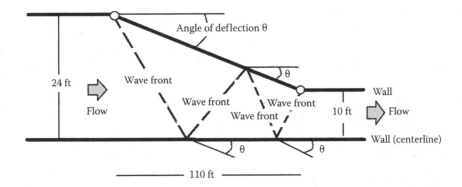

Figure Q8.5 Top view of contraction section.

The height of freeboard for this channel must be >2.96 ft.

Q8.4 If a drop structure can be used to reduce the bottom slope for the concrete channel in *Q8.2*, what is the maximal channel slope to maintain a stable supercritical flow? (*Solution:* $S_o \leq 0.011$)

Q8.5 A transition is built to convert the rectangular channel from 48 ft wide at the upstream station to 20 ft wide at the downstream station as shown in Figure Q8.5. The approach flow velocity at the upstream station is 36.4 fps with a depth of 2.75 ft. (1) Determine the angle of wall deflection. (2) Analyze the cross waves through a length of 110 ft.

Bibliography

ADOWR. (2000). *Design Guideline for Diversion Channels*, Flood Warning and Dam Safety Section, Arizona Department of Water Resources, Phoenix, AZ.

Brock, R.R. (1969). "Development of Roll Wave Trains in Open Channels," *Proceedings of the ASCE Journal of the Hydraulics Division*, Vol. 95, No. HY4, pp. 1401–1427.

CCFCD. (1995). "Open Channels," *Hydrologic Criteria and Drainage Design Manual*, Clark County Regional Flood Control District, Clark County, Las Vegas, NV.

Chow, V.T. (1959). *Open Channel Hydraulics*, McGraw-Hill Civil Engineering Series, New York.

Escoffier, F.F., and Boyd, M.B. (1962). "Stability Aspects of Flow in Open Channels," *Journal of the Hydraulics Division*, ASCE, Vol. 88, No. HY6, pp. 145–166.

French, R.H. (1985). *Open Channel Hydraulics*, McGraw Hill, New York.

Guo, J.C.Y. (1999a). *Channel Design and Flow Analysis*, Water Resources Publication, Littleton, CO.

Guo, J.C.Y. (1999b). "Roll Waves in High Gradient Channels," *International Water Resources Association's International Journal of Water*, Vol. 3, No.1.

Guo, J.C.Y. (2009). "Grade Control for Urban Channel Design," *Journal of Hydro-Environmental Research*, Vol. 2, No. 4, pp. 239–242, doi: 10.1016/j.jher.2009.01.001.

Guo, J.C.Y., and Blackler, G. (2009). "Least Cost and Most Efficient Channel Cross Sections," *ASCE Journal of Irrigation and Drainage Engineering*, Vol. 135, No. 2, pp. 248–251.

Guo, J.C.Y., and Hughes, W. (1984). "Optimal Trapezoidal Cross Sections for Open Channel with Freeboard," *ASCE Journal of Irrigation and Drainage Engineering*, Vol. 110, No. 3.

Ippen, A.T. (1949). "Design of Channel Contractions," *Transactions of the American Society of Civil Engineers*, Vol. 116, pp. 326–346.

Ippen, A.T. (1951). "Mechanics of Supercritical Flow," *Transactions of the American Society of Civil Engineers*, Vol. 116, pp. 268–295.

Manning, R. (1891). "On the Flow of Water in Open Channel and Pipes," *Transactions of the Institution of Civil Engineers of Ireland*, Vol. 20, pp. 161–207.

Mayer, P.G.W. (1957). " A Study of Roll Waves and Slug Flows in Inclined Open Channels," PhD Dissertation, Cornell University, Ithaca.

Rouse, H., Bhoota, B.V., and Hsu, E.Y. (1949). "Design of Channel Expansions," *Transactions of the American Society of Civil Engineers*, Vol. 116, pp. 347–400.

Shunkry, A. (1950). "Flow Around Bends in an Open Flume," *ASCE Transactions*, Paper No. 2411, Vol. 115, pp. 751–779.

Stonstreet, S.E. (1996). "Stochsatic Determination of Wave Heights for Flood Control Channels," ASCE Conference Proceedings of North American Water and Environment Congress.

Thorsky, G.N., and Haggman, D.C. (1970). Discussion of "Development of Roll Wave Trains in Open Channels," *ASCE Journal of Hydraulic Division*, Vol. 96, No. HY4, April.

UDFCD. (2010). *Urban Drainage and Stormwater Design Criteria*, Volumes 1 and 2, Urban Drainage and Flood Control District, Denver, CO.

USCOE. (2000). *Hydraulic Design of Flood Control Channels*, Department of the Army, US Army Corps of Engineers, Washington, DC, 20314–1000, EM 1110-2-1601.

Vedernikov, V.V. (1945). "Conditions at the Front of a Translatory Wave Distributing a Steady Motion of a Real Fluid," C.R. USSR Academy of Sciences, Vol. 48, No. 4.

Vedernikov, V.V. (1946) "Characteristic Features of a Liquid Flow in an Open Channel," C. R. USSR Academy of Sciences, Vol 52, pp. 207–210.

Chapter 9

Street hydraulic capacity

The primary function of a street is to facilitate traffic movement. However, during a storm event, a street provides an emergency passage for flood flows. A street drainage system consists of *collection, conveyance,* and *storage* facilities. A collection system includes street inlets and culverts. A conveyance system consists of street gutters and storm drains (sewers). The storage capacity in a street drainage system consists of sump areas, surface detention, and basin detention. As illustrated in Figure 9.1, street inlets are placed at the street intersection where storm runoff is concentrated and has to be collected into storm sewers. All inlets at a street intersection must be connected to the manhole placed at the center of the street interception. Between two manholes, the water flow is conveyed by the sewer line. Storm sewer lines are designed as a network to jointly transport storm runoff from the streets to the downstream natural waterway. For the purpose of stormwater quality enhancement, a storage basin shall be placed at the exit of the sewer system. Storm runoff will be temporarily stored for trash, debris, and sediment controls, and gradually released at a permissible rate.

In this chapter, both (1) *street hydraulic conveyance capacity* (SHCC) and (2) *street hydraulic storage capacity* (SHSC) in the street drainage system are discussed. As demonstrated in Figure 9.2, SHCC is related to the stormwater flowing capacity on a street that has a positive slope, whereas SHSC is related to the storage capacity at a depressed session or sump area. It is important that designs of street drainage take advantage of the SHCC as the rules and regulations permit. For instance, the water spread shall not encroach into the emergency traffic lane during the major event, and the gutter flow depth shall not exceed the curb height during the minor event (Anderson, 1993). The optimal use of SHCC can significantly reduce the numbers and sizes of street inlets and storm sewers. Obviously, a poorly designed street drainage system will degrade the level of service and even impose serious safety issues to the public.

9.1 Street classifications

Although a street drainage system shall be designed to quickly remove stormwater from the traffic areas, care has to be taken because water flows on pavements can impose hydroplaning effects on cars to the level of traffic function becoming paralyzed.

It is important to keep in mind that the primary purpose of a street is for traffic movement; therefore, the drainage function is considered secondary. As the rule of thumb, the street drainage facilities do not interfere with the traffic functions and safety. A street drainage system (Figure 9.3) is a joint operation of the *minor and major drainage systems*. A minor system consists of *street inlets* and *storm sewers*. During a major event, all street gutters operate like wide and shallow channels that carry the storm runoff in excess of the capacity of the minor drainage system.

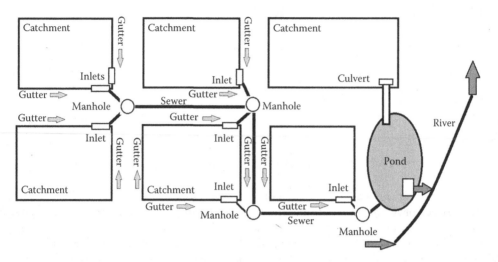

Figure 9.1 Layout of street drainage system.

Figure 9.2 Street conveyance capacity versus storage capacity. (a) Gutter flow (conveyance) and (b) water spread (storage).

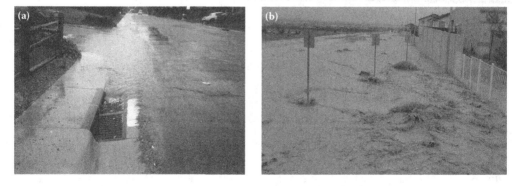

Figure 9.3 Minor and major events on street. (a) Minor event—gutter flowing full and (b) major event—street flowing full.

From traffic engineers' point of view, streets and roadways are classified into *local, collector, principal, and freeway systems,* depending on the *Average Daily Traffic* (ADT). Details are discussed as follows:

1. *Local streets* are placed in a small residential or industrial community. They are designed to provide local services with no through traffic. They may have stop signs on two moving traffic lanes. Parking lane along each side of the street is allowed.
2. *Collector streets* collect the traffic flows from local streets and intersect with an arterial street. Stop signs and traffic signals are placed at the intersections. Collectors may be two to four moving traffic lanes with car parking adjacent to the curbs.
3. *Arterial streets* have four to six traffic lanes equipped with traffic signals at major intersections. Car parking adjacent to the curbs is prohibited on an arterial.
4. *Freeways* are designed for rapid and efficient movement of traffic. Access is controlled with ramps at an interchange. A freeway may be as many as eight lanes. Car parking along the curbs of a freeway is prohibited along a freeway.

From a drainage engineer's view point, streets are classified based on the allowable water spread width on the street and water depth in the street gutter. As illustrated in Figure 9.4, from the aspect of stormwater drainage, streets are classified into *urban, rural, and semiurban* streets.

Figure 9.4 Comparison between urban and rural streets. (a) Urban street, (b) semiurban street, and (c) rural street.

1. *Urban streets* are usually characterized with curbs, gutters, and inlets.
2. *Rural streets* are equipped with well-defined grass-lined ditches and swales.
3. *Semiurban streets* drain stormwater into shallow riprap or concrete-lined trickle channels.

The demarcation between rural and urban streets depends on whether the drainage system is formed with curbs and gutters. Although an urban drainage system is efficient for extreme events, it does increase erosion potentials and solid loads into the downstream water bodies.

9.2 Curb and gutter types

In general, a street cross-section consists of sidewalks, curbs, gutters, and traffic lanes. As illustrated in Figure 9.5, curbs and gutters collect the overland flows and provide shallow and wide channels to transport the concentrated flows into the downstream inlets.

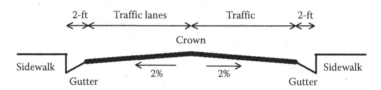

Figure 9.5 Typical urban street sections.

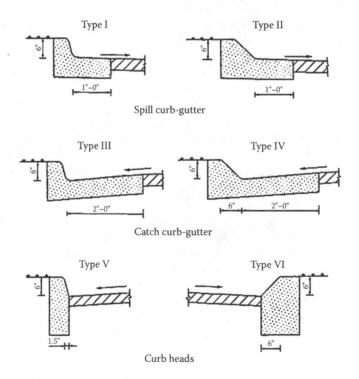

Figure 9.6 Types of street gutter and curbs.

A curb-gutter unit is designed to guide the overland flows from the pavement into a concentrated flow in the gutter. Curb and gutter can be attached or detached to the sidewalk. A *combined curb-gutter-sidewalk unit* is used only on residential streets because its capacity is limited to a 4-in. depth at the flow line. The *vertical curb-gutter units* have a depth of 6 in. at the flow line and are often used on local, collector, or major arterial because of a higher drainage capacity. Figure 9.6 shows six types of vertical curb-gutter units available for engineering uses. Type I and II are the prefabricated *spill curb-gutter units*, which are installed on a curved ramp to drain surface runoff from the elevated outer shoulder to the inner gutter along the ramp. Type III and IV are the typical *catch curb-gutter units*, which collect the surface overland flows from the pavements and create a concentrated flow at the location of downstream inlet. In comparison, type V and VI are the curb units without a prefabricated gutter. The depth of the curb head can be tailored for the local needs. For a narrow neighborhood street with sidewalks, a low curb head allows the cars to roll over the curb. On the contrary, a high curb head >6 in. may be installed along a wide street to increase the SHCC.

9.3 Street hydraulic conveyance capacity

As indicated in Hydraulic Engineering Circulars No. 12 and No. 22, entitled, "Drainage of Highway Pavements," SHCC depends on street geometry and hydraulic parameters such as pavement surface roughness coefficient. Figure 9.7 illustrates the flow condition in a triangular street gutter cross-section. Stormwater flowing through such a gutter section can be described by the revised Manning's equation as

$$Q = \frac{K}{n} S_x^{1.67} T^{2.67} \sqrt{S_o}$$ (9.1)

in which Q = SHCC in cfs or cms, K = 0.56 for the foot-second unit system or 0.376 for the meter-second unit system, n = Manning's roughness of street surface such as 0.016, S_x = street transverse slope in [L/L], S_o = street longitudinal slope in [L/L], and T = water spread width in [L] on the street. A milder transverse slope <2% will noticeably reduce the SHCC. On the contrary, a steeper slope >2% will compromise driver's comfort. In practice, the standard transverse slope for street is recommended to be 2%.

Referring to Figure 9.7, the gutter flow depth, Y in [L], along the flow line in a triangular street section is calculated as

$$Y = TS_x$$ (9.2)

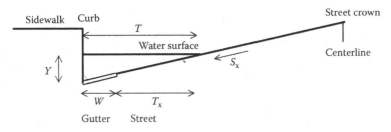

Figure 9.7 Triangular cross-section of gutter flow.

The flow cross-sectional area, A in $[L^2]$, for a street flow is calculated as

$$A = 0.5YT \tag{9.3}$$

The cross-sectional average flow velocity, V in $[L/T]$, is

$$V = \frac{Q}{A} \tag{9.4}$$

As illustrated in Figure 9.8, a straight street cross-section is often improved with a standard gutter depression of 2 in. introduced at the street curb in order to increase the SHCC. With a gutter depression, the stormwater flow in a street gutter is divided into *gutter flow* and *side flow*. The gutter flow is the amount of flow carried within the gutter width, W, while the side flow is the amount of flow carried by the water spread, T_x, encroaching into the traffic lanes. In practice, the standard gutter width is recommended to be 2 ft. As a result, the transverse slope across the gutter width is increased to

$$S_w = S_x + \frac{D_s}{W} \tag{9.5}$$

in which S_w = cross slope in $[L/L]$ over gutter width, S_x = street transverse slope in $[L/L]$ over the traffic lanes, and D_s = gutter depression in $[L]$.

The water depth at the curb, D in $[L]$, is the sum of the flow depth, Y, and the gutter depression, D_s:

$$D = Y + D_s \tag{9.6}$$

The water depth in a gutter must not exceed the curb height or the maximum allowable depth equal to the gutter full depth, D_m, under a safety concern. For convenience, the water spread width, T_s in $[L]$, across the gutter width is calculated as

$$T_s = \frac{D}{S_w} \tag{9.7}$$

The total water spread, T, is the sum as

$$T = W + T_x \tag{9.8}$$

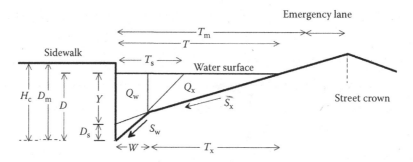

Figure 9.8 Composite street cross-section.

in which T_x = water spread in [L] for side flow into the traffic lanes. Under the assumption that the internal friction is negligible, both the gutter and side flows are analyzed independently. Applying Equation 9.1 to the gutter and side flows separately yields

$$Q_x = \frac{K}{n} S_x^{1.67} T_x^{2.67} \sqrt{S_o} \tag{9.9}$$

$$Q_w = \frac{K}{n} S_w^{1.67} [T_s^{2.67} - (T_s - W)^{2.67}] \sqrt{S_o} \tag{9.10}$$

in which Q_x = side flow in [L³/T], Q_w = gutter flow in [L³/T], W = gutter width in [L], and T_s = water spread in [L] as defined in Figure 9.8.
The total flow, Q in [L³/T], on the street is

$$Q = Q_x + Q_w \tag{9.11}$$

The flow cross-sectional area, A in [L²], for a composite street cross-section is

$$A = 0.5YT + 0.5WD_s \tag{9.12}$$

EXAMPLE 9.1

A street section in Figure 9.9 has $n = 0.016$, $W = 2$ ft, $D_s = 2$ in., $S_o = 1.50\%$, and $S_x = 2\%$. The gutter full depth, D_m, for this section is equal to the cub height, H_c, of 6 in. Determine the *gutter full capacity* for this street.

Solution: The *gutter full capacity* for a street cross-section is defined as the gutter water depth equal to the curb height. For this case, $D = D_m = H_c = 6.0$ in. First, let us calculate the cross slope over the gutter width as

$$S_w = S_x + D_s / W = 0.02 + (2/12)/2 = 0.103 \text{ ft/ft,}$$

The gutter full water spread width, T, is calculated as

$$T = \frac{(D_m - D_s)}{S_x} = \frac{(6-2)/12}{0.02} = 16.67 \text{ ft}$$

$$T_x = T - W = 14.67 \text{ ft}$$

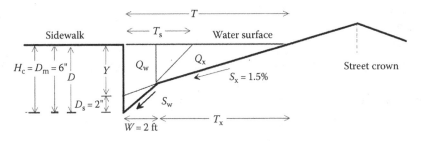

Figure 9.9 Street cross-section used in Example 9.1.

Table 9.1 Impact of gutter depression with $S_x = 0.02$ and $S_o = 2.0\%$

Water spread width (ft)	With depression			Without depression			Difference in capacity (%)
	Capacity (cfs)	Gutter depth (in.)	VD (cfs/ft)	Capacity (cfs)	Gutter depth (in.)	VD (cfs/ft)	
10.00	4.78	4.40	1.50	3.37	4.20	2.42	29.50
20.00	23.56	6.80	3.20	21.43	6.60	3.80	9.04
28.00	55.25	8.72	5.01	52.63	8.52	4.90	4.74

$$T_s = 6.0/(12 \times 0.103) = 4.85\,\text{ft}$$

$$Q_w = \frac{0.56}{0.016}(0.103)^{1.67}[4.85^{2.67} - (4.85 - 2.0)^{2.67}]\sqrt{0.015} = 4.95\,\text{cfs}$$

$$Q_x = \frac{0.56}{0.016}0.02^{1.67}14.67^{2.67}\sqrt{0.015} = 8.11\,\text{cfs}$$

$$Q = Q_x + Q_w = 13.06\,\text{cfs}$$

Table 9.1 presents a comparison of SHCC with and without a gutter depression. On a slope of $S_o = 2\%$, the increase of SHCC due to the 2 in. depression can be as much as 29.5% for a 10-ft water spread and then gradually diminishes to 4.74% for a 28-ft water spread.

9.4 Water spread and curb height

The key parameter for determining the SHCC is the water spread. The *allowable water spread* (as shown in Figure 9.8) shall be the smaller one between the *available water spread*, T_m, and the *gutter full depth*, D_m. For instance, in a business district, the gutter water depth shall not exceed the curb height, and the water spread is limited to keeping at least one traffic lane free from water for emergency use. Aided by Equation 9.2, the gutter full depth, D_m, can be converted into its water spread. The *design water spread* for determining the SHCC is selected as

$$T = \min\left(T_m, \frac{D_m - D_s}{S_x}\right) \tag{9.13}$$

EXAMPLE 9.2

Consider the street section in Example 9.1. The gutter full depth is set to be the curb height of 6 in., and the water spread into the traffic lanes is limited to 15 ft. Determine the SHCC.

Solution: First, let us determine the design water spread for this street section as

$$T = \min\left(15, \frac{(6-2)/12}{0.02}\right) = \min(15, 16.7) = 15\,\text{ft}$$

The street hydraulic capacity is calculated for a 15-ft water spread as

$$T_x = T - W = 15.0 - 2.0 = 13.0 \, \text{ft}$$

$$D = T \times S_x + D_s = 15.0 \times 0.02 + \frac{2.0}{12} = 0.466 \, \text{ft}$$

$$S_w = 0.02 + \frac{(2/12)}{2} = 0.103$$

$$T_s = \frac{D}{S_w} = 4.52 \, \text{ft}$$

$$Q_x = \frac{0.56}{0.016} \times 0.02^{1.67} \times 13.0^{2.67} \times \sqrt{0.015} = 47.98 \times \sqrt{0.015} = 5.88$$

$$Q_w = \frac{0.56}{0.016} \times 0.103^{1.67} \times [4.52^{2.67} - (4.52 - 2)^{2.67}] \times \sqrt{0.015} = 39.13 \times \sqrt{0.015} = 4.79$$

The total flow is as follows: $Q = Q_w + Q_x = 10.67 \, \text{cfs}$

It is noted that the earlier approach takes the geometry of a street gutter into consideration but not the safety concerns. For instance, a steep street can carry more flow, but its high flow velocity has to be subject to a reduction due to the safety concern. Among various recommendations on street flow safety, two different criteria have been developed for the SHCC designs, including (1) *permissible VD product* and (2) *discharge reduction factor*.

9.5 Permissible VD product

Traffic accidents during a storm event are caused by the hydroplaning acted on cars. As shown in Figure 9.10, water splashes occur to the cars on the flooded pavements. Dynamic hydroplaning effects were studied based on the condition that the brake force coefficient is reduced to zero when analyzing the initiation of hydroplaning (Agrawal et al., 1977). The vehicle speed at incipient hydroplaning was analyzed by the change in the rotational speed of a wheel due to the loss of contact with the pavement surface

Figure 9.10 Examples of hydroplaning. (a) Hydroplaning effect and (b) possible hydroplanning.

(Gallaway et al., 1979). An empirical equation between vehicle speed at incipient hydroplaning and the water film thickness exhibits a hyperbolic relationship between the vehicle velocity and the water depth (Huebner et al., 1986).

From the previous studies, the most sensitive parameters in hydroplaning effects are the water depth and flow velocity on the pavement. As a result, the hazards from a high-speed flood flow on the street are directly proportional to the flow momentum in the gutter flow. Similar to the concept used in the hydroplaning analysis, a hyperbolic relationship between the water flow velocity and depth in a street gutter is developed as a basis to determine the maximum allowable runoff discharge on the street. For simplicity, let us assume that $D_s = 0$. The VD product is derived from Equation 9.1 as

$$VD = \frac{2K}{n}(TS_x)^{1.67}\sqrt{S_o} \tag{9.14}$$

$$V = \frac{Q}{A} \tag{9.15}$$

Aided with Equations 9.1 and 9.14, Equation 9.15 is rearranged as

$$Q = \frac{1}{2}(VD)T \tag{9.16}$$

Equation 9.16 describes the relationship between the SHCC and the VD product in a street flow. The *Hydrologic Criteria and Drainage Design Manual* (Clark County, 1999) used in Las Vegas, NV, has suggested that the VD product in a street gutter flow be <6 for a minor event or 8 for a major event. According to Equation 9.16, the street stormwater conveyance capacity, q, per unit width of water spread is

$$q = \frac{Q}{T} = \frac{VD}{2} \tag{9.17}$$

Equation 9.17 indicates that the SHCC in a street is proportional to its VD product. A reduction on the VD product imposes a limit on both the unit-width capacity and the flow momentum force in a gutter flow. Consider that the VD product of a gutter flow shall not exceed a limit defined by safety as

$$VD \leq L \tag{9.18}$$

in which L = permissible VD product. Substituting Equation 9.14 into Equation 9.18 yields the permissible water spread, T_L, for the specified L as

$$T_L \leq \frac{1}{S_x}\left(\frac{nL}{2K\sqrt{S_o}}\right)^{0.6} \tag{9.19}$$

By Equations 9.16 and 9.18, the allowable SHCC, Q_L, is limited to

$$Q_L \leq \frac{1}{2}LT_L \tag{9.20}$$

in which Q_L = allowable SHCC subject to the limiting VD product. For a given limiting VD product, Equation 9.13 is revised to

$$T = \min\left(T_m, \frac{D_m}{S_x}, T_L\right)$$ (9.21)

Equation 9.12 assists the engineer in selecting design water spread with consideration of both street gutter geometry and safety. Substituting the design water spread determined by Equation 9.21 into Equation 9.1, the allowable SHCC can be determined.

EXAMPLE 9.3

An arterial highway (as shown in Figure 9.11) is to be widened to an 80-ft bus route. The street geometry is described with $S_o = 0.01$, $S_x = 0.02$, and $W = 2.0$ ft. During the 100-year event, the street must maintain a 12-ft traffic lane free from storm runoff, and the gutter water depth is not to exceed the curb height of 8 in. Determine the SHCC subject to $VD \leq 2$.

Solution: With $D_s = 0$, we have $T_m = 40-12 = 28$ ft (maximum available spread width). According to Equation 9.20, the water spread for $L = 2$ is calculated as

$$T_L = \frac{1}{0.02}\left(\frac{0.016 \times 2}{2 \times 0.56 \times \sqrt{0.01}}\right)^{0.6} = 23.6 \, \text{ft}$$

The selection of water spread is

$$T = \min\left(28, \frac{8/12}{0.02}, 23.6\right) = 23.6 \, \text{ft.}$$

The corresponding allowable flow is calculated as

$$Q_s = \frac{0.56}{0.016}(0.02)^{1.67} \times 23.6^{2.67} \times \sqrt{0.01} = 23.52 \, \text{cfs}$$

The water depth in the gutter is $D = 23.6 \times 0.02 = 0.47$ ft

The flow area is $A = 0.5 \times 0.47 \times 23.6 = 5.55 \, \text{ft}^2$

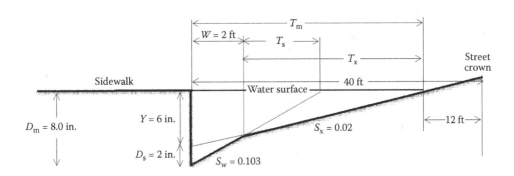

Figure 9.11 Design example for street hydraulic conveyance capacity.

The flow velocity is $V = 23.52/5.55 = 4.24\,\text{fps}$

The VD produce is $L = 4.24 \times 0.47 = 1.99 < 2.0$ (Ok!)

Remember that Equation 9.20 was derived with $D_s = 0.0$. If $D_s > 0$, Equation 9.20 will provide only an approximation. Repeat Example 9.3 with $D_s = 2.0\,\text{in}$.

$$T = \min\left(28, \frac{(8-12)}{0.02}, 23.6\right) = 23.6\,\text{ft}$$

$$S_w = 0.02 + \frac{2/12}{2} = 0.103$$

$$Y = TS_x = 23.6 \times 0.02 = 0.47\,\text{ft}$$

$$D = Y + D_s = 0.472 + 2/12 = 0.64\,\text{ft}$$

$$T_x = T - W = 23.6 - 2.0 = 21.6\,\text{ft}$$

$$T_s = D/S_w = 0.64/0.103 = 6.20\,\text{ft}$$

$$Q_w = \frac{0.56}{0.016} \times 0.103^{1.67} \times [6.20^{2.67} - (6.20 - 2.0)^{2.67}] \times \sqrt{0.01} = 6.63\,\text{cfs}$$

$$Q_x = \frac{0.56}{0.016} \times 0.02^{1.67} \times 21.6^{2.67} \times \sqrt{0.01} = 18.61\,\text{cfs}$$

The total flow is the sum as : $Q_s = 6.63 + 18.61 = 25.24\,\text{cfs}$

The flow area is : $A = 0.5 \times 0.472 \times 23.6 + 0.5 \times 2 \times 2/12 = 5.74\,\text{ft}^2$

The flow velocity is : $V = \dfrac{Q_s}{A} = \dfrac{25.24}{5.74} = 4.40\,\text{fps}$

Check on the VD product as : $VD = 4.40 \times 0.64 = 2.82 \leq L$ (close, but not exactly correct!)

The above discrepancy was caused by the difference between that with and without a gutter depression. For this case, the water spread of 23.6 ft serves only as a reference to find the solution. After a couple of iterations, the water spread of 18 ft is chosen to produce a flow rate of 12.85 cfs with $VD = 2.0$.

9.6 Discharge reduction method

In addition to the approach of permissible VD product, the *discharge reduction method* is also developed as recommended for street drainage designs. For instance, the City of Denver, CO, converts the limiting VD product into a set of discharge reduction factors. The allowable SHCC is equal to the *gutter full capacity* multiplied by a *discharge reduction factor* that is defined as

$$R = \frac{Q_L}{Q_{Full}} \tag{9.22}$$

in which R = discharge reduction factor. As mentioned previously, when the water spread is wide, the flow condition on a street is converged to a straight section. Setting the water depth equal to the gutter full depth, Equation 9.16 is converted to

$$Q_{Full} = \frac{1}{2} V D_m T \tag{9.23}$$

Substituting Equations 9.19, 9.20, and 9.23 into Equation 9.22 yields

$$R = \frac{\frac{1}{2} L T_L}{V\left(\frac{1}{2} D_m T\right)} = \frac{1}{(TS_x)^{2.67}} \left[\frac{nL}{2K\sqrt{S_o}}\right]^{1.60} \quad 0 \le R \le 1.0 \tag{9.24}$$

Equation 9.24 has four variables: R, T, S_o, and L. In order to produce a design chart, let us select a representative water spread, T_R, and accept the design criterion, L_R, for the VD product. Equation 9.24 is rewritten as

$$R = \frac{1}{(T_R S_x)^{2.67}} \left(\frac{nL_R}{2K\sqrt{S_o}}\right)^{1.60} \tag{9.25}$$

in which T_R = regional water spread in [L] and L_R = regional VD limit in [L^2/T]. For instance, the City of Denver, CO, accepts $T_R = 12.5$ ft and $L_R = 1.0$ cfs/ft for the minor event and $T_R = 20.5$ ft and $L_R = 2.0$ cfs/ft for the major event. Figure 9.12 was developed as the discharge reduction factors versus street longitudinal slopes. It shows that the

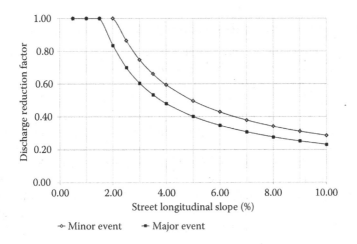

Figure 9.12 Reduction factors used in Denver, CO.

steeper the street, the higher the discharge reduction factor. Both curves in Figure 9.12 have been adopted by the City of Denver for street drainage designs.

As shown in Equation 9.25, it requires a pair of regional values for L_R and T_R in order to derive a set of consistent discharge reduction factors. In practice, the first step is to determine the gutter full capacity. Next, the discharge reduction factor should be chosen from Figure 9.12 based on the street longitudinal slope. The allowable SHCC is determined by Equation 9.22.

EXAMPLE 9.4

Considering the street section in Example 9.1, apply regional parameters $T_R = 12.5\,\text{ft}$ and $L_R = 1$ to analyze the allowable SHCC for $S_o = 0.04$.

Solution: Referring to Example 9.1, the design parameters are as follows: $H_c = 6\,\text{in.}$, $W = 2.0\,\text{ft}$, $D_s = 2.0\,\text{in.}$, $S_x = 0.02$, $n = 0.016$, $T = 16.67\,\text{ft}$, $T_s = 4.85\,\text{ft}$, $S_w = 0.103$, and $S_o = 0.04$; the gutter full capacity is calculated with the following:

$$Q_x = \frac{0.56}{0.016} \times 0.02^{0.67} \times 16.7^{2.67} \times \sqrt{0.04} = 8.08\,\text{cfs}$$

$$Q_w = \frac{0.56}{0.016} \times 0.103^{1.67} \times [4.84^{2.67} - (4.84 - 2.0)^{2.67}] \times \sqrt{0.04} = 13.24\,\text{cfs}$$

The gutter full flow is the sum of the above two flows: $Q_{Full} = 21.32\,\text{cfs}$.
Based on the given condition, the discharge reduction factor is calculated as

$$R = \frac{1}{(12.5 \times 0.02)^{2.67}} \left(\frac{0.016 \times 1.0}{2 \times 0.56 \times \sqrt{0.04}} \right)^{1.6} = 0.59$$

With a discharge reduction factor of 0.59, the allowable SHCC is reduced to

$$Q_L = 0.59 \times 21.32 = 12.58\,\text{cfs}$$

9.7 Street hydraulic storage capacity

Water ponding at a street intersection (as shown in Figure 9.13) may be caused by inadequate maintenance or operational malfunctions such as backing up the sewer system. Whenever the incoming runoff is greater than the outflow at an inlet, the street is flooded with standing water.

The concept of SHCC was derived from *a conveyance-based method*. At a depressed section, the runoff volume is accumulated through the increase of the water depth. The performance of a sump inlet is directly related to the local street storage capacity rather than the street conveyance capacity along the gutter. As a result, *a volume-based method* needs to be developed to predict the water spread at a sump area. At a street corner, crossing water flows are prevented by raising both street crowns on a side slope of 2%. As a result, stormwater entrained at a street corner can only be drained out by a sump inlet. As a result, the water spread at a depression street section is dictated by the volume balance between (1) *stormwater detention volume* and (2) *storage capacity* around the sump inlet (Figure 9.14).

Figure 9.13 Water accumulated at sump. (a) Clogged inlet and (b) backed-up inlet.

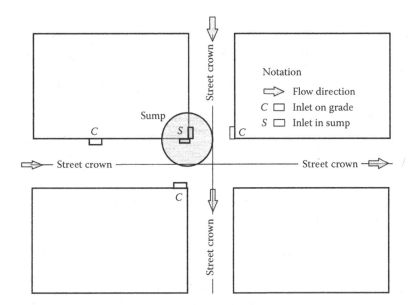

Figure 9.14 Sump at street intersection.

9.7.1 Stormwater detention volume

In an urban area, the distance between two adjacent street inlets is approximately 300–400 ft. The catchment area for a sump inlet is approximately 3–4 acres (1 ha = 2.5 acres). To predict the peak runoff flow rate and volume from a small urban catchment, the rational method states

$$Q_p = K_r CIA \tag{9.26}$$

$$V_I = K_r CT_d I \, A \tag{9.27}$$

in which C = runoff coefficient, A = watershed area in acres (hectare), I = rainfall intensity in in./h (mm/h), T_d = rainfall duration in minutes, Q_p = peak runoff rate in cfs or cms, V_I = runoff volume in cubic feet or cubic meters, and K_r = 1 for acre-in./h or 1/360 for hectare-mm/h. The rainfall intensity is described as

$$I = \frac{a}{(b+T_d)^m} \tag{9.28}$$

where a, b, and m are constants in the local intensity–duration–frequency (IDF) formula. For a given event, T_d, Equation 9.27 represents the inflow volume entering the sump area. The outflow volume is estimated by the sump inlet capacity as

$$V_o = (1-C\log)Q_s T_d \tag{9.29}$$

in which Q_s = sump inlet capacity in $[L^3/T]$ such as 5–10 cfs designed to pass the minor event and Clog = clogging factor between zero and unity. The storage volume at the sump area is the difference between the inflow and outflow volumes as

$$V_D = V_I - V_o \tag{9.30}$$

The detention volume in Equation 9.30 varies with the rainfall duration. In practice, we shall test a range of rainfall duration until the volume in Equation 9.30 becomes maximized (Department of the Army and the Air Force, 1977; Guo, 1999). Care must be taken about the unit conversions when using the *rational volume-based method*. It is advisable to convert all variable to foot and second or meter and second to avoid potential calculation errors.

9.7.2 Street hydraulic storage capacity

As illustrated in Figure 9.15, the depression storage volume at a street corner is similar to a conic volume with the street transverse slope as the conic side slope. The *conic volume* between two adjacent depths from h_1 to h is calculated as

$$V_h = \frac{1}{3}h_1 A_1 + \frac{(h-h_1)}{3}\left(A+A_1+\sqrt{AA_1}\right) \tag{9.31}$$

in which h = water depth in $[L]$, A = water surface area in $[L^2]$, and V_h = conic volume in $[L^3]$. The subscript 1 represents the variables at depth, h_1. When $h_1 = 0$, i.e., on the ground, its area, $A_1 = 0$. Equation 9.31 is reduced to

$$V_h = \frac{1}{3}hA \tag{9.32}$$

As illustrated in Figure 9.15, the storage volume around a curb is divided into two portions: *(a) volume below the curb height, H_c, and (b) the volume above the curb height.*

1. Volume for $h \le H_c$
   When the depth, h, is below the curb height, H_c, the water surface area (Figure 9.16) can be approximated as a fraction of a circle, depending on the configuration of the street section. For instance, k = 1/4 for a 90° depressed street section, k = 1/2 for a straight depressed street section, k = 3/4 for a 270° intersection area of two sloping

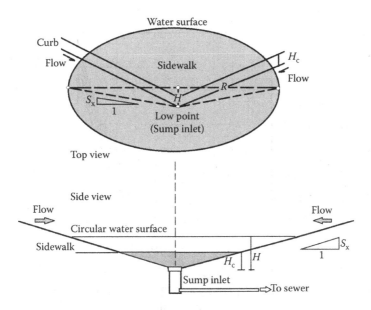

Figure 9.15 Illustration of street depression storage volume.

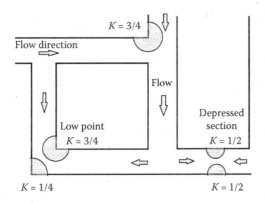

Figure 9.16 Storage capacity parameter at street sump.

streets, and $k = 1$ for the condition of no curb. With a specified k, the water surface area at a depth, $h < H_c$, is

$$A_h = k\pi R_h{}^2 \text{ for } h < H_c \tag{9.33}$$

$$R_h = h/S_x \tag{9.34}$$

in which k = a fraction of a circle and R_h = radius in [L] of the circular water area at a sump. Substituting Equations 9.33 and 9.34 into Equations 9.32 yields

$$V_h = \frac{k\,\pi h^3}{3\,S_x{}^2} \text{ for } h \le H_c \tag{9.35}$$

Equation 9.35 is the storage volume with a depth below the curb height.

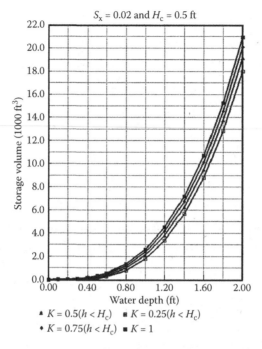

Figure 9.17 Street storage capacity with $S_x = 0.02$ and $H_c = 0.5$ ft.

2. **Volume for $h > H_c$**

    Above the curb height, the additional volume between h and H_c is estimated by Equation 9.31 as

    $$V_h = \frac{\pi(h - H_c)}{3S_x^2}(H_c^2 + h^2 + H_c h)$$ (9.36)

    The total storage volume for a depth above the curb height is the sum of Equations 9.35 and 9.36 as

    $$V_h = \frac{k}{3}\frac{\pi H_c^3}{S_x^2} + \frac{\pi(h - H_c)}{3S_x^2}(H_c^2 + h^2 + H_c h) \text{ for } h > H_c$$ (9.37)

    When the ponding depth exceeds 18 in., Equation 9.37 can be an approximation for Equation 9.36. Having known the detention volume by Equation 9.30, the water depth can be predicted by Equation 9.35 or 9.37, depending on the water depth relative to the curb height. The corresponding spread, R, can then be predicted by Equation 9.34. In general, the transverse slope of 0.02 is used for street designs. Figure 9.17 represents the plot of Equations 9.35 and 9.37 with $S_x = 0.02$.

EXAMPLE 9.5

Determine the water spread at a street intersection as illustrated in Figure 9.18. The design information includes the drainage area of 3.0 acres and runoff coefficient of 0.65. The IDF formula at the site is described as

$$I = \frac{28.5P_1}{(10 + T_d)^{0.789}}$$

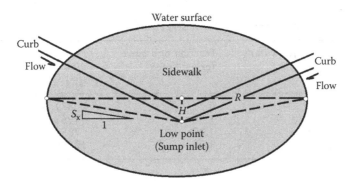

Figure 9.18 Illustration of water pool at a 3/4-circle intersection.

in which I = rainfall intensity in in./h, T_d = duration in minutes, and P_1 = 1-h rainfall depth in inches. Knowing that the 100-year 1-h rainfall depth is 2.6 in., determine the water spread for the 100-year event.

Solution: For this case, a Type 16 grate inlet is installed at the sump. During a 100-year event, the sewer system is backed up. The capacity of this inlet is estimated to be 10.0 cfs without clogging. For the IDF curve, $a = 28.5 \times 2.6 = 74.1$, $b = 10$, and $m = 0.789$. To find the storm water volume, let $K_r = 1.0$, $C = 0.65$, $A = 3.0$ acre, $C_{log} = 0.3$, and $Q = 10.0$ cfs. With $T_d = 12$ min, Equation 9.30 becomes

$$V_d = 1.0 \times 0.65 \times 3.0 \times \frac{28.5 \times 2.6}{(12.0+1.0)^{0.789}} \times 12.0 \times 60 - (1-0.3) \times 10.0 \times 12.0 \times 60 = 4114 \, \text{ft}^3$$

The stormwater volume of 4114 ft³ will enter the sump area when $T_d = 12$ min. Table 9.2 illustrates the calculations of stormwater detention volumes for this example. The increment of 1 min applies to the rainfall duration. The maximal volume of 4115.2 ft³ is obtained when $T_d = 13$ min.

Next is to estimate the available storage volume at the site. Assuming that $S_x = 0.02$, the water depth is found by Equation 9.37 to be

$$4115 = \frac{0.75 \, \pi \times 0.5^3}{3} \frac{1}{0.02^2} + \frac{\pi(h-0.5)}{3 \times 0.02^2}(0.5^2 + h^2 + 0.5 \times h)$$

Substituting a range of depths into the above equation yields a ponding depth of 1.15 ft or 13.8 in. (as shown in Table 9.3).

The corresponding radius or water spread at this water pool is

$$R_h = \frac{1.15}{0.02} = 57.5 \, \text{ft}$$

Table 9.2 Calculation of storage volume at street sump

Duration (min)	Rainfall intensity (in./h)	Inflow volume (ft³)	Outflow rate (cfs)	Outflow volume (ft³)	Detention volume (ft³)
11.0	6.708	8705.022	7.000	4620.000	4085.0217
12.0	6.466	9154.148	7.000	5040.000	4114.1483
13.0	6.243	9575.209	7.000	5460.000	4115.2095
14.0	6.037	9971.249	7.000	5880.000	4091.2490

Table 9.3 Calculations of storage volumes at street intersection

Water depth (in.)	Radius for water pool (ft)	Fraction of a circle for pool area	Pool surface area (ft²)	Cumulative volume (ft³)
4.00	16.67	0.75	872.7	97.0
6.00	25.00	1.00	1963.5	327.3
12.00	50.00	1.00	7854.0	2618.0
14.00	58.33	1.00	10690.2	4157.3

For this case, the upstream sloping street sections have been examined and confirmed that it can pass the 100-year stormwater within the spread limit of 30 ft. However, the water pool at the street intersection has a radius of 57.5 ft, or the sidewalk and traffic lanes at the intersection will be inundated. This example presents a typical situation that at a street intersection, the critical water spread is dominated by the SHSC rather than the SHCC.

9.8 Closing

The street drainage system must be evaluated by both the SHCC for sloping street sections and the SHSC for depressed street sections. When the section of street flow extended into neighborhood front yard, the SHCC may be calculated by the elements in the flow section (as shown in Figure 9.19).

The standard gutter depression of 2 in. (Figure 9.20) creates a gutter crossing slope as steep as 0.103. To be safe for bicycle riders, it is also recommended that the gutter depression be reduced to 1.52 in. (as illustrated in Figure 9.20) (City of Denver, 2010). With $D_s = 1.52$ in., the gutter crossing slope is 0.083.

In the case of severe water encroachments into the emergency traffic lane at a depressed section, the mitigation to such a problem may include the following:

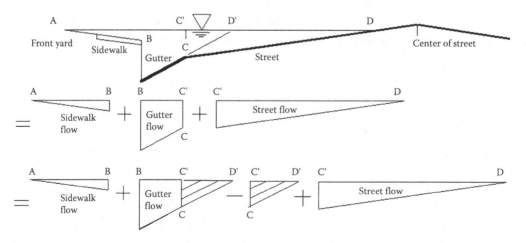

Figure 9.19 Extended flow section in street.

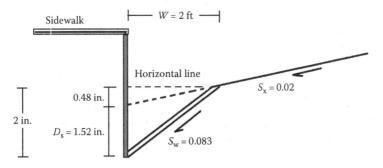

Figure 9.20 Gutter depression of 1.52 in. for traffic safety.

1. *To reduce the tributary catchment area by adding an inlet upstream*
   The rule of thumb of having one inlet per every 300–400 ft along the street may apply to sloping street sections. This practice may result in too much water accumulated at a sump area. To alleviate the inundation at a sump area, it is necessary to reduce the tributary area by adding one more inlet on the upstream sloping street section.

2. *To upsize both the sump inlet and the downstream storm sewers*
   Increasing the capacity of the sump inlet will reduce the accumulated stormwater volume. However, the efficiency of an inlet is not only determined by its size but also influenced by the energy grade line in the downstream storm sewer. In case that a higher inlet capacity is desired, it is necessary to upsize both the inlet and the sewer line downstream. If upsizing the collection drainage facility is not feasible, stormwater shall be diverted before the sump area.

3. *To increase the depression storage by applying a steep transverse slope*
   The stormwater storage capacity at a street intersection is directly related to the street transverse slope. The steeper the transverse slope, the larger the storage volume. Increasing a transverse slope from 2.0% to 3.0% can effectively reclaim the traffic lanes from the water spread, but it will compromise driver's comfort.

9.9 Homework

Q9.1 A street cross-section in Figure Q9.1 has a width of 30 ft from the gutter flow line to the street crown. The emergency use requires a 12-ft traffic lane. Knowing that the design parameters of gutter width, W, curb height, H_c, and Manning's N are 2 ft, 6 in., and 0.016, respectively, determine the SHCC and conduct comparisons among the following cases.

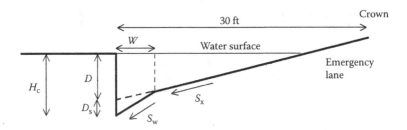

Figure Q9.1 Street cross-section.

1.   Sensitivity test of Q versus D_s and S_o

Available water spread	$T = 30 - 12 = 18\,ft$
Maximum gutter depth	$H_c = 6\,in.$

Street slope	Discharge (cfs)	Discharge (cfs)
S_o	$D_s = 1.52\,in.$ and $S_x = 0.02$	$D_s = 2\,in.$ and $S_x = 0.02$
0.01		
0.03		
0.05		

2.   Sensitivity test of Q versus S_x and S_o

Street slope	Discharge (cfs)	Discharge (cfs)
S_o	$D_s = 2\,in.$ and $S_x = 0.01$	$D_s = 2\,in.$ and $S_x = 0.02$
0.01		
0.03		
0.05		

Q9.2 A wide street has a transverse slope of 0.02 ft/ft, Manning's roughness of 0.016, gutter width of 2.0 ft, and gutter depression of 2.0 in. Without considering flow reduction due to the steep slope, establish a design chart for the stormwater conveyance capacity for the range of street slope from 0.01 to 0.10 ft/ft and the water spread from 15 to 40 ft.

Q9.3 As illustrated in Figure Q9.3, the section of south Speer Blvd. between Lincoln Avenue and Broadway Boulevard in the City of Denver is designed to have a 2000-ft depressed street section on a slope of 3%. The two inlets at the low point have a total

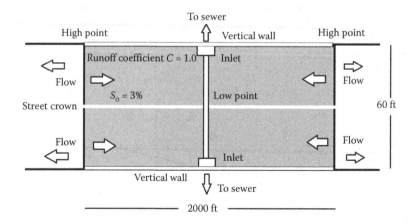

Figure Q9.3 Stormwater at depressed roadway.

capacity of 10 cfs. Considering a clogging factor of 0.5, estimate the water depth during the 100-year event. The rainfall IDF formula is given as

$$I(\text{in./h}) = \frac{74.1}{(10+T_d)^{0.784}} \text{ in which } T_d = \text{duration in minutes}$$

3. Find the rainfall duration to maximize the stormwater detention volume.

Solution: $T_d = 25$ min

Duration (min)	Rainfall intensity (in./h)	Inflow volume (acre-ft)	Outflow rate (cfs)	Outflow volume (acre-ft)	Storage volume (acre-ft)	
5.00	8.75	0.17	5.00	0.03	0.13	
15.00	5.85	0.34	5.00	0.10	0.23	
25.00	4.48	0.43	5.00	0.17	0.26	Max
35.00	3.68	0.49	5.00	0.24	0.25	

4. Determine the stormwater detention volume.
   Solution: 0.26 acre-ft
5. Estimate the maximum water depth around the inlet.
   Solution: a depth of 3.35 ft

Q9.4 Determine the SHCC for the 30-ft street sections in Figure Q9.4 with the following parameters: $W = 2$ ft, $D_s = 2$ in., $S_x = 0.02$, $n = 0.016$, and $H_c = 6$ in., which includes a gutter depression of 2 in. The longitudinal slope for each street section is determined by the elevation contours.

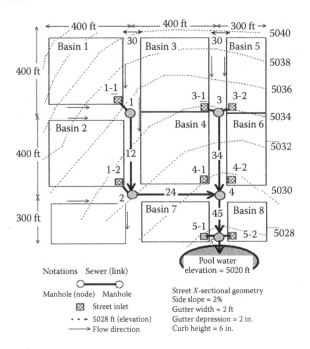

Figure Q9.4 Layout of street map.

Bibliography

Agrawal, S.K., and Henry, J.J. (1980). "Experimental Investigation of the Transient Aspect of Hydroplaning," Proceedings of the Annual Meeting of Transportation Research Board, Washington, DC.

Agrawal, S.K., Henry, J.J., and Mayer, W.E. (1977). "Measurement of Hydroplaning Potential," Final Report to Pennsylvania Department of Transportation, Report No. FHWA-PA-72-6.

Anderson, D.A. (1993). *"Improved Surface Drainage of Pavement,"* Research Report prepared by the Pennsylvania State University for National Cooperative Highway Research Program, National Research Council, Transportation Research Board, Washington, DC.

CDOT. (1990). *Hydraulic Design Criteria for Highways,* Colorado Department of Transportation, Denver, CO.

City of Denver. (2010). *Stormwater Design Criteria,* City of Denver, CO.

Clark County, Nevada. (1999). *Hydrologic Criteria and Drainage Design Manual,* Montgomery Watson Inc., Las Vegas, NV.

Department of the Army and the Air Force. (1977). "Surface Drainage Facilities for Airfields and Heliports," Technical Manual No. 5-820-1, Washington, DC.

Gallaway, B.M., Ivey, D.L., Hayes, G.G. Ledbetter, W.G., Olson, R.M., Woods, D.L., and Schiller, R.E. (1979). *"Pavement and Geometric Design Criteria for Minimizing Hydroplaning,"* Federal Highways Administration, Research Report No. FHWA-RD-79-31.

Guo, J.C.Y. (1997). *Street Hydraulics and Inlet Sizing,* Water Resources Publications, LLC, Littleton, CO.

Guo, J.C.Y. (1999). "Design of Detention Basins for Small Urban Catchments," *ASCE Journal of Water Resources Planning and Management,* Vol. 122, No. 1.

Guo, J.C.Y. (2000a). "Design of Grate Inlet with a Clogging Factor," *Journal of Advances in Environmental Research,* Vol. 4, pp. 181–186.

Guo, J.C.Y. (2000b). "Street Storm Water Conveyance Capacity," *ASCE Journal of Irrigation and Drainage Engineering,* Vol. 126, No. 2, pp. 119–123.

Guo, J.C.Y. (2000c). "Street Storm Water Storage Capacity," *Journal of Water Environment Research,* Vol. 27, No. 6.

Guo, J.C.Y. (2006). "Decay-Based Clogging Factor for Curb Inlet Design," *ASCE Journal of Hydraulic Engineering,* Vol. 132, No. 11, pp. 1237–1241.

Guo, J.C.Y., McKenzie, K., and Mommandi, A. (2008). "Sump Inlet Hydraulics," *ASCE Journal of Hydraulic Engineering,* Vol. 135, No. 1.

HEC12. (1984). "Drainage of Urban Highway Drainage Pavement." Hydraulic Engineering Circular No. 12, US Department of Transportation, Federal Highways Administration.

HEC22. (2010). Urban *Drainage Manual,* Hydraulic Engineering Circular No. 22, US Department of Transportation, Federal Highways Administration.

Huebner, R.S., Reed, J.R., and Henry, J.J. (1986). "Criteria for Predicting Hydroplaning Potential," *ASCE Journal of Transportation Engineering,* Vol. 112, No. 5, pp. 549–553.

USWDCM. (2010). *Urban Stormwater Drainage Design Criteria Manual,* Volumes 1 and 2, Urban Drainage and Flood Control District, Denver, CO.

Waugh, P.D., Jones, J.E., Urbonas, B.R., MacKenzie, K.A., and Guo, J.C.Y. (2002). "Denver Urban Storm Drainage Criteria Manual," *ASCE Journal of Urban Drainage,* Vol. 112, No 56.

Chapter 10

Street inlet hydraulics

Stormwater is removed from the streets through inlets. Frequently, street geometry dictates the location of inlets. Inlets are placed at low points (sumps), highway median breaks, and street intersections. Additional inlets should be placed at the points where the design peak flow on the street is approaching the allowable street hydraulic conveyance capacity (SHCC). Traffic may be impeded if inlets are not properly installed on the street. The *Hydraulic Engineering Circular No.12* (HEC 12, 1984), entitled "*Drainage of Highway Pavements,*" was published by the US Department of Transportation. Later on, HEC 12 was updated by the report of HEC 22. These two reports provide semitheoretical methods developed to estimate the SHCC for sizing street inlets. Although the HEC 22 procedure is recommended for inlet designs, it does fall short when the inlet is subject to potential clogging. Because urban debris is always one of the major parameters in street drainage designs, the HEC 22 procedure is expanded into street inlet designs under clogging effects in this chapter.

10.1 Types of inlet

As illustrated in Figure 10.1, an inlet unit collects surface stormwater into an inlet box that has a connector to drain stormwater into the nearby manhole. There are four major types of inlets used for street drainage. They are *grate inlet*, *curb-opening inlet*, *combination inlet*, and *slotted inlet*. Each one has its pros and cons when considering debris clogging, hydraulic efficiency, and traffic interference.

10.1.1 Grate inlet

As shown in Figure 10.2, an inlet grate is described by its length, L, width, W, type of steel bars, and spacing between steel bars. Grate inlets are often installed horizontally along the flow line within the gutter width. A grate can function well in the places where debris clogging and bike interference are not a problem. As illustrated in Figure 10.3, based on the layout and shape of steel bars, grates are further divided into *bar* and *vane* grates. In comparison, a vane grate is safer for bikes and more efficient to collect stormwater. Specifics of each type of grate can be found from industrial manufactory catalogues.

10.1.2 Curb-opening inlet

A curb-opening inlet (Figure 10.4) comprises a vertical opening on the street curb and a horizontal depression pan in the street gutter. Curb-opening inlets (shown in Figure 10.5) are hydraulically efficient and also much less susceptible to debris clogging.

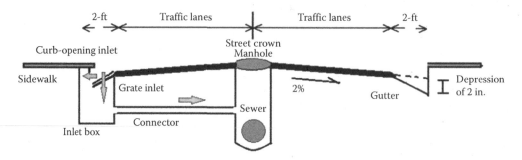

Figure 10.1 Inlet–sewer–gutter system.

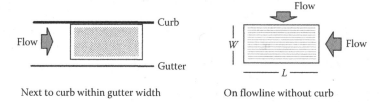

Figure 10.2 Grate inlet.

Figure 10.3 Bar and vane grates. (a) Bar grate and (b) vane grate with a 3-ft curb-opening inlet.

A curb-opening inlet is described by its height, H, and length, L, of the vertical opening, and the wing width, W_p, of the depression pan. Although a depression pan can enhance the hydraulic efficiency, it does interfere with bikes in the summer and snow plow operations in the winter. Therefore, a depression pan is optional to a curb-opening inlet.

10.1.3 Combination inlet

As shown in Figures 10.6 and 10.7, a combination inlet comprises a vertical curb-opening inlet and a horizontal grate inlet. Obviously, a combination inlet has a higher capacity of flow interception. The depression pan can increase the hydraulic efficiency for the grate, but it is optional.

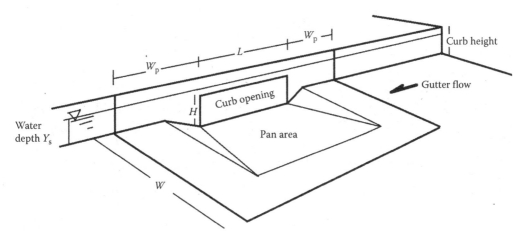

Figure 10.4 Curb-opening inlet.

Figure 10.5 Curb-opening inlet. (a) Four units of 10-ft curb-opening inlet and (b) one unit of 5-ft curb-opening inlet.

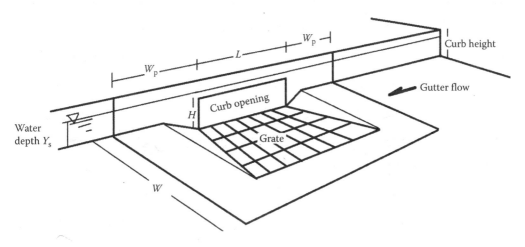

Figure 10.6 Combination inlet.

Figure 10.7 Combo inlets. (a) Combo inlet with double units and (b) 3-ft curb-opening and bar grate.

During a storm event, in case that the grate was clogged, the vertical opening would still function well. Therefore, a combination inlet is the best choice for sump areas and low points on the street.

10.1.4 Slotted inlet

A slotted inlet (as shown in Figures 10.8 and 10.9) is a strip of horizontal grate placed on top of an underground trench. A slotted inlet can be either parallel to the curb or perpendicular to the curb, depending on stormwater spread, traffic interference, snow plowing, and other maintenance considerations. On a steep slope where storm runoff is widely spread out, a slotted inlet can efficiently intercept the wide and shallow flow. A slotted inlet is often placed across a ramp at a highway interchange or at the entrance to an underground parking area.

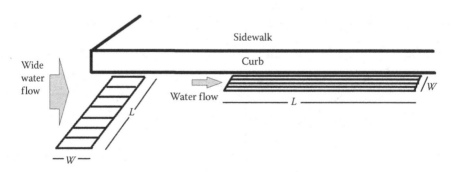

Figure 10.8 Slotted inlet.

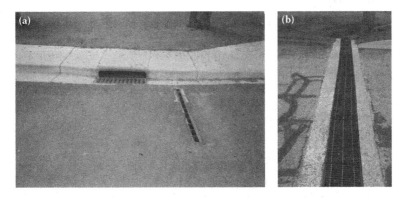

Figure 10.9 Slotted inlets. (a) Slotted inlet on steep street and (b) slotted trench on slope.

10.2 Inlet hydraulics

There are two major hydraulic factors to determine the inlet performance. They are
(1) *water depth at the inlet* and (2) *local street slope at the inlet*. In a sump, the inlet operates
like a weir when the water depth is so shallow such that the inlet opening area is not submerged. On the other hand, the inlet operates like an orifice when the inlet opening is completely covered under water. When an inlet is placed on a continuous grade, the stormwater
spreads out from the curb toward the street crown. The interception efficiency of an inlet on
a continuous grade depends on the water spread and gutter flow hydraulics. In general, the
steeper the street is, the less interception an inlet can have. When the inlet is placed in a depressed area, the deeper the water is, the more interception a sump inlet will have. Therefore,
it is important to identify whether the inlet is located on a grade or in a sump before sizing
the inlet. Often, a street consists of sloping and depression sections. Figure 10.10 shows an
example to identify whether the inlet is on a grade or in a sump.

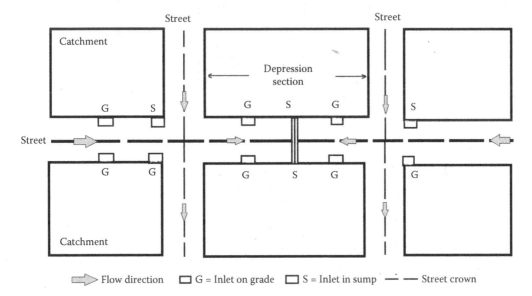

Figure 10.10 Inlets on a grade versus inlets in a sump.

10.3 Inlet spacing

If cross flows are not allowed for safety concerns, inlets must be placed at a street intersection. A sump demands an inlet by all means. On a continuous slope, the average distance between two adjacent inlets is approximately 300–400 ft. As illustrated in Figure 10.11, each individual inlet is placed using the following steps:

1. Prepare a street map at the project site.
2. Identify the street classifications.
3. Select a design storm event.
4. Choose a location for the inlet.
5. Calculate the peak flow from the local tributary area.
6. Add the carryover flow from the upstream inlets.
7. Determine the allowable SHCC based on water spread and gutter flow depth.
8. Check if the design flow is close to, but does not exceed, the SHCC. If not, move the inlet further upstream to reduce the tributary area and repeat the above steps until the design flow is comparable to the SHCC.
9. Choose a type of inlet.
10. Compute the interception capacity under a clogging condition. Inlets on a continuous grade are often sized to collect 70%–90% of the water flow on the street. Residual stormwater is termed *carryover flow* that moves toward the downstream inlet. Go to step 4 to size the next inlet.

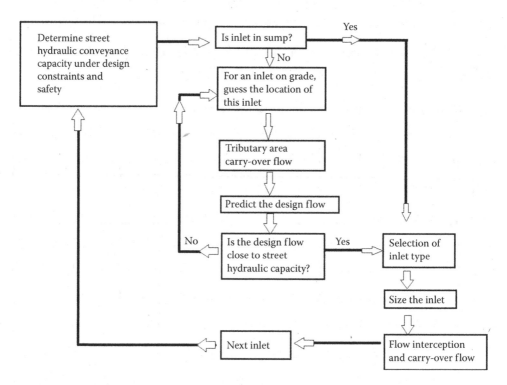

Figure 10.11 Procedure to determine inlet spacing.

10.4 Design discharge

At the design point, the design flow consists of the local flow generated from the local tributary area and the carryover flow from upstream inlet(s). Two methods are applicable to combine these two flows together as follows:

1. **Summation method**

   When the carryover flow is a small amount, it is suggested that the design discharge, Q, be the sum of the carryover flow, Q_c, and the local peak discharge, Q_b. This approach is valid if the times of concentration from various flow paths to reach the design point are about the same. Such an assumption is generally applicable to a series of inlets with tributary catchments of similar size.

$$Q = Q_b + Q_c \tag{10.1}$$

   in which Q = design flow in $[L^3/T]$, Q_b = local peak flow in $[L^3/T]$, and Q_c = carryover flow in $[L^3/T]$.

2. **Time shift method**

   With a large amount of carryover flow or off-site runoff, the design discharge on the street shall be either *the local flow* or *the combined flow*—whichever is higher. As illustrated in Figure 10.12, the local peak flow has a smaller area and a shorter time of concentration. The combined flow has a larger area and a longer time of concentration. To combine these two flows, the carryover flow shall be converted to its equivalent drainage area. Rearranging the rational method, the equivalent tributary area for the carryover flow is

$$C_c A_c = \frac{Q_c}{K_Q I_c} \tag{10.2}$$

   in which A_c = equivalent drainage area in acre or hectare, Q_c = carryover flow in cfs or cms, C_c = runoff coefficient of the upstream catchment from which the carryover was generated, I_c = rainfall intensity for carryover flow in in./h or mm/h, and $K_Q = 1$ for cfs-acre-in./h or 1/360 for cms-hectare-mm/h. The time of concentration is the accumulated flow time for the carryover flow to reach the design point:

$$T_T = T_c + T_f \tag{10.3}$$

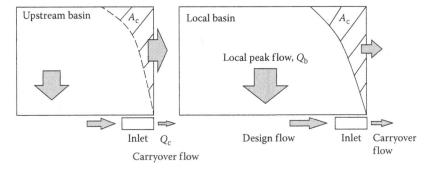

Figure 10.12 Carryover flow on the street.

in which T_T = flow time in minutes, T_c = time of concentration of the upstream catchment, and T_f = gutter flow time through the local catchment. The combined flow is

$$Q_T = K_Q I_T (C_b A_b + C_c A_c) \tag{10.4}$$

in which Q_T = combined flow, C_b = local runoff coefficient, A_b = local area, and I_T = rainfall intensity based on T_T. The design discharge shall be

$$Q = \text{Max}(Q_b, Q_T) \tag{10.5}$$

The design discharge in Equation 10.5 must be close to, but not exceeding, the street hydraulic capacity; otherwise, the abovementioned procedure shall be repeated with a reduced local area until the proper design discharge is achieved.

EXAMPLE 10.1

Use the street section in Example 9.1 to determine the design discharge for the situation shown in Figure 10.13. The local catchment has a runoff coefficient of 0.85 (or imperviousness of 80%). The overland flow length is 200 ft on a slope of 0.02 ft/ft. The gutter flow length is 500 ft on a slope 0.01 ft/ft. The gutter flow is described by the Soil Conservation Service (SCS) upland method with a conveyance parameter of 20.0. The carryover flow is 4.0 cfs that has a time of concentration of 10.0 min. The rainfall intensity in in./h is given as

$$I = \frac{76.5}{(10 + T_c)^{0.786}} \text{ in which } T_c = \text{time of concentration in minutes.}$$

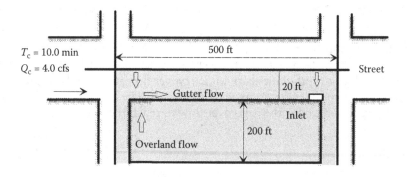

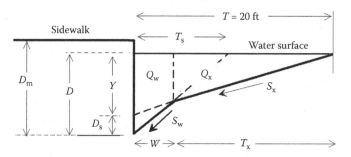

Figure 10.13 Catchment layout and street cross-section for inlet design.

1. **Analysis of local flow**

For this case, the overland flow time is calculated as

$$T_o = \frac{0.39(1.1-C)L_o^{0.5}}{S_o^{0.33}} = \frac{0.39(1.1-0.85)\sqrt{200}}{0.02^{0.33}} = 5.06 \text{ min}$$

The gutter flow time through the local catchment is estimated as

$$T_f = \frac{L_f}{60 \times 20 \times \sqrt{S_f}} = \frac{500}{60 \times 20 \times \sqrt{0.01}} = 4.17 \text{ min}$$

The computed time of concentration for the local flow is

$$T_c - \text{computed} = 5.06 + 4.17 = 9.23 \text{ min}$$

Next, we shall check with the regional time of concentration. First, let us start with the average slope, S_a, along the waterway as

$$S_a = \frac{S_oL_o + S_fL_f}{L_o + L_f} = \frac{0.02 \times 200 + 0.01 \times 500}{700} = 0.013$$

$$T_c - \text{check} = (18 - 15I_a) + \frac{L}{60(24I_a + 12)\sqrt{S_a}} = 6.0 + \frac{700}{60 \times 31.2\sqrt{0.013}} = 9.74 \text{ min}$$

$$T_c = \text{Min}(9.23, 9.74) = 9.23 \text{ min}$$

$$I_b = \frac{76.5}{(10 + 9.23)^{0.785}} = 7.51 \text{ in/h}$$

The local peak flow is

$$Q_b = C_bA_b = 0.85 \times 7.51 \times (220 \times 500)/43,560 = 15.96 \text{ cfs}$$

2. **To combine with the carryover flow**

The design rainfall intensity for the carryover flow is

$$I_c = \frac{76.5}{(10 + 10.0)^{0.785}} = 7.28 \text{ min}$$

The equivalent tributary area for the carryover flow is

$$C_cA_c = \frac{Q_c}{I_c} = \frac{4.0}{7.28} = 0.55 \text{ acre}$$

The travel time for the carryover flow to reach the design point is

$$T_T = 10 + \frac{500}{60 \times 20 \times \sqrt{0.01}} = 14.17 \text{ min}$$

With the consideration of the carryover flow, the design rainfall intensity is

$$I_T = \frac{76.5}{(10+14.17)^{0.785}} = 6.28 \text{ min}$$

The combined flow is

$$Q_T = I_T(C_c A_c + C_b A_b) = 6.28 \times [0.55 + 0.85 \times (220 \times 500 / 43,560)] = 16.8 \text{ cfs}$$

The design discharge shall be the higher one between the local and combined flows as

$$Q = \text{Max}(Q_b, Q_T) = 16.8 \text{ cfs}$$

Next, place $Q = 16.8$ cfs on the street that has a street longitudinal slope $S_o = 0.01$, street transverse slope $S_x = 0.02$, gutter depression, D_s, = 2 in., curb height, H, = 7 in., Manning's roughness, n, = 0.016, and available water spread, T, = 20 ft. Referring to Figure 10.13, the SHCC is determined by the available spread as

Try: $T = 20$ ft (available water spread width)

$$Y = T_m S_x = 20 \times 0.02 = 0.4 \text{ ft}$$

$$D = Y + D_s = 0.4 + (2.0/12) = 0.57 \text{ ft} < 7 \text{ in. (Ok)}$$

$$T_s = \frac{D}{S_w} = 5.48 \text{ ft}$$

$$T_x = T_m - W = 18 \text{ ft}$$

$$Q_x = \frac{K}{n} S_x^{1.67} T_x^{2.67} \sqrt{S_o} = 11.5 \text{ cfs (side flow)}$$

$$Q_w = \frac{K}{n} S_w^{1.67} [T_s^{2.67} - (T_s - W)^{2.67}] \sqrt{S_o} = 5.2 \text{ cfs (gutter flow)}$$

Check: $Q_s = 5.2 + 11.5 = 16.7$ cfs close to the design flow. So, $T = 20$ ft is accepted. The flow velocity is determined to be

$$A = 4.17 \text{ ft}^2 \text{ (flow area)}$$
$$V = 4.03 \text{ fps (flow velocity)}$$

The flow condition in Example 10.1 will be used to illustrate the inlet sizing procedures for different types of street inlets.

10.5 Clogging factor

The performance of an inlet is subject to debris clogging. Selection of a clogging factor depends on the debris amount and types of trash on the street. As a common practice for street drainage (CDOT, 1990), a clogging factor of 50% is recommended for the design of a single grate inlet, whereas a clogging factor of 10% is recommended for sizing a single curb-opening inlet. In practice, it often takes more than a single unit to collect stormwater on the street. The clogging factor applied to a multiple-unit inlet shall be decreased with respect to the length of the inlet. As shown in Figure 10.14, linearly applying a clogging factor to a multiple-unit inlet leads to an excessive length. For instance, assuming a clogging factor of 50%, the length of an inlet should be doubled, or it implies that a six-unit inlet would intercept the same amount of storm runoff as a three-unit inlet under no clogging.

As expected, the first grate is mostly vulnerable to debris clogging. As the number of inlet units increases, the clogging potential decays to none. Therefore, a decay function on clogging factor is derived as (Guo, 2000a)

$$C_G = \frac{1}{N_n}\left(C_o + eC_o + e^2C_o + e^3C_o + \cdots + e^{Nn-1}C_o\right) = \frac{C_o}{N_n}\sum_{i=1}^{i=N_n} e^{i-1} \tag{10.6}$$

in which C_G = multiple-unit clogging factor, C_o = single-unit clogging factor, e = decay ratio less than unity, and N_n = number of units. As shown in Figure 10.15, the amount of

Figure 10.14 Street inlet with excessive length. (a) Vane grate of 45-ft long and (b) curb-opening inlet of 55 ft long.

Figure 10.15 Decay of debris amount on multiple grates. (a) Single grate, (b) double grates, and (c) triple grate.

debris carried in the gutter flow is diminished as the flow moves downstream. It implies that the second unit may catch only 25% of the debris volume landed on the first unit. For a single-unit inlet, Equation 10.6 results in $C_G = C_o$. For a multiple-unit inlet, the value of C_o decays with respect to the number of units or the length of the inlet. The decay ratio, e, can be estimated by the incremental amount of debris captured by every single unit added to the inlet. When N_n becomes large, Equation 10.6 converges to

$$C_G = \frac{C_o}{N_n(1-e)} \tag{10.7}$$

For instance, when $e = 0.25$ and $C_o = 50\%$, then $C_G = 0.67/N_n$ for a large number of N_n. Equation 10.6 was evaluated by field data. Table 10.1 indicates that the predicted clogging factors for both multiple-unit curb-opening inlet and grate inlet closely agree with field experience when $e = 0.25$.

A shallow flow overtops the sides around a grate. After the grate is submerged, it operates like an orifice. Under a clogging condition, the flow interception is proportional to the clear side length for overtopping flows or the clear grate area for submerged flows. Table 10.2 presents the lengths, widths, and heights for both bar and vane grates. These dimensions are the key factors when calculating orifice and weir flows through the grate. The *area-opening ratio, m*, represents the ratio of the net flow-through area to the total grate area after subtracting steel or vane area from the grate area. Similarly, the *length-opening ratio, n*, represents the ratio of the net overtopping flow length to the side length of a grate inlet.

As shown in Figure 10.16, under a shallow water condition, the clear length for overtopping flow is calculated as

$$L_e = (1 - C_G)L \tag{10.8}$$

Table 10.1 Clogging factors predicted and field experience

Number of unit	Curb-opening inlet		Grate inlet	
	Observed	Predicted with e = 0.25	Observed	Predicted with e = 0.25
1.00	0.12	0.12	0.50	0.50
2.00	0.08	0.08	0.35	0.31
3.00	0.05	0.05	0.25	0.21
4.00	0.03	0.04	0.15	0.16

Table 10.2 Dimensions of various types of grate and vane inlets

Dimension	Bar grate Type 13	Vane grate Type 16	5-ft curb-opening	3-ft curb-opening
Length (ft)	2.96	2.96	5.00	3.00
Width (ft)	1.58	1.65	—	—
Height of curb opening (ft)	—	—	0.50	0.50
Length-opening ratio (n^*)	0.70	0.73	1.00	1.00
Area-opening ratio (m^*)	0.43	0.32	1.00	1.00

Figure 10.16 Area clogging versus wetted perimeter clogging. (a) Area clogging and (b) perimeter clogging.

in which L_e = effective length (unclogged) and L = side length of grate in [L]. Similarly, under a submerged case, a clogging factor shall be applied to the grate area as

$$A_e = (1 - C_G)A \tag{10.9}$$

in which A_e = clear (unclogged) opening area in [L^2] and A = grate area in [L^2].

10.6 Grate inlet on grade

As illustrated in Figure 10.17, a street gutter flow is divided into the *gutter flow*, which is carried by the street gutter within the gutter width, and the *side flow*, which is the water spread into the traffic lanes. The ratio of the gutter flow to the total runoff flow on a street is defined as

$$E_w = \frac{Q_w}{Q} \tag{10.10}$$

$$S_w = S_x + \frac{D_s}{W} \tag{10.11}$$

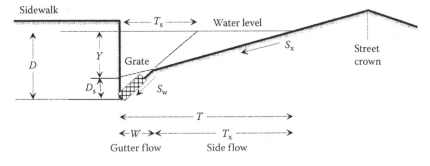

Figure 10.17 Illustration of street flow.

in which E_w = ratio of gutter flow to total water flow, Q_w = gutter flow in $[L^3/T]$ carried within the gutter width, Q = street flow in $[L^3/T]$, S_w = gutter transverse slope, S_x = street transverse slope, D_s = gutter depression in $[L]$ such as 2 in., and W = gutter width in $[L]$ such as 2 ft. Equation 10.10 is applicable to both composite and straight street sections. For a straight triangular cross section, $D_s = 0.0$, $S_w = S_x$, and $T = T_s$. As a result, Equation 10.10 is reduced to

$$E_w = 1 - \left(1 - \frac{W}{T}\right)^{2.67} \text{ only for } D_s = 0 \tag{10.12}$$

The ratio of side flow, Q_x, to street flow, Q, is

$$E_x = \frac{Q_x}{Q} = 1 - E_w \tag{10.13}$$

in which E_x = ratio of side flow to total runoff flow on the street.

The interception of the gutter flow by a grate is determined by the length of the grate, average cross-sectional water velocity, and water splash velocity due to the interference of the grate. A regression analysis using the laboratory data reported by HEC-12 results in the following empirical formula for determining the splash-over velocity, V_o, as a function of grate length and type of grate bars.

$$V_o = \alpha + \beta L_e - \gamma L_e^2 + \eta L_e^3 \tag{10.14}$$

in which V_o = water splash velocity in ft/s, as shown in Figure 10.18, L_e = effective length for each unit in feet of grate inlet, and α, β, γ, and η = constants, depending on the type of steel bars, as shown in Table 10.3.

The interception capacity of a grate is separately determined for the gutter and side flows. The interception ratio of a gutter flow, Q_w, is estimated as

$$R_w = 1 - 0.09(V - V_o) \text{ if } V \geq V_o, \text{ otherwise } R_w = 1 \tag{10.15}$$

in which R_w = interception ratio of gutter flow, $0 \leq R_w \leq 1$, and V = average cross-sectional water velocity in ft/s, which can be obtained from the street runoff flow analysis. *For*

Figure 10.18 Splash velocity over inlet grate on grade. (a) Splash velocity at on-grade grate and (b) splash velocity at sump grate.

Table 10.3 Splash velocities for various inlet grates

Type of grate	A	β	Γ	H
Bar P-1-7/8	2.22	4.03	0.65	0.06
Bar P-1-1/8	1.76	3.12	0.45	0.03
Bar P-1-7/8-4	0.74	2.44	0.27	0.02
45° Bar	0.99	2.64	0.36	0.03
30° Bar	0.51	2.34	0.20	0.01
Vane Grate	0.30	4.85	1.31	0.15
Reticuline	0.28	2.28	0.18	0.01
Type 13 Valley Grate	0.00	0.680	0.060	0.0023
Type 16 Valley Grate	0.00	0.815	0.074	0.003
Type C Standard Grate	2.22	4.03	0.65	0.06
Type C Close Mesh Grate	0.74	2.44	0.27	0.02

most cases, the condition, $V < V_o$, prevails or $R_w = 1.0$. The interception ratio of the side flow, Q_x, is expressed as

$$R_x = \frac{1}{\left(1 + \dfrac{K_R V^{1.8}}{S_x L_e^{2.3}}\right)} \tag{10.16}$$

in which $K_R = 0.15$ for foot-second units or 0.083 for meter-second units and $R_x =$ interception ratio of side flow, $0 \le R_x \le 1$. As a result, the total interception capacity, Q_a in $[L^3/T]$, for a grate inlet is calculated as

$$Q_a = R_w Q_w + R_x Q_x = [R_w E_w + R_x (1 - E_w)]Q \tag{10.17}$$

EXAMPLE 10.2

A unit of vane grate in Figure 10.19 has a width of 1.85 ft and a length of 3.25 ft. Considering a clogging factor of 0.5 for a single unit, determine the number of vane grates to intercept 80% of the design flow in Example 10.1.
   The gutter flow to street flow ratio is

$$E_w = Q_w/Q = 5.22/16.80 = 0.31$$

Consider four vane grates. The total grate length is

$$L_g = 4 \times 3.25 = 13.00 \, \text{ft}$$

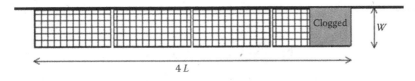

Figure 10.19 Interception capacity for vane grate on grade.

The clogging factor is 0.5 for a single unit. With e = 0.25, the clogging factor for four units is computed by Equation 10.7 as

$$C_G = \frac{0.5}{4 \times (1 - 0.25)} = 0.17$$

The effective grate length free from clogging is

$$L_e = (1 - 0.17) \times 13.00 = 10.79 \, \text{ft}$$

As illustrated in Example 10.1, the average water flow velocity is 4.30 fps. Referring to Table 10.3, the splash velocity is estimated as

$$V_o = 0.30 + 4.85 L_e - 1.31 L_e^2 + 0.15 L_e^3 = 88.50 \, \text{fps} > V, \text{ so}, R_f = 1.0$$

$$R_x = \frac{1}{\left(1 + \dfrac{0.15 V^{1.8}}{S_x L_e^{2.3}}\right)} = \frac{1}{\left(1 + \dfrac{0.15 \times 4.03^{1.8}}{0.02 \times 10.79^{2.3}}\right)} = 0.72$$

The flow interception for these four grates is calculated as

$$Q_c = [R_f E_w + R_x (1 - E_w)] Q = [1.0 \times 0.31 + 0.72 \times (1 - 0.31)] \times 16.8 = 13.6 \, \text{fps}$$

For this case, the interception ratio is 13.6/16.8 = 81%, which is acceptable. The carryover flow is 3.2 cfs.

10.7 Grate inlet in a sump

The performance of a sump grate (Figure 10.20) depends on whether the grate surface area is submerged or not. Under a shallow water depth, the flow interception by the grate is estimated using the weir flow formula as

$$Q_W = \frac{2}{3} C_d \sqrt{2g} P_e D^{1.5} \tag{10.18}$$

Figure 10.20 Grate inlet in sump. (a) Orifice flow—submerged grate and (b) weir flow—around grate.

or

$$Q_W = C_w P_e D^{1.5} \tag{10.19}$$

in which Q_W = interception capacity in $[L^3/T]$ as a weir flow, C_d = orifice discharge coefficient varied from 0.6 to 0.7, C_w = weir coefficient varied from 3.2 to 3.6 for English units or 1.8–2.1 for SI units, D = water depth in $[L]$, and P_e = effective weir length in $[L]$ around the inlet grate, which is defined as

$$P_e = n^*[(1-C_G)L+2W] \tag{10.20}$$

where n^* = length-opening ratio varied from 0.70 for bar grate to 0.73 for vane grate, L = grate length $[L]$, and W = grate width in $[L]$. Equation 10.20 is applicable to a grate placed next to a curb as shown in Figure 10.20, i.e., flow entering into the grate from three sides.

After the grate is submerged, it operates like an orifice. Its capacity is estimated as

$$Q_o = C_d A_e \sqrt{2gY_s} \tag{10.21}$$

$$A_e = (1-C_G)m^*WL \tag{10.22}$$

in which Q_o = orifice flow in $[L^3/T]$, C_d = orifice discharge coefficient such as 0.60–0.70, g = gravitational acceleration in $[L/T^2]$, Y_s = water depth in $[L]$, and m^* = area-opening ratio varied from 0.32 for vane grate to 0.43 for bar grate. Obviously, the flow interception at a grate is varied from weir to orifice flow. Transition between weir and orifice flows is not clearly understood. In practice, for a specified water depth, the interception capacity, Q_a, on a grate is evaluated by both Equations 10.18 and 10.21, and the smaller one dictates.

$$Q_a = Min(Q_w, Q_o) \tag{10.23}$$

EXAMPLE 10.3

A bar grate (shown in Figure 10.21) has a length of $L = 3.25\,ft$ and width of $W = 1.85\,ft$. The length- and area-opening ratios are $n^* = 0.70$ and $m^* = 0.45$. The clogging factor of 0.50 is recommended for a single unit. Calculate the interception capacity for a vane grate under a water depth of 0.5 ft.

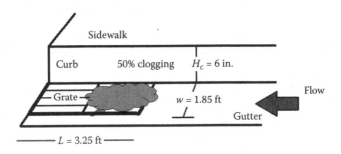

Figure 10.21 Interception capacity for vane grate in sump.

When the inlet operates like a weir, the capacity is

$$P_e = n^*[2 \times W + (1-C_G)L] = 0.70[2 \times 1.85 + (1-0.5) \times 3.25] = 2.59\,\text{ft}$$

With $C_w = 3.3$ for vane grate in English units, the weir capacity is

$$Q_W = 3.3 \times 2.59 \times 0.5^{1.5} = 3.02\,\text{cfs}$$

The net opening area for a grate is

$$A_e = (1-0.5) \times 0.45 \times 3.25 \times 1.85 = 1.35\,\text{ft}^2$$

With $C_d = 0.60$, the interception capacity is

$$Q_o = 0.60 \times 1.35 \times \sqrt{64.4 \times 0.5} = 4.60\,\text{cfs}$$

For this case, the weir flow dictates the interception capacity as

$$Q_a = \min(Q_w, Q_o) = 3.02\,\text{cfs}.$$

10.8 Curb opening on a grade

The performance of a curb-opening inlet is similar to a side weir. The flow interception capacity of a curb-opening inlet is empirically developed using English units. To install a curb-opening inlet on grade, as shown in Figure 10.22, the required *curb-opening length* to have a 100% runoff interception, L_t, to completely intercept the design flow on the street is computed by the empirical formula as

$$L_t = K_t Q^{0.42} S_o^{0.30} \left(\frac{1}{n S_e} \right)^{0.6} \tag{10.24}$$

$$S_e = S_x + S_w E_w \tag{10.25}$$

in which L_t = required curb-opening length in [L] for a 100% runoff interception, $K_t = 0.60$ for feet-second units or 0.82 for meter-second units, n = Manning's roughness

Figure 10.22 Curb-opening inlet on grade and in sump. (a) In sump and (b) on grade.

of 0.016, Q = design flow in $[L^3/T]$ on the street, S_o = street longitudinal slope, S_x = street transverse slope, S_w = gutter side slope, and S_e = equivalent transverse street slope. The curb-opening inlet shall have a length less than, but close to, L_t. The interception capacity, Q_a, in $[L^3/T]$, for the selected effective length, L_e in $[L]$, is calculated as

$$Q_a = Q\left[1-\left(1-\frac{L_e}{L_t}\right)^{1.80}\right]$$

(10.26)

in which Q_a = interception capacity for inlet in $[L^3/T]$ and L_e = effective $[L]$ of the curb-opening inlet.

EXAMPLE 10.4

A unit of curb-opening inlet has a length of 5 ft. Considering a clogging factor of 0.12 for a single unit, determine how many units of curb-opening inlet is required for the design flow in Example 10.1.

For this case, the gutter side slope is

$$S_w = \frac{D_s}{W} = \frac{(2/12)}{2.0} = 0.083\,\text{ft/ft}$$

The equivalent transverse slope is calculated as

$$S_e = S_x + S_w E_w = 0.02 + 0.083 \times 0.31 = 0.0458\,\text{ft/ft}$$

With $N = 0.016$ for asphalt surface, the required length of the curb-opening inlet to have a 100% runoff interception is

$$L_t = 0.60 \times 16.80^{0.42} \times 0.01^{0.30} \times \left(\frac{1}{0.016 \times 0.0458}\right)^{0.6} = 37.48\,\text{ft}$$

Try five units. The total length of the inlet is

$$L = 5.0 \times 5.0 = 25\,\text{ft}$$

Aided by Equation 10.7, the clogging factor for these five units of curb-opening inlet is

$$C_G = \frac{0.12}{5 \times (1-0.25)} = 0.027$$

The selected effective length of this curb-opening inlet is

$$L_e = (1-0.027) \times 25.0 = 24.33\,\text{ft}$$

Substituting the selected effective length into Equation 10.26 yields

$$Q_a = 16.80 \times \left[1-\left(1-\frac{24.33}{37.48}\right)^{1.80}\right] = 14.25\,\text{cfs}$$

This inlet has an interception ratio of 84.8% or a carryover flow of 2.55 cfs.

10.9 Curb-opening inlet in a sump

Referring to Figure 10.23, when a curb-opening inlet in a sump operates like a weir, its interception capacity is estimated as

$$Q_w = \frac{2}{3}C_d\sqrt{2g}P_eY_s^{1.5} = C_wP_eY_s^{1.5} \qquad (10.27)$$

$$P_e = (1-C_G)(L+kW_p)+2W_p \qquad (10.28)$$

in which P_e = effective weir length in [L] around the depressed pan in front of curb-opening inlet, W_p = width in [L] of depressed pan, and k = 1.8 to 2.0 for two sides of the pan, as illustrated in Figure 10.23. When the cub-opening gets submerged, it operates like an orifice and can be modeled as

$$Q_o = C_dA_e\sqrt{2g(Y_s-0.5H)} \qquad (10.29)$$

$$A_e = (1-C_G)HL \qquad (10.30)$$

in which H = height of curb opening in [L]. In practice, for a given water depth, the interception capacity for a curb-opening inlet in sump is determined by the smaller one between Equations 10.27 and 10.29.

EXAMPLE 10.5

Given a clogging factor of 12%, weir flow coefficient of 3.0, orifice flow coefficient of 0.60, and the water depth of 0.5 ft, determine the interception capacity for the curb-opening inlet shown in Figure 10.23.

For this case, the water depth is given as

$$Y_s = 0.5 \text{ ft}$$

Consider $k = 2.0$. Aided by Equation 10.28, the effective weir length for the depression pan is

$$P_e = (1-0.12)\times(3.0+2.0\times1.0)+2\times1.85 = 8.10 \text{ ft}$$

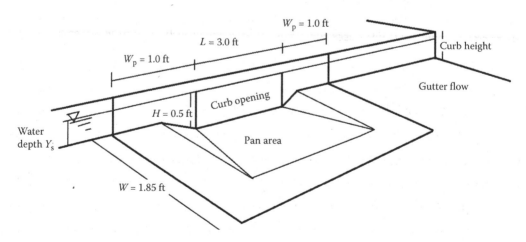

Figure 10.23 Example of curb-opening inlet in sump.

The weir flow capacity is estimated by Equation 10.27 as

$$Q_w = 3.0 \times 8.10 \times 0.5^{1.5} = 17.2\,\text{cfs}$$

The unclogged curb-opening area by Equation 10.30 is

$$A = (1 - 0.12) \times 3 \times 0.5 = 1.32\,\text{ft}^2$$

The orifice flow capacity is estimated by Equation 10.29 as

$$Q_o = 0.60 \times 1.32 \sqrt{2g(0.5 - 0.5/2)} = 3.18\,\text{cfs}$$

The interception capacity for this curb opening is

$$Q_a = \min(Q_w, Q_o) = 3.18\,\text{cfs}$$

10.10 Slotted inlet

A slotted inlet is hydraulically similar to a curb-opening inlet. As a result, all design formulas developed for curb-opening inlet are also applicable to slotted drain inlets.

10.11 Combination inlet

A combination inlet (Figure 10.24) consists of a horizontal grate placed in the gutter and a vertical curb-opening inlet on the curb face. When water flows through a combination inlet, the grate intercepts the shallow flow. The curb opening will not function until the grate is submerged. Different approaches were developed to size a combination inlet. For instance, it has been recommended that the capacity of a combination inlet be the higher interception between the grate and the curb opening. Or the street flow shall be applied to the grate first, and the remaining water depth is then applied to the curb-opening inlet (Guo, 1997). The laboratory test and data analyses indicate that the algebraic sum consistently overestimates the capacity of a combination inlet. A modification to the sum of the two inlets is introduced as (Guo et al. 2008):

$$Q_t = Q_g + Q_c - K_c \sqrt{Q_g Q_c} \tag{10.31}$$

Figure 10.24 Combinations of grate and curb-opening for combo inlet. (a) Middle grate in combo inlet and (b) end grate in combo inlet.

where Q_t = interception capacity in $[L^3/T]$ for combination inlet, Q_g = interception capacity in $[L^3/T]$ for grate, Q_c = interception capacity in $[L^3/T]$ for curb opening, K_c = reduction factor, $K_c = 0.37$ for bar grate, and $K_c = 0.21$ for vane grate (Guo et al., 2008).

10.12 Carryover flow

The actual amount of flow intercepted by an inlet affected by debris clogging is equal to

$$Q_c = Q - Q_a \tag{10.32}$$

in which Q_c = carryover flow.

10.13 Case study

Task 1 A street section in the City of Denver, CO, is illustrated in Figure 10.25. The hydraulic parameters are $n = 0.016$, $W = 2$ ft, $D_s = 2$ in., $H = 6$ in., $S_o = 1.0\%$, $S_x = 2\%$, and $T_m = 20$ ft. The design criteria for the Denver area recommends that the SHCC is the smaller one between the capacity for the maximum allowable water spread and the gutter full capacity subject to a safety reduction. Determine the SHCC for this case.

1. **Street capacity for the maximum water spread of $T_m = 20$ ft**

$$S_w = S_x + D_s/W = 0.103 \text{ ft/ft}$$

$$Y = T_m S_x = 20 \times 0.02 = 0.4 \text{ ft}$$

$$D = Y + D_s = 0.4 + (2.0/12) = 0.57 \text{ ft} > 6 \text{ in.}$$

$$T_s = \frac{D}{S_w} = 5.48 \text{ ft}$$

$$T_x = T_m - W = 18 \text{ ft}$$

$$Q_x = \frac{K}{n} S_x^{1.67} T_x^{2.67} \sqrt{S_o} = 11.44 \text{ cfs (side flow)}$$

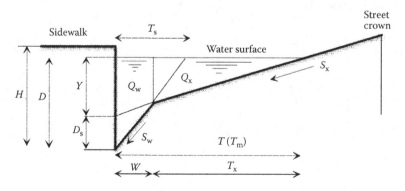

Figure 10.25 Street cross-section for case study.

$$Q_{\mathrm{w}} = \frac{K}{n} S_{\mathrm{w}}^{1.67} [T_{\mathrm{s}}^{2.67} - (T_{\mathrm{s}} - W)^{2.67}] \sqrt{S_{\mathrm{o}}} = 5.22\,\text{cfs} \;\; (\text{gutter flow})$$

$Q_{\mathrm{T}} = 5.22 + 11.44 = 16.66$ cfs for the maximum water spread

2. Gutter full capacity, Q_{F}

$$T_{\mathrm{X}} = \frac{(H - D_{\mathrm{s}})}{S_{\mathrm{X}}} = \frac{(6-2)/12}{0.02} = 16.7\,\text{ft}$$

$$T_{\mathrm{s}} = \frac{H}{S_{\mathrm{s}}} = \frac{6/12}{0.103} = 4.85\,\text{ft}$$

$$Q_{\mathrm{X}} = \frac{K}{n} S_{\mathrm{X}}^{1.67} T_{\mathrm{X}}^{2.67} \sqrt{S_{\mathrm{o}}} = 9.31\,\text{cfs}$$

$$Q_{\mathrm{w}} = \frac{K}{n} S_{\mathrm{w}}^{1.67} [T_{\mathrm{s}}^{2.67} - (T_{\mathrm{s}} - W)^{2.67}] \sqrt{S_{\mathrm{o}}} = 4.04\,\text{cfs}$$

$Q_{\mathrm{F}} = 4.04 + 9.31 = 13.35$ cfs for the gutter-full flow condition

3. Street hydraulic capacity for a minor storm event
Based on the street slope, the reduction factor for this case is $R = 1.0$. Therefore, the street capacity is determined as

$$Q_{\mathrm{s}} = \min{(Q_{\mathrm{T}}, R_{100} \times Q_{\mathrm{F}})} = \min{(16.66, 1.0 \times 13.35)} = 13.35\,\text{cfs}$$

Task 2 The catchment in Figure 10.26 has a drainage area of 1.87 acres, runoff coefficient of 0.70, and time of concentration of 10.6 min. The peak discharge generated from the catchment is 4.66 cfs. Knowing that the carryover flow is 2.5 cfs, determine the design discharge at the inlet.

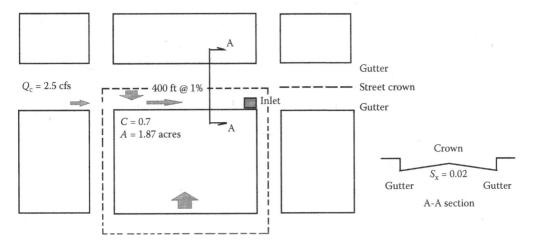

Figure 10.26 Tributary catchment.

According to the rational method, the local tributary discharge is determined to be 4.66 cfs. To combine with the carryover flow, Q_c, of 2.5 cfs, the design flow for the inlet is summed up as

$$Q_s = 2.50 + 4.66 = 7.16 \, \text{cfs} < \text{street capacity}$$

The design discharge is less than the street hydraulic capacity of 13.35 cfs. Therefore, the inlet location for this case is acceptable.

Task 3 Conduct the hydraulic analysis for the design flow of 7.16 cfs on the street shown in Figure 10.27.

Try: $T = 14.0 \, \text{ft}$

$$T_x = T - W = 12.0 \, \text{ft}$$

$$Y = TS_x = 14.0 \times 0.02 = 0.28 \, \text{ft and } D = Y + D_s = 0.45 \, \text{ft}$$

$$T_s = \frac{D}{S_w} = \frac{0.45}{0.103} = 4.34 \, \text{ft}$$

$$Q_x = \frac{K}{n} S_x^{1.67} T_x^{2.67} \sqrt{S_o} = 3.96 \, \text{cfs (side flow)}$$

$$Q_w = \frac{K}{n} S_w^{1.67} [T_s^{2.67} - (T_s - W)^{2.67}] \sqrt{S_o} = 3.22 \, \text{cfs (gutter flow)}$$

The total flow on the street is

The flow area, A_s, and flow velocity, V_s, on the street are

$$A_s = 0.5W(D + Y) + 0.5YT_x = 0.5 \times 2 \times (2/12 + 0.28) + 0.5 \times (0.45 + 0.28) \times 12.0 = 2.47 \, \text{ft}^2$$

$$V_s = \frac{Q_s}{A_s} = 2.96 \, \text{fps}$$

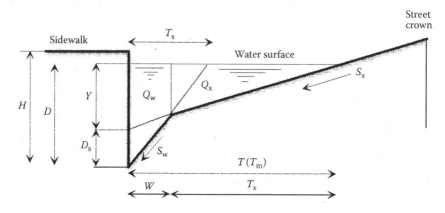

Figure 10.27 Design discharge on the street.

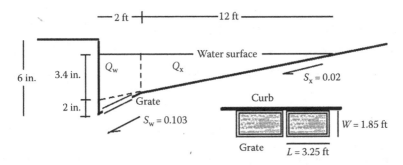

Figure 10.28 Grate inlet.

Task 4 Size a grate inlet on a continuous grade in Figure 10.28 for the design flow on the street.

$$E_o = \frac{Q_w}{Q_s} = \frac{3.22}{7.17} = 0.45$$

Try three vane grates. The dimensions for a vane grate include $W = 1.85$ ft and $L = 3.25$ ft. Consider a clogging factor of 0.50 for a single grate. With three grates or $N_n = 3$, use a decay coefficient of 0.5 to calculate the clogging factor for three grates as

$$C_G = \frac{1}{3}(0.5 + 0.5 \times 0.5 + 0.5 \times 0.5 \times 0.5) = 0.29$$

So, the selected effective length for three grates is

$$L_e = (1 - C_G) \times N_n \times L = (1 - 0.29) \times 3 \times 3.25 = 6.91 \text{ ft}$$

The interception percentage to the side flow is calculated as

$$R_s = \frac{1}{\left(1 + \dfrac{0.15 V_s^{1.8}}{S_x L_e^{2.3}}\right)} = \frac{1}{\left(1 + \dfrac{0.15 \times 2.96^{1.8}}{0.02 \times 6.91^{2.3}}\right)} = 0.62$$

Let $R_f = 1.0$. The interception for three inlet grates is

$$Q_a = [R_f E_o + R_s(1 - E_o)]Q_s = [1.0 \times 0.45 + 0.62 \times (1 - 0.45)] \times 7.17 = 5.65 \text{ cfs}$$

The carryover flow is

$$Q_c = Q_s - Q_a = 7.17 - 5.65 = 1.52 \text{ cfs on the street.}$$

Task 5 Size a bar grate inlet in a sump with a ponding depth of $Y = 0.5$ ft. The grate's dimension is shown in Figure 10.29.

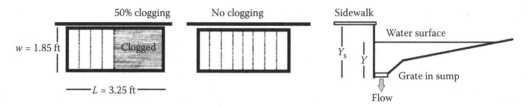

Figure 10.29 Grate inlet in sump.

Let the clogging factor for a single grate be 0.50. Try two grates for this case, $N_n = 2$. The clogging factor for two grates is

$$C_G = \frac{1}{2}(0.5 + 0.5 \times 0.5) = 0.375$$

When the grates operate like a weir, the unclogged wetted perimeter is

$$P = n^*[(1 - C_G)N_n L + 2W] = 0.7 \times [(1 - 0.375) \times 2 \times 3.25 + 2 \times 1.85] = 5.43\,\text{ft}$$

With a weir coefficient of 3.0 and a headwater of 0.5 ft, the weir capacity is

$$QW = 3.0 \times 5.43 \times 0.5^{1.50} = 5.76\,\text{cfs}$$

When the grates operate like an orifice, the unclogged opening area, A_t, is

$$A_t = (1 - C_G)N_n WL = (1 - 0.375) \times 2 \times 1.85 \times 3.25 = 7.52\,\text{ft}^2$$

With an area-opening ratio of 0.43 (after subtracting the steel bars), the net grate-opening area is

$$A_g = 0.43 A_t = 3.23\,\text{ft}^2$$

With an orifice discharge coefficient of 0.60 and a headwater of 0.5 ft, the orifice capacity is

$$Q_o = 0.60 \times 3.23 \times (2.0 \times 32.2 \times 0.5)^{0.50} = 11.0\,\text{cfs}$$

The interception capacity is determined as

$Q_a = \min(Q_w, Q_o) = 5.76\,\text{cfs} >$ design flow of 7.17 cfs. This is an undersized sump inlet. The pounding depth will be increased (>0.5 ft) and the drain time becomes longer.

Task 6 Size a curb-opening inlet on a grade for the design discharge. The layout of the curb-opening inlet is shown in Figure 10.30.

Let the clogging factor for a single unit be 0.10. Try three units, $N_n = 3.0$, subject to a decay coefficient of 0.25. The clogging coefficient for three curb-opening inlets is

$$C_G = \frac{1}{3}(0.1 + 0.1 \times 0.25 + 0.1 \times 0.25 \times 0.25) = 0.04$$

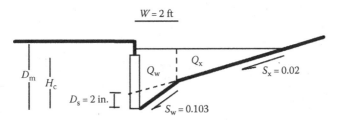

Figure 10.30 Curb-opening inlet sizing.

The unit length for a curb-opening inlet is 5 ft. The total unclogged curb-opening length for three units is

$$L_c = (1-0.04) \times 5.0 \times 3.0 = 14.4 \, \text{ft}$$

The required curb-opening length for a 100% runoff interception is calculated as

$$E_o = \frac{Q_w}{Q_s} = 0.45$$

$$S_e = S_x + S_w E_o = 0.02 + 0.103 \times 0.45 = 0.066$$

$$L_t = 0.60 Q_s^{0.42} S_0^{0.30} \left(\frac{1}{nS_e}\right)^{0.6} = 0.60 \times 7.17^{0.42} \times 0.01^{0.30}$$

$$\times \left(\frac{1}{0.016 \times 0.066}\right)^{0.60} = 21.05 \, \text{ft}$$

The interception capacity for the inlet with a length of 14.4 ft is determined as

$$Q_a = \left\{ 1 - \left[1 - \left(\frac{L_c}{L_t} \right) \right]^{1.8} \right\} Q_s = \left\{ 1 - \left[1 - \left(\frac{14.2}{21.05} \right) \right]^{1.80} \right\} \times 7.17 = 6.26 \, \text{cfs}$$

The carryover flow is

$$Q_c = 7.17 - 6.26 = 0.91 \, \text{cfs}$$

Task 7 Size a curb-opening inlet in a sump. The layout of this inlet is shown in Figure 10.31.

Given the parameters $W_g = 2 \, \text{ft}$, $L = 5 \, \text{ft}$, angle of throat = 1.05 rad. or 60°, $C_o = 0.6$, and clogging factor, $C_G = 0.1$, and considering that the curb opening operates like an orifice, determine the ponding water depth, Y, for the design flow.

Solution: With a bottle neck inclined with an angle of 60°, the flow area is

$$A = (H \sin \theta) \, L = 0.5 \sin 60° \times 5.0 = 2.16 \, \text{ft}^2$$

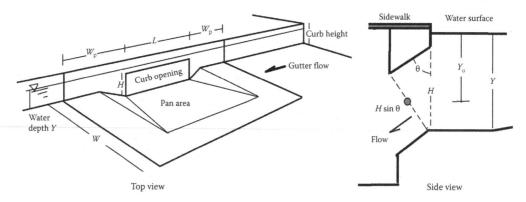

Figure 10.31 Sizing a curb-opening inlet in sump.

With $C_d = 0.6$, the orifice flow for this curb-opening inlet is

$$Q_o = (1 - C_G) C_d A \sqrt{64.4 Y_o} = (1 - 0.1) \times 0.60 \times 2.16 \times \sqrt{2 \times 32.2 Y_o} = 7.17 \, \text{cfs},$$

So, $Y_o = 0.59 \, \text{ft}$

$$Y_o = Y - 0.5 H \sin \theta = Y - 0.50 \times 0.50 \times \sin 60° \, \text{ft} = 0.59 \, \text{ft}$$

So, the ponding depth $Y = 0.81 \, \text{ft}$.

Next, check if the curb-opening inlet operates like a weir. The effective wetted perimeter is

$$P = (1 - C_G) \times L + 1.80 \times W_g = (1 - 0.1) \times 5.0 + 1.8 \times 2 = 8.1 \, \text{ft}$$

With a weir coefficient of 3.0, the weir capacity, Q_W, is

$$Q_W = 3.0 P Y^{1.5} = 3.0 \times 8.1 \times 0.81^{1.5} = 21.8.0 \, \text{cfs}$$

The interception capacity for this single curb-opening inlet in a sump shall be determined as

$$Q_a = \min (Q_o, Q_W) = 7.17 \, \text{cfs}$$

10.14 Homework

Q10.1 Figure Q10.1 shows the layout of street inlets. The design information is given as follows:

1. The 5-year design rainfall intensity in in./h is calculated using T_c in minutes as

$$I(\text{in./h}) = \frac{38.5}{(10 + T_c)^{0.786}}$$

where T_c = time of concentration set to be rainfall duration = time of concentration in minutes

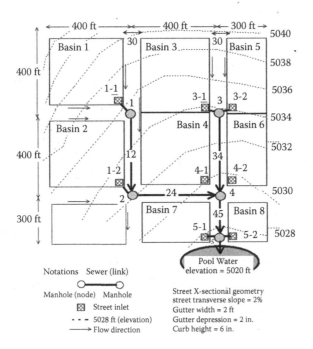

Figure Q10.1 Layout of inlets.

2. The overland flow time, T_o in minutes, shall be estimated by the airport formula using C = 5-year runoff coefficient, L_o = overland flow length in feet, and S_O = overland slope in ft/ft.

$$T_o = \frac{0.395(1.1-C)\sqrt{L_o}}{S_O^{0.33}} \, (\text{min})$$

The gutter flow time through the local catchment, T_f, in minutes, shall be estimated by the SCS upland method using L_f = gutter length in ft and S_f = gutter slope in ft/ft.

$$T_f = \frac{L_f}{60 \times 20\sqrt{S_f}} \, (\text{min})$$

3. The catchment hydrologic parameters are listed in Table Q10.1.

Table Q10.1 Catchment hydrologic parameters for inlet designs

Basin ID number	Area (acres)	Runoff coefficient and Imp%	Overland slope (%)	Overland length (ft)	Gutter slope (%)	Gutter length (ft)
1.00	3.67	0.55/60	1.25	300.00	1.25	550.00
2.00	3.67	0.62/65	1.50	300.00	1.50	525.00
3.00	3.67	0.85/80	1.50	250.00	1.50	500.00
4.00	3.67	0.85/80	1.00	300.00	1.00	500.00
5.00	2.75	0.85/80	1.50	300.00	1.50	450.00
6.00	2.75	0.85/80	1.00	300.00	1.00	400.00
7.00	2.75	0.45/40	0.75	150.00	0.75	400.00
8.00	2.11	0.45/40	0.75	150.00	0.75	350.00

Q10.2 Produce design charts for determining the interception capacity for a bar grate inlet on a grade under water spread of 10, 20, or 30 ft.

Q10.3 Produce design charts for determining the interception capacity for a bar grate inlet in a sump under headwater from 0.25 to 1.0 ft.

Q10.4 Produce design charts of determining the interception capacity for a 5-ft curb opening on a grade under water spread of 10, 20, or 30 ft.

Q10.5 Produce design charts for determining the interception capacity for a 5-ft curb-opening inlet in a sump under headwater from 0.25 to 1.0 ft.

Bibliography

CDOT. (1990). *Hydraulic Design Criteria for Highways*, Colorado Department of Transportation, Denver, CO.

Clark County Regional Flood Control District. (1990). *Hydrologic Criteria and Drainage Design Manual*, Las Vegas, NV.

Guo, J.C.Y. (1997). *Street Hydraulics and Inlet Sizing*, Water Resources Publications, LLC, Littlton, CO.

Guo, J.C.Y. (2000a). "Design of Grate Inlet with a Clogging Factor," *Journal of Advances in Environmental Research*, Vol. 4, pp. 181–186.

Guo, J.C.Y. (2000b). "Street Storm Water Conveyance Capacity," *ASCE Journal of Irrigation and Drainage Engineering*, Vol. 126, No. 2, pp. 119–123.

Guo, J.C.Y. (2000c). "Street Storm Water Storage Capacity," *Journal of Water Environment Research*, Vol. 27, No. 6.

Guo, J.C.Y. (2006). "Decay-Based Clogging Factor for Curb Inlet Design," *ASCE Journal of Hydraulic Engineering*, Vol. 132, No. 11, pp. 1237–1241.

Guo, J.C.Y., McKenzie, K., and Mommandi, A. (2008). "Sump Inlet Hydraulics," *ASCE Journal of Hydraulic Engineering*, Vol. 135, No. 1.

HEC12. (1984). "Drainage of Urban Highway Drainage Pavement." Hydraulic Engineering Circular No. 12, U.S. Department of Transportation, Federal Highways Administration.

HEC22. (2010). *Urban Drainage Manual*, Hydraulic Engineering Circular No. 22, U.S. Department of Transportation, Federal Highways Administration.

Waugh, P.D., Jones, J.E., Urbonas, B.R., MacKenzie, K.A., and Guo, J.C.Y. (2002). "Denver Urban Storm Drainage Criteria Manual," *ASCE Journal of Urban Drainage*, Vol. 112, No. 56.

USWDCM. (2010). *Urban Storm Water Design Criteria Manual*, Volumes 1 and 2, Urban Drainage and Flood Control District, Denver, CO.

Chapter 11

Culvert hydraulics

A culvert is designed to pass a stream flow under a barrier such as roadways. The hydraulic capacity of a culvert varies with respect to the headwater depth at the entrance and the tailwater depth at the exit. A culvert system operates like a small detention basin. When the inflow to the culvert is greater than the outflow through the culvert, the excess water will be temporarily stored in the entrance pool upstream of the culvert. As soon as the accumulated water depth at the entrance is adequate to pass the inflow, the stored water begins to be released. To be conservative, a culvert is sized to pass the peak discharge and then evaluated by a range of small to high flows. According to the headwater and tailwater depths, the culvert may carry open-channel flow, surcharge flow, or pressure flow. Design of a culvert involves the determination of design discharge, selection of barrels, evaluation of culvert performance, and overall considerations of construction, safety, maintenance, and esthetics.

11.1 Functions of culvert

The major function of a culvert is to maintain the continuity of stream flow. Figure 11.1 presents culverts serving for different purposes. A *crossing culvert* is installed underneath the roadway to maintain the continuity of a waterway. A *roadside culvert* is placed along the roadway to collect storm runoff. Culverts are also utilized for short-span bridges that can be prefabricated or constructed in field.

11.2 Culvert elements

As shown in Figure 11.2, a culvert system includes three basic elements: *entrance, barrel,* and *exit.* The entrance pool is shaped to collect inflows, and the exit pool is designed as a transition to tie into the downstream stream bed. A culvert system can be laid with multiple barrels that can be box, oval, arch, or circular in shape. The performance of a culvert depends on *orifice hydraulics* at the entrance and *conduit hydraulics* through the barrel, whichever is smaller. The orifice hydraulic capacity is dominated by the *headwater depth*, whereas the conduit conveyance capacity depends on the balance of energy principle subject to the specified *tailwater depth.*

11.2.1 Culvert entrance

At the entrance, the *headwater depth* is referred to as the vertical distance between the culvert invert (Figure 11.2) and the water surface. The *available depth* at the entrance is the vertical distance from the culvert invert to the top of the road. The entrance of a culvert shall form a pool for water to build up the required headwater depth. Design factors for

Figure 11.1 Crossing and roadside culverts. (a) Crossing culvert as bridge and (b) roadside culvert as storm drain.

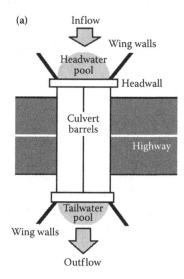

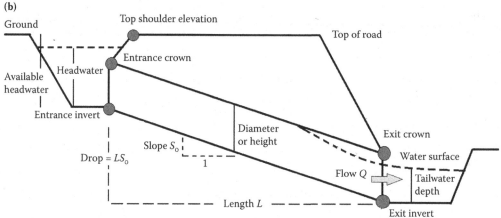

Figure 11.2 Elements of culvert. (a) Top view and (b) side view.

Figure 11.3 Culvert entrance.

an entrance pool include sediment deposit, trash rack for debris control, head walls for hydraulic efficiency, and cutoff wall to stabilize the structure from the groundwater seepage flow and uplift force. *Head walls* shall maintain a streamlined angle, 35°–75°, to guide the water into the barrel. The *cutoff wall* must be deep to avoid seepage water flowing through the foundation of the culvert. As shown in Figure 11.3, when the alignment of the culvert is laid on a steep slope, the hydraulic efficiency of a culvert can be increased with an *improved inlet*, which is a vertical drop created at the entrance. An improved inlet increases the headwater depth and decreases the culvert slope. The steep bottle neck at the entrance accelerates the water flow into the barrel. If the culvert is also designed to intercept the runoff from the roadway ditches, a *drop inlet* can be installed on the top of the barrel. If the barrel is larger than 45 mm (18 in.) in diameter, a *trash rack* must be installed in front of the entrance for the sake of safety and maintenance. It is critically important that the surface area of the rack is at least four times the cross-sectional area of the culvert entrance.

11.2.2 Culvert barrels

Culvert barrels are made up of different materials in different shapes, including circular, box, elliptical, and arch. Design of a culvert is often a question with multiple solutions. When a single large barrel is not suitable, a number of smaller barrels can be adopted under the assumption that each barrel evenly shares the design flow and works independently. Selections of culvert shape and material depend on the site constraints and design criteria. An arch or elliptical culvert shall be considered when encountering a narrow clearance, which is the vertical distance between the invert at the entrance and the top of the road. As shown in Figure 11.4, a *reinforced concrete pipe* (RCP) shall be considered when the soil coverage is inadequate. On a natural waterway, a *concrete box culvert* (CBC) is preferable because it introduces the least backwater effect to the waterway. *Corrugated metal pipes* (CMP) are rougher and require a higher headwater. A CMP is preferred for temporary uses. Concrete pipes are smooth and efficient for passing stream flows, but they also produce high exit velocities for potential scours. During construction, most detour culverts are built with used pipes for a short and temporary service. At the exit, a high tailwater demands flood gates to protect the culvert from the potential of reverse flows.

11.2.3 Culvert exit (outlet)

As illustrated in Figure 11.5, the configuration of a *culvert exit* shall be designed to provide an efficient transition to tie into the existing waterway downstream. The outlet work is an overtopping flow system, including erosion control using a concrete or riprap apron for energy dissipation and flare walls for hydraulic efficiency.

Figure 11.4 Types of culvert material. (a) Corrugated metal pipes used in rural area, (b) reinforced concrete pipe with flood gate, and (c) concrete box culvert with wing walls.

Figure 11.5 Erosion protection at culvert exit. (a) Riprap protection and (b) plunging pool.

11.3 General design considerations

11.3.1 Culvert alignment

As shown in Figure 11.6, the proper location to place the culvert under design depends on the stream alignment and the layout of the roadway. From the hydraulic standpoint, a culvert shall be lined with the existing stream alignment through a smooth wing-wall pool at the entrance. An abrupt change in the direction of a stream flow is discouraged. If a change in the flow direction cannot be avoided, it should be made at the culvert outlet rather than at the entrance.

During the construction of the permanent bridge, a detour culvert and diversion dam shall be used to bypass the stream flow from the construction site. A skewed alignment requires a longer culvert pipe, improvements at the entrance, and riprap protections at the exit. Scours may become the major potential problem in a skewed culvert.

11.3.2 Culvert slope and flow line

An ideal grade for a culvert is the one that produces neither sediment silting at the entrance nor scour at the exit, and the one that gives the shortest length and makes the replacement simplest. As shown in Figure 11.7, the preferable grade for the culvert is to

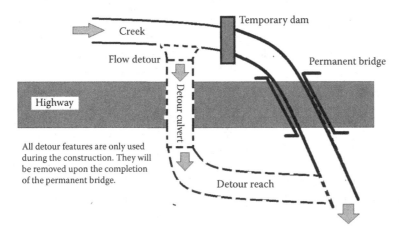

Figure 11.6 Detour culvert for construction of bridge.

Figure 11.7 Flow line and slope for culver alignment. (a) Laid on existing slope and (b) filled by sediment.

lay the barrels on the existing stream bed because the existing stream morphology represents the long-term equilibrium of sediment transport and stream bed scour. Lowering or elevating the culvert invert below or above the existing streambed will invite changes in sediment transport.

11.3.3 Culvert permissible velocities

At the entrance of a culvert, the water flow becomes concentrated and accelerated. When the water flow velocity is faster than 18.0 fps, it tends to be destructive because the flow may jump upstream of the culvert entrance (as shown in Figure 11.8). At the exit, an accelerated outflow increases the potential scour to the stream bed. On the other hand, when the water flow velocity is slower than 2.0 fps, it becomes incapable of self-cleaning. From an overall consideration, the culvert may be laid on an alignment slightly deviated from the existing stream bed and may be justifiable for a good purpose of maintaining the allowable flow velocity.

11.3.4 Available headwater

During a severe event, the accumulated water depth at the entrance pool may exceed the *available headwater depth* and produce an overtopping flow across the roadway. Of course, an inundated culvert (as shown in Figure 11.9) will be washed away as soon as a soil piping failure is developed. The induced potential damage to the adjacent buildings, roads, and bridges must be considered when setting the available headwater depth.

11.3.5 Allowable headwater

The headwater depth at the entrance of a culvert is determined by the energy grade line. Although a high headwater increases the hydraulic capacity, the excessive hydrostatic force and seepage uplift force may damage the barrels. Therefore, a culvert is designed to function under the *allowable headwater depth*, not the available headwater, determined by maximizing the flow loading as well as safety. Table 11.1 is recommended by the Colorado Department of Transportation on the allowable headwater (CDOT, 2015).

Figure 11.8 Fast flow through sharply curved alignment. (a) Sharp curved alignment at culvert entrance and (b) hydraulic jump of 5 m high.

Figure 11.9 Failure of culvert. (a) Overtopping flow, (b) piping failure, and (c) barrel washed away.

Table 11.1 Maximum ratios of headwater to culvert height

Culvert diameter or height	Max headwater/height
Less than 3 ft	1.5
From 3 to 5 ft	1.3
From 5 to 7 ft	1.2
Larger than 7 ft	1.0

11.3.6 Culvert barrels

The vertical profile of a culvert is often subject to inadequate clearance due to the conflicts among the underground utilities. Among all feasible alternatives, pipes with various shapes shall be compared, based on hydraulic capacity, durability, cost, maintenance, and service life. For instance, a CMP is more economic than a concrete pipe, but it has a much rougher surface or lesser flow capability. The final selection of the barrel shape depends on the construction constraints and weight loading at the project site. For instance, a narrow passage needs an elliptical barrel as a replacement of the circular pipe. Selection of the culvert size also depends on the construction convenience. For instance, at a remote mining site, one uniform size for all crossing culverts is preferred. Trash rack is always an important element for maintenance and safety. Information concerning the

Table 11.2 Recommended minimum culvert sizes

Type of culvert	Minimum diameter (in.)
Cross culvert	24
Side drain	18
Median drain	18
Storm sewer trunk line	18
Storm sewer connections	15
Irrigation crossing	18

type and the amount of debris are important for the engineer to design the debris-control structure. Owing to the potential for clogging, Table 11.2 presents recommendation on the minimum pipe sizes for various uses.

For maintenance, a minimum culvert size of 18-in. is applied to storm drains, while 24-in. culverts are applied to roadside drains with sediment-laden flows. Other factors in sizing a culvert also include, but are not limited to, public safety and upstream and downstream inundation. The length of culvert barrel used for highway crossing shall be in multiples of 2 ft, and the length of a CBC shall be to the nearest foot.

11.3.7 Headwall and wing walls

Without *a headwall*, a large metal projecting or sharply skewed mitered inlet will be easily collapsed under the hydrostatic forces of high headwater depths. The headwall should be placed perpendicular to the centerline along the culvert alignment. Concrete headwall is recommended on metal culverts of >30 in. in diameter. A *full headwall* and *wing-wall* systems (Figure 11.10) are recommended for culverts ≥96 in. in diameter. As illustrated with flow stream lines in Figure 11.11, a *square edge* is not as hydraulically efficient as a *beveled edge* because of the contraction of flow area at the entrance.

11.3.8 End treatment

Culverts shall be blended into the embankment for esthetics. Projecting culverts should not be used on interstate or urban highways. End treatments reduce potential erosion on the slope areas and improve the hydraulic efficiency. Concrete aprons or grouted riprap

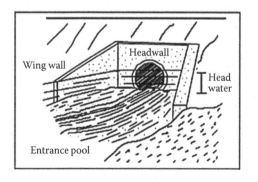
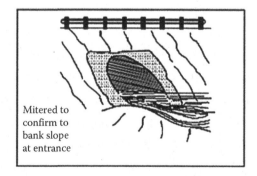

Figure 11.10 Head wall and wing walls.

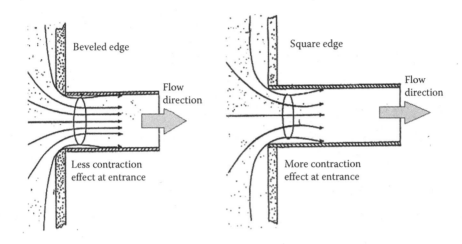

Figure 11.11 Beveled and square edge at entrance.

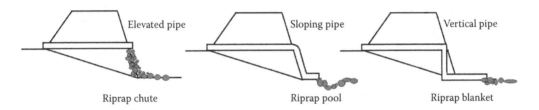

Figure 11.12 Treatments of culvert end.

should be applied to an *elevated outlet* for erosion control. Riprap rocks for a chute (Figure 11.12) may not be the best solution at an elevated end. An alternative may be a *sloping pipe*, which reduces the erosion potential on the slope, but the day-lighted concrete pipe is not blended with the environment well. The third option can be a *vertical manhole*, which provides a maintenance access and also serves the purposes of erosion control and esthetics.

11.4 Culvert sizing

Sizing a culvert depends on the *level of protection*. As a rule of thumb, a roadside culvert shall be designed to pass the 2- to 5-year event as part of the minor drainage system. A crossing culvert under a two-lane highway in rural areas shall be sized to pass the 25-year event. Culverts under a four-lane highway in rural areas must be able to pass the 50-year event. Freeways and Interstate Highways in urban areas must be equipped with culverts to pass the 100-year event.

A culvert is sized for the design event under specified constraints and safety criteria. The hydraulic resistance in a culvert depends on the surface materials and treatments. For instance, a CMP has a Manning's roughness coefficient of 0.025 versus 0.014 for RCPs. Without considering inlet and outlet effects, the normal flow capacity is assumed to approximate the flow condition in the culvert. As illustrated in Figure 11.13,

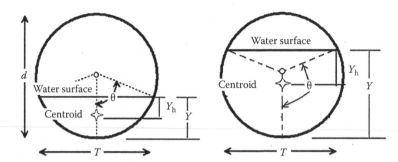

Figure 11.13 Hydraulic parameters in circular conduit.

the flow parameters of a *partially full flow* are directly related to *a half of the central angle*, θ, as

$$A = \frac{d^2}{4}(\theta - \sin\theta\cos\theta) \qquad (11.1)$$

$$P = d\theta \qquad (11.2)$$

$$R = \frac{A}{P} = \frac{d(\theta - \sin\theta\cos\theta)}{4\theta} \qquad (11.3)$$

$$Y = \frac{d}{2}(1 - \cos\theta) \qquad (11.4)$$

$$T = d\sin\theta \qquad (11.5)$$

$$Y_h = Y - \frac{d}{2} + \frac{2d(\sin\theta)^3}{3[2\theta - \sin(2\theta)]} \qquad (11.6)$$

$$Q = \frac{k}{N}P^{-\frac{2}{3}}A^{\frac{5}{3}}\sqrt{S_o} \qquad (11.7)$$

$$F_r = \sqrt{\frac{Q^2 T}{gA^3}} \qquad (11.8)$$

in which A = flow area in $[L^2]$, d = diameter of pipe in $[L]$, P = wetted perimeter in $[L]$, R = hydraulic radius in $[L]$, T = top width in $[L]$, θ = central angle in radians varied from zero to π shown in Figure 11.13, Y = flow depth in $[L]$, Q = design discharge in $[L^3/T]$, N = Manning's roughness, S_o = slope of culvert, F_r = flow Froude number, g = gravitational acceleration in $[L/T^2]$, Y_h = depth in $[L]$ to centroid of flow area, and k = 1.486 for foot-second units or 1.0 for meter-second units. Noted that when the pipe is full, the top width of the flow area is reduced to zero or the value of Froude number vanishes.

For a given design flow, Q, the hydraulically required pipe size is determined under *the full-flow condition*. Aided with Equations 11.1 and 11.2, the hydraulic radius under a full-flow condition is

$$R = \frac{d}{4} \quad \text{when } \theta = \pi \qquad (11.9)$$

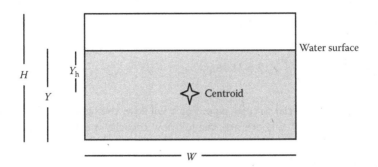

Figure 11.14 Hydraulic parameters in box conduit.

Substituting the full-flow condition into Equation 11.7 yields

$$d = \left(\frac{NQ}{K\sqrt{S_o}}\right)^{\frac{3}{8}}$$ (11.10)

in which d = hydraulically required circular diameter in [L] and $K = 0.462$ for feet-second units or 0.311 for meter-second units. In practice, the next larger commercial pipe shall be used for design. As a result, the design discharge is partially full in the commercial pipe. The corresponding angle, θ, in Figure 11.13, is calculated as

$$\frac{Q}{Q_f} = \frac{1}{\pi}\left(\frac{1}{\theta}\right)^{\frac{2}{3}}(\theta - \sin\theta\cos\theta)^{\frac{5}{3}}$$ (11.11)

in which Q = design flow in [L³/T] and Q_f = flowing full capacity in [L³/T] for the commercial pipe. Equation 11.11 is solved by trial and error. Having known the angle, θ, Equations 11.1 to 11.8 provide the normal flow condition in the commercial pipe.

The hydraulic performance of a box conduit (Figure 11.14) depends on its cross-sectional elements, including its height, H, and width, W.

The hydraulic parameters in a box conduit can be computed by

$$A = WY$$ (11.12)

$$P = W + 2Y$$ (11.13)

$$R = \frac{WY}{W + 2Y}$$ (11.14)

where W = width in [L] of box pipe and H = height in [L] of box pipe. The maximum capacity for a closed conduit is not at $Y = H$ or $Y = d$ but approximately at Y/H or $Y/d = 0.95$ because of the additional friction force from the top lid.

EXAMPLE 11.1

Size a circular concrete pipe to deliver a discharge of 40 cfs on a slope of 1.0% with a Manning's roughness coefficient of 0.015.

Solution:

1. Find the hydraulically required pipe size as

$$d = \left(\frac{0.015 \times 40.0}{0.462\sqrt{0.01}}\right)^{\frac{3}{8}} \times 12 = 31.36 \text{ in.}$$

2. Use a 36-in. commercial circular pipe. For a full flow, the hydraulic radius is 0.75 ft and the flow area is 7.07 ft². Therefore, the full-flow capacity is calculated as

$$Q_f = \frac{1.486}{0.015} \times 0.75^{\frac{-2}{3}} \times 0.707^{\frac{5}{3}} \times \sqrt{0.01} = 57.9 \text{ cfs for } d = 36 \text{ in.}$$

3. Determine the design flow condition in the 36-in. pipe as

$$\frac{Q}{Q_f} = \frac{40}{57.92} = 0.69 = \frac{1}{\pi}\left(\frac{1}{\theta}\right)^{\frac{2}{3}}(\theta - \sin\theta\cos\theta)^{\frac{5}{3}}$$

By trial and error, the central angle is found to be 1.79 rad or 102.8°. The partially full-flow condition for the design discharge in the 36-in. pipe is calculated as

$$Y = \frac{d}{2}(1 - \cos\theta) = \frac{3}{2}(1 - \cos 1.79) = 1.83 \text{ ft}$$

$$A = \frac{d^2}{4}(\theta - \sin\theta \cos\theta) = \frac{3^2}{4}(1.79 - \sin 1.79 \times \cos 1.79) = 4.52 \text{ ft}^2$$

$$V = \frac{Q}{A} = \frac{40.0}{4.52} = 8.85 \text{ fps}$$

$$T = d\sin\theta = 3 \times \sin 1.79 = 2.93 \text{ ft}$$

$$F_r = \sqrt{\frac{Q^2 T}{gA^3}} = \sqrt{\frac{40.0^2 \times 2.93}{32.2 \times 4.52^3}} = 1.26$$

11.5 Culvert hydraulics

Performance of a culvert depends on the culvert length, cross-sectional geometry, hydraulic resistance, and inlet and outlet conditions. There are four distinct sections (as shown in Figure 11.15) when flow passing through a culvert. They are as follows:

1. Section 1 at the upstream entrance pool (approaching section)
2. Section 2 at the culvert entrance (inlet section)
3. Section 3 at the culvert exit (outlet section)
4. Section 4 at the downstream channel (tailwater section)

Because the energy principle dictates the capacity of a culvert, the energy grade line (HGL) has to be established between sections 1 and the control section. A control section in a culvert is defined as the section whose location and flow depth are known. For

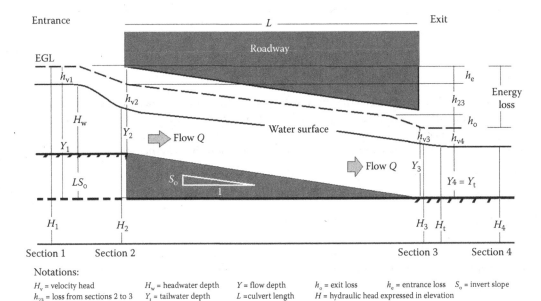

Figure 11.15 Culvert flow sections.

instance, when a culvert is placed on a steep slope with neither the entrance nor the exit being submerged, the control section will be section 2 (Figure 11.15) because the critical depth will be developed at the entrance. When the culvert is placed on a mild slope with a drop exit, the control section will be section 3 (Figure 11.15) because the critical depth will occur at the exit. If the culvert is laid under a submerged condition, the downstream tailwater depth at section 4 will serve as the control section for backwater analyses using the principle of energy.

At the entrance of a culvert, the headwater depth is measured relative to the entrance invert as

$$H_w = Y_1 - LS_o \qquad (11.15)$$

in which H_1 = elevation head in [L], Y_1 = water depth in [L] or hydraulic head at section 1, and L = culvert length in [L]. The ratio, H_R, of headwater depth to culvert height is defined as

$$H_R = \frac{H_w}{d} \qquad (11.16)$$

At section 1, the flow is diffused into the entrance pool; the Bernoulli's sums at sections 1–4 are

$$E_1 = H_1 + \frac{U_1^2}{2g} \approx H_1 \quad \text{because } U_1 \approx 0 \text{ for most of cases} \qquad (11.17)$$

$$E_2 = H_2 + \frac{U_2^2}{2g} \qquad (11.18)$$

$$E_3 = H_3 + \frac{U_3{}^2}{2g} \tag{11.19}$$

$$E_4 = H_4 + \frac{U_4{}^2}{2g} \tag{11.20}$$

in which E = Bernoulli's sum in [L], Y = flow depth in [L], U = flow velocity in [L/T], and H = hydraulic head in [L] including flow depth and elevation head. The subscript 1 represents the flow parameters at section 1, etc. As discussed earlier, the energy principle shall be applied between section 1 and the control section that can be one of sections 2–4. The Bernoulli's sum at the control section is

$$E_x = H_x + \frac{U_x{}^2}{2g} \tag{11.21}$$

in which the subscript of x represents the flow parameters at the control section. Applying the energy principle between section 1 and the control section yields

$$H_1 = E_x + h_f \tag{11.22}$$

in which h_f = energy losses between two sections. Substituting Equation 11.21 into Equation 11.22 yields

$$U_x = \sqrt{2g\left(H_1 - H_x - h_f\right)} \tag{11.23}$$

According to the principle of continuity equation, the discharge in the culvert is

$$Q = C_o A_x \sqrt{2g\left(H_1 - H_x - h_f\right)} \tag{11.24}$$

in which Q = culvert capacity in [L^3/T], g = gravitation acceleration in [L/T^2], A_x = flow area at the control section in [L^2], and C_o = discharge coefficient. Notice that Equation 11.22 only includes the friction losses. In order to compensate the contraction, bend, entrance, and exit losses, a discharge coefficient is introduced to Equation 11.24. The value of C_o varies between 0.60 and 0.70, depending on the culvert entrance geometry and the ratio of headwater to culvert height. A value of 0.65 is recommended for design. Equation 11.24 indicates that the capacity of a culvert depends on where the control section is located: at the entrance or at the exit. Therefore, the complicated culvert hydraulics is divided into two categories:

1. Inlet-control culvert
2. Outlet-control culvert

Inlet-control headwater depth is the headwater depth required by the orifice formula applied to the entrance. On the other hand, *outlet-control headwater depth* is determined by the energy balance between the entrance and the known tailwater depth at the exit. When the inlet-control headwater depth is greater than the outlet-control headwater depth, the culvert operates like an orifice or under inlet control; otherwise, it operates like a pipe flow under the backwater effect from the given tailwater depth or under outlet control. A culvert may behave like inlet control to pass one discharge and then switches to outlet control for passing another discharge. Therefore, for a given discharge, the culvert must be examined by both the orifice formula and the energy balance. The higher headwater depth dictates the culvert hydraulics.

11.6 Inlet-control culvert hydraulics

A culvert can be arranged such that it has a control section at the entrance. As a result, its capacity is determined by the entrance geometry and the headwater depth at the entrance. Such a culvert is termed *inlet-control culvert*. In other words, the length, barrel slope, and roughness do not affect the hydraulic performance if the culvert is under the condition of inlet control. The Sections 11.6.1 and 11.6.2 are two cases for inlet-control culverts:

11.6.1 Culvert on a steep slope with unsubmerged entrance

Often, the length of a highway crossing culvert is approximately 50–100 ft. They are considered a short culvert. When a short culvert is placed on a steep slope (i.e., $F_r > 1$) with an unsubmerged entrance, the capacity of the culvert is solely determined by the energy difference between section 1 in Figure 11.16 and the critical depth at section 2. With a sufficient tailwater depth downstream, this flow may have a hydraulic jump near the culvert exit. Setting the critical flow condition to be the control section and ignoring the minor friction loss, Equation 11.21 becomes

$$H_1 = H_c + \frac{U_c^2}{2g} \tag{11.25}$$

Substituting Equation 11.25 into Equation 11.24 yields

$$Q = A_c\sqrt{2g(H_1 - H_c)} = A_c\sqrt{2g(H_w - Y_c)} = A_c\sqrt{2g(Y_1 - LS_o - Y_c)} \tag{11.26}$$

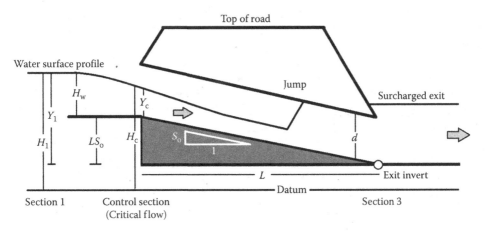

Figure 11.16 Critical depth at culvert entrance.

in which H_1 = hydraulic head in [L] at section 1, H_c = hydraulic head in [L] for critical flow at the entrance, H_w = headwater depth in [L], Y_1 = water depth in [L] relative to the exit invert, L = length of culvert in [L], S_o = slope of culvert in [L/L], and A_c = crossing area for critical flow at entrance section in [L²]. Equation 11.26 requires the critical flow area, A_c in [L²], and depth, Y_c in [L]. The critical flow condition is defined by the flow Froude number equal to unity as

$$\frac{Q^2 T_c^2}{g A_c^3} = 1 \tag{11.27}$$

The subscript, c, is referenced to the variables associated with the critical depth. Equation 11.27 is applicable to all shapes of conduit. The only unknown in Equation 11.27 is the critical flow depth, which is the function of the central angle, as illustrated in Figure 11.13.

11.6.2 Culvert with high headwater and unsubmerged exit

When the ratio of headwater depth to culvert height is between 1.0 and 1.5, the flow passing through the entrance becomes rapid. Under such a submerged entrance and an unsubmerged exit (as depicted in Figure 11.17), the culvert capacity is controlled by section 2 that acts like an orifice. The capacity of the culvert is estimated by the orifice formula with a discharge coefficient, C_o, of 0.6–0.7

$$\begin{aligned}
Q &= C_o A \sqrt{2g(H_1 - H_o)} \\
&= C_o A \sqrt{2g\left(H_w - \frac{d}{2}\right)} \\
&= C_o A \sqrt{2g\left(Y_1 - LS_o - \frac{d}{2}\right)}
\end{aligned} \tag{11.28}$$

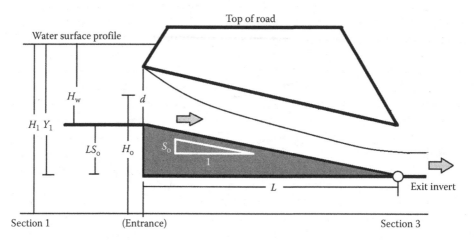

Figure 11.17 Culvert with submerged entrance and clear exit.

in which H_o = elevation in [L] at the center of the entrance area, d = diameter or height in [L] of the entrance area, and A = crossing area at entrance section in [L^2].

11.7 Outlet-control culvert hydraulics

The capacity of an outlet-control culvert depends on outlet geometry, tailwater depth, barrel roughness, flow line slope, and culvert length. Cases of outlet-control culverts are discussed as follows.

11.7.1 Culvert on mild slope with a drop exit

With a drop exit, water flowing through a culvert on a mild slope produces a draw-down water surface profile that is ended with the critical depth at the exit. As shown in Figure 11.18, the layout creates a critical depth at the exit. The energy balance in Equation 11.22 between sections 1 and 3 is written as

$$H_1 = H_c + \frac{U_c^2}{2g} + h_e + h_{23}$$ (11.29)

$$\begin{aligned}
Q &= A_c \sqrt{2g(H_1 - H_c - h_e - h_{23})} \\
&= A_c \sqrt{2g(Y_1 - Y_c - h_e - h_{23})} \\
&= A_c \sqrt{2g(H_w + LS_o - Y_c - h_e - h_{23})}
\end{aligned}$$ (11.30)

The value of h_{23} has to be determined by the backwater profile computations through the length of the culvert, and the critical flow depth and area are solved with Equation 11.27.

11.7.2 Culvert with both entrance and exit submerged

When both exit and entrance of a culvert are submerged, the culvert is pressurized. As illustrated in Figure 11.19, the full-flow capacity of the culvert is dictated by the energy grade line across the entire culvert.

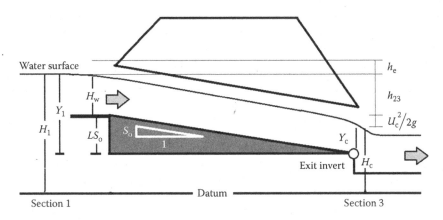

Figure 11.18 Culvert on mild slope with drop exit.

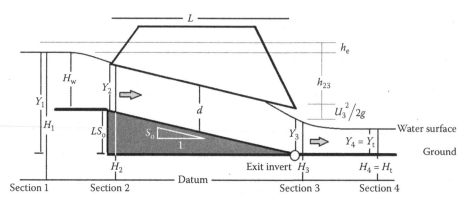

Figure 11.19 Culvert with submerged inlet and outlet.

Applying the energy principle between sections 1 and 3 in Figure 11.19 yields

$$Y_1 = H_w + LS_o = Y_3 + \frac{U_3^2}{2g} + h_e + h_{23} \text{ (use the exit invert as the datum)} \tag{11.31}$$

The entrance loss can be related to the flow velocity head as

$$h_e = K_e \frac{U_2^2}{2g} \tag{11.32}$$

The friction loss through the barrel is calculated by Manning's formula as

$$h_{23} = S_f L = \left[\frac{N^2 U_f^2}{k^2 R^{\frac{4}{3}}} \right] L = \left[\frac{2g}{k^2} \frac{N^2}{R^{\frac{4}{3}}} L \right] \frac{U_f^2}{2g} = K_N \frac{U_f^2}{2g} \tag{11.33}$$

in which h_{23} = friction loss in [L] between sections 2 and 3, K_N = roughness coefficient, and U_f = full-flow velocity in [L/T] in which K_e = contraction loss coefficient at the entrance such as 0.3. Because the culvert is flowing full, we have

$$U_2 = U_3 = U_f \tag{11.34}$$

Aided with Equation 11.34, substituting Equations 11.32 and 11.33 into Equation 11.31 yields

$$Q = A \left[\frac{2g(Y_1 - Y_3)}{1 + K_e + K_N} \right]^{0.5}$$

$$= A \sqrt{\frac{1}{1 + K_e + K_N}} \sqrt{2g(H_w + LS_o - Y_3)} \tag{11.35}$$

As shown in Figure 11.19, the flow condition, H_3, has to be determined by the tailwater depth, Y_t, at section 4, where the culvert is tied into downstream channel. If the downstream channel information is insufficient, it is recommended that the flow depth at section 3 be approximated by the average of the critical depth and the height of the culvert. In practice, the tailwater depth is determined as

$$Y_3 = \max\left[Y_t, 0.5(Y_c + d)\right] \tag{11.36}$$

in which Y_t = known tailwater depth in [L], Y_c = critical depth in [L], and d = height in [L] of the culvert. Having known Y_3, the discharge is further calculated by Equation 11.35.

11.8 Culvert design

For a given discharge, the design headwater depth is examined by both inlet- and outlet-control hydraulics; the higher of these will dictate the flow. Equation 11.28 is used for *the inlet-control condition*, and Equation 11.35 is recommended for *the outlet-control condition*. Often, the tailwater information is not known; as a result, Equation 11.36 provides an acceptable estimate of tailwater depth.

EXAMPLE 11.2

The 36-in. circular concrete culvert shown in Figure 11.20 is laid on a slope of 2% and designed to carry a flow of 50 cfs. The tailwater is not given. Determine the headwater depth.

1. **Headwater depth under inlet control**
   With $C_o = 0.6$ and $d = 3\,\text{ft}$, the inlet-control headwater depth is calculated using Equation 11.28 as

$$Q = C_o A \sqrt{2g\left(H_w - \frac{d}{2}\right)} = 0.60 \times \frac{3.1416 \times 3^2}{4} \sqrt{2 \times 32.2 \times \left(H_w - \frac{3}{2}\right)} = 50$$

   So, $H_w = 3.66\,\text{ft}$

2. **Headwater depth under outlet control**
   The given design information includes $N = 0.015$, $L = 162\,\text{ft}$, $K_e = 0.2$, and $S_o = 0.02$. The tailwater depth is not specified. Therefore, the critical flow depth is needed. Referring

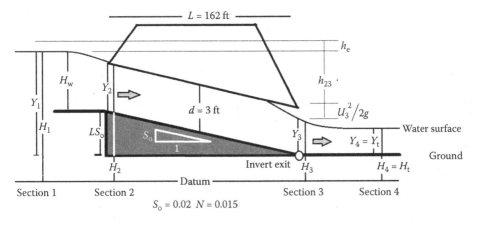

Figure 11.20 Determination of headwater depth.

Table 11.3 Critical flow in circular pipe

Flow variables	Critical flow condition
Central angle	$\theta = 2.13$ rad
Flow depth	$Y = 2.30$ ft
Top width	$T = 2.54$ ft
Wetted perimeter	$P = 6.40$ ft
Flow area	$A = 5.82$ ft^2
Flow variable	$V = 8.59$ fps
Check on Froude number	$F_r = 1.00$

to Figure 11.13, try the central angle, $\theta = 2.13$ rad; the critical flow condition is tabulated in Table 11.3.

Set the datum to be the exit invert at section 3. The hydraulic head at section 3 is equal to its flow depth as

$$Y_3 = \max\left[Y_t, 0.5(Y_c + d)\right] = \left[0, 0.5(2.3 + 3.0)\right] = 2.65\,\text{cfs}$$

Next, the hydraulic radius for the flowing full condition through the culvert is

$$R = \frac{A}{P} = \frac{d}{4} = \frac{3}{4} = 0.75$$

The entrance loss and friction loss coefficients are

$$K_e = 0.02$$

$$K_N = \frac{2 \times 32.2}{1.486^2} \frac{0.015^2 \times 162}{0.75^{4/3}} = 1.55$$

The headwater depth at section 1 is determined as

$$Q = \frac{3.1416 \times 3^2}{4}\left[\frac{2 \times 32.2 \times (Y_1 - 2.65)}{1 + 0.02 + 1.55}\right]^{0.5} = 50$$

Solution: $Y_1 = 4.73$ ft relative to the exit invert at section 3. So, the headwater depth is

$$H_w = Y_1 - LS_o = 4.73 - 162 \times 0.02 = 1.49\,\text{ft under the outlet-control condition.}$$

For this case, the design headwater depth is the higher between the inlet and outlet-control conditions as

$$H_w = \max(3.66, 1.49) = 3.66\,\text{ft. This is a case of inlet control.}$$

Check the headwater-to-diameter ratio: $H_w/d = 1.22 < 1.30$. The 36-in. pipe is acceptable for this case.

EXAMPLE 11.3

Repeat Example 11.2 using a corrugated pipe with $N = 0.027$. Determine the headwater depth. A higher Manning's N will increase the friction coefficient as

$$K_N = \frac{2 \times 32.2}{1.486^2} \frac{0.025^2 \times 162}{0.75^{4/3}} = 5.03$$

$$Q = \frac{3.1416 \times 3^2}{4} \left[\frac{2 \times 32.2 \times (Y_1 - 2.65)}{1 + 0.02 + 5.03} \right]^{0.5} = 50, \text{ So } Y_1 = 7.43 \text{ ft}$$

$$H_w = Y_1 - LS_o = 7.43 - 162 \times 0.02 = 4.19 \text{ ft}$$

$$H_w = \max(3.66, 4.19) = 4.19 \text{ ft. This is a case of outlet control}$$

Check the headwater-to-diameter ratio: $H_w/d = 1.40 > 1.3$. The 36-in. pipe is too small for this case. Next, try a 42-in. pipe.

The earlier examples reveal the fact that the higher the roughness coefficient in the pipe, the more the culvert to become an outlet-control culvert. Similarly, the steeper the pipe, the more the culvert to become an inlet-control culvert.

11.9 Stilling basin at culvert outlet

At the exit, the concentrated flow released from the culvert has to be spread out over an energy dissipater. As shown in Figure 11.21, three major categories of energy dissipaters

Figure 11.21 Dissipater at culvert exit. (a) Grouted riprap, (b) drop steps, and (c) stilling basin.

are commonly applied to culvert exits, including (1) *rough surfaces* to create tumbling flows such as grouted riprap and blocks, (2) *continuous drops* to create impingement flows such as a series of steps, and (3) *plunging pools* to create diffusive flows such as a stilling basin. Selection of a dissipater depends on the constraints at the site, pollutants in storm runoff, and the flow regime: subcritical or supercritical flow.

Although a grouted riprap apron is considered as durable as a concrete, its effectiveness of energy dissipation is not comparable to stilling basin. A stilling basin is laid out to trigger a hydraulic jump under the design condition. As illustrated in Figure 11.22, the basin is expanded in width, starting from the same width as the culvert exit, and then expended using a length-to-width ratio of 2–3. Concrete floor and walls are preferred for convenience of maintenance.

Hydraulic jump involves a tremendous amount of energy dissipation through rolls and eddies in the turbulent flow. However, a jump only occurs if the required tailwater depth exists in the pool. Therefore, placing a weir at the end of the stilling basin is a common practice to raise the tailwater depth. Referring to Figure 11.22, the hydraulic jump is analyzed by the balance of specific forces between sections 1 and 2 as

$$F_1 = F_2 \tag{11.37}$$

where F = specific force in pound or newton. The subscript of "1" means the flow variables at section 1, etc. The specific force associated with a flow consists of flow static and dynamic forces that are directly related to the flow depth as

$$F = \rho Q U + \gamma Y_h A \tag{11.38}$$

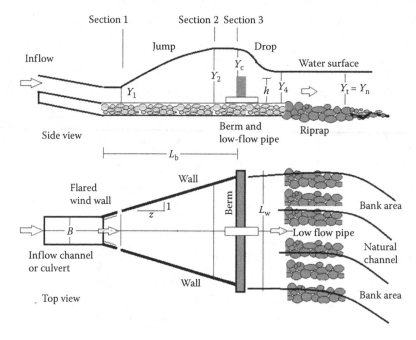

Figure 11.22 Stilling basin at culvert outlet.

where ρ = density of water such as $1.94\,\text{slug/ft}^3$ or $1000\,\text{kg/m}^3$, γ = specific weight of water such as $62.4\,\text{lb/ft}^3$ or $9800\,\text{N/m}^3$, Y_h = depth in [L] to the centroid of the flow area, and A = flow area in $[\text{L}^2]$.

The specific energy associated with a flow consists of hydraulic head and dynamic head as

$$E = Y + \frac{U^2}{2g} \text{ relative to the basin floor} \qquad (11.39)$$

At the exit of the culvert, the hydraulic force is associated with the flow momentum. To induce a hydraulic jump, the required conjugate depth, Y_2, is determined by Equation 11.37. To pass the design flow, the weir section is designed to carry the critical depth for the overtopping flow. The principle of energy between section 2 and weir section is written as

$$E_2 = E_c + h \qquad (11.40)$$

where E_2 = specific energy in [L] at section 2, E_c = specific energy in [L] for critical flow at weir section, and h = height of weir in [L].

The length of jump, L_b in [L], is approximated as

$$L_b = 10Y_1\left(Fr_1 - 1\right) \qquad (11.41)$$

where L_b = length of jump set to be the minimum length for the pool in [L], and Fr_1 = flow Froude number at section 1.

EXAMPLE 11.4

A culvert system is designed to pass the 100-year peak flow of 500 cfs into a stilling basin. At the windwall section, the flow condition is described by a rectangular section with $B = 10\,\text{ft}$, $N = 0.014$, and $S_o = 0.025$. Determine the height of the weir and the minimal length of the basin.

Solution: Firstly, we need to find the incoming flow condition. Trying the normal depth, $Y = 2.23\,\text{ft}$, the incoming flow condition is calculated as

$$A = WY = 10 \times 2.23 = 22.3\,\text{ft}^2$$

$$P = W + 2Y = 10 + 2 \times 2.23 = 14.5\,\text{ft}$$

$$R = \frac{WY}{W + 2Y} = \frac{22.3}{14.5} = 1.54\,\text{ft}$$

$$Q = \frac{k}{N} P^{-\frac{2}{3}} A^{\frac{5}{3}} \sqrt{S_o} = \frac{1.486}{0.014} \times 14.5^{-0.67} \times 22.3^{1.67} \times \sqrt{0.025} = 500\,\text{cfs}$$

The flow depth, $Y = 2.23\,\text{ft}$, is accepted as the normal depth for the design discharge of 500 cfs.

$$U = \frac{Q}{A} = \frac{500}{22.3} = 22.43\,\text{fps}$$

$$F_r = \sqrt{\frac{Q^2 T}{gA^3}} = \sqrt{\frac{500^2 \times 10}{32.2 \times 22.3^3}} = 2.65$$

The incoming flow is supercritical and expected to have a hydraulic jump in the pool. The depth to the centroid of the flow area is $Y_h = 0.5Y$ for the rectangular channel. The specific force is calculated as

$$F = \rho QU + \gamma Y_h A = 1.94 \times 500 \times 22.43 + 62.4 \times (2.23 / 2) \times 22.3 = 23,310 \text{ lb of force.}$$

Table 11.4 presents a summary of the normal flow condition. The incoming flow is supercritical with a specific force of 23,310 lb.

The width of this stilling basin is expanded to $L_W = 25$ ft at the weir section. The conjugate depth, Y_2, is the one that produces the same specific force as

$$F = \rho QU + \gamma Y_h A = 1.94 \times 500 \times \frac{500}{25 \times Y_2} + 62.4 \times (Y_2 / 2) \times (25 \times Y_2) = 23,310$$

So, $Y_2 = 4.99$ ft

The critical depth is determined by the flow Froude number equal to unity as

$$F_r = \sqrt{\frac{Q^2 T}{gA^3}} = \sqrt{\frac{500^2 \times 25}{32.2 \times (25 \times Y_c)^3}} = 1.0$$

So, $Y_c = 2.32$ ft, which is the overtopping depth on the weir. Table 11.5 summarizes the conjugate and critical flow conditions as

According to Equation 11.40, the height of the weir is determined as

$$h = E_2 - E_c = 5.24 - 3.47 = 1.77 \text{ ft}$$

Aided with Equation 11.41, the minimal length of the basin is

$$L_b = 10 Y_1 (F_{r1} - 1) = 10 \times 2.23 \times (2.65 - 1) = 38.5 \text{ ft.}$$

Table 11.4 Incoming flow to stilling basin

Flow depth, Y (ft)	Flow area, A (ft^2)	Wetted P-meter, P (ft)	Hydraulic radius, R (ft)	Flow velocity, U (fps)	Flow rate, Q (cfs)	Froude number, F_r	Specific energy, E (ft)	Depth to centroid, Y_h (ft)	Specific force, F (klb)
2.23	22.3	14.5	1.54	22.43	500.0	2.65	10.02	1.11	23.31

Table 11.5 Conjugate and critical depth in stilling basin

Section	Flow depth, Y (ft)	Basin width, T (ft)	Flow area, A (ft^2)	Flow velocity, U (fps)	Kinetic energy, $U^2/2g$ (ft)	Specific energy, E (ft)	Froude number, F_r	Depth to centroid, Y_h (ft)	Specific force, F (klb)
1	2.23	10.00	22.3	22.43	7.81	10.04	2.65	1.11	23.31
2	4.99	25.00	124.8	4.01	0.25	5.24	0.32	2.50	23.31
3 (weir)	2.32	25.00	57.9	8.63	1.16	3.47	1.00	1.16	12.56

This system will be built for the 100-year peak discharge of 500 cfs. During a 10-year event, the overtopping flow on the as-built weir will provide the required tailwater depth equal to the conjugate depth, Y_2, to trigger the hydraulic jump. The stilling basin and the end weir are sized to pass extreme events. Add two circular drains of 6 in. in diameter at the base of the end weir to pass frequent, small flows.

EXAMPLE 11.5

A broken-back culvert system (Figure 11.23) comprises two pipes connected at the break point on the invert slope. The upstream pipe is 4 ft in diameter and 100 ft long on a slope of 0.016. The downstream pipe is 6 ft in diameter for a length of 200 ft on a slope of 0.001. The design discharge for this culvert system is 95.7 cfs. It is expected that the critical flow will occur at the entrance. Considering both pipes will have their normal depths at the break point on the invert. Determine the headwater depth at the entrance and verify that the hydraulic jump occur at the break point.

Solution: For this case, the culvert flow is controlled by the critical flow at the entrance. Referring to Figure 11.13, trying $\theta = 2.07$ rad or $118.85°$, the critical flow condition is solved as

$$T_c = 4 \times \sin 2.07 = 3.50 \, \text{ft}$$

$$A_c = \frac{4^2}{4}(2.07 - \sin 2.07 \times \cos 2.07) = 9.99 \, \text{ft}^2$$

$$F_r = \sqrt{\frac{95.7^2 \times 3.50}{32.2 \times 9.99^3}} = 1.0. \text{ So, the central angle of 2.07 rad is accepted.}$$

The critical flow condition is summarized in Table 11.6.

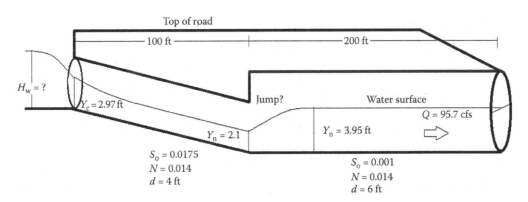

Figure 11.23 Broken-back culvert.

Table 11.6 Critical flow at entrance

Central angle (rad)	Depth, Y_c (ft)	Area, A_c (ft²)	W parameter, P_c (ft)	H radius, R_c (ft)	Top width, T_c (ft)	Centroid, Y_h (ft)	Velocity, U_c (fps)	Flow, Q_s (fps)	Check Froude number
2.07	2.97	9.99	8.30	1.20	3.50	1.32	9.58	95.70	1.00

Aided with Equation 11.26, the headwater at the entrance is determined as

$$Q = A_c \sqrt{2g(H_W - Y_c)} = 9.99 \times \sqrt{2 \times 32.2 \times (H_W - 2.97)} = 95.7 \text{ cfs}$$

The headwater depth at the entrance is $H_w = 4.39$ ft. Next, let us calculate the normal depth in the 4-ft pipe using Equations 11.1 through 11.8:
Try $\theta = 1.62$ rad $= 92.84°$.

$$A = \frac{4^2}{4}(1.62 - \sin 1.62 \times \cos 1.62) = 6.68 \text{ ft}^2$$

$$P = d\theta = 4 \times 1.62 = 6.48 \text{ ft}$$

$$R = \frac{4(1.62 - \sin 1.62 \times \cos 1.62)}{4 \times 1.62} = 1.03 \text{ ft}$$

$$Y = \frac{4}{2}(1 - \cos 1.62) = 2.10 \text{ ft}$$

$$T = 4 \times \sin 1.62 = 4 \text{ ft}$$

$$Y_h = 2.1 - \frac{4}{2} + \frac{2 \times 4 \times (\sin 1.62)^3}{3[2 \times 1.62 - \sin(2 \times 1.62)]} = 0.89 \text{ ft}$$

$$Q = \frac{1.486}{0.014} 6.48^{-\frac{2}{3}} 6.68^{\frac{5}{3}} \sqrt{0.0175} = 96.0 \text{ cfs close to the design flow.}$$

So, the central angle is accepted.

$$U = \frac{Q}{A} = \frac{95.7}{6.68} = 14.33 \text{ fps}$$

$$F_r = \sqrt{\frac{95.7^2 \times 4}{32.2 \times 6.68^3}} = 1.95 > 1.0, \text{ so the inflow is supercritical.}$$

$$F = \rho QU + \gamma Y_h A = 1.94 \times 95.7 \times 14.33 + 62.4 \times 0.89 \times 6.68 = 3030 \text{ lb.}$$

Repeat the same process for the normal flow in the 6-ft pipe. The flow conditions in these two pipes are summarized in Table 11.7.

As shown in Table 11.7, at the break point, the upstream flow carries a specific force of 3030 lb, whereas the downstream normal flow matches with the same force. Therefore, the hydraulic jump occurs at the break point on the invert.

Table 11.7 Hydraulic jump analysis in broken culvert

Pipe diameter (ft)	Central angle (rad)	Depth, Y (ft)	Area, A (ft²)	Top width, T (ft)	Centroid, Y_h (ft)	Velocity, U (fps)	Froude, F_r	Specific energy, E (ft)	Specific force, F (lb × 10³)
4	1.62	2.10	6.68	4.00	0.89	14.33	1.95	5.29	3.03
6	1.89	3.95	19.73	5.69	1.73	4.85	0.46	4.31	3.03

11.10 Detour culvert

According to the cost data published by the *Colorado Department of Transportation* (CDOT, 1992), approximately one dollar out of four was spent for drainage structures in highway constructions. This is also true at a national level. During the construction of a permanent highway drainage structure, it often requires a temporary culvert to maintain the continuity of traffic flow and runoff flow. In general, a detour drain serves only through the construction period, which is often less than a year in most cases. Design of such an interim drainage structure must begin with the selection of the design flood. The temporary nature of the detour drainage structure makes this task difficult. Because of its short service, the cost of an interim drainage facility should be kept as economical as possible. However, the failure of an undersized interim drainage structure may cause as much damage as losing the permanent structure. Based on the fact that the cost of a drainage structure increases with respect to its capacity, the cost of installing a drainage culvert, C, can be related to its capacity, Q, as

$$C = F(Q) \tag{11.42}$$

in which $F(Q)$ = functional relationship between cost and capacity of highway drainage structures. During an interim phase, a temporary culvert will only serve for several months. The cost-capacity ratio for culverts can be approximated by a linear relationship as

$$\frac{C_d}{C_P} = a\frac{q}{Q} \tag{11.43}$$

in which C_d = cost of detour culvert, C_p = cost of permanent drainage structure, q = detour culvert capacity, Q = capacity of permanent drainage structure, and a = the cost-capacity coefficient. Analysis of highway cost data revealed that the cost-capacity coefficient varies between 0.35 and 0.57 (Guo 1987, 1998). Of course, the cost ratio in Equation 11.43 can also be described by other nonlinear forms as long as they fit local cost data.

One of the primary drawbacks to many risk-cost analysis procedures is the degree of difficulty in capitalizing the monetary damage resulting from a failure. The seriousness of any traffic delay is proportional to the highway site, traffic volume, availability of alternate routes, and the overall importance of the route. As far as the losses due to the discontinuity of traffic are concerned, we may conservatively consider that the failure of a detour drainage structure may result in the same damage as that incurred in the failure of the permanent structure. The chance of failure of a detour culvert can be assessed by the joint probability that includes the following:

1. The exceedance probability, P_T, of having a flood exceeding the capacity of the detour culvert.
2. The occurrence probability, P_m, of such a flood to occur during the service period of the detour drain.

Assuming that the detour culvert will fail when a flood exceeds the design capacity and the two events mentioned previously are independent, the expected damage associated with the failure of a detour culvert can be written as

$$C_r = P_T P_m D_P \tag{11.44}$$

in which C_r = expected damage due to the failure of the detour culvert and D_p = losses due to the discontinuity of traffic. When the detour culvert is designed to pass a flood with a frequency of T-year, the exceedance probability, P_T, is

$$P_T = \frac{1}{T} \tag{11.45}$$

The occurrence probability, P_m, of having such a flood during the construction months can be approximated by either the monthly rainfall distribution or the monthly runoff distribution normalized by its annual amount. This normalization process produces an approximation of monthly occurrence probability curve that has a total area of unity. For instance, when the detour culvert is to serve from the ith month to jth month in a year, the occurrence probability, P_m, can be estimated as

$$P_m = \sum_{k=i}^{k=j} \frac{\Delta P_k}{P_o} \quad \text{for } i = 1 \text{ to } 12, \ j = i + m, \text{ and } m < 12 \tag{11.46}$$

in which ΔP_k = kth monthly average rainfall or runoff amount, i = beginning month of construction, j = end month of construction, m = construction span in months, and P_o = annual rainfall or runoff amount. By definition, the total risk cost of a detour culvert can then be written as

$$C_T = C_d + C_r \tag{11.47}$$

in which C_T = total risk cost for a detour culvert. Substituting Equations 11.43 through 11.45 into Equation 11.47 yields

$$C_T = a \frac{q}{Q} C_P + P_T P_m D_P = a \frac{q}{Q} C_P + \frac{P_m D_P}{T} \tag{11.48}$$

As illustrated in Figure 11.24, the least total cost in Equation 11.48 can be achieved in terms of the selection of return period, T, which is the design flood frequency for the detour culvert.

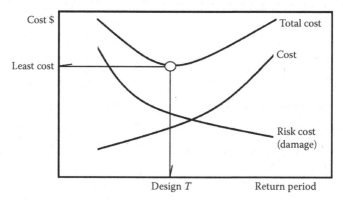

Figure 11.24 Minimization of total cost.

Mathematically, the first derivative of Equation 11.48 for the least cost must be equal to zero. Therefore, we have

$$\frac{dq}{dT} = \frac{QP_m}{aT^2} \frac{D_P}{C_P} \tag{11.49}$$

Equation 11.49 implies that the solution can be achieved by identifying the optimal slope on the flood flow frequency curve. According to the flood frequency analysis, the variable, q, with a return period, T, can be statistically related to its mean and standard deviation as

$$q = Q_m + K_T S \tag{11.50}$$

in which Q_m = mean flood flow in $[L^3/T]$, S = standard deviation of the flood variable in $[L^3/T]$, and K_T = frequency factor of the flood variable. Taking the first derivative of Equation 11.50 with respect to the variable, T, yields

$$\frac{dq}{dT} = S \frac{dK_T}{dT} \tag{11.51}$$

Substituting Equation 11.51 into Equation 11.49 yields

$$\frac{dK_T}{dT} = \frac{P_m Q}{aST^2} \frac{D_P}{C_P} \tag{11.52}$$

Equation 11.52 applies to any probability distribution as long as it fits the runoff data. Runoff data used in the hydrologic peak flow frequency analysis can be structured as either annual maximum series or annual exceedance series. Application of Equation 11.52 to these two types of data series is discussed as follows:

1. **Annual maximum series**
   The Gumbel distribution may be considered. Its frequency factor is defined as (Chow et al. 1988)

$$K_T = \frac{\sqrt{6}}{\pi}\left[0.5772 + \ln\left(\frac{T}{T-1}\right)\right] \tag{11.53}$$

   in which $\pi = 3.1416$ and $\ln$ = the natural logarithmic function.
   Taking the first derivative of Equation 11.53 with respect to T yields

$$\frac{dK_T}{dT} = \frac{\sqrt{6}}{\pi}\left[\frac{1}{\ln\left(\frac{T}{T-1}\right)(T-1)T}\right] \tag{11.54}$$

Substituting Equation 11.54 into Equation 11.52 yields

$$P_m = B\left[\frac{\dfrac{T}{T-1}}{\ln\left(\dfrac{T}{T-1}\right)}\right]\frac{C_P}{D_P} \tag{11.55}$$

$$B = \frac{a\sqrt{6}}{\pi}\left(\frac{S}{Q}\right) \tag{11.56}$$

2. **Annual exceedance series**
    The exponential distribution is considered. Its frequency factor is defined as

$$K_T = \frac{\sqrt{6}}{\pi}\left(\ln T - 0.5772\right) \tag{11.57}$$

Taking the first derivative of Equation 11.57 with respect to T yields

$$\frac{dK_T}{dT} = \frac{\sqrt{6}}{\pi}\frac{1}{T} \tag{11.58}$$

Substituting Equation 11.58 into Equation 11.52 yields a linear relation

$$P_m = BT\frac{C_P}{D_P} \tag{11.59}$$

The cost-to-damage ratio in Equations 11.55 and 11.59 can be determined by economic and social considerations. For instance, $C_p/D_p = 1$ represents the economic break-even point. On the other hand, if the failure of the detour culvert is not tolerable for the site, a conservative design can be achieved by using a smaller ratio of C_p/D_p. Figure 11.25 is produced using Equation 11.55 for obtaining the design flood frequency, T, when B, P_m, and C_p/D_p are specified.

EXAMPLE 11.6

The existing bridge, Number N-10-C, is located at State Highway 160 and the South Fork River near Creed, CO. This bridge is to be replaced with concrete culverts. It will take 3 months to complete the construction. As a four-lane highway in a rural area, the new bridge is designed to pass the 50-year flood flow. The mean and standard deviation of the annual maximum peak flow database are found to be 1516.6 and 754.4 cfs, respectively. The magnitude of a 50-year flood is then determined to be 3472 cfs using the Gumbel distribution. Suggest the sizes of detour culvert during the 3-month construction periods.

Solution: The *USGS Water Resources Data* for Colorado provides the monthly runoff records from 1961 through 1997 near the project site. The monthly average runoff rates are listed in Table 11.8 as follows:

The sum of these monthly runoff rates, P_o, is 5656 cfs, which is then used to normalize the monthly runoff rates to approximate the flood occurrence probability distribution. Applying

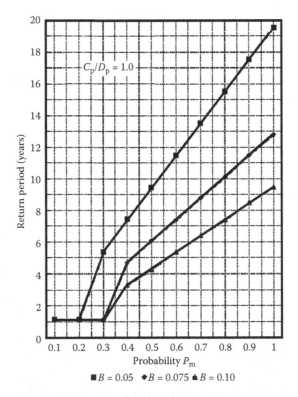

Figure 11.25 Detour culvert design using Gumbel distribution.

Table 11.8 Flood occurrence probability by monthly runoff distribution

Month	Runoff rate (cfs)	P_k/P_o
January	56	0.010
February	62	0.011
March	453	0.027
April	574	0.101
May	1550	0.274
June	1470	0.260
July	770	0.136
August	375	0.066
September	284	0.050
October	198	0.035
November	89	0.016
December	75	0.013

Table 11.8 to the construction period from April through June as an example, the value of P_m in Equation 11.55 is

$$P_m = 0.101 + 0.274 + 0.260 = 0.635$$

It implies a chance of 63.5% to have a flood flow exceeding the culvert capacity within the selected 3 months in a year. Substituting a value of 0.5 for the cost-capacity ratio into

Table 11.9 Selection of design flows for various construction periods

Start	End month											
	Jan.	Feb.	Mar.	Apr.	May	June	July	Aug.	Sept.	Oct.	Nov.	Dec.
Jan.	1.1	1.1	1.1									
Feb.		1.1	1.1									
Mar.			4.5	4.5	4.5							
Apr.				6.6	6.6	6.6						
May					7.0	7.0	7.0					
June						4.6	4.6	4.6				
July							1.9	1.9	1.9			
Aug.								1.1	1.1	1.1		
Sept.									1.1	1.1	1.1	
Oct.										1.1	1.1	1.1
Nov.	1.1										1.1	1.1
Dec.	1.1	1.1										1.1

Note: The value of 1.1 represents that the recommended return period ≤1 year.

Equation 11.56, we produce $B = 0.0847$. Substituting the values of B and P_m into Equation 11.55 results in a recommended capacity of 6.6-year flood flow for the detour culvert when $C_p/D_p = 1.0$ (see Figure 11.25). Table 11.9 is the summary of design floods for 12 different construction periods in a year for $C_p/D_p = 1.0$.

In practice, the cost of detour culvert depends on engineer's experience about the project. Without adequate local knowledge, the dry season is preferred for construction. Table 11.9 provides a basis to estimate the cost and size of the required detour culverts for all possible construction periods in a year. In practice, used culverts are recommended for detour application because of the short service in time.

11.11 Homework

Q11.1 A dual culvert system is designed to pass 90 cfs. The layout of the culvert is given in Figure Q11.1. Try $D = 36$ in. and $Q = 45$ cfs per barrel. (1) Determine the tailwater

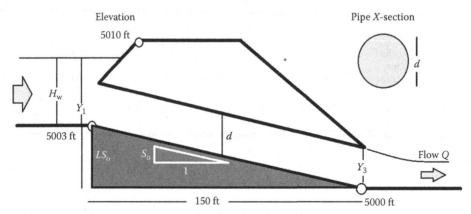

Figure Q11.1 Culvert design without specified tailwater depth.

depth. (2) Estimate the inlet-control and outlet-control headwater depths. (3) Determine the headwater depth for design.

Solution:

Design information (input)		
Pipe diameter (in.)	$d = 36.0$ in.	914.4 mm
Pipe diameter (ft)	$d = 3.0$ ft	0.91 m
Invert elevation at entrance	E-in = 5003.0 ft	1525.30 m
Invert elevation at exit	E-out = 5000.0 ft	15.24 m
Pipe length	$L = 150.0$ ft	45.7 m
Slope of culvert	$S_o = 0.0200$ ft/ft	33.0200 m/m
Manning's roughness	$N = 0.0140$	0.0140
Entrance loss coefficient	$K_e = 0.50$	0.5000
Valve or additional loss	$K_x = 0.00$	0.0000
Bend loss coefficient	$K_b = 0.00$	0.0000

Design discharge		
Total design discharge	$Q_t = \mathbf{90.00}$ cfs	2.55 cms
Number of barrels	$n = 2.00$	2.00
Total culvert inner width	Pipe width = 6.00 ft	0.91 m
Discharge per barrel	$Q = \mathbf{45.00}$ cfs	1.28 cms

Flow condition		
Flow variables	Critical flow	Normal flow
Half central angle	$\theta{-}c = 2.04$ rad	$\theta{-}n = 1.59$ rad
Flow depth	$Y_c = 2.18$ ft	$Y_n = 1.53$ ft
Flow area	$A_c = 5.51$ ft^2	$A_n = 3.61$ ft^2
Wetted perimeter	$P_c = 6.13$ ft	$P_n = 4.76$ ft
Check on Froude number or design flow	$F_r = 1.00$	$d_Q = 0.00$ cfs
Flow variable	$V_c = 8.16$ fps	$Fr_n = 2.00$

Headwater depth by inlet control		
Orifice coefficient	$C_o = 0.52$	0.52
Headwater depth—inlet control	H_w-inlet = **3.83** ft	1.18 m
Pipe cross-sectional area	$A = 7.07$ ft^2	0.67 m^2
Pipe flow velocity	$V_f = 6.37$ fps	1.96 mps

Headwater depth by outlet control 0.00		
If known, enter the tailwater depth	$Y_t = 0.00$ ft	0.00 m
$Y_3 = \max[Y_t, 0.5(D + Y_c)]$	$Y_3 = 2.59$ ft	0.80 m
Friction loss coefficient	$K_N = 1.25$	
Sum of all loss coefficients	$K_e + K'_N = 1.75$	
Headwater depth—outlet control	H_w-out = **1.32** ft	0.41 m

Design headwater depth ratio		
Headwater for design	$H_w = \mathbf{3.83}$ ft	1.18 m
H_w/D ratio	$H_w/d = \mathbf{1.28}$	1.28

Q11.2 The 36-in. culvert in *Q11.1* is modified with an improved inlet that has a drop slope on 1V:10 H at the entrance as shown in Figure Q11.2. During the design event, the headwater surface elevation remains the same as *Q11.1*, but its capacity will be calculated relative to elevation 5002 ft. Determine the capacity of the improved-inlet culvert.

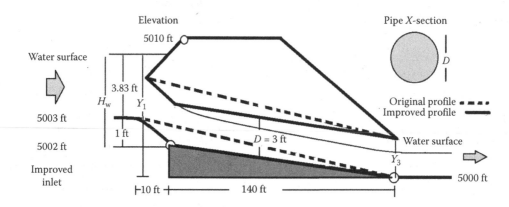

Figure QII.2 Improved inlet and known tailwater depth.

Bibliography

ASCE. (1991). "Design and Construction of Urban Stormwater Management Systems," Manuals and Reports of Engineering Practice No 77 or Water Environmental Federation WEF Manual of Practice FD-20.

CDOT. (1992). "1986–1992 Cost Data," Cost Estimates Squad of the Staff Design Branch, Colorado Department of Transportation, Denver, CO.

CDOT. (2000). "Standard Specifications for Road and Bridge Construction," Colorado Department of Transportation, Denver, CO.

CDOT. (2015). "*Roadway Design Manual*," Colorado Department of Transportation, Denver, CO.

Chow, V.T., Maidment, D.R., and Mays, L.M. (1988). "*Applied Hydrology*," McGraw-Hill Company, New York.

Guo, J.C.Y. (1987). "Development of Risk-Cost Methodology for Detour Culvert Design," the Research Contract Number 1503A, CDOH-UCD-R-88-5, Colorado Department of Highway, Denver, CO.

Guo, J.C.Y. (1998). "Risk-Cost Approach to Interim Drainage Structure Design," *ASCE Journal of Water Resources Planning and Management*, Vol. 124, No. 6, pp. 330–333.

Guo, J.C.Y. (1999). "Sand Recovery for Highway Drainage Designs," *ASCE Journal of Drainage and Irrigation Engineering*, Vol. 125, No. 6, pp. 380–384.

HEC14. (1975). "Hydraulic Design of Energy Dissipaters for Culverts and Channels," U.S. Department of Transportation, Federal Highway Administration, Washington, DC.

USGS. (1977). "Water Resources Data Colorado," U.S. Geological Survey, Water Year 1961 to 1977.

USWDCM. (2010). "*Urban Stormwater Design Criteria Manual*," Urban Drainage and Flood Control District, Denver, CO.

Chapter 12

Storm sewer system design

A storm sewer system consists of a series of inlets, manholes, and pipes that collect and convey storm runoff from streets to the downstream collector. Storm sewers are placed where storm runoff exceeds the street gutter capacity. In general, storm sewer systems are designed to pass the *minor storm flows* such as 2- to 5-year events. Investigation of the performance of a sewer system is not limited only to the design event but is also done under the *major storm flows* such as 10- to 100-year events. During the major storm event, a storm sewer system will be surcharged and the excess runoff will be carried by the street gutters. Street drainage relies on a dual flow system in which the minor and major flow systems coexist. Utilizing a dual system, the underground sewers provide the conveyance capacity for 2- to 5-year events, and the street gutters provide an overland flow system to convey the 10- to 100-year events.

12.1 Layout of sewer system

The importance of a storm drainage plan is to incorporate the existing natural waterways and man-made drainage facilities into the drainage system. Therefore, the storm drainage plan shall be undertaken prior to the finalization of the street layout in order to effectively incorporate the major and minor drainage concepts into a storm drainage plan. Storm sewers are usually located within the right-of-way such as streets, roadways, and easements for ease of access during repair and maintenance operations. To be economic, a sewer system shall follow the natural topography as closely as possible. Topographic maps, aerial photographs, and drawings of existing utilities are required before the sewer system can be laid out. The layout of a storm sewer system is governed by many factors, including the following:

1. Existing utility locations
2. Street alignment
3. Inlet placement
4. Outfall location
5. Surface topography

These conditions impose the inherent constraints to the layout of a storm sewer system. In addition, the storm sewer system often takes priority when other conflicts or limits cause undesirable hydraulic conditions. For instance, a sewer can be designed around a water line or the water line has to be relocated, depending on the hydraulic grade line. Such limits as a result of other service utilities on sewer vertical and horizontal alignments are discussed in the following sections. The layout of a sewer system is depicted by its *plan view* and *vertical view*. A *plan view* shows the connectivity of the sewer network in terms of streets, manholes, sewers, and buildings. A *vertical view* is the plot of the vertical profile along the sewer line. As illustrated in Figures 12.1 and 12.2, all sewers are marked with their length, slope, and identification numbers as S-1, S-2, etc. All manholes are marked by their identification numbers as M-1, M-2, etc. A manhole is covered by a metal cap. The *rim elevation* of the metal cap is set to be at the local ground elevation. Each incoming sewer at the manhole is marked with its *crown* and *invert elevations*, so is the outgoing sewer. The difference between the incoming and outgoing sewer inverts is termed *manhole drop*.

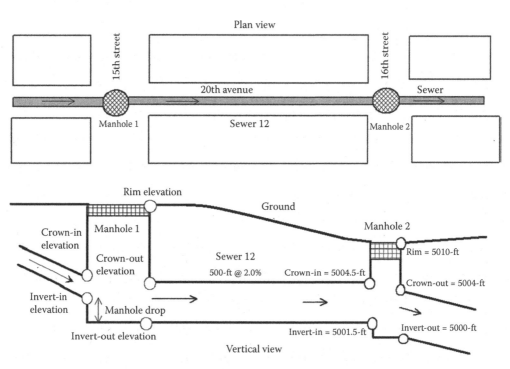

Figure 12.1 Flow conditions in a sewer–manhole system.

Figure 12.2 Construction of sewer line. (a) Dual concrete box sewer and (b) single concrete circular sewer.

Figure 12.3 Manhole systems. (a) Manhole and inlet on sewer line and (b) branch lateral into sewer line.

12.1.1 Manholes

Manholes and *junctions* (Figure 12.3) provide efficient transitions in the storm sewer system and also serve as the access passages to storm sewer lines for maintenance and cleaning. Therefore, to maintain hydraulic efficiency and adequate maintenance access, manholes shall be located at the locations where we have to make changes to the following:

1. Pipe size
2. Sewer line alignment
3. Invert grades along incoming and outgoing sewer lines
4. Manhole drop for energy dissipation
5. Design discharge due to laterals
6. Access for cleaning or maintenance and
7. Spacing between manholes

12.1.2 Sewer line vertical alignment

The required soil cover on top of a storm sewer pipe is dependent on many factors, including pipe strength, pipe size, and cover material. For practical purposes, the storm sewer should be protected from potential surface disturbances and displacements. Therefore, the minimum allowable cover over the storm sewer pipe shall be 2 ft or greater at any point along the pipe. If there is less than 2 ft of cover, the pipe shall be concrete encased. The maximum cover is also contingent upon the pipe strength.

12.1.3 Lateral connectors

In general, the angle of confluence between main line and lateral (shown in Figure 12.4) shall not exceed 45°. A connector pipe from an inlet box may join the main line at an angle greater than 45° up to a maximum of 90°. Care must be taken when the backwater effects from the main line may severely impact the flow conditions in the lateral. A smooth transition can significantly reduce the surcharge condition under backwater effect.

Figure 12.4 Laterals in sewer system. (a) Inlet connected to sewer line and (b) manhole on sewer line.

12.1.4 Utility clearances

Storm sewers shall be located to minimize potential contamination and disturbance from water supply lines and sanitary sewers. This goal can be accomplished by distancing the storm sewer from the surrounding utilities. The engineer must collect adequate information about the existing underground utilities and be responsible for using proper design criteria for safety.

Water mains

In a place where a storm sewer or storm inlet run crosses a water main or comes within 10 horizontal feet (clear distance) of a water main, the storm sewer pipe shall be located a minimum of 18 in. clear distance vertically below the water main. If this clear distance cannot be obtained, then the storm sewer pipe section must be designed and constructed so as to protect the water main. Minimum protection shall consist of a 20-ft section of storm sewer centered over the water main being encased in concrete at least 4 in. thick. In addition, watertight joints shall be used within the 20-ft section. In no case shall the clearance between the water main and the storm sewer be less than 12 in.

Sewer mains

In a place where a storm sewer or storm inlet run crosses a sanitary sewer main or comes within 10 horizontal feet (clear distance) of each other, the storm sewer pipe shall be located a minimum of 12 in. clear distance vertically above or below the sanitary sewer main. If this clear distance cannot be obtained, then the sanitary sewer pipe section must be designed or improved to provide a structurally sound sewer main. For instance, the sanitary sewer is encased with concrete at least 4 in. thick and extending to a distance of 10 ft on either side of the storm sewer.

12.2 Design constraints

The impingement of a supercritical flow at a bend or manhole juncture can damage the pipe walls and possibly result in cavitations. The maximum allowable water flow velocity in a sewer pipe depends on pipe material, flow condition, pipe joints, manhole drops, and connections with the laterals. Considering the abovementioned factors, the maximum flow velocity in a sewer system shall be limited to 25 ft/s. Such a flow velocity control can be achieved by adding manhole drops in a sewer system. Self-cleaning capability is also recognized as a goal to minimize the costs for maintenance of storm sewer facilities. Sediment deposits, once established, are generally difficult to remove without pressure-cleaning equipment. In general, a minimum flow velocity is suggested to be 2 fps under a flowing full condition. The slope of a sewer line is often dictated by the surface gradient. The minimum storm sewer slope shall be 0.25%, because it becomes difficult to construct a sewer with a slope less than 0.25%. The hydraulic capacity of a sewer is estimated by open-channel hydraulics using Manning's formula. To be conservative, Manning's roughness should be selected to account for the pipe material, the debris and sediment in storm water, and the deterioration of pipe interior surface condition over the entire service of the pipe. The minimum allowable pipe size for storm sewers depends on the convenience of maintenance and inspection during the service period. As a rule of thumb, the minimum pipe size for street inlet connectors to a manhole is 15 in. Storm sewers on the trunk line shall not be smaller than 18 in. in diameter for a round pipe or shall have a minimum flow area of 2.2 ft^2 for other pipe shapes. In summary, a sewer system must satisfy the design criteria and constraints. Although every project has its own limitations, the general design criteria for storm sewers are summarized as follows:

1. Permissible flow velocity in a sewer: between 20 and 3 ft/s
2. Minimum earth coverage of 2 ft
3. Minimum sewer diameter of 18 in.
4. Minimum manhole drop of 0.20 ft
5. Minimum of 2 ft used for sewer trench bottom width
6. Earth side slope of 1V:1H used in sewer trench excavation
7. Maximum manhole spacing of 400 ft
8. The maximum ratio of normal flow depth to sewer diameter or height to be 0.8

To accommodate the potential backwater effects, a sewer shall be sized to have the normal depth for the design discharge not exceeding 80% of the diameter for a circular sewer or the height of a box sewer. Because the design discharge in a sewer system increases downstream, sewer sizes in a system must also increase downstream. Decrease in sewer size due to steep invert slope or smooth pipe roughness must be avoided.

12.3 Design discharge at street inlet

Referring to Figure 12.5, street inlets are connected to manholes, and manholes are connected by sewers. Determination of design flows in a sewer system starts from the catchment analysis at the upstream inlet. The design rainfall intensity depends on the local time of concentration:

$$I = \frac{k_1 H_1}{\left(k_2 + T_c\right)^{k3}} \tag{12.1}$$

in which k_1, k_2, and k_3 = empirical coefficients, T_c = time of concentration in minutes, H_1 = index rainfall depth such as 1-h precipitation depth in inches or millimeters, and

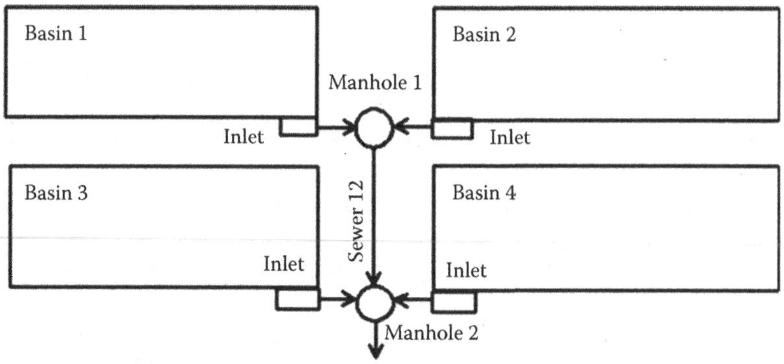

Figure 12.5 Design flows for inlets and manhole.

I = rainfall intensity in in./h or mm/h. The design peak runoff rate at a street inlet depends on the runoff coefficient and the tributary area:

$$A_e = CA \tag{12.2}$$

$$Q = KIA_e \tag{12.3}$$

in which Q = peak discharge in $[L^3/T]$ such as cfs or cms, I = rainfall intensity in in./h or mm/h, A_e = effective area in $[L^2]$ such as acres or hectare, C = runoff coefficient, A = local tributary area in $[L^2]$ such as acres or hectare, and K = 1 using cfs-acre-in./h or 1/360 using cms-ha-mm/h.

12.4 Sewer sizing

A considerable effort has been devoted to the development of storm sewer computer modeling techniques. Although many flood routing methods have been developed to numerically simulate flood wave movement in a sewer network, the most widely used design method for storm sewer designs has not gone beyond the concept of the rational method. Sizing a sewer system starts from the most upstream manhole. Referring to Figure 12.5, there are two flow paths to reach manhole 1. As a result, the time of concentration at manhole 1 is the longer one as shown below:

$$T_{M1} = \max\left(T_{c1}, T_{c2}\right) \tag{12.4}$$

in which T_{M1} = time of concentration in minutes at manhole 1, T_{c1} = time of concentration in minutes for basin 1, and T_{c2} = time of concentration in minutes for basin 2. The cumulative area at a manhole is calculated as

$$A_e = \sum_{i=1}^{i=n} C_i A_i \tag{12.5}$$

in which A_e = accumulated effective tributary area. The subscript, i, represents the variables at the ith manhole upstream of the nth manhole. For instance, at manhole 1 in. Figure 12.5, the total effective area is

$$A_e = C_1 A_1 + C_2 A_2 \tag{12.6}$$

The design discharge, Q_{M1}, at manhole 1 is

$$Q_{M1} = KI(C_1 A_1 + C_2 A_2) \tag{12.7}$$

Next, the pipe diameter for sewer 12 can be calculated by Manning's formula under the full-flow condition. As illustrated in Figure 12.5, manhole 1 has three incoming flows: two local inlet flows and one sewer flow from manhole 1. For comparison, the flow times along these three flow paths must be calculated because the longest one is chosen for the design rainfall duration. For instance, from manhole 1 to manhole 2, the travel time is

$$T_s = \frac{L_s}{(60 \times V_s)} \tag{12.8}$$

in which T_s = travel time in minutes, L_s = length for sewer 12, and V_s = flow velocity in fps or mps through sewer 12.

For simplicity, the sewer flow velocity is assumed to be the flowing full velocity. Often, the flow time through a sewer is not numerically sensitive to flow velocity. For this case, the time of concentration at manhole 2 is determined as

$$T_{M2} = \max(T_{c3}, T_{c4}, T_{M1} + T_s) \tag{12.9}$$

in which T = time of concentration. The subscript, $c3$, represents the flow time of basin 3, etc. The design rainfall intensity derived at manhole 2 applies to basins 1–4. The design discharge, Q_{M2}, at manhole 2 is

$$Q_{M2} = KI(C_1 A_1 + C_2 A_2 + C_3 A_3 + C_4 A_4) \tag{12.10}$$

Repeat the abovementioned procedure to each manhole in the system until the exit sewer pipe is sized. Having all sewers sized individually, the sewer system is ready for the evaluation of tailwater effect starting from the system exit. The energy and hydraulic grade lines will detect if any manholes and sewers are surcharged. To avoid the reverse flows to the street from a surcharged manhole, it is necessary to continue modifying the sewer vertical profile until the hydraulic grade line for the design flow is kept below the ground at all manholes in the system.

EXAMPLE 12.1

Consider the 5-year storm runoff for the three catchments in the city of Denver, CO. Determine the design discharge at point B in Figure 12.6.

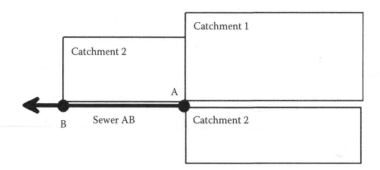

Figure 12.6 Layout of example sewer system.

Catchment parameters	Catchment 3	Catchment 2	Catchment 1
Tributary area (acres)	4.8	5.2	10.5
T_c to basin outlet (min)	12	15	23
Flow time through sewer AB	0	2.5	2.5
Runoff coefficient, C	0.65	0.75	0.51

Solution: There are three flows concentrated at manhole B. Their flow paths are as follows:

1.   Catchment 1 through sewer AB
2.   Catchment 2 through sewer AB
3.   Catchment 3 as the local flow

The longest time of concentration among these three flow paths is calculated by Equation 6.2 as

$$T_c = \max(23.0+2.5, 15+2.5, 12.0) = 25.5\,\text{min}$$

The 5-year 1-h precipitation is 1.35 in. for the project site. As a result, the corresponding rainfall intensity is

$$I = \frac{28.5 \times 1.35}{(25.5+10)^{0.789}} = 2.30\,\text{in./h}$$

At manhole B, the total effective drainage area is

$$A = 0.65 \times 4.8 + 0.75 \times 5.2 + 0.51 \times 10.5 = 12.38\,\text{acres}$$

The design peak discharge at manhole B is

$$Q = 2.30 \times 12.38 = 28.46\,\text{cfs}$$

Considering the flowing full condition for $Q_f = 28.46\,\text{cfs}$, $N = 0.014$, and $S_o = 0.01$, the required diameter of the sewer pipe is

$$d = \left(\frac{NQ_f}{K_d\sqrt{S_o}}\right)^{\frac{3}{8}} \text{ in ft or } m = 12 \times \left(\frac{0.014 \times 28.46}{0.462 \times \sqrt{0.01}}\right)^{\frac{3}{8}} \text{ in in.} = 26.85 \text{ in.}$$

in which N = Manning roughness, Q_f = flowing full capacity equal to the design flow, S_o = conduit invert slope, d = required culvert diameter, and K_d = 0.462 for ft-s units or 0.311 for m-s units (see Chapter 11). For this case, a 27- or 30-in. pipe is recommended.

12.5 Sewer system design

Design of a sewer system begins with its plan layout and vertical profile. As a rule of thumb, the ground slopes provide guidance for the first approximation. Manhole drops shall be introduced into the system to control the flow condition. The sewer sizing process shall start from the most upstream manhole in the trunk line and then march downstream as the flow rate and time of concentration are accumulated downstream. At a manhole, the flow times along various flow paths are calculated, and the longest one shall be considered as the design rainfall duration. The flow time through a sewer can be computed by its flowing full velocity.

The design peak discharge at a manhole is determined by the accumulated effective drainage area. The downstream sewer at each manhole is then sized by the open-channel uniform flow condition. The calculated hydraulic pipe size may not be available in manufactory. Therefore, the next larger commercially available pipe size shall be used. For instance, the available commercial circular pipe sizes in diameter are as follows: 6, 8, 10, 12, 15, 18, 21, 24, 27, 30, 36, 42, 48, 54, 60, 66, 72, 78, 84-in., etc.

The vertical profile of a sewer system may have to be adjusted several times in order to satisfy the design criteria and constraints. For instance, a sewer slope shall be adjusted until the flow velocity is within the low and high permissible velocities and also meets the required minimum soil cover. As a common practice, a manhole drop of 0.2 ft is preferred. After all sewers are sized by open-channel hydraulics under the normal flow condition, the sewer system is further subject to the performance evaluation under tailwater effects at the system exit. The energy and hydraulic grade lines will predict if any manholes in the sewer system are surcharged. Further modifications on the vertical profile may be required until the hydraulic grade line for the design flow is kept below the ground at all manholes in the system.

EXAMPLE 12.2

As illustrated in Figure 12.7, the layout of sewer system A is given. Use the following information to size the sewer pipes.

1. The 5-year design rainfall intensity, I (in./h), is given as

$$I (\text{in./h}) = \frac{28.5 P_1}{(10+T_d)^{0.789}} \text{ in which } P_1 = 1.35 \text{ in. and } T_d = \text{duration in minutes.}$$

2. The overland flow time shall be calculated by the airport formula with the maximum overland flow length of 300 ft. The swale and gutter flow velocities, can be estimated by the Soil Conservation Service (SCS) upland method using a conveyance factor of 20.
3. Design parameters for subareas are listed in Table 12.1.
4. Design parameters for sewer pipes are listed in Table 12.2.

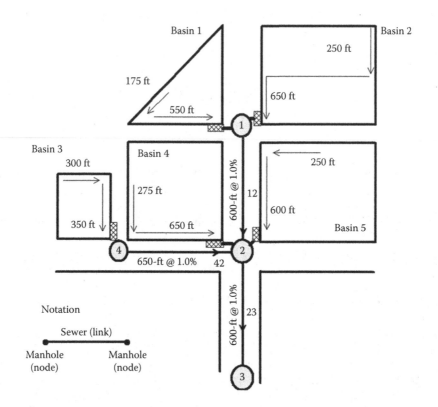

Figure 12.7 Layout of example storm sewer system.

Table 12.1 Hydrologic parameters for sewer system A

Basin number	Area (acres)	Runoff coefficient and Imp%	Overland		Channel	
			Slope (%)	Length (ft)	Slope (%)	Length (ft)
1	4.0	0.80/85%	1	175	1	550
2	8.0	0.60/65%	2	250	2	650
3	3.0	0.75/80%	2	300	1.5	350
4	8.5	0.80/85%	1	275	1	650
5	7.0	0.85/95%	2	250	1	600

Table 12.2 Sewer pipe parameters for sewer system A

Sewer ID	Length (ft)	Slope (%)	Roughness
12	600	1	0.014
42	650	1	0.014
23	600	1	0.014

Solution: At design point 1, both flows from basins 1 and 2 will be combined together. The analysis of the combined flow is summarized as follows:

1. Flow time analysis for basin 1
    a. The overland flow time is computed as

    $$T_o = \frac{0.393 \times (1.1 - 0.8) \times \sqrt{175}}{0.01^{0.33}} = 7.12 \, min$$

    b. The gutter flow time is calculated as

    $$T_2 = \frac{550}{60 \times 20 \times \sqrt{0.01}} = 4.58 \, min$$

    c. The computed time of concentration is the sum as shown below:

    $$T_{comp} = 7.12 + 4.58 = 11.70 \, min$$

    d. The Denver regional time of concentration is computed as

    $$T_{Check} = (18 - 15 l_a) + \frac{L}{60 \times (24 l_a + 12) \sqrt{S_a}}$$

    $$= (18 - 15 \times 0.85) + \frac{175 + 550}{60 \times (24 \times 0.85 + 12) \sqrt{0.01}} = 6.15 \, min$$

    The time of concentration for basin 1 is

    $$T_c = min(11.70, 6.15) = 6.15 \, min$$

2. Flow time analysis for basin 2
    Repeating the same procedure as basin 1, the time of concentration for basin 2 is found to be 15.0 min.
3. Design peak flow at design point 1

    $$T_c = max(6.15, 15.0) = 15.0 \, min$$

    $$I(in./h) = \frac{28.5 \times 1.35}{(10 + 15.0)^{0.789}} = 3.04 \, in./h$$

    The design peak flow from basins 1 and 2 is

    $$Q_A = I(C_1 A_1 + C_2 A_2) = 3.04 \times (3.2 + 4.8) = 24.28 \, cfs$$

4. Diameter for sewer 12

    $$d = 12 \times \left(\frac{0.014 \times 24.28}{0.462 \times 0.01^{0.5}} \right)^{\frac{3}{8}} = 25.34 \, in.$$

Use a 27-in. circular pipe. The flowing full velocity in a 27-in. pipe is

$$V_s = \frac{24.28}{\frac{\pi}{4} \times \left(\frac{27}{12}\right)^2} = 6.37\,fps$$

The flow time to the immediately downstream manhole, design point 2, is

$$T_s = \frac{600}{60 \times 6.37} = 1.57\,min$$

5. At point 2, the cumulative flow time from sewer 12 is

$$15.0 + 1.57 = 16.57\,min$$

Similarly, the cumulative flow time from sewer 42 to design point 2 is 13.10 min. In comparison, the longer one shall be used for the design rainfall duration. As a result, the rainfall intensity at design point 2 is

$$I = \frac{28.5 \times 1.35}{(10 + 16.57)^{0.789}} = 2.89\,in./h$$

And the peak runoff flow for sewer 23 is

$$Q = I \sum_{i=1}^{i=5} C_i A_i = 2.89 \times (3.20 + 4.80 + 2.25 + 6.80 + 5.95) = 66.54\,cfs$$

The required pipe diameter for a flow of 66.54 cfs is 36.99 in.; therefore, a 42-in. pipe is recommended. Solutions for this case are summarized in Tables 12.3 through 12.6.

12.6 Sewer–manhole element

The water surface profile represents the hydraulic gradient line (HGL) in a sewer line. The difference between the HGL and its energy gradient line (EGL) is the flow kinetic energy or velocity head. Having all sewers sized in a system, it is necessary to ensure that the HGL for the design condition is kept below the ground elevations at all manholes. Figure 12.8 depicts examples of pressurized manholes.

To analyze the energy and HGLs in a sewer system, a sewer system is divided into sewer–manhole elements. As shown in Figure 12.9, a *sewer–manhole element* has four distinct sections. They are as follows:

Section 1 is located immediately downstream of the sewer exit.
Section 2 is located immediately upstream of the sewer exit.
Section 3 is located immediately downstream of the sewer entrance.
Section 4 is located immediately upstream of the sewer entrance.

Applying the energy principle to sections 1 and 2 needs to include the exit loss. From section 2 to 3, the flow is subject to the friction loss determined by the sewer roughness

Table 12.3 Basin hydrology

Catchment data				Computation of time of concentration						Computed T_c (min)	Regional T_c (min)	Storm duration (min)	Rainfall intensity (in./h)	Peak flow (cfs)
			Effective area (acres)	Overland flow			Gutter flow							
				Slope (%)	Length (ft)	Time (min)	Slope (%)	Length (ft)	Time (min)					
ID	A	C	CA	S_o	L_o	T_o	S_2	L_2	T_2	T_c-comp	T_c-reg	T_d	I	Q_p
Basin ID	Area (acres)	Runoff coefficient												
1	4.00	0.80	3.20	1.00	175.0	7.14	1.00	550.0	4.58	11.73	14.03	11.73	3.39	10.85
2	8.00	0.60	4.80	2.00	250.0	11.32	2.00	650.0	3.83	15.15	15.00	15.00	3.46	16.63
3	3.00	0.75	2.25	1.50	300.0	9.55	1.50	350.0	2.38	11.93	13.61	11.93	3.83	8.61
4	8.50	0.80	6.80	1.00	250.0	8.54	1.00	600.0	5.00	13.54	14.72	13.54	3.63	24.66
5	7.00	0.85	5.95	1.50	250.0	6.22	1.00	600.0	5.00	11.22	14.72	11.22	3.92	23.34

Table 12.4 Manhole hydrology

Sewer pipe ID	Sewer length (ft) L_s	Sewer slope (ft/ft) S_s	Local catchment			Cumulative parameters at upstream manhole			Travel time through sewer			Sum of effective area (acres) CA
			Basin ID	Effective area (acres)	Rainfall duration (min)	Upstream manhole	Effective area (acre)	Rainfall duration (min)	Length (ft)	Velocity (fps)	Time (min)	
ID	L_s	S_s	ID	CA	T_d	ID	CA	T_d	L_s	V_s	T_s	CA
12	600	0.01	1.0	3.20	11.7	—	0.0	15.0	600.0	6.37	1.57	8.00
42	650	0.01	2.0	4.80	15.0	4	0.0	11.9	650.0	9.27	1.17	2.25
23	600	0.01	3.0	2.25	11.9							
			4.0	6.80	13.5							
			5.0	5.95	11.2							23.0

Table 12.5 Sewer sizing

Sewer pipe ID	Sum of effective area (acres) CA	Storm duration (min) T_d	Rainfall intensity (in./h) I	Peak flow (cfs) Q_p	Pipe diameter required (in.) D-required	Pipe diameter used (in.) D-used	Sewer flow velocity (fps) V_s
12	8.00	15.00	3.04	24.28	25.34	27.00	6.37
42	2.25	11.93	3.37	7.57	16.37	18.00	9.27
23		13.54					
		11.22					
		16.57					
		13.10					
	23.00	16.57	2.89	66.54	36.99	42.00	3.84

Table 12.6 Sewer profile

Sewer ID	Length (ft)	Slope (ft/ft)	Downstream ground elevation (ft)	Downstream manhole drop (ft)	Downstream crown elevation (ft)	Downstream invert elevation (ft)	Design flow rate (cfs)	Pipe diameter required (in.)	Pipe diameter used (in.)	Pipe wall thick (in.)	Upstream crown elevation (ft)	Upstream invert elevation (ft)	Upstream ground elevation (ft)	Downstream soil depth (ft)	Upstream soil depth (ft)
Trunk															
23	600.0	0.010	5028.0	0.0	5023.0	5019.50	66.54	36.99	42.00	4.50	5029.0	5025.5	5033.0	5.00	4.00
12	600.0	0.010	5035.0	0.2	5028.0	5025.7	24.28	25.34	27.00	3.25	5034.0	5031.7	5040.0	7.05	6.05
Branch															
42	650.0	0.010	5035.0	0.2	5027.2	5025.70	7.57	16.37	18.00	2.50	5033.7	5032.2	5042.0	7.80	8.30

Figure 12.8 High pressure buildup in sewer line. (a) Pressurized manhole and (b) pressurized inlet.

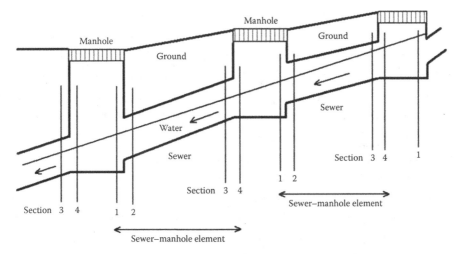

Figure 12.9 Manhole–sewer element.

and the flow conditions. From section 3 to 4, the entrance losses must be included. From section 4 to 1, the manhole configurations dictate the manhole loss.

As shown in Figure 12.9, the EGL through a sewer–manhole element starts from the outfall point. The system exit is the most downstream section 1 where the water surface elevation must be known. Analysis of energy balance for a sewer–manhole element is divided into two steps:

1. *Sewer hydraulics* for the energy losses through the sewer, including

   a. Exit losses from section 1 to 2
   b. Friction losses from section 2 to 3
   c. Entrance losses from section 3 to 4

2. *Manhole hydraulics* for the energy losses through the manhole, including

   a. Lateral losses due to incoming lateral
   b. Bend losses from section 4 to 1

12.7 Sewer hydraulics

12.7.1 Energy analysis from section 1 to 2

With the known energy, E_1 in [L], at section 1, the energy balance between sections 1 and 2 is written as

$$E_{12} = E_1 + K_x \frac{V_2^2}{2g} \tag{12.11}$$

in which E_{12} = energy in [L] at section 2 due to E_1, V_2 = velocity in [L/T] at sewer exit, and K_x = exit energy loss coefficient ranging from 0.5 to 1.0. In practice, the exit loss can be treated as part of the bend losses because both are expressed as a fraction of the flow kinetic energy. Therefore, Equation 12.11 is simplified to

$$E_{12} = E_1 \tag{12.12}$$

As illustrated in Figure 12.10, the EGL at section 2 is not solely determined by E_{12} because a significant manhole drop at section 2 can introduce a discontinuity to the flow line or establish a new EGL to replace E_{12}. Of course, the energy difference due to the discontinuity on the flow line is interpreted as dissipation. Therefore, it is necessary to examine the flow condition and manhole drop at the sewer exit for choosing the proper one to continue the EGL calculations. The flow at the exit may be one of the cases shown in Figure 12.10.

For a subcritical flow with a significant manhole drop, we expect an M-2 profile with its critical depth at the sewer exit. Therefore, the EGL at the sewer exit is

$$F_r < 1.0 \quad E_e = Z_e + \frac{V_c^2}{2g} + Y_c \quad \text{for a free fall exit} \tag{12.13}$$

in which F_r = Froude number, E_e = EGL in [L] at sewer exit, Z_e = sewer invert elevation in [L] at exit, V_c = critical flow velocity in [L/T], and Y_c = critical depth in [L]. For a subcritical flow with a submerged exit, the flowing full condition shall have an EGL as

$$F_r = 1.0 \quad E_e = Z_e + \frac{V_f^2}{2g} + D \quad \text{for a submerge exit} \tag{12.14}$$

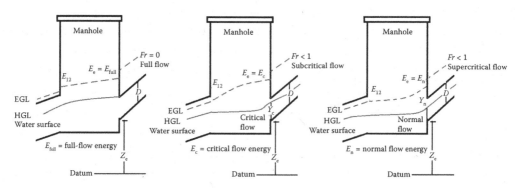

Figure 12.10 Flow conditions at sewer exit.

in which D = sewer diameter in [L] and V_f = full-flow velocity in [L/T]. For a supercritical flow, the critical depth occurs at the upstream entrance, and the normal depth is assumed to be developed at the exit. As a result, the EGL at the sewer exit is

$$F_r > 1.0. \quad E_e = Z_e + \frac{V_n^2}{2g} + Y_n \quad \text{for a drop exit with supercritical flow} \tag{12.15}$$

in which Y_n = normal depth [L] and V_n = normal flow at the sewer exit. Equations 12.13 through 12.15 provide the possible EGL dictated by the sewer exit condition.

From the earlier discussion, the EGL at section 2 is subject to the following: (1) the downstream tailwater condition, E_{12}, at section 1 and (2) the exit configuration, E_e, at the sewer exit. Of course, the higher one dictates the EGL at section 2 as

$$E_2 = \max(E_{12}, E_e) \tag{12.16}$$

If E_{12} is selected, the EGL is a continuous curve. If E_e is selected, it means the manhole drop has created a discontinuity on the EGL. After the value of E_2 is selected, the corresponding water surface elevation, W_2, at section 2 is calculated as

$$W_2 = E_2 - \frac{V_2^2}{2g} \tag{12.17}$$

in which V_2 = flow velocity in [L/T] at the sewer exit. There are several possibilities discussed as follows:

$V_2 = V_f$ if the exit is submerged
$V_2 = V_n$ if the exit is not submerged with a supercritical flow
$V_2 = V_c$ if the exit is not submerged with a subcritical flow

in which V_n = normal flow velocity in [L/T], and V_c = critical flow velocity in [L/T]. Using a proper flow velocity, Equation 12.17 calculates the corresponding water surface elevation, W_2 in [L], at section 2.

EXAMPLE 12.3

A 24-in. circular sewer in Figure 12.11 carries a discharge of 24 cfs. The downstream manhole creates a tailwater EGL at $E_1 = 90$ ft. The sewer exit may be set at elevations of 96 or 100 ft. Under the flowing full condition, estimate the EGL at the sewer exit for both cases.

Solution: Under the flowing full condition, the flow velocity is calculated as

$$V_f = \frac{Q}{A} = \frac{24.0}{3.14 \times \dfrac{2.0^2}{4}} = 7.64 \, \text{fps}$$

When $Z_e = 96$ ft, the flow condition at the exit is

$$E_e = Z_e + D + \frac{V_f^2}{2g} = 96.0 + 2.0 + \frac{7.64^2}{2.0 \times 32.2} = 98.9 \, \text{ft}$$

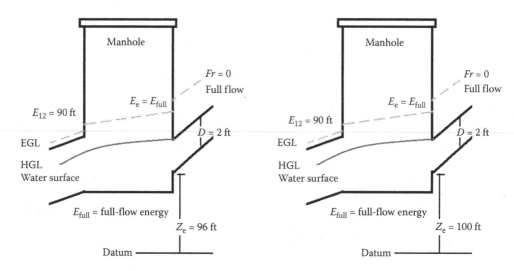

Figure 12.11 Examples for flow conditions at sewer exit.

As a result, the EGL at section 2 is dominated by the EGL at section 1 as

$$E_2 = \max(E_{12}, E_e) = \max(100.0, 98.9) = 100.0\,\text{ft}$$

The corresponding water surface at the exit is

$$W_2 = 100.0 - \frac{7.64^2}{2.0 \times 32.2} = 99.1\,\text{ft}$$

Because W_2 is greater than the sewer crown elevation at the exit, it is a submerged case that produces a surcharge flow at the downstream end of the sewer. On the contrary, when $Z_e = 100\,\text{ft}$, the EGL at section 2 becomes dominated by the manhole drop as

$$E_2 = \max(E_{12}, E_e) = \max(100.0, 102.9) = 102.9\,\text{ft}$$

The corresponding water surface at the exit is

$$W_2 = 102.9 - \frac{7.64^2}{2.0 \times 32.2} = 102.0\,\text{ft}$$

12.7.2 Energy analysis from section 2 to 3

The water elevation, W_2, relative to the exit sewer crown is critical to the flow condition through the sewer. For instance, when $W_2 > C_w$ in which C_w is the sewer crown elevation at the exit, the sewer has a surcharge flow as illustrated in Figure 12.12. If the surcharged length exceeds the length of the sewer, the entire sewer carries a pressure flow that is independent of Froude number. Otherwise, only the downstream portion of the sewer

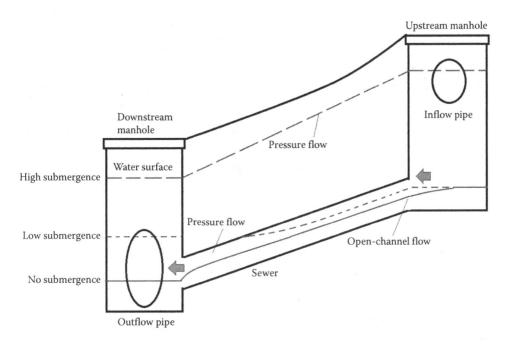

Figure 12.12 Effects of downstream water elevation on sewer hydraulics.

is pressurized, and the upstream portion of the sewer still remains aerated. The energy analysis from section 2 to 3 is discussed separately for the following five cases:

1. Subcritical flow with unsubmerged exit
2. Supercritical flow with unsubmerged exit
3. Subcritical flow with submerged exit
4. Supercritical flow with submerge exit
5. Undersized and flat sewers

Case 1. Subcritical flow with unsubmerged exit

When the sewer exit is free from submergence as shown in Figure 12.13, the sewer carries open-channel flows. Therefore, Froude number dictates the water surface profiles. For instance, a subcritical flow with a manhole drop at the sewer exit will produce an M-2 curve that has the critical depth at the exit. Otherwise, an M-1 curve will be developed when the tailwater depth is greater than the normal depth.

Using the finite difference approach, the sewer length between sections 2 and 3 can be divided into several segments. The water surface profile illustrated in Figure 12.14 can be computed by the standard step method or the direct step method. For each segment, the friction loss is calculated by Manning's equation in order to balance the energy principle between two adjacent sections.

The energy principle between sections a and b is written as

$$Y_a + Z_a + \frac{V_a^2}{2g} = Y_b + Z_b + \frac{V_b^2}{2g} + H_f \qquad (12.18)$$

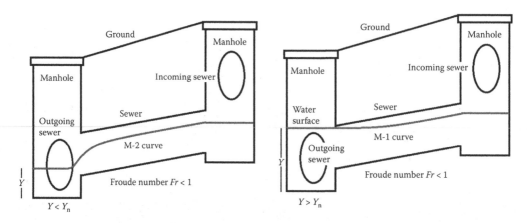

Figure 12.13 Subcritical flows with unsubmerged sewer exit.

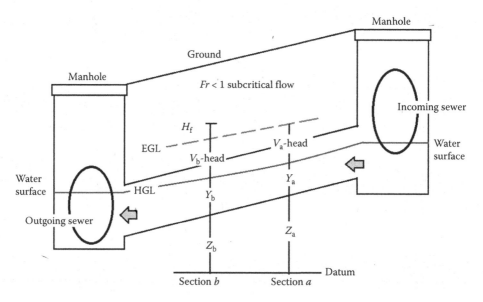

Figure 12.14 Energy grade line in subcritical flow.

$$H_f = \bar{S}_f X \tag{12.19}$$

$$S_a = \frac{N^2 V_a^2}{K_N^2 R_a^{4/3}} \tag{12.20}$$

$$S_b = \frac{N^2 V_b^2}{K_N^2 R_b^{4/3}} \tag{12.21}$$

$$\bar{S}_f = \frac{S_a + S_b}{2} \tag{12.22}$$

in which Y = flow depth in [L], V = flow velocity in [L/T], V-head = kinetic energy in [L], Z = invert elevation in [L], H_f = friction losses in [L], N = Manning's roughness, K_N = 2.22 for feet-second units or 1.0 for meter-second units, R = hydraulic radius in [L], S_f = friction slope, X = distance in [L] between sections a and b, and the subscript "a" represents variables associated with section a that is upstream of section b. The total loss is obtained by accumulating individual loss through the segments from section 2 to 3.

$$E_3 = E_2 + \sum_{i=1}^{i=m} H_{f_i} \quad \text{for a subcritical flow} \tag{12.23}$$

in which E_3 = Bernoulli energy at section 3, m = number of divisions on sewer length, and i = ith segment.

Case 2. Supercritical flow with unsubmerged exit

When the sewer carries a supercritical flow with an unsubmerged exit, an S-2 water profile will be developed in Figure 12.15. The water depth varies from the critical depth at the entrance toward the normal depth further downstream.

The S-2 profile is a typical flow under inlet control. Therefore, it is not necessary to compute the entire water surface profile, because the critical depth occurs at the entrance.

$$E_3 = E_c = Y_c + \frac{V_c^2}{2g} + Z_u \tag{12.24}$$

in which E_c = energy of the critical flow and Z_u = upstream sewer invert elevation.

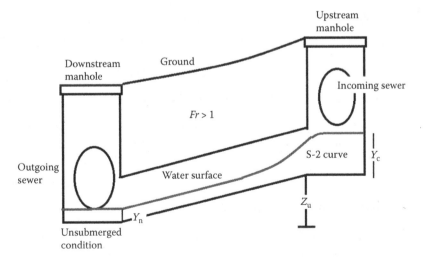

Figure 12.15 Supercritical flow with unsubmerged sewer exit.

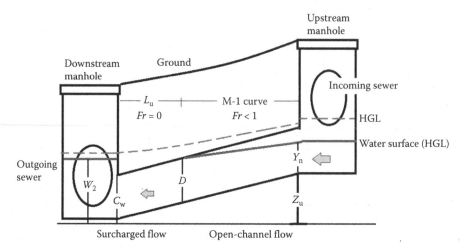

Figure 12.16 Subcritical flow with submerged sewer exit.

Case 3. Subcritical flow with submerged exit

As shown in Figure 12.16, a submerged exit results in a surcharge flow near the downstream end of the sewer. The length of the surcharge flow depends on the sewer slope. In case the surcharged length is shorter than the sewer length, the flow becomes an open-channel flow again further upstream. Figure 12.16 presents a case on a mild slope. The surcharge length, L_u, can be approximated by the direct step method as

$$L_u = \frac{W_2 - C_w}{S_s - S_f} \tag{12.25}$$

Use the flowing full velocity to calculate the friction slope as

$$V_f = \frac{Q}{A_f} \tag{12.26}$$

$$S_f = \frac{N^2 V_f^2}{K_N^2 (D/4)^{\frac{4}{3}}} \tag{12.27}$$

The friction loss through the surcharged length is

$$H_f = S_f \times L_u \tag{12.28}$$

In case the surcharged length is shorter than the sewer length, an M-1 curve will be developed with the downstream water depth equal to the sewer diameter. If the sewer is long enough, the EGL at section 3 will be the normal flow condition. Detailed computations are similar to Case 1. The sum of friction losses through the M-1 profile is calculated as

$$E_3 = E_2 + H_f + \sum_{i=1}^{i=m} H_{f_i} \quad \text{for subcritical flow} \tag{12.29}$$

in which H_{f_i} = friction loss for ith segment in M-1 profile and m = number of segments under M-1 profile.

Case 4. Supercritical flow with submerged exit

Similar to Case 3, when the flow is carried in a steep sewer with a submerged exit, the surcharged flow produces a hydraulic jump as shown in Figure 12.17.

For this case, the flow profile begins with a critical depth at the entrance and then follows an S-2 water profile toward the surcharged section. If the pipe is long enough, the normal flow may be developed before the jump. A hydraulic jump consumes a large amount of energy that cannot be easily estimated by theories. The analysis of HGL at the entrance for this case includes two possibilities:

1. The flow goes through the critical depth.
2. The flow has to overcome the friction loss through the surcharged length.

As a result, we can compare these two conditions and choose the higher one for the HGL at the upstream manhole. This approach is similar to culvert hydraulics that checks the headwater depth for both conditions of inlet and outlet control and then selects the higher one. As a result, the HGL at section 3 is determined as

$$E_3 = \max(E_c, E_2 + H_f) \quad \text{for } F_r > 1.0 \tag{12.30}$$

in which E_c = energy of the critical flow by Equation 12.24.

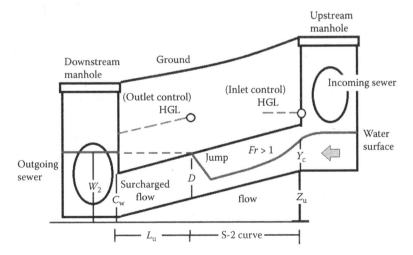

Figure 12.17 Supercritical flow with submerged sewer exit.

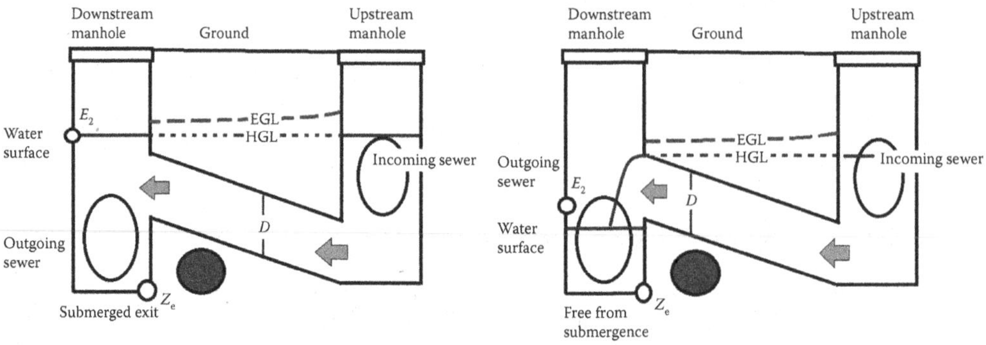

Figure 12.18 Flow in sewer with flat or negative slope.

Case 5. Undersized or flat sewer

Undersized, flat, or negatively sloped sewers result in a flowing full condition. As illustrated in Figure 12.18, the minimum energy required for water to pass through the sewer is the friction loss. However, the sewer exit configuration may also impose a requirement on the headwater depth. Therefore, it is necessary to compare both conditions and use the higher one for EGL.

Similar to Equation 12.14, the HGL at section 3 for a negatively sloped pipe is determined as

$$E_3 = \max\left(E_2 + H_f, Z_e + D + \frac{V^2}{2g} + H_f \right) \tag{12.31}$$

12.7.3 HGL analysis from section 3 to 4

The HGL analysis from section 3 to 4 in a sewer–manhole element essentially deals with the entrance loss as

$$E_4 = E_3 + K_e \frac{V_o^2}{2g} \tag{12.32}$$

in which K_e = entrance loss coefficient between 0.25 and 0.5, depending on the configurations of the entrance, and V_o = velocity at the sewer entrance, which is the critical velocity if it is a supercritical flow, or the full-flow velocity if the entrance of the outgoing sewer is submerged. Considering that the entrance loss can be included in the bend losses, Equation 12.33 is simplified to

$$E_4 = E_3 \tag{12.33}$$

12.8 Manhole hydraulics

A manhole in Figure 12.19 may have several incoming sewers but only one outgoing sewer. The energy analysis across a manhole starts with the known energy, E_4, which

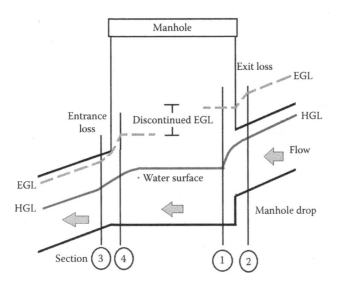

Figure 12.19 Sections I and 4 in manhole hydraulic analysis.

has been determined by Equation 12.33 by the downstream EGL analysis. There are two types of juncture losses at a manhole. They are as follows:

1. *Bend losses* due to the change in the flow direction caused by the change of the sewer alignment
2. *Lateral losses* caused by incoming lateral sewers that produce additional losses to the trunk line.

Because the flow length across a manhole is so short that the friction loss is negligible, the energy balance across the manhole between the entrance of the outgoing sewer and the exit of an incoming sewer is

$$E_1 = E_4 + H_b + H_m \quad \text{(for trunk line)} \tag{12.34}$$

$$E_1 = E_4 + H_b \quad \text{(for lateral line)} \tag{12.35}$$

in which E_1 = HGL at section 1. Bend loss, H_b, is often estimated as a fraction of the full-flow velocity head in the incoming sewer. Bend loss equation states

$$H_b = K_b \frac{V_f^2}{2g} \tag{12.36}$$

in which V_f = full-flow velocity in the sewer coming to the manhole. The value of K_b is determined by the angle between the incoming flow direction and the outgoing flow direction at the manhole. Table 12.7 lists bend loss coefficients. It is noticed that a straight-through alignment still has a bend loss coefficient of 0.05.

Lateral losses are only applicable to the truck (main) line sewers. Lateral losses count for the additional turbulence caused by the branch sewers. The value of lateral loss

Table 12.7 Loss coefficients at manholes K_m

Angle (°)	Bend loss coefficient for curved deflector in the manhole	Bend loss coefficient for nonshaping deflector in the manhole	Lateral loss coefficient on main line sewer
Straight through	0.050	0.050	Not applicable
22.500	0.080	0.100	0.750
45.000	0.280	0.400	0.500
60.000	0.460	0.640	0.350
90.000	1.000	1.320	0.250

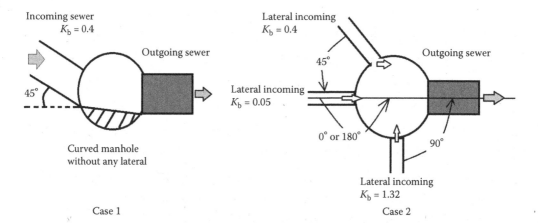

Figure 12.20 Lateral losses and bend losses at a manhole.

coefficient, K_m, is determined by the angle between the branch sewer line and the main line. Lateral loss, H_m, is estimated by

$$H_m = \frac{V_{fo}^2}{2g} - K_m \frac{V_{fi}^2}{2g} \tag{12.37}$$

in which V_{fo} = full-flow velocity of outgoing sewer at manhole and V_{fi} = full-flow velocity of incoming lateral at manhole.

As illustrated in Figure 12.20, the manhole in case 1 has only one incoming lateral. The bend loss is estimated by a coefficient of 0.28 determined by the angle of 45°. Case 2 involves three incoming sewers. Each incoming sewer has its own bend loss coefficient due to its own angle relative to the outgoing sewer. The main line has a bend loss coefficient of 0.05 (straight through) plus an additional loss due to the two laterals. According to the angles of the two laterals, we have two different lateral loss coefficients. To be conservative, the smaller one is used because it results in a higher loss according to Equation 12.37.

The previous discussion gives a general guideline as to how to select loss coefficients. It is important to understand that the engineer must take the manhole configuration into consideration when choosing the values of K_m and K_b. For instance, the lateral introduces much less disturbance to the main line when it comes into the manhole with a significant drop. As a result, the value of K_m can be reduced to reflect this arrangement. Also, the value of K_b needs to include the sewer exit and entrance losses.

Aided by Equations 12.36 and 12.37, Equation 12.34 is solved across the manhole to obtain the EGL at section 1, i.e., E_1 in Equation 12.12. Having known E_1, we can repeat the aforementioned procedure until we finish the HGL for the most upstream manhole.

EXAMPLE 12.4

Refer to the case 2 in Figure 12.20. The full-flow velocity is 12.5 fps in the outgoing sewer, 10.0 fps in the incoming main line sewer, 8.5 fps in the lateral one sewer, and 6.5 fps in the lateral two sewer. The known Bernoulli energy at the entrance of the outgoing sewer is 5020 ft. The value of E_1 at each sewer exit is derived as follows:

1. For lateral sewer one—$(E_1)_1$

$$H_b = K_b \times \frac{V_f^2}{2g} = 0.41 \times \frac{6.5^2}{2 \times 32.2} = 0.27 \text{ ft}$$

$$(E_1)_1 = E_4 + H_b = 5020 + 0.27 = 5020.27 \text{ ft}$$

2. For lateral sewer two—$(E_1)_2$

$$H_b = K_b \times \frac{V_f^2}{2g} = 1.32 \times \frac{8.5^2}{2 \times 32.2} = 1.48 \text{ ft}$$

$$(E_1)_2 = E_4 + H_b = 5020 + 1.48 = 5021.48 \text{ ft}$$

3. For incoming sewer on the main line—$(E_1)_3$

$$H_b = K_b \times \frac{V_f^2}{2g} = 0.05 \times \frac{10.0^2}{2 \times 32.2} = 0.08 \text{ ft}$$

$$H_m = \min(0.50, 0.25) = 0.25$$

$$H_m = \frac{V_{fo}^2}{2g} - K_m \frac{V_{fi}^2}{2g} = \frac{12.5^2}{2 \times 32.2} - 0.25 \times \frac{10^2}{2 \times 32.2} = 2.03 \text{ ft}$$

$$(E_1)_3 = E_4 + H_b + H_m = 5020 + 1.48 + 2.03 = 5023.83 \text{ ft}$$

A sewer system often has discontinuity points on its EGL because of manhole drops. Having completed the EGL and HGL analyses, the energy balance between two adjacent manholes is recalculated as

$$(E_1)_{\text{upstmanhole}} = (E_1)_{\text{dnstmanhole}} + H_m + H_b + HF \qquad (12.38)$$

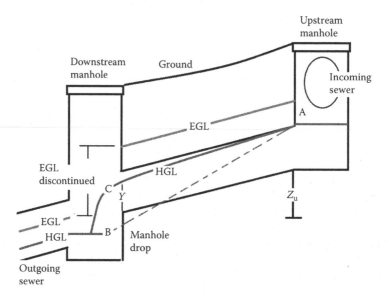

Figure 12.21 Plots of hydraulic gradient line and energy gradient line.

The unknown in Equation 12.38 is the term *HF* that is the lump sum of friction loss, manhole loss, and the dissipation caused by manhole drops and hydraulic jumps. Rearrange Equation 12.38 to yield

$$\text{Or } HF = (E_l)_{\text{upstmanhole}} - (E_l)_{\text{dnstmanhole}} - H_m - H_b \tag{12.39}$$

To transfer the calculated HGL and EGL to the sewer vertical profile, it is critically important that both HGL and EGL must be plotted through the sewer exit. For instance, between the two manholes in Figure 12.21, the line **AB** could be misinterpreted that the HGL falls below the sewer invert. In fact, the HGL must go through the water depth at the exit similar to the line **AC**. In doing so, we properly plot the HGL between these two manholes. It is important to know that the discontinuity of EGL occurs only across a manhole because of the manhole drop. The difference between EGL and HGL is always the flow kinetic energy.

12.9 Sewer trench

From the plan view and vertical profile, we can calculate the lengths of sewers and the heights of manhole tubes. As illustrated in Figure 12.22, the height of a manhole tube is calculated from the ground elevation to the lowest invert elevation among incoming and outgoing sewers at the manhole. According to the manufactory formula for ASTM C76 Reinforced Concrete Circular Sewer Type Wall B, the pipe wall thickness is calculated as

$$T = \frac{D}{12} + 1 \tag{12.40}$$

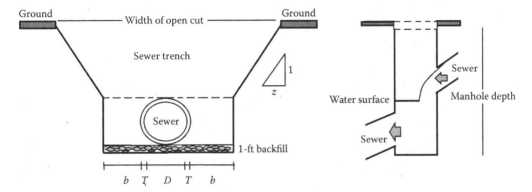

Figure 12.22 Sewer trench cross section.

where T = wall thickness in inches and D = pipe equivalent diameter in inches. For instance, a 48-in. circular pipe shall have a wall thickness of 5 in., or the external diameter for a 48-in. pipe is 58 in.

Along a sewer trench, we can calculate the excavated earth volumes. The engineer has to specify the side slope and minimum width for the sewer trenches. The bottom width of a sewer trench is set to be the outer diameter of the sewer plus 1–2 ft if the sewer is smaller than 48 in. in diameter, or 2–4 ft if the sewer is greater than or equal to 48 in. The additional space along the two sides of the sewer line is for convenience of on-site machinery operations. The excavated earth volume of a sewer trench is estimated by the trench cross-sectional area times the length of the sewer.

$$V_S = \frac{(A_U + A_D)}{2} L_S \tag{12.41}$$

where V_S = soil-excavated volume in $[L^3]$, A_U = sectional area at upstream end of trench in $[L^2]$, A_D = sectional area at downstream end of trench in $[L^2]$, and L_S = length of sewer line in $[L]$. Applying Equation 12.41 to the entire sewer system provides a basis for cost estimations.

EXAMPLE 12.5

A 24-in. circular sewer line has a length of 400 ft. The sewer line is buried 2 ft below the ground. Determine the soil-excavated volume.

Solution: A 24-in. circular pipe has a wall thickness of 3 in. As a result, the outer diameter of this 24-in. pipe is 30 in. or 2.5 ft. The excavated section is shown in Figure 12.23. Adding an additional space of 1.0 ft to both sides of the sewer line, the bottom width of the trench is 4.5 ft. Considering that the soil stable slope is IV:IH for a depth of 2 ft, the top width of the trench is 8.5 ft.

$$A_U = 4.5 \times 3.5 + \frac{(8.5 + 4.5)}{2} \times 2 = 28.75 \, \text{ft}^2$$

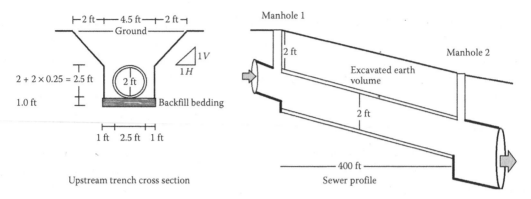

Figure 12.23 Excavation of sewer trench.

$$A_U = A_D$$

$$V_S = \frac{(28.75 + 28.75)}{2} \times 400 = 11,500\,\text{ft}^3.$$

12.10 Sewer HGL and EGL analyses

This section presents an example as shown in Figure 12.24. Detailed HGL and EGL analyses are summarized from Examples 12.1 through 12.6 and plotted in Figures 12.25 through 12.30.

EXAMPLE 12.6

Hydraulic analysis for sewer 4799 is shown in Figure 12.25.

Design information

Q (cfs)	Y_n (ft)	V_n (fps)	S_s (ft/ft)	Y_c (ft)	V_c (fps)	S_c (ft/ft)	N	F_r	L_s (ft)
56.7	2.34	6.04	0.25%	1.84	7.71	0.48%	0.013	0.70	410

where Q = design flow, Y_n = normal depth, V_n = normal velocity, S_s = sewer slope, Y_c = critical depth, V_c = critical flow velocity, S_c = critical flow slope, N = Manning's roughness, F_r = Froude number, and L_s = sewer length.

Manhole 99 is the system exit. The water surface elevation at the receiving lake is 87 ft, or the sewer exit is not submerged. The flow Froude number in sewer 4799 is 0.7 or subcritical flow. As a result, an M-2 curve is expected with the critical depth at the exit. If sewer 4799 is long enough, the normal depth can be developed at the entrance or manhole 47. The following sections provide the EGL analysis.

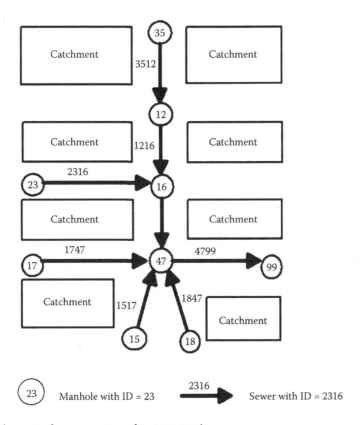

Figure 12.24 Layout of sewer system for case study.

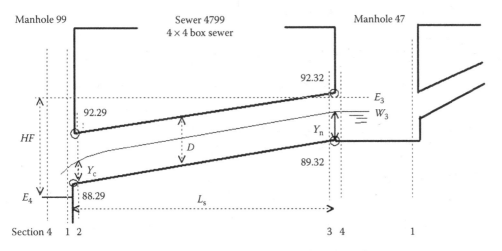

Figure 12.25 Hydraulic gradient line and energy gradient line along sewer 4799.

Manhole hydraulics

The lake water surface elevation is given at 87 ft or section 4. Assuming that the exit loss is negligible, the energy at section 1 is set to be

$$E_1 = 87.0 \, ft$$

Sewer hydraulics

1. At section 2, the critical flow depth occurs at the exit. Let $Y_2 = Y_c$ and $V_2 = V_c$. The energy sum for the critical flow is

    From the critical flow at the exit, we have

    $$E_e = Z_2 + Y_2 + \frac{V_2^2}{2g} = 91.05 \, ft$$

    From the downstream water surface, we have

    $$E_{22} = 87 \, ft$$

    In comparison, the higher one dictates the HGL at section 2 as

    $$E_2 = \max(E_e, E_{22}) = 91.05 \, ft$$

2. From section 2 to 3, it is an M-2 drawdown water surface profile. Assuming that the normal flow depth can be developed at the entrance, let $Y_a = Y_n = 2.34 \, ft$, $V_a = V_n = 6.04$ fps, and $S_a = S_s = 0.0025 \, ft/ft$. The EGL for the normal flow condition is

    $$E_a = Z_a + Y_a + \frac{V_a^2}{2g} = 89.32 + 2.34 + \frac{6.04^2}{2 \times 32.2} = 92.23 \, ft$$

    At section 2, we have $Y_b = Y_c = 1.84 \, ft$, $V_b = V_c = 7.71$ fps, and $S_b = S_c = 0.0048 \, ft/ft$. The EGL for the critical flow condition is

    $$E_b = Z_b + Y_b + \frac{V_b^2}{2g} = 88.29 + 1.84 + \frac{7.71^2}{2 \times 32.2} = 91.05 \, ft$$

    Using the direct step method, the distance required between these two sections is

    $$X = \frac{E_a - E_b}{0.5 \times (S_a + S_b)} = 322.45 < 410 \, ft$$

    It is concluded that this sewer is long enough to develop the normal depth at the entrance. Therefore, we have

    $$E_3 = Z_3 + Y_n + \frac{V_n^2}{2g} = 92.23 \, ft$$

    $$W_3 = E_3 - \frac{V_3^2}{2g} = 92.23 - \frac{6.04^2}{64.4} = 91.66 \, ft$$

3. At section 4, let $E_4 = E_3$ and $W_4 = W_3$ when neglecting entrance losses. It is concluded that the water surface elevation at manhole 47 is 91.66 ft, and its EGL is 92.23 ft.

Between manholes
Between manholes 47 and 99, the energy balance is calculated as

$$92.23 = 87.0 + HF \quad \text{so, } HF = 5.23 \, \text{ft}$$

It is noted that the value of 5.23 ft includes the manhole drop from 88.9 to 87.0 ft at the system outfall.

EXAMPLE 12.7

Hydraulic analysis for sewer 1847 is shown in Figure 12.26.

Design information

Q (cfs)	Y_n (ft)	V_n (fps)	Y_c (ft/ft)	V_c (ft)	V_f (fps)	N	F_r	S_s (ft/ft)	K_b
1.85	0.46	4.05	0.53	3.33	1.05	0.013	1.05	0.75%	1.0

From Example 12.1, the EGL at manhole 47 is found to be $E_4 = 92.23\,\text{ft}$ and $W_4 = 91.66\,\text{ft}$. Sewer 1847 carries a supercritical flow because the flow Froude number is 1.05.

Manhole hydraulics
At section 4, i.e., manhole 47, $E_4 = 92.23\,\text{ft}$ and $W_4 = 91.66\,\text{ft}$. Crossing manhole 47, i.e., from section 4 to 1, the bend loss is calculated as

$$H_b = K_b \frac{V_f^2}{2g} = 1.0 \times \frac{1.05^2}{2 \times 32.2} = 0.017\,\text{ft}$$

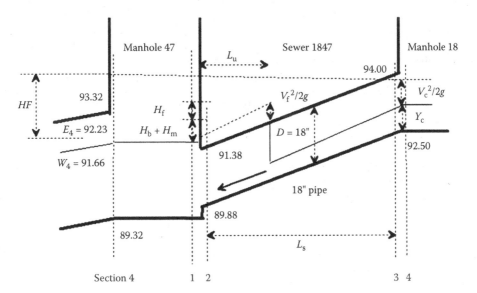

Figure 12.26 Hydraulic gradient line and energy gradient line along sewer 1847.

Note that sewer 1847 is not on the main line. Therefore, it does not have a lateral loss, i.e., $K_m = 0.0$. The HGL across manhole 47 is calculated as

$$E_l = E_4 + H_b + H_m = 92.23 + 0.017 + 0.0 = 92.25 \, \text{ft}$$

$$W_l = E_l - \frac{V_f^2}{2g} = 92.25 - \frac{1.05^2}{2 \times 32.2} = 92.23 \, \text{ft}$$

Sewer hydraulics

1. The EGL at section 2 takes consideration of tailwater depth at manhole 47 and the flow condition at the exit of sewer 1847.

    From the downstream water elevation, we have

    $$E_{22} = E_l = 92.25 \, \text{ft}$$

    At the sewer exit, we have

    $$E_e = \text{Crown elevation} + \frac{V_f^2}{2g} = 91.38 + \frac{1.05^2}{2 \times 32.2} = 91.39$$

    In comparison, the higher one dictates as

    $$E_2 = \max(E_e, E_{22}) = 92.25 \, \text{ft}$$

    The corresponding water surface elevation at section 2 is

    $$W_2 = E_2 - \frac{V_f^2}{2g} = 92.25 - \frac{1.05^2}{2 \times 32.2} = 92.23 \, \text{ft}$$

    Note that the case has a submerged exit because W_2 is above the sewer exit crown elevation at 91.38 ft. Therefore, the sewer is surcharged at its downstream end.

2. Because the exit is submerged, the surcharged length is

    $$S_f = \frac{N^2 V_f^2}{K_N R^{\frac{4}{3}}} = \frac{0.013^2 \times 1.05^2}{2.22 \times \left(\frac{1.5}{4}\right)^{\frac{4}{3}}} = 0.0003 \, \text{ft/ft}$$

    $$L_u = \frac{W_2 - C_w}{S - S_f} = \frac{(92.23 - 91.38)}{(0.0075 - 0.0003)} = 118.1 \, \text{ft}$$

3. From section 2 to 3, the flow is subject to the surcharged exit and the critical flow condition at the upstream end. The higher dictates the headwater at the sewer entrance or HGL at manhole 18. The energy required to overcome the friction losses due to the surcharged length is calculated as

    $$H_f = S_f \times L_u = 0.0003 \times 118.1 = 0.035 \, \text{ft}$$

$$E_{33} = E_2 + H_f = 92.23 + 0.035 = 92.27 \text{ ft}$$

The energy associated with the critical flow at the entrance is calculated as

$$E_c = Z_c + Y_c + \frac{V_c^2}{2g} = 92.50 + 0.53 + \frac{3.33^2}{2 \times 32.2} = 93.20 \text{ ft}$$

In comparison, the higher one is selected as

$$E_3 = \max(E_{33}, E_c) = 93.20 \text{ ft and } W_3 = Y_c + Z_c = 93.03 \text{ ft}$$

4. At section 4, $E_4 = E_3$ and $W_4 = W_3$ when neglecting entrance losses.

Between manholes
The energy balance between manholes 18 and 47 is

$$(W_4)_{18} = (W_4)_{47} + H_m + H_b + HF$$

$$93.20 = 92.23 + 0.017 + 0 + HF \quad \text{so, } HF = 0.96 \text{ ft}$$

EXAMPLE 12.8

Hydraulic analysis for sewer 1747 is shown in Figure 12.27.

Design information

Q (cfs)	Y_n (ft)	V_n (fps)	Y_c (ft/ft)	V_c (ft)	K_b	N	F_r	S_s (ft/ft)	L_s (ft)
2.1	0.38	5.99	0.58	3.37	1.21	0.013	2.02	2.0%	200

The HGL at manhole 47 is $E_4 = 92.23$ ft and $W_4 = 91.66$ ft. Sewer 1747 carries a supercritical flow.

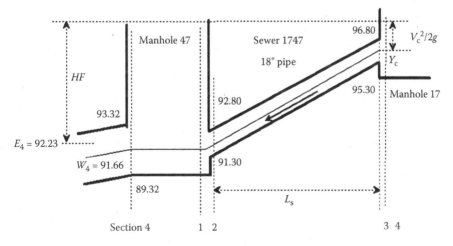

Figure 12.27 Hydraulic gradient line and energy gradient line along sewer 1747.

Manhole hydraulics

From section 4 to 1, we have

$$H_b = K_b \frac{V_f^2}{2g} = 1.0 \times \frac{1.21^2}{2 \times 32.2} = 0.023 \, \text{ft}$$

Sewer hydraulics

The sewer carries a supercritical open-channel flow with an unsubmerged exit. As a result, it is a case of upstream control. The energy at the entrance is dictated by the critical flow condition.

$$E_c = Z_{entrance} + Y_c + \frac{V_c^2}{2g} = 95.30 + 0.58 + \frac{3.37^2}{2 \times 32.2} = 96.06 \, \text{ft}$$

Therefore, the energy at section 3 is

$$E_3 = 96.06 \, \text{ft}$$

The water surface elevation is calculated as

$$W_3 = E_3 - \frac{V_c^2}{2g} = 95.88 \, \text{ft}$$

For this case, $W_4 = W_3$ and $E_4 = E_3$.

Between manholes

The energy balance between manholes 47 and 17 is

$$(E_4)_{17} = (E_4)_{47} + H_b + HF$$

$$96.06 = 92.23 + 0.023 + HF \quad \text{so,} \, HF = 3.81 \, \text{ft}$$

EXAMPLE 12.9

Hydraulic analysis of sewer 1647 is shown in Figure 12.28.

Design information

Q (cfs)	V_f (fps)	K_b	K_m	N	F_r	S_s (ft/ft)	L_s (ft)
35.6	8.94	0.05	0.25	0.013	0	−0.1%	200

This is a case of flow in a flat sewer. Sewer 1647 is on the main line. As a result, the manhole is subject to both bend and lateral losses.

Manhole hydraulics

From section 4 to 1, the bend and lateral losses are

$$H_b = K_b \frac{V_f^2}{2g} = 0.05 \times \frac{8.94^2}{2 \times 32.2} = 0.06 \, \text{ft}$$

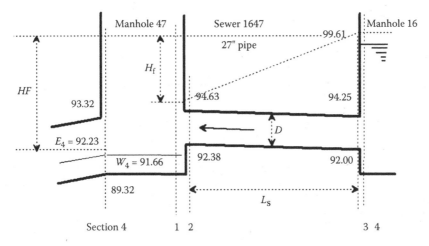

Figure 12.28 Hydraulic gradient line and energy gradient line along sewer 1647.

$$H_m = \frac{V_{fo}^2}{2g} - K_m \frac{V_{fi}^2}{2g} = \frac{3.52^2}{2 \times 32.2} - 0.25 \times \frac{8.94^2}{2 \times 32.2} = -0.12 \text{ or } 0.0 \text{ ft}$$

$$E_1 = E_4 + H_b + H_m = 92.23 + 0.06 + 0.0 = 92.29 \text{ ft}$$

Sewer hydraulics

1. At section 2, the downstream water surface elevation suggests the EGL as

    $$E_{22} = E_1 = 92.29 \text{ ft}$$

    The configuration of the sewer exit requests the EGL as

    $$E_e = Z_e + D + \frac{V_f^2}{2g} = 92.38 + 2.25 + \frac{8.94^2}{2 \times 32.2} = 95.87 \text{ ft}$$

    In comparison, the higher dictates

    $$E_2 = \max(E_{22}, E_e) = 95.87 \text{ ft}$$

    Under the flowing full condition, the water surface elevation at section 3 is

    $$W_2 = E_2 - D = 95.87 - 2.25 = 93.63 \text{ ft}$$

2. From section 2 to 3, the friction loss is

    $$H_f = S_f \times L_S = \frac{N^2 V_f^2}{K_N R^{\frac{4}{3}}} \times L_s = \frac{0.013^2 \times 8.94^2}{2.22 \left(\frac{2.25}{4}\right)^{\frac{4}{3}}} \times 200 = 4.98 \text{ ft}$$

The energy at section 3 is

$$E_3 = E_2 + H_f = 95.87 + 4.98 = 100.86 \, \text{ft}$$

The water surface elevation at section 3 is

$$W_3 = E_3 - \frac{V_f^2}{2g} = 100.86 - \frac{8.94^2}{2 \times 32.2} = 99.61 \, \text{ft}$$

Ignoring the entrance losses, let $E_4 = E_3$ and $W_4 = W_3$.

Between manholes

Energy balance between manholes 47 16 is

$$(E_4)_{16} = (E_4)_{47} + H_m + H_b + HF$$

$$100.86 = 92.23 + 0.25 + 0.06 + HF \quad \text{so,} \, HF = 8.32 \, \text{ft}$$

EXAMPLE 12.10

Hydraulic analysis of sewer 1547 is shown in Figure 12.29.

Design information

Q (cfs)	Required D (in.)	Existing D (in.)	V_f (fps)	N	S_s (ft/ft)	L_s (ft)	K_b
13.4	21	18	7.59	0.013	1.5%	295	0.4

This is an undersized lateral that carries full flow.

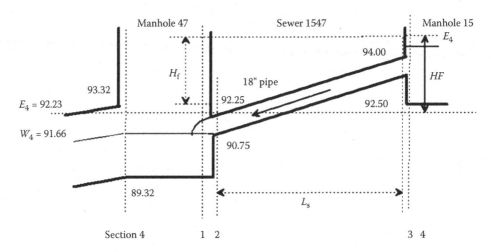

Figure 12.29 Hydraulic gradient line and energy gradient line along sewer 1547.

Manhole hydraulics

Crossing manhole 47, $K_m = 0$ and $K_b = 0.4$. The bend loss is

$$H_b = K_b \frac{V_f^2}{2g} = 0.40 \times \frac{7.59^2}{2 \times 32.2} = 0.36 \, \text{ft}$$

Sewer hydraulics

1. The HGL at section 2 is dictated by the manhole water surface or the sewer exit configuration, whichever is higher. For this case, it is obvious that the manhole drop at the sewer exit dictates the HGL at section 2 as

$$W_2 = 96.67 \quad \text{and} \quad E_2 = 96.67 + \frac{7.59^2}{2 \times 32.2} = 97.56 \, \text{ft}$$

2. From section 2 to 3, the friction loss is calculated as

$$H_f = S_f \times L_s = \frac{N^2 V_f^2}{K_N R^{\frac{4}{3}}} \times L_s = \frac{0.013^2 \times 7.59^2}{2.22 \times \left(\frac{1.5}{4}\right)^{\frac{4}{3}}} \times 295 = 4.78 \, \text{ft}$$

$$E_3 = E_2 + H_f = 97.56 + 4.78 = 102.34 \, \text{ft}$$

3. Neglecting the entrance losses, $E_4 = E_3$.

Between manholes

Energy balance between manholes 47 and 15 is

$$(E_4)_{15} = (E_4)_{47} + H_m + H_b + HF$$

$$102.34 = 92.23 + 0.36 + 0.0 + HF \quad \text{so,} \, HF = 9.75 \, \text{ft}$$

EXAMPLE 12.11

Hydraulic analysis of sewer 1216 is shown in Figure 12.30.

Q (cfs)	V_f (fps)	K_b	K_m	N	S_s (ft/ft)	L_s (ft)
23.5	7.48	0.05	0.25	0.013	0.8%	360

Sewer 1216 is completely surcharged. The arch pipe is 28-in. by 20-in. in dimension. Its equivalent diameter for this sewer is 24 in. Because sewer 1216 is on the main line, the exit is subject to both bend and lateral losses.

Manhole hydraulics

The bend and lateral losses are

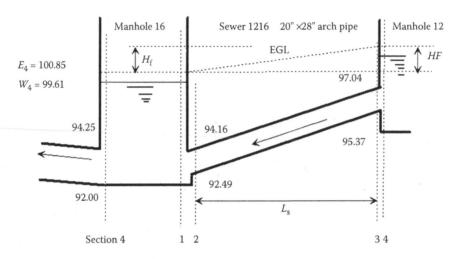

Figure 12.30 Hydraulic gradient line and energy gradient line along sewer 1216.

$$H_b = K_b \frac{V_f^2}{2g} = 0.05 \times \frac{7.48^2}{2 \times 32.2} = 0.043$$

$$H_m = \frac{V_{fo}^2}{2g} - K_m \frac{V_{fi}^2}{2g} = \frac{8.94^2}{2 \times 32.2} - 0.25 \times \frac{7.48^2}{2 \times 32.2} = 1.02 \text{ ft}$$

$$E_1 = E_4 + H_m + H_b = 100.85 + 1.02 + 0.043 = 101.91 \text{ ft}$$

Sewer hydraulics

1. The HGL at the sewer exit is dictated by either the full flow at the exit or the downstream manhole EGL, whichever is higher. So, we have

$$E_{22} = E_1 = 101.91 \text{ ft}$$

$$E_e = Z_e + \frac{V_f^2}{2g} = 94.16 + \frac{7.48^2}{2 \times 32.2} = 95.03 \text{ ft}$$

In comparison, the EGL at section 2 is determined as

$$E_2 = \max(E_{22}, E_e) = 101.91 \text{ ft}$$

$$W_2 = E_2 - \frac{V_f^2}{2g} = 101.91 - \frac{7.48^2}{2 \times 32.2} = 101.04 \text{ ft}$$

2. The downstream sewer end is under such high submergence that the entire sewer is surcharged. The friction loss through the sewer pipe is

$$S_f = \frac{N^2 V_f^2}{K_N R^{\frac{4}{3}}} = \frac{0.013^2 \times 7.48^2}{2.22 \times 2.0^{\frac{4}{3}}} = 0.0107$$

$$H_f = S_f \times L_s = 0.0107 \times 360 = 3.85 \, \text{ft}$$

$$E_3 = E_2 + H_f = 101.91 + 3.85 = 105.76 \, \text{ft}$$

$$W_3 = E_3 - \frac{V_f^2}{2g} = 104.92 \, \text{ft}$$

3. Ignoring entrance loss, we have $E_4 = E_3$ and $W_4 = W_3$.

Between manholes
Energy balance between manholes 12 and 16 is

$$(E_4)_{12} = (E_4)_{16} + H_m + H_b + HF$$

$$104.92 = 100.85 + 1.02 + 0.0043 + HF \quad \text{so,} \, HF = 3.0 \, \text{ft}$$

12.11 Homework

Q12.1 A sewer system shown in Figure Q12.1 consists of three sewers and five catchments. Design information is given as follows:

1. Design rainfall intensity formula

$$I = \frac{40}{(10 + T_d)^{0.74}}$$

in which T_d = duration in minutes, and I = intensity in in./h.

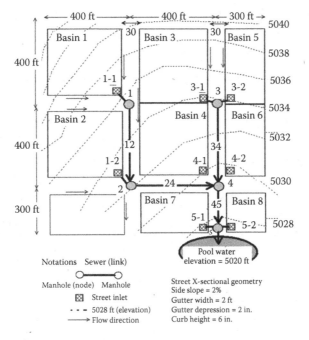

Figure Q12.1 Layout of sewer system.

2. Design rainfall duration for estimating the peaking runoff from a catchment is the time of concentration of the catchment. The overland flow time shall be estimated by the airport formula, and the gutter flow velocity is estimated by the SCS upland method using a conveyance factor of 20.

3. Design constraints include the permissible flow velocity in a sewer between 20 and 3 ft/s, a minimum coverage of 2 ft, a minimum sewer diameter of 18 in., and a minimum manhole drop of 0.50 ft.

4. Information of subbasins for inlet design:

Basin ID number	Area (acres)	Runoff coefficient	Overland		Channel	
			Slope (%)	Length (ft)	Slope (%)	Length (ft)
1	3.67	0.55	1.25	300.00	1.25	550.00
2	3.67	0.62	1.50	300.00	1.50	525.00
3	3.67	0.85	1.50	250.00	1.50	500.00
4	3.67	0.85	1.00	300.00	1.00	500.00
5	2.75	0.85	1.50	300.00	1.50	450.00
6	2.75	0.85	1.00	300.00	1.00	400.00
7	2.75	0.45	0.75	150.00	0.75	400.00
8	2.11	0.45	0.75	150.00	0.75	350.00

5. Information for manholes:

Manhole ID number	Ground elevation (ft)	Tributary area (acres)	Runoff coefficient	Overland	Overland	Gutter	Gutter
				Slope (%)	Length (ft)	Slope (%)	Length (ft)
1 (Bsn1)	5037.0	3.67	0.55	1.25	300.00	1.25	550.00
2 (Bsn2)	5032.0	3.67	0.62	1.50	300.00	1.50	525.00
3 (Bsn3 + 5)	5034.0	6.42	0.85	1.50	300.00	1.50	450.00
4 (Bsn4 + 6)	5029.5	6.42	0.85	1.00	300.00	1.00	500.00
5 (Bsn7 + 8)	5027.0						

6. Sewer information

Sewer ID	Length (ft)	Slope (%)	Upstream crown elevation (ft)	Diameter (in.)	Height or rise (in.)	Width or span (in.)	Bend loss coefficient	Lateral loss coefficient
12 (round)	400	1.00	5032.0				1	
24 (arch)	400	1.00	5027.5				1	0.25
34 (round)	400	1.35	5029.0				0.05	
45 (round)	300	1.00	5023.0				0.05	

Bibliography

AISI. 1980a. *Modern Sewer Design*, American Iron and Steel Institute, Washington, DC.

AISI. 1980b. *Handbook of Steel Drainage and Highway Construction Product*, American Iron and Steel Institute, Washington, DC.

ASCE. 1979. *Design and Construction of Sanitary and Storm Sewer*, American Society of Civil Engineers, New York.

Guo, J.C.Y. (1985). "Technical Manual for UDSEWER Computer Model." Research Report, Department of Civil Engineering, University of Colorado at Denver, Denver, CO.

Guo, J.C.Y. (1989a). "Energy Dissipation in Storm Sewer System," Proceedings of ASCE Conference on Stormwater Modeling and Management, held at Denver, CO.

Guo, J.C.Y. (1989b). "Auto Sizing Techniques for Storm Sewer System Design Using UDSWMM," Proceedings for the National EPA Conference, Denver, CO.

Mays, L. (2005). *Stormwater Collection Tools*, McGraw Hills, New York.

USWDCM. (2010). *Urban Storm Water Design Criteria Manual*, Urban Drainage and Flood Control District, Denver, CO.

Chapter 13

Detention basin design

One of the major tasks in stormwater management is to reduce peak flows after the development (ASCE, 1984, 1992). A stormwater drainage system consists of conveyance and storage facilities. Detention and retention basins are the major storage facilities designed for stormwater quantity and quality controls (EPA, 1986, 1994). Storage facilities in a drainage network should be placed at the strategic locations in order to effectively attenuate peak flood flows. During the preliminary studies, it is necessary to evaluate all feasible combinations of basin locations, storage volumes, and allowable release rates. Decision making relies on the impact assessments by numerical simulations for the entire watershed with and without the proposed detention basin. A detention basin may operate as an on stream facility if the inflow channel directly drains into the basin or as an off stream facility if stormwater is diverted into the basin from the inflow channel. Flow diversion is triggered when the flood flow exceeds the predetermined channel capacity.

This chapter presents the methods to calculate the required detention storage volume. *The rational volumetric method* is applicable to small watersheds using the continuity principle among inflow, outflow, and storage volumes. The *hydrograph method* is recommended if the detailed inflow hydrograph is available. At the preliminary stage, the basin geometry can be approximated by a regular cross-section and then refined with the detailed grading plan at the final stage. The outflow structure comprises low-flow and high-flow inlets as well as outfall pipes. With the proposed outflow structure, the basin's characteristic curve is formulated using orifice, weir, and culvert hydraulics applied to the outflow structure. The performance of the basin under design can then be evaluated by hydrograph routing to confirm that the flow release does not exceed the allowable.

13.1 Basics in stormwater detention

Urbanization results in more impervious areas. Reduction in soil infiltration leads to the increases in runoff volumes and peak flows. Pavements allow stormwater to move faster and to become more concentrated. Solids and pollutants in stormwater flow are increased when surface runoff washes urban streets. Pollutant sources include debris, dirt, and chemicals and contaminants from streets, open areas, and domestic and industrial areas. The increase of storm runoff in an urban area is closely related to the areal ratio of imperviousness (EPA, 1983). Using the Colorado Urban Hydrograph Procedure (CUHP), a sensitivity study of the areal imperviousness on the 100-year storm runoff was conducted. Considering that the baseline case has an area imperviousness of 5%, as shown in Figure 13.1, the peak discharge increases 3.84 times, and the runoff volume increases 1.61 times when the watershed's imperviousness increases to 90%. High peak flows on the streets result in public safety issues, and increased runoff volumes cause the

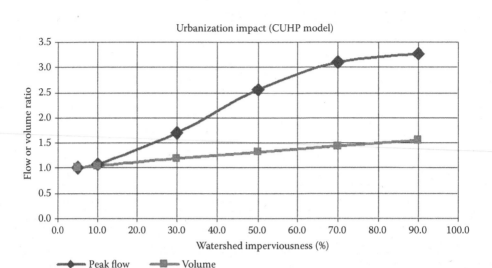

Figure 13.1 Sensitivity of runoff flows and volumes to watershed imperviousness.

deterioration of the water environment. Stormwater management is aimed at reducing the negative impact of urbanization on receiving water bodies. Stormwater best management practices (BMPs) include the following: (1) *source controls*, (2) *collection system controls*, (3) *storage and treatment*, and (4) *complex controls* incorporating the above elements together (Athayde, 1976). Stormwater detention reduces peak flows and also enhances stormwater quality. Since 1970, stormwater detention has been widely used as an effective drainage facility for flow release control.

A *flood-control detention basin* is designed to store the excess storm runoff associated with the increased watershed imperviousness. As shown in Figure 13.2, the stormwater detention storage volume is the volume difference between the inflow and outflow hydrographs. Between two adjacent storm events, the detention basin remains dry. During a major event, the operation of a flood detention basin is divided into a *filling period,* when the inflow rate is greater than the outflow rate, and a *depletion period,* when the outflow rate is greater than the inflow rate. These two distinct periods are separated by the peak outflow, which is selected based on the allowable flow release. Figure 13.2 shows that the higher the peak outflow, the lesser the storage volume. The peak outflow is an important design parameter and shall be determined according to the allowable flow released from the tributary watershed. According to the continuity principle, the total stored (detention) volume during the filling period is equal to the total released volume during the depletion period. In order to provide a large storage volume, a stormwater detention system is often blended into floodplains, depressed areas, recreational parks, and/or sport fields.

Design of a stormwater system shall not transfer any on site flooding problems to downstream properties. The allowable stormwater released at a design point is often determined by the following considerations:

1. Peak discharge under the predevelopment condition
2. Critical capacity of the downstream existing drainage facility
3. Allowable flow release published on the regional master drainage plan
4. Recommended flow release by the local design criteria

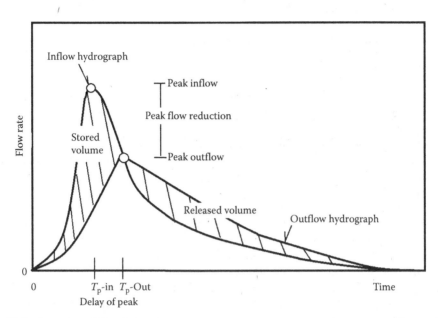

Figure 13.2 Concept of flood detention.

Table 13.1 Allowable flow release rates recommended for Denver area

Design frequency (year)	Type A soil (cfs/acre)	Type B soil (cfs/acre)	Type C or D soil (cfs/acre)
2	0.02	0.03	0.04
5	0.07	0.13	0.17
10	0.13	0.23	0.30
25	0.24	0.41	0.52
50	0.33	0.56	0.68
100	0.50	0.85	1.00

In addition to the above considerations, any design constraints associated with the project must also be reviewed. Among all considerations, the smallest release shall be adopted for design. Table 13.1 is a typical example in which the recommended flow releases are defined based on the imperviousness of 5% (USWSCM, 2010).

EXAMPLE 13.1

A lot of 100 acres is ready for development. The soil texture at the site is classified as type B soil. From the regional master drainage study, the flow release from this site was set to be 90 cfs. The downstream existing storm sewer line can take no more than 80 cfs from the site. Determine the 100-year peak flow release from the site.

Solution: According to Table 13.1, the 100-year release for type B soil is 0.85 cfs/acre or 85 cfs for 100 acres. Compared with the critical capacity for the existing sewer, the design release for this site is set to be

$$Q\text{-allowable} = \min\,(85, 90, 80) = 80\,\text{cfs}$$

13.2 Types of detention basins

13.2.1 Classification based on functionality

Detention basins are designed to meet different needs and also to achieve multiple purposes. Using the functionality as the criterion, detention basins are classified into three major categories: (1) *flood-control detention basin*, (2) *stormwater retention basin, and* (3) *infiltrating basin and trench.*

1. Flood-control detention basin (dry basin)
    A flood-control detention basin is placed at the major outfall point to temporarily store the excess storm runoff and then to discharge the stored water volume at a rate not more than the allowable. Between two sequential storm events, a flood-control detention basin remains dry and can be accessible as an open space for the public. As shown in Figure 13.3, a *local detention basin* serves a small, local tributary area for flow release control, and a regional detention basin is designed to mitigate the flood flows collected along the major waterway.
2. Stormwater retention basin (wet basin)
    A *retention basin* (as shown in Figure 13.4) is installed at the low point and operated with a permanent wet pool. The basin is sized to capture stormwater for the purposes of groundwater recharge, water quality enhancement, and/or local runoff volume disposal. A retention basin is often mixed with wetland features to settle solids and pollutants in stormwater (Shueler and Helfrich, 1989).
3. Infiltrating basin and trench (porous basin)
    Infiltrating basins and trenches are utilized as the common low-impact development (LID) devices to reduce the increased runoff volume. The infiltrating basins (as shown in Figure 13.3) include rain gardens, infiltration pools, riprap trenches, vegetation beds, etc. An infiltrating basin consists of an on-surface storage volume, landscaping vegetation bed, highly porous bottom, and overflow weir. They are often located at the outlet of an industrial park, a business district, or a highway intersection (see Figure 13.3).

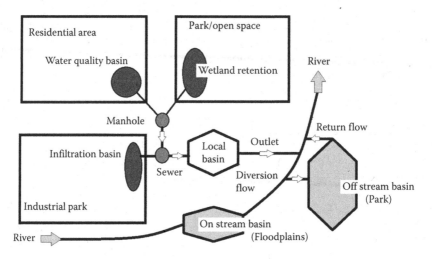

Figure 13.3 Types and locations of detention basins.

Figure 13.4 Examples of storage basins. (a) Detention basin, (b) retention basin, (c) infiltration basin, and (d) detention and infiltration basin.

13.2.2 Classification based on location

Basin location is an important factor in determining the collection of stormwater. Upstream locations serve the purpose of source control of pollutants, whereas downstream locations are more favorable to the reduction of peak flows. Based on the location, detention basins are classified into the following:

1. Upstream and downstream basins
2. On stream and off stream basins
3. Local and regional basins
4. On site and off site basins

An *upstream basin* shall be placed to focus on solids removal, whereas a *downstream basin* shall be installed for the purpose of peak flow reduction. For instance, pollutants carried in storm runoff from an industrial park shall be collected into an *upstream basin*, which is placed at the outfall point of the industrial park for water quality enhancement. A trunk line of a sewer system needs a *downstream basin* to control the flow release. Widening the floodplain width and constructing an embankment across the floodplain bottom create *on stream detention* (see Figure 13.5). Diverting the excess flood flow from a waterway into the adjacent open area such as depressed parks and sport fields to control stormwater release is termed *off stream detention* (Figure 13.5). An *on site detention* is implemented to dispose the increased stormwater at a building site. When the easement is available, the increased stormwater may be safely conveyed to a downstream *off*

Figure 13.5 Offstream and onstream detention basins. (a) Offstream detention basin (sport field) and (b) onstream detention basin (floodplain).

site detention basin for flow release control. An on site detention is often recommended as a source control when the high concentration of pollutants is the major problem in stormwater.

Storage of stormwater becomes *regional* when the facility is sized for a large tributary area and located on the major waterway. A *local detention basin* serves a small residential subdivision, industrial park, or business district to store the runoff flows before entering the major waterway. All upstream, local detention basins and the downstream regional detention system have to be simulated altogether in a numerical model to ensure that the flow releases are well coordinated to collectively reduce the peak flows.

13.3 Design considerations

Design of a detention system is an integration of functional integrity, land value, esthetics, recreation, and safety to merge into the urban setting. From the engineering aspect, the design of stormwater detention basin shall take the following factors into consideration.

13.3.1 Location

In an urban area, parking lot, parks, sport fields, road embankments, and depressed areas provide stormwater storage volumes. Selection of basin site depends on costs, public safety, and maintenance. It is important to impose the concept of multiple land uses to blend a stormwater detention feature into a park, ball field, and/or green belt. A large detention system shall also provide recreational functions, including jogging, walking, bicycling, playground, skating, golfing, etc. Different specialists need to work together as a joint effort to develop desirable and acceptable criteria that fit the community recreational needs as well as the flood mitigation purposes.

13.3.2 Basic layout

The basic elements (see Figure 13.6) for a detention basin include *inlet structure* to collect runoff flows, *energy dissipation system* for erosion control at the entrance, *fore bay* for sediment settlement, *trickle channel* to pass frequent nuisance flows, *storage basin* for

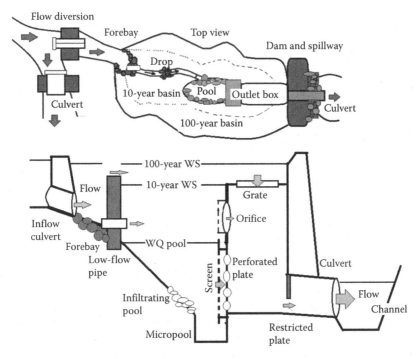

Figure 13.6 Layout of detention basin for multiple storm events.

mitigation of design events, and *outlet structure* to control flow releases. At the entrance, a proper energy dissipater shall be designed for erosion protection. A trickle channel or a low-flow channel shall be installed through the bottom of the basin to pass frequent nuisance flows. In general, the capacity of a trickle channel is 1.0%–3.0% of the 100-year peak discharge, and the low-flow channel shall pass 1/3 to 1/2 of the 2-year peak flow. Proper drop structures shall be placed along the trickle channel to reduce erosion. The trickle channel drains into the permanent pool for stormwater quality control. The permanent pool is directly connected to the outlet structure (USWSCM, 2010). An outlet structure is formed with perforated plate, riser, orifices, and weirs to collect low to high flows into the concrete vault, and the outfall pipes discharge the water flow from the concrete vault into the downstream receiving water body.

The basin width to length ratio must be >2 so that the flood flows can be sufficiently expanded and diffused into the water body to enhance the sedimentation process. Slopes on embankments have to maintain the bank slope stability. As a rule of thumb, slopes on earthen embankments shall not be steeper than 1V:4H and on riprap embankments shall not be steeper than 1V:2H. The cross-sectional geometry of the basin shall be designed for multiple events. As shown in Figure 13.6, the lower storage volume in a basin is shaped from the water quality control volume to 10-year storage volumes. From the 10-year water surface up to the weir crest shall provide an additional storage volume to accommodate the 100-year event. From the weir crest up to the brimful of the basin is the height of the freeboard. To mimic the predevelopment watershed hydrologic condition, it is preferred to drain the low storage volume over 6–48 h, whereas the 100-year storage volume shall be emptied out over no more than 24–72 h.

13.3.3 Groundwater impacts

The detention basin operates with a dry ground between two adjacent events. It is necessary to ensure that the drain time of the basin is not to exceed the average time interval between two adjacent storm events (average interevent time). On the contrary, a retention basin is designed to be a wet pool. Care must be taken in assessing infiltration to and exfiltration from the local ground water table. It is necessary to carefully evaluate the water budget among groundwater, surface water, and associated hydrologic losses. To design a retention basin without an outlet, soil infiltration tests must be carefully investigated. Although vertical drainage wells backfilled with aggregate gravel can be installed to increase the infiltration rate, the subsurface soil hydraulic conductivity must be carefully examined to ensure that the subsurface geometry sustains the infiltration rate on the land surface. Otherwise, the soil medium below the basin will become saturated and results in water mounding to the local groundwater system (Guo, 2007).

13.3.4 Inlet and outlet works

Inlets and outlets of a detention basin shall be protected from erosion and deposition of sediments. Design parameters shall be selected with consideration of trash racks, inlet grates, backwater surcharge, etc. Orifice and weir coefficients must be selected according to the postconstruction operations. Trash rack is critically important with regard to the public safety. As a rule of thumb, a trash rack is absolutely needed at the entrance of any outfall pipe larger than 18 in. (450 mm) in diameter. The outlet system for a basin must be designed with the full understanding of the downstream tailwater effects. The performance of the outfall culverts must be evaluated for a range of headwater depths at the entrance and tailwater depths at the exit. It is preferable that the outfall pipe is designed under the condition of inlet control.

13.3.5 Others

Operations of a stormwater detention system also involve many institutional issues, including the infrastructure needed to ensure proper planning, design, construction, operation, and maintenance. A monitoring or regulatory mechanism is required to ensure that the approved design is constructed, the operational integrity is implemented, and the maintenance is regularly provided. Other considerations also include public safety, access facilities, landscaping, and esthetics.

13.4 Design procedure

The main objective of stormwater detention is to mitigate the increased storm runoff peak flow rates. Although the design event for a detention basin is often specified to be the major event such as 50- to 100-year storm, the operations of a detention basin need to accommodate events of all kinds. Inflows to the detention basin shall be studied for both the existing and future conditions. For each project development, it is necessary to identify the changes and mitigation measures between the predevelopment and postdevelopment conditions. This effort is an attempt to identify the existing and future flood problems. It will serve as a basis for impact evaluations and alternative selections.

Figure 13.7 outlines the design steps beginning with the basin site selection. During the stage of preliminary design, little information is available. Therefore, it is suggested that the basin geometry be approximated by a triangular, rectangular, or circular shape

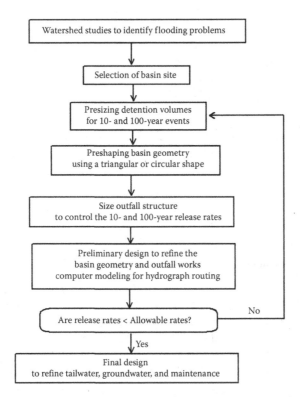

Figure 13.7 Procedure for detention basin design.

and that the basin operation be approximated by the inlet-control capacity determined by weir and orifice hydraulics only. Of course, when the project moves to the final stage, the preliminary design can be further refined with more information. For instance, the tail-water and backwater effects shall be considered to refine the basin characteristic curves, and the basin performance must be evaluated by hydrologic routing techniques. The abovementioned procedure is an iterative process until all design criteria and safety concerns are satisfied.

13.5 Detention volume

Detention volume is defined as the difference between the inflow and outflow volumes. The inflow volume is generated from the tributary area under a specified design storm, whereas the outflow volume is determined according to the flow release control. There are empirical methods developed to calculate detention volumes. In general, the *hydrograph method* is recommended for large watersheds, and the *volumetric method* is suitable for urban watersheds less than 150 acres (Aron and Kibler, 1990).

13.5.1 Hydrograph method

The inflow hydrograph to the detention site shall be predicted for the future developed condition. The allowable release is determined by not exceeding the existing flow release or the critical capacity of the downstream drainage facilities. At the planning stage, the

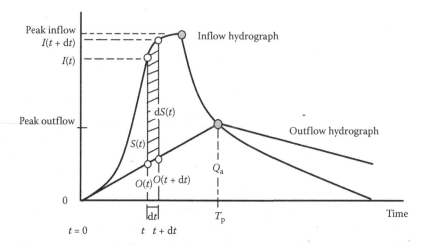

Figure 13.8 Detention volume by hydrograph method.

detailed knowledge of the outlet structure is not known. As a result, the after-detention hydrograph (as shown in Figure 13.8) is approximated by the linear rising hydrograph from the beginning of the event to the allowable release rate on the inflow recession hydrograph (Guo, 1999). The required detention storage volume is then calculated by the volume difference between the inflow and outflow hydrographs (McCuen, 1998). As shown in Figure 13.8, the outflow rate, $O(t)$, at time t on the linear rising limb is estimated as

$$O(t) = \frac{O_a}{T_p} t \quad \text{for } 0 \le t \le T_p \tag{13.1}$$

in which $O(t)$ = outflow rate in [L^3/T] at time t, Q_a = allowable flow release in [L^3/T], T_p = time to peak on after-detention hydrograph in [T], and t = elapsed time in [T]. The accumulated storage volume, $S(t)$, is the volume difference between the inflow hydrograph and the rising outflow hydrograph:

$$S(t) = \sum_{t=0}^{t=T_p} [I(t) - O(t)]\Delta t \tag{13.2}$$

in which $S(t)$ = cumulative storage volume in [L^3], $I(t)$ = inflow rate in [L^3/T] at time t, and Δt = time increment in [T].

The design storage volume is then calculated by Equation 13.2 from $t = 0$ to $t = T_p$ as

$$S_m = S(T_p) \tag{13.3}$$

in which S_m = detention storage volume in [L^3]. The hydraulic performance of a detention basin is described by its characteristic curve, i.e., *storage-outflow curve*. For instance, HEC HMS (2005), and US EPA SWMM (2005) computer models define the performance of a detention basin by such a curve. During the preliminary study, design information is not adequately available; the pairs (S, O) in Equations 13.1 and 13.2 can serve as the preliminary *storage-outflow curve* for the basin under design. Of course, such a preliminary relationship can be refined after the detailed information becomes available (Guo, 2004).

EXAMPLE 13.2

As shown in Table 13.2, the inflow hydrograph to a detention basin has a peak flow of 750 cfs. The allowable flow released from the basin is set to be 250 cfs. On the recession curve of the inflow hydrograph, the flow of 250 cfs occurs at 60 min. Under the assumption of a linear rising outflow hydrograph, Equation 13.1 becomes

$$O(t) = \frac{250}{60}t \tag{13.4}$$

With a time increment of 5 min, the cumulative storage volume, $S(t)$, is computed as

$$S(t) = \sum_{t=0}^{t=60}\left[I(t) - \frac{250}{60}t\right] \times (10 \times 60)/43,560 \tag{13.5}$$

Note that $Q(t)$ is expressed in cfs, and $S(t)$ is expressed in acre-ft. As shown in Table 13.2, the cumulative volume $S(T_p = 60) = 17.56$ acre-ft. The pairs of (O, S) in Table 13.2 can serve as the preliminary storage-outflow curve for the detention basin under design.

EXAMPLE 13.3

A detention basin is sized for a tributary watershed of 40 acres, located in Denver, CO. The soil texture in the tributary area is classified as type B soil. According to Table 13.1, the 10- and 100-year allowable flow release rates are $Q_{10} = 0.23 \times 40 = 9.20$ cfs and $Q_{100} = 0.85 \times 40 = 34.0$ cfs. The 10- and 100-year inflow hydrographs are given in Table 13.3. The detention volumes for the 10- and 100-year events are determined as shown in Table 13.3.

Table 13.2 Preliminary storage-outflow curve by hydrograph method

Time (min)	Given inflow, I(t) (cfs)	Linear outflow, O(t) (cfs)	Incremental volume (acre-ft)	Cumulative volume, S(t) (acre-ft)
0.0	0.00	0.00	0.00	0.00
10.0	50.00	41.67	0.11	0.11
20.0	250.00	83.33	2.30	2.41
30.0	750.00	125.00	8.61	11.02
40.0	500.00	166.67	4.59	15.61
50.0	350.00	208.33	1.95	17.56
60.0	250.00	250.00	0.00	17.56
70.0	200.00	250.00	–	–

Table 13.3 Detention volumes for detention basin

Time (min)	10-year detention volume				100-year detention volume			
	Peak time = 90.0 min				Peak time = 70.0 min			
	10-year hydrograph (cfs)	Linear outflow (cfs)	Incremental volume (acre-ft)	Accumulated volume (acre-ft)	100-year hydrograph (cfs)	Linear outflow (cfs)	Incremental volume (acre-ft)	Accumulated volume (acre-ft)
0.00	0.00	0	0.00	0.00	0.00	0.00	0.00	0.00
10.00	19.00	1.02	0.25	0.25	17.00	4.86	0.17	0.17
20.00	46.00	2.04	0.61	0.85	64.00	9.71	0.75	0.91
30.00	51.00	3.07	0.66	1.51	108.00	14.57	1.29	2.20
40.00	40.00	4.09	0.49	2.01	103.00	19.43	1.15	3.35
50.00	29.00	5.11	0.33	2.34	78.00	24.29	0.74	4.09
60.00	22.00	6.13	0.22	2.56	56.00	29.14	0.37	4.46
70.00	17.00	7.16	0.14	2.69	37.00	34.00	0.04	4.50
80.00	13.00	8.18	0.07	2.76	22.00	0.00	0.00	4.50
90.00	10.00	9.20	0.01	2.77	13.00	0.00	0.00	4.50
100.00	9.00	0.00	0.00	2.77	10.00	0.00	0.00	4.50
110.00	7.00	0.00	0.00	2.77	7.00	0.00	0.00	4.50
120.00	4.00	0.00	0.00	2.77	4.00	0.00	0.00	4.50

100-year storage volume in acre-ft = 2.77

10-year storage volume in acre-ft = 4.50

13.5.2 Volumetric method

To model watersheds less than 150 acres, the assumption of uniform rainfall is acceptable for runoff volume predictions. Therefore, the required storage volume for a small watershed can directly be estimated by the volume difference between the rainfall volume on the tributary watershed and the runoff volume released from the basin. For simplicity, trapezoidal hydrographs in Figure 13.8 are considered for volume calculations (FAA, 1970, 1977).

On stream detention volume

As shown in Figure 13.9, the inflow hydrograph has a linear rising limb over the time of concentration of the tributary watershed, and the peaking portion of the inflow hydrograph is a plateau from the time of concentration, T_c, to the end of the rainfall event. For the specified rainfall duration, the effective rainfall volume into the basin is represented by the area *abce* in Figure 13.9.

$$V_i = CAI_dT_d \tag{13.6}$$

in which V_i = effective rainfall volume in $[L^3]$, C = runoff coefficient, A = tributary area in $[L^2]$, I_d = rainfall intensity in $[L/T]$, and T_d = rainfall duration in $[T]$.

Care must be taken when using variables in different units. Ensure all variables in Equation 13.6 are converted into feet-second or meter-second for further computations.

The outflow volume is

$$V_o = \frac{1}{2}Q_a\left(T_d + T_c\right) \tag{13.7}$$

The detention volume is the difference between inflow and outflow volumes as

$$V_d = V_i - V_o \tag{13.8}$$

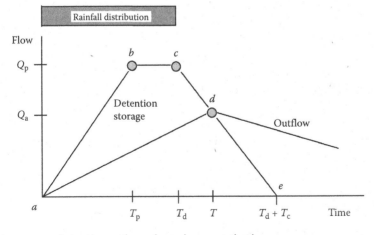

Figure 13.9 In stream detention volume by volume method.

For convenience, the outflow volume is calculated by the average outflow over the rainfall duration (FAA, 1970) as

$$V_o = QT_d \tag{13.9}$$

The average flow release can be related to the allowable or peak outflow rate as

$$Q = mQ_a \tag{13.10}$$

Aided by Equations 13.7, 13.9, and 13.10, the value of m is found to be (Guo, 1999)

$$m = \frac{1}{2}\left(1 + \frac{T_c}{T_d}\right) \text{ for } T_d \geq T_c \tag{13.11}$$

The basic concept used in the volume method is to find the maximum volume difference between the inflow and outflow volumes for a range of storm events in terms of rainfall duration. The design detention storage, S_m in [L^3], is set to be the maximal volume difference

$$S_m = \max(V_i - V_o) = \max(CAI_dT_d - mQ_aT_d) \quad \text{for } T_d \geq T_c \tag{13.12}$$

Equation 13.12 will be operated over a range of rainfall duration. Start from rainfall duration equal to T_c, and then use an increment of 5, 10, or 15 min for storm duration to compute the inflow and outflow volumes until the maximum storage volume is identified (Guo, 1999; Urbonas and Stahre, 1992).

EXAMPLE 13.4

The watershed to the Denver detention basin in Example 13.2 has developed with its imperviousness area ratio of 0.60. The time of concentration is 15 min. The runoff coefficients are 0.61 and 0.76 for the 10- and 100-year events. Allowable release rates have been determined to be 9.2 and 34 cfs for the 10- and 100-year events. The design rainfall intensity–duration–frequency (IDF) curve is

$$I(\text{in./h}) = \frac{28.5P_1}{(10+T_d)^{0.789}}$$

in which $P_1 = 1.61$ in. for 10-year event and 2.60 in. for the 100-year event and $T_d = $ rainfall duration in minutes. Determine the detention storage volume by the rational volumetric method.

Solution:

1. Let us start with the 100-year 30-min storm event as an example. The rainfall IDF curve at the project site is described as

$$I_d = \frac{28.5 \times 2.6}{(10+T_d)^{0.789}} = \frac{74.1}{(10+30)^{0.789}} = 4.02 \text{ in./h}$$

$$V_i = CI_aAT_d = 0.76 \times (4.02/12) \times 40 \times (30/60) = 5.10 \text{ acre-ft}$$

Table 13.4 Detention volume by volumetric method

Duration (min)	Rainfall intensity	Inflow volume (acre-ft)	Peak runoff (cfs)	Adjustment factor (m)	Outflow volume (acre-ft)	Storage volume (acre-ft)
30.00	4.03	5.11	122.65	0.75	1.05	4.06
40.00	3.38	5.71	102.85	0.69	1.29	4.43
70.00	2.33	6.90	70.98	0.61	1.99	4.91
80.00	2.13	7.19	64.68	0.59	2.22	4.96
90.00	1.96	7.44	59.52	0.58	2.46	4.98
100.00	1.82	7.67	55.21	0.58	2.69	4.98
110.00	1.70	7.88	51.55	0.57	2.93	4.95
120.00	1.59	8.07	48.39	0.56	3.16	4.90

2.   Outflow runoff volume is calculated as

$$V_o = m Q_a T_d = \frac{1}{2}\left(1 + \frac{15}{30}\right) \times 34 \times (30 \times 60)/43,560 = 1.05\,\text{acre-ft}$$

3.   Stormwater storage volume, V_d, for the 30-min rain storm is the volume difference as

$$V_d = V_i - V_o = 5.10 - 1.05 = 4.05\,\text{acre-ft}$$

Repeating this process (as shown in Table 13.4), the maximized storage volume for this example is 4.98 acre-ft for the event with a duration of 100 min.

Repeating the same process, the 10-year detention storage volume is determined to be 2.52 acre-ft. It is noted that both the rational volumetric method and hydrograph method produce good agreement on the detention storage volumes for this case.

Off stream detention volume

As illustrated in Figure 13.10, an off stream detention allows the channel to carry its base flow, Q_1, and only diverts the peak runoff volume into the basin, which is located outside of the floodplain. In practice, a diversion weir is installed along the channel bank to divert the peak flow into the basin. The basin is designed to release its peak flow, Q_2. The sum of Q_1 and Q_2 must not exceed the allowable release, Q_a, based on the predevelopment condition. In practice, the straight-through capacity, Q_1, is the maximum allowable in the downstream channel. Flow diversion begins at the preset flow rate, Q_1. The outflow volume, the area of *abefg* in Figure 13.10, can be calculated as two trapezoids, *bef* and *abfg*, as follows (Guo, 2012; Guo and Clark, 2006):

$$V_o = \frac{Q_2}{2}(T_d + T_c - 2T_1) + \frac{Q_1}{2}[(T_d + T_c - 2T_1) + (T_d + T_c)] \qquad (13.13)$$

in which V_o = outflow volume in $[L^3]$ and T_1 = time to begin flow diversion in $[T]$.

On the linear rising limb, the peak inflow occurs at the time of concentration, T_c, and the diversion flow is triggered at

$$T_1 = \frac{Q_1}{Q_p} T_c \qquad (13.14)$$

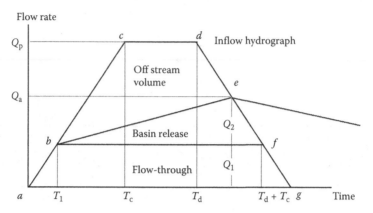

Figure 13.10 Off stream detention volume.

in which Q_1 = downstream channel capacity equal to the allowable flow-through capacity in $[L^3/T]$ and Q_p = peak inflow in $[L^3/T]$ at T_c. The peak inflow is calculated as

$$Q_p = CIA \tag{13.15}$$

in which C = runoff coefficient and A = watershed tributary area in $[L^2]$.
For mathematical convenience, the outflow volume, V_o in $[L^3]$, is expressed by the average release over the rainfall duration, T_d, as

$$V_o = mQ_aT_d \tag{13.16}$$

in which m = volume adjustment factor. Equating Equation 13.13 through Equation 13.16 yields

$$m = \frac{1}{2}\left[1 + \frac{(T_c - 2T_1)}{T_d}\right] + \frac{Q_1}{Q_2}\left[1 + \frac{(T_c - T_1)}{T_d}\right] \text{ for } T_d > T_c \tag{13.17}$$

For an in stream detention basin, $Q_2 = 0$ and $T_1 = 0$. As a result, Equation 13.17 is reduced to Equation 13.11. When using Equations 13.13 through 13.17, convert all variables to the units of feet-second or meter-second for calculating water volumes.

EXAMPLE 13.5

The inflow channel collects stormwater generated from a tributary area of 62 acres in Denver, CO. The runoff coefficient for the tributary area is $C = 0.68$. The time of concentration of the watershed is $T_c = 20$ min. The total allowable stormwater release is set to be 62 cfs. The channel is so undersized that the downstream capacity is limited to 15 cfs. The detention basin

Table 13.5 Example for off stream detention volume

Duration (min)	Rainfall intensity (in./h)	Inflow volume (acre-ft)	Peak runoff (cfs)	Diversion time, T_I (min)	Coefficient, m	Outflow volume (acre-ft)	Storage volume (acre-ft)
40.00	3.42	8.02	144.32	2.08	1.16	3.00	5.01
50.00	2.97	8.68	125.05	2.40	1.08	3.51	5.18
60.00	**2.63**	**9.23**	**110.78**	**2.71**	**1.03**	**4.01**	**5.22**
70.00	2.37	9.70	99.75	3.01	1.00	4.52	5.18
80.00	2.16	10.10	90.93	3.30	0.97	5.02	5.08

Note: 1 in. = 25.4 mm, 1 ft = 0.305 m, 1 acre = 0.4 hectare.

is sized to receive the excess stormwater from the channel. Determine the 100-year detention volume for this off stream detention basin.

Solution: For this case, the allowable release from the basin is 47 cfs (62–15 = 47). Try $T_d = 50$ min using Denver rainfall IDF formula with $P_1 = 2.6$ in. The calculations are summarized as follows:

1. Inflow volume

$$I = \frac{28.5 \times 2.60}{(10+50)^{0.789}} = 2.97 \, \text{in./h}$$

$$Q_p = CIA = 0.68 \times 2.97 \times 62 = 125.2 \, \text{cfs}$$

$$V_i = CIAT_d = 0.68 \times 2.97 \times 62 \times 50 = 8.68 \, \text{acre-ft}$$

2. Outflow volume

$$T_I = \frac{Q_I}{Q_p} T_c = \frac{15}{125.2} \times 20 = 2.40 \, \text{min}$$

$$m = \frac{1}{2} \left[1 + \frac{20 - 2 \times 2.4}{50} \right] + \frac{15}{47} \left(1 + \frac{20 - 2.4}{50} \right) = 1.1$$

$$V_o = mQ_a T_d = 1.1 \times 47 \times 50 \times 60/43,560 = 3.51 \, \text{acre-ft}$$

3. Stormwater detention volume, S_d, for the 50-min rain storm is

$$V_d = 8.68 - 3.15 = 5.18 \, \text{acre-ft}$$

Repeating this process for the range of rainfall duration from 40 to 80 min, Table 13.5 summarizes the variation of detention storage volumes. The maximum storage volume is identified to be 5.22 acre-ft with a storm duration of 60.0 min.

13.6 Preliminary shaping

Hydraulic structures are designed to process flows generated from small to extreme events. As a result, a detention basin is built with multiple layers of storage volume, starting from the bottom layer for the 2-year storage volume, the mid layer to store up to the 10-year detention volume, and the additional top layer to accommodate the 100-year storage volume. After knowing the detention volumes, the cross-sections of the basin can be approximated by a truncated cone with a base of circular, triangular, or rectangular shape. Refinements to the basin shape can always be added to the future grading plan after the detailed information becomes finalized.

13.6.1 Rectangular basin

The basic geometric parameters are the width and length of the cross-sectional area for each layer in a rectangular basin as shown in Figure 13.11. The cumulative storage volume between two cross-sectional areas is calculated as

$$L_2 = L_1 + 2zH \tag{13.18}$$

$$B_2 = B_1 + 2zH \tag{13.19}$$

$$A_1 = L_1 B_1 \tag{13.20}$$

$$A_2 = L_2 B_2 \tag{13.21}$$

$$V = \frac{1}{3}\left(A_1 + A_2 + \sqrt{A_1 A_2}\right) \approx 0.5 \times (A_1 + A_2)H \tag{13.22}$$

in which L = length in [L], B = width in [L], z = average side slope, H = vertical distance in [L], and V = storage volume in [L^3]. The subscript 1 represents the variables at the bottom layer, and 2 represents the variables at the top layer.

Usually, a basin is divided into multiple layers, and the slope of the basin's embankment varies with respect to the water depth, starting from as steep as 1V:2H at the bottom to as flat as 1V:10H on the top surface.

13.6.2 Elliptical basin

To calculate the volume in an elliptical basin (as shown in Figure 13.12), an inverted cone with a truncated bottom is used to calculate the storage volume for the basin under design. The bottom area and side slope are the required geometric parameters. The long and short radii, B_2 and L_2, of the upper layer can be estimated by Equations 13.18 and 13.19. The cross-sectional area is calculated as

$$A_2 = \frac{1}{4}\pi L_2 B_2 \tag{13.23}$$

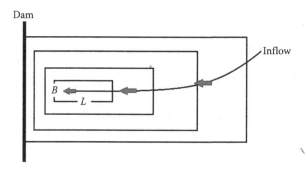

Figure 13.11 Preshaping for rectangular basin.

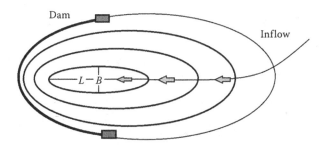

Figure 13.12 Preshaping for elliptical basin.

After the top and bottom areas are known, Equation 13.22 is then used to estimate the storage volume between two layers of a basin.

13.6.3 Triangular basin

Repeating the abovementioned procedure, the cross-sectional area in a triangular basin (as shown in Figure 13.13) is calculated by the base width, B_2, and height, L_2, as

$$A_2 = 0.5B_2L_2 \tag{13.24}$$

The volume between two triangular layers is estimated by Equation 13.22.

Applying Equations 13.18 through 13.24 to the 2-, 10-, and 100-year storage volumes, the basin shape can be approximated at various stages. Between two adjacent layers, the required side slope can be incorporated into the storage volume calculation. The *stage-storage curve* and *stage-contour area curve* can then be established. Upon completion of preshaping and presizing a basin, the engineer can begin to work on the hydrologic and hydraulic modeling and evaluations. During the stage of final design, the detailed shape and contours of the detention basin can be further outlined.

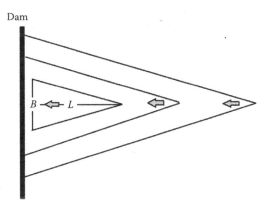

Figure 13.13 Preshaping for triangular basin.

Table 13.6 Example for preshaping of detention basin

Water surface elevation (ft)	Basin side slope (ft/ft)	Width of cross-section (ft)	Length of cross-section (ft)	Cross pond (acres)	Sectional area (ft²)	Accumulated pond (acre-ft)	Storage volume (cfs)	Identify design water elevation
5000.00	3.00	200.00	350.00	0.80	35,000.0	0.00	0.0	
5000.50	3.00	203.00	353.00	0.82	35,829.5	0.41	17,707.4	
5001.00	3.00	206.00	356.00	0.84	36,668.0	0.82	35,831.8	
5001.50	5.00	211.00	361.00	0.87	38,085.5	1.25	54,520.1	
5002.00	5.00	216.00	366.00	0.91	39,528.0	1.70	73,923.5	
5002.50	5.00	221.00	371.00	0.94	40,995.5	2.16	94,054.4	
5003.00	5.00	226.00	376.00	0.98	42,488.0	2.64	114,925.3	
5003.50	5.00	231.00	381.00	1.01	44,005.5	3.13	136,548.6	WS-10
5004.00	10.00	241.00	391.00	1.08	47,115.5	3.66	159,328.9	
5004.50	10.00	251.00	401.00	1.16	50,325.5	4.22	183,689.1	
5005.00	10.00	261.00	411.00	1.23	53,635.5	4.81	209,679.4	
5005.50	10.00	271.00	421.00	1.31	57,045.5	5.45	237,349.6	WS-100
5006.00	10.00	281.00	431.00	1.39	60,555.5	6.12	266,749.9	Freeboard
5006.50	10.00	291.00	441.00	1.47	64,165.5	6.84	297,930.1	Overflow

EXAMPLE 13.6

The 10- and 100-year detention volumes for the basin are 2.77 acre and 4.98 acre-ft. Distribute this volume using a triangular basin. The bottom triangle has a width of 200 ft and height of 350 ft. The side slope varies from IV:3H for water depths less than 1 ft and IV:5H for depths between 1 and 3.5 ft, IV:10H. The solution is summarized in Table 13.6.

Table 13.6 shows that for this case, the 10-year water surface elevation is 5003.4 ft, whereas the 100-year water depth is 5.5 ft. The 1-ft freeboard provides a brim-full capacity of 6.12 acre-ft.

13.7 Outlet works

Outlet works for a detention basin consist of *extended outlet, low-flow outlet, high-flow outlet*, and *emergency outlet*. As shown in Figure 13.14, the outlet structure is formed by risers, perforated plates, orifices, weirs, and culverts (Akan, A.O., 1990; ASCE, 1984). A riser (Figure 13.15a) has a perforated vertical pipe with a cap on

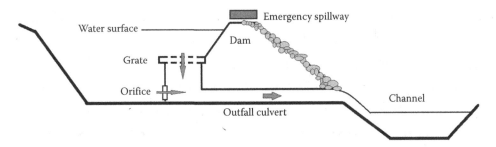

Figure 13.14 Example outlets for detention basin.

Figure 13.15 Concrete vault with low- and high-flow outlets. (a) Riser attached to concrete vault, (b) perforated plate, (c) concrete vault with multiple outlets, and (d) screen and trash rack.

Figure 13.16 Trash rack and metal gate. (a) Trash rack at pipe entrance and (b) metal gate for culvert.

the top. Holes on the riser pipe function as multiple orifices. The density of holes on the riser depends on the diameter of the vertical pipe. A perforated plate (Figure 13.15b) is similar to a riser, except that all holes are drilled on a flat plate. In comparison, a plate is easier to be installed than a riser. Both are designed to have an extended release for the purpose of stormwater quality control. Orifices and weirs (Figures 13.15c and d) are usually installed on a concrete vault to serve as *low-flow outlets* and *high-flow outlets*. A low-flow outlet is sized for 2- or 10-year flow release, whereas a high-flow outlet is designed to pass the 100-year flow release. The *emergency outlet* is used for events greater than the major storm.

A concrete vault (Figure 13.14) collects flows from the orifices and weirs, and then discharges flows through the outfall pipes. The capacity of an outfall pipe (Figure 13.14) is determined by culvert hydraulics under headwater and tailwater effects. The operation of the concrete vault reflects its culvert hydraulics under either inlet or outlet control. When a vault collects more water flows than it can release, the water depth in the vault increases. At a given depth, both collection and discharge capacities must be calculated; the smaller of these dominates the operation of the outlet structure.

A trash net shall be installed around a riser. Orifices shall be protected with a screen (Figure 13.15d) in front. If the vault is designed to have an open top (Figure 13.15c), a metal grate must be installed on the top of the vault opening. When water is directly to drain into an outfall pipe, a trash rack (Figure 13.16a) must be installed at the pipe entrance. A metal gate (Figure 13.16b) must be installed on the entrance for pipes greater than 72 in. (180 cm) in diameter.

It is important to understand that a trash rack in front of an orifice is not only for debris control but also a life saver for public safety. For a large detention basin, its emergency bypass such as a spillway must be constructed with an adequate capacity. Water released through a spillway often has a high potential for downstream erosion; therefore, a baffle system or a stilling basin shall be installed for energy dissipation.

13.7.1 Orifice hydraulics

The capacity of a horizontal orifice depends on the elevation and size of the orifice opening. The collection rate through an orifice varies with respect to the water surface elevation. When the depth above the orifice opening area is so shallow that the orifice opening area is not completely submerged, the horizontal orifice operates like a weir with a crest length equal to the circumference of a circular orifice. As shown in Figure 13.17, the horizontal orifice collects flows from three sides of the steel grate. The collection flow into a horizontal orifice under a low head is described as

$$Q_w = \frac{2}{3} C_d \sqrt{2g} P_e Y^{1.5} = \frac{2}{3} n C_d \sqrt{2g}(B+2L)Y^{1.5}$$ (13.25)

$$Y = H - E_o \text{ for } H > E_o$$ (13.26)

in which, Q_w = collection capacity as a weir in [L³/T], C_d = discharge coefficient between 0.60 and 0.65 as shown in Table 13.7, H = head in [L] of approach flow or water surface elevation in a reservoir, E_o = elevation of orifice center in [L], g = gravitational acceleration in [L/T²], P_e = effective weir length in [L], n = net opening area ratio such as 0.60, and Y = effective headwater depth in [L].

When the water depth is deep enough to submerge the entire orifice opening area, the collection rate is described by the orifice equation as

$$Q_o = C_d n A_o \sqrt{2gY} = C_d n (B \times L)\sqrt{2gY}$$ (13.27)

in which, Q_o = collection capacity of orifice in [L³/T] and A_o = orifice area in [L²].

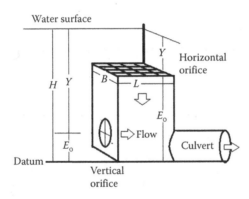

Figure 13.17 Orifice hydraulics.

Table 13.7 Orifice coefficients

Shape of orifice	C_d
Circular orifice	0.614
Square orifice	0.616
Rectangular orifice with L to B ratio of 4:1, long side in vertical direction	0.626
Rectangular orifice with L to B ratio of 4:1, long side in horizontal direction	0.627
Rectangular orifice with L to B ratio of 10:1, long side in vertical direction	0.637
Rectangular orifice with L to B ratio of 10:1, long side in horizontal direction	0.637
Triangular orifice	0.615

Source: Brate, E.F., King, H.W., Handbook of Hydraulics, McGraw-Hill Book Company, New York, 1976.

Both Equations 13.25 and 13.27 are dimensionally consistent. In practice, the collection capacity, Q_c, of an orifice under a given water depth shall be the smaller one between Equations 13.25 and 13.27 as

$$Q_c = \min(Q_w, Q_o) \tag{13.28}$$

A vertical orifice can be installed on a vertical wall. When the entire opening area is submerged, the capacity of a vertical orifice is calculated by applying Equation 13.27 with the headwater depth from the water surface to the center of the vertical orifice.

EXAMPLE 13.7

As shown in Figure 13.18, a riser has nine rows of 1-in. holes between elevations 5002 to 5004 ft. Each row has 12 1-in. holes. The orifice coefficient is 0.60. Determine the collection capacity of this riser under water surface elevations from 5002 to 5006 ft.

Solution: The total opening area of holes at each layer is

$$A = 12 \times \frac{3.1416 \times (1/12)^2}{4} = 0.065 \, \text{ft}^2 \text{ for 12 one-inch holes per row}$$

The collection discharge for each row is

$$Q = 0.60 \times 0.065 \times \sqrt{2.0 \times 32.2 \times (H - E_o)}$$

The total collection capacity for this riser is summarized in Table 13.8.

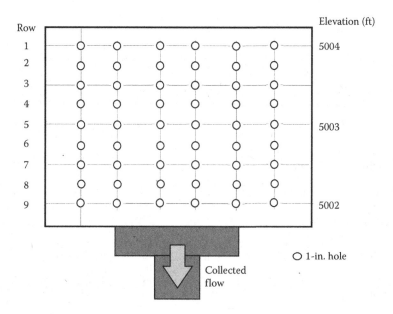

Figure 13.18 Collection capacity of a riser.

Table 13.8 Collection capacity of a riser

Stage (ft)	Riser collection capacity (cfs) Center elevation for holes (ft)									Total flow rate (cfs)
	5002.00	5002.25	5002.50	5002.75	5003.00	5003.25	5003.50	5003.75	5004.00	
	Flow (cfs)									
5002.00	0									0.00
5002.25	0.16	0								0.16
5002.50	0.22	0.16	0							0.38
5002.75	0.27	0.22	0.16	0						0.65
5003.00	0.32	0.27	0.22	0.16	0					0.97
5003.25	0.35	0.32	0.27	0.22	0.16	0.00				1.32
5003.50	0.39	0.35	0.32	0.27	0.22	0.16	0.00			1.71
5003.75	0.42	0.39	0.35	0.32	0.27	0.22	0.16	0.00		2.12
5004.00	0.45	0.42	0.39	0.35	0.32	0.27	0.22	0.16	0.00	2.57
5004.25	0.47	0.45	0.42	0.39	0.35	0.32	0.27	0.22	0.16	3.04
5004.50	0.50	0.47	0.45	0.42	0.39	0.35	0.32	0.27	0.22	3.38
5004.75	0.52	0.50	0.47	0.45	0.42	0.39	0.35	0.32	0.27	3.68
5005.00	0.55	0.52	0.50	0.47	0.45	0.42	0.39	0.35	0.32	3.96
5006.00	0.63	0.61	0.59	0.57	0.55	0.52	0.50	0.47	0.45	4.88

13.7.2 Weir hydraulics

Weirs are classified by their cross-sectional shapes, such as rectangular, triangular, and trapezoidal weirs.

Rectangular weir

The capacity of a *rectangular weir* (Figure 13.19) is determined by its crest width as

$$Q_w = \frac{2}{3} C_d \sqrt{2g} L_c Y^{\frac{3}{2}}$$ (13.29)

$$L_e = L_w - 0.1mY$$ (13.30)

$$Y = H - H_w$$ (13.31)

in which, Q_w = collection capacity in $[L^3/T]$, C_d = discharge coefficient such as 0.60–0.65, L_w = crest width in $[L]$, L_e = effective weir width in $[L]$, E_w = weir crest elevation in $[L]$, m = number of end contractions, and g = gravitational acceleration in $[L/T^2]$. Equation 13.29 is dimensionally consistent. For English units, Equation 13.29 is reduced to

$$Q_w = C_w L_e Y^{\frac{3}{2}}$$ (13.32)

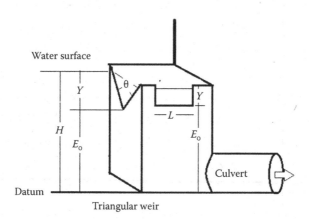

Figure 13.19 Triangular and rectangular weirs.

Table 13.9 Weir coefficients for broad-crested weirs

Breath of weir crest (ft)	Headwater 1.0 ft	Headwater 2.0 ft	Headwater 3.0 ft	Headwater 4.0 ft	Headwater 5.0 ft
5.00	2.68	2.65	2.66	2.70	2.79
10.00	2.68	2.64	2.64	2.64	2.64
15.00	2.63	2.63	2.63	2.63	2.63

Source: Brate, E.F., King, H.W., *Handbook of Hydraulics*, McGraw-Hill Book Company, New York, 1976.

Table 13.10 Weir coefficients for triangular weirs

Headwater (ft)	H:V 1.0:1.0	H:V 2.0:1.0	H:V 3.0:1.0	H:V 5.0:1.0	H:V 10.0:1.0
0.50	3.85	3.49	3.22	3.05	2.84
1.0	3.85	3.50	3.40	3.13	2.91

Source: Brate, E.F., King, H.W., *Handbook of Hydraulics*, McGraw-Hill Book Company, New York, 1976.

in which C_w = rectangular weir coefficient between 2.6 and 3.8 for English units in Table 13.9.

Triangular weir

The release rate of a *triangular weir* is governed by its V-notch angle, θ, as illustrated in Figure 13.19.

$$Q_w = \frac{8}{15} C_d \sqrt{2g} \tan\left(\frac{\theta}{2}\right) Y^{\frac{5}{2}} \tag{13.33}$$

For English units, Equation 13.33 is further reduced to

$$Q_w = C_t \tan\left(\frac{\theta}{2}\right) Y^{\frac{5}{2}} \tag{13.34}$$

in which C_t = triangular weir coefficient. Theoretically speaking, under the same hydraulic condition, a rectangular weir coefficient, C_t, is approximately 20% less than a triangular weir coefficient, C_w (Table 13.10).

Trapezoidal weir

As illustrated in Figure 13.20, a *trapezoidal weir* is composed of a rectangular weir with a crest length equal to the trapezoidal bottom width and a triangular weir with the notch angle equal to the trapezoidal side slope, Z, as

$$\frac{\theta}{2} = \frac{\pi}{2} - \tan^{-1}\left(\frac{1}{Z}\right) \tag{13.35}$$

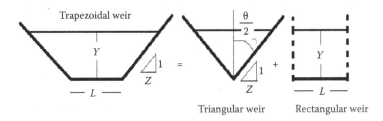

Figure 13.20 Trapezoidal weir hydraulics.

The weir coefficient for various side slopes is summarized in Table 13.10. Typically, a weir discharge coefficient, C_w, for a broad- or sharp-crested weir ranges from 2.65 to 3.10. Without knowing the weir specifics and downstream tailwater conditions, the value of $C_w = 3.0$ is recommended for sharp weirs and 2.65 is recommended for broad-crested weirs. However, care must be taken in selecting the appropriate discharge coefficient after the downstream tailwater condition becomes well understood.

EXAMPLE 13.8

Figure 13.21 presents the outlet concrete box designed for the Denver Detention Basin discussed in Example 13.3. The 10- and 100-year water depths are 3.5 and 5.5 ft, respectively, above the basin floor at an elevation of 5000 ft. The high-flow orifice is formed by a horizontal 3-ft by 3-ft square grate installed on top of the concrete box at an elevation of 5004 ft. This square orifice is protected by a steel grate with a net area ratio of 0.60. The low-flow orifice is a 1-ft by 2-ft opening with its center located at 5001.0 ft. The orifice coefficient is 0.60, and the weir coefficient is 3.0. Determine the collection capacity of the low- and high-flow orifices.

Solution: For a given stage of water surface >5004 ft, the collection capacity of the high-flow orifice shall be evaluated by both weir and orifice hydraulics, and the smaller one dictates. The weir collection capacity for the top square grate is calculated as

$$Y = H - 5004.0$$

$$Q_w = 3.0 \times 2 \times (3.0 + 3.0) \times Y^{1.5} = 36Y^{1.5} \text{ cfs}$$

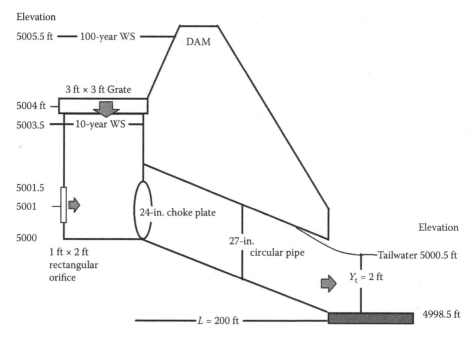

Figure 13.21 Outlet structure for example detention basin.

Table 13.11 Collection capacity for concrete vault in Denver Detention Basin

Water stage (ft)	Storage volume (acre-ft)	Flow collection		
		Low-flow V orifice (cfs)	High-flow H orifice (cfs)	Collection capacity (cfs)
5000.00	0.00	0.00	0.00	0.00
5000.50	0.41	0.00	0.00	0.00
5001.00	0.82	1.20	0.00	1.20
5001.50	1.25	4.09	0.00	4.09
5002.00	1.70	5.78	0.00	5.78
5002.50	2.16	7.08	0.00	7.08
5003.00	2.64	8.17	0.00	8.17
5003.50	3.13	9.14	0.00	9.14
5004.00	3.66	10.01	0.00	10.01
5004.50	4.22	Submerged	12.73	12.73
5005.00	4.81	Submerged	28.17	28.17
5005.50	5.45	Submerged	34.50	34.50
5006.00	6.12	Submerged	39.84	39.84
5006.50	6.84	Submerged	44.54	44.54

The orifice collection capacity for the top square grate is calculated as

$$Q_o = 0.60 \times (0.6 \times 3.0 \times 3.0) \times \sqrt{2.0 \times 32.2 \times Y} = 26.0\sqrt{Y} \text{ cfs}$$

At a stage $H \geq 5004$ ft, the collection capacity of the grate is

$$Q_c = \min\left(36.0Y^{1.5}, 26.0\sqrt{Y}\right) \text{cfs}$$

For a high stage >5004 ft, the concrete box is submerged. As a result, the low orifice becomes hydraulically connected or it passes a negligible flow. For a low stage <5004 ft, the low-flow opening functions as an orifice. As shown in Table 13.11, the flow collection curve is a combination of low- and high-stage flows.

13.8 Culvert hydraulics

Water enters the concrete vault (Figure 13.22) through the low and high orifices and then discharges into the downstream receiving system through the outlet pipes that are usually short enough, 100–200 ft, to act like a culvert. During the preliminary design, the tailwater information is not yet available. The capacity of the outlet structure in a detention basin is approximated by orifice and weir hydraulics. At the stage of final design, the outflow capacity must be refined with culvert hydraulics under the tailwater effects.

The conveyance capacity of a culvert is dictated by its inlet and outlet conditions. Under inlet control, the capacity of a culvert is independent of the tailwater at the culvert outlet. For instance, a culvert laid on a steep slope can be operated under inlet control because its capacity is solely determined by the critical depth at the entrance. On the contrary, when

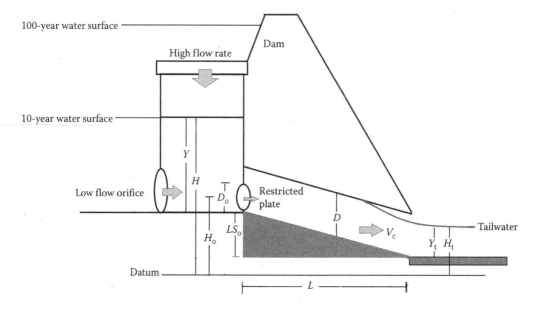

Figure 13.22 Culvert hydraulics.

a pipe has a submerged exit, the flow capacity in such a pipe may become dictated by the tailwater depth at the exit, namely under outlet control. Therefore, the flow capacity of a culvert needs to be examined by both inlet and outlet controls. For a given discharge, the one that requires a higher headwater depth at the entrance dictates the culvert's operation. For a given headwater depth, the one that passes smaller discharge dictates the culvert's capacity (HEC5, 1965).

In practice, the design tailwater condition is not always warranted during the design event. As a result, it is preferable to design the outfall pipe under so much inlet control that the flow release from a detention basin is independent of the downstream tailwater condition. To achieve such an inlet-control operation, it is recommended that a restricted plate be installed at the entrance of the outfall pipe. A restricted plate requires a higher headwater depth and leads to an inlet-control operation. For instance, as shown in Figure 13.21, a 24-in. plate is inserted at the entrance of the 27-in. pipe.

13.8.1 Outlet-control culvert hydraulics

As illustrated in Figure 13.22, under outlet control, the culvert capacity is determined by the balance of energy between sections 1 and 2 as

$$H = Y + LS_o = (K_e + K_x + K_b + K_n)\frac{V_c^2}{2g} + Y_t + \frac{V_c^2}{2g} \tag{13.36}$$

$$K_n = \alpha \frac{N^2 L}{D^{\frac{4}{3}}} \text{ for a circular pipe} \tag{13.37}$$

$$K_n = \beta \frac{N^2 L}{R^{\frac{4}{3}}} \text{ for a noncircular pipe} \qquad (13.38)$$

in which H = water surface elevation in [L] at the entrance, Y = headwater depth in [L] at the entrance, L = length of the pipe in [L], S_o = pipe slope in [L/L], K_e = entrance loss coefficient (Tables 13.12 and 13.13), K_x = exit loss coefficient between 0.5 and 1.0, K_b = bend loss coefficient as shown in Table 13.14, K_n = friction coefficient, V_c = flowing full velocity in [L/T], N = Manning's roughness coefficient such as 0.025

Table 13.12 Entrance loss coefficients for box culverts

Structure of box culvert and entrance	Coefficient-K_e
Headwall parallel to embankment (no wing wall)	
a Square-edged on three edges	0.50
b Three edges rounded	0.20
Headwall with wing walls at 15°–45° to barrel	
a Square-edge top corner	0.40
b Top corner rounded	0.20

Source: HEC5, *Hydraulic Charts for the Section of Highway Culverts*, US Department of Commerce, Bureau of Roads, Washington, DC, 1965.

Table 13.13 Entrance loss coefficient for circular culverts

Structure of circular culvert and entrance	Coefficient-K_e
Concrete pipe projecting from fill (no headwall)	
a Socket end of pipe	0.20
b Square cut end of pipe	0.50
Concrete pipe with headwall or headwall and wing walls	
a Socket end of pipe	0.10
b Square cut end of pipe	0.50
c Rounded entrance	0.10
Corrugated metal pipe	
a Projecting from fill (no headwall)	0.80
b Headwall or headwall and wing walls	0.50

Source: HEC5, *Hydraulic Charts for the Section of Highway Culverts*, US Department of Commerce, Bureau of Roads, Washington, DC, 1965.

Table 13.14 Bend loss coefficients

Angle (°)	0	20	40	60	80	90
Bend loss coefficient	0	0.1	0.2	0.45	0.80	1.0

for metal pipes and 0.015 for concrete pipes, D = diameter of circular pipe in [L], R = hydraulic radius in [L], and Y_t = tailwater depth in [L]. It is noted that $\alpha = 184$ for foot-second units or 124 for meter-second units, and $\beta = 29$ for foot-second units or 19.5 for meter-second units. For convenience, let K be the sum of all the loss coefficients as

$$K = K_e + K_x + K_b + K_n \tag{13.39}$$

The culvert capacity under outlet control, Q_O in [L^3/T], is calculated using water depths as

$$Q_O = V_c A_c = A_c \sqrt{\frac{1}{K+1}} \sqrt{2g(Y + LS_o - Y_t)} \tag{13.40}$$

Or, it may be calculated using elevations as

$$Q_O = V_c A_c = A_c \sqrt{\frac{1}{1+K}} \sqrt{2g(H - H_t)} \tag{13.41}$$

in which A_c = pipe cross-sectional area in [L^2] and H_t = elevation in [L] for tailwater depth, Y_t. Equation 13.41 is similar to the orifice equation except that the orifice coefficient is computed by the sum of all the loss coefficients.

13.8.2 Inlet-control culvert hydraulics

Under inlet control, the culvert hydraulics is independent of the tailwater effect and energy losses. The culvert capacity under inlet control shall operate like an orifice as

$$Q_I = C_d \frac{\pi D_o^2}{4} \sqrt{2g\left(Y - \frac{D_o}{2}\right)} = C_d A_o \sqrt{2g(H - H_o)} \tag{13.42}$$

in which Q_I = culvert capacity under inlet control in [L], D_o = equivalent diameter in [L] for the restricted plate installed at the culvert entrance, and H_o = elevation in [L] at the center of the restricted plate. Equation 13.42 is valid if the headwater elevation is above the tailwater elevation.

13.8.3 Discharge capacity of concrete vault

In practice, the range of headwater depth at the culvert entrance needs to be identified first. For a given headwater, H in [L], the discharge capacity, Q_C in [L^3/T], from the concrete vault is dictated by Equations 13.41 and 13.42, whichever is smaller.

$$Q_C = \min(Q_O, Q_I) \tag{13.43}$$

EXAMPLE 13.9

The outfall pipe in Figure 13.21 has a diameter of 27 in. The length of the pipe is 200 ft laid on a slope of 0.75%. The loss coefficients are 0.20 at the entrance, 0.50 at the exit, and 0.014 for Manning's roughness. A 24-in. restricted plate is installed at the entrance. The orifice coefficient for this plate is 0.60. Construct the stage-outflow curve for this culvert under the tailwater of $d = 2$ sft or at an elevation of $H_t = 5000.5$ ft.

Solution: Applying Equations 13.36 and 13.39 to the friction loss yields

$$K_n = 184.1 \frac{0.014^2 \times 200}{2.25^{4/3}} = 2.45$$

$$K = 0.2 + 0.5 + 2.45 = 3.15$$

$$LS_o = 200 \times 0.0075 = 1.5 \, \text{ft}$$

Substituting the variables into Equations 13.41 and 13.42 yields

$$Q_O = \sqrt{\frac{1}{3.15+1}} \times \frac{3.14 \times 2.25^2}{4} \times \sqrt{2 \times 32.2 \times (H - 5000.5)} = 15.65\sqrt{H - 5000.5}$$

The above equations provide the stage-outflow relationship for the culvert. For a given H, the discharge capacity from the concrete vault is dictated by the smaller one.

$$Q_C = \min(15.65\sqrt{H - 5000.5}, \ 15.12\sqrt{H - 5001})$$

As shown in Table 13.15, the restricted plate effectively dominates the culvert hydraulics as inlet control.

Table 13.15 Discharge capacity from concrete vault in Denver Detention Basin

Water stage at entrance (ft)	Headwater depth above entrance (ft)	Outlet control flow rate (cfs)	Inlet control flow rate (cfs)	Discharge capacity (cfs)	Design water surface elevation
5000.00	0.00	0.00	0.00	0.00	
5000.50	0.50	0.00	6.61	0.00	
5001.00	1.00	11.08	11.34	11.08	
5001.50	1.50	15.67	14.17	14.17	
5002.00	2.00	19.19	15.12	15.12	
5002.50	2.50	22.16	18.52	18.52	
5003.00	3.00	24.77	21.38	21.38	
5003.50	3.50	27.13	23.91	23.91	10-year WS
5004.00	4.00	29.31	26.19	26.19	
5004.50	4.50	31.33	28.29	28.29	
5005.00	5.00	33.23	30.24	30.24	
5005.50	5.50	35.03	32.07	32.07	100-year WS
5006.00	6.00	36.74	33.81	33.81	
5006.50	6.50	38.37	35.46	35.46	

13.9 Characteristic curve

Hydrologic analyses on basin shaping produce the stage-storage curve (SS curve), while hydraulics analyses on outlet concrete vault produce the stage-outflow curve (SO curve). These two curves are then merged into the stage-storage-outflow (SSO) curve, which is termed the *characteristic curve* for the detention basin under design. Reservoir routing requires the SSO curve to perform the numerical simulation to confirm that the detention basin under design will produce the allowable flow release and meet the required storage volume. Development of the SO curve takes both stage-collection and stage-discharge relationships into consideration. At a specified (S)tage, H in [L], the (O)utflow in [L^3/T] from the detention basin is determined as

$$\text{Outflow} = \min \{\text{collection capacity, discharge capacity}\} \text{ for a given } H \qquad (13.44)$$

For instance, the SO, SS, and SSO curves developed for the outlet structure in Figure 13.21 are summarized in Table 13.16.

The SSO curve was developed using a triangular basin, and the tailwater effect was removed with a restricted plate. As shown in Table 13.16, during the 10-year event, the water surface elevation would be at 5003.5 ft with a storage volume of 3.13 acre-ft and the allowable peak flow of 9.14 < 9.2 cfs. During the 100-year event, the water surface elevation would be at 5005.5 ft with a storage volume of 5.45 acre-ft and the allowable peak flow of 32.07 < 34 cfs. As a preliminary design, this SSO characteristic curve developed for the basin under design has met the goals of peak flow reduction. Furthermore, a reservoir routing process may be performed to understand the flow movement through the basin.

13.10 Underground detention

Underground detention is an option for stormwater mitigation when the on-ground open space is not available or the local land cost is expensive. Usually, the tributary area is rather small, and the storage volume is manageable using pipes or vaults under parking lots or pavements. The major components for an underground basin include (1) a *sump intake* to prime the system, (2) a *storage unit* to temporarily store water volume, and (3) an *outflow structure* to control flow release. Figure 13.23 presents an example. The required detention storage volume for an underground basin is calculated using the rational volumetric method. The storage volume is provided by multiple pipes and/or vaults. The sump intake is shaped like a swimming pool that receives runoff flows from the surrounding surface areas and also controls the flow release through an outflow structure. As soon as the sump intake is loaded, water will flow into the underground pipes by the hydraulic grade line. During the period of recession, the stored water volume gradually flows back into the sump intake by the gravity and is then drained out through the outflow box.

EXAMPLE 13.10

A 2-acre catchment in Figure 13.24 has a runoff coefficient of 0.76 and the time of concentration of 10 min. The design rainfall IDF curve is formulated as: $I(\text{in./h}) = 76/(10 + T_d)^{0.789}$ in which T_d = rainfall duration in minutes. Using the allowable flow release of 2 cfs, determine the detention storage volume, and size the number of pipes.

Table 13.16 Characteristic stage-storage-outflow curve developed for Denver detention basin

Water stage (S) (ft)	Storage volume (S) (acre-ft)	Flow collection by orifice and weir			Outflow released from concrete box		
		Low-flow V orifice (cfs)	High-flow H orifice (cfs)	Collection by O and W (cfs)	Discharge by culverts (cfs)	Outflow (O) (cfs)	Allowable flow release
5000.00	0.00	0.00	0.00	0.00	0.00	0.00	
5000.50	0.41	0.00	0.00	0.00	0.00	0.00	
5001.00	0.82	1.20	0.00	1.20	11.08	1.20	
5001.50	1.25	4.09	0.00	4.09	14.17	4.09	
5002.00	1.70	5.78	0.00	5.78	15.12	5.78	
5002.50	2.16	7.08	0.00	7.08	18.52	7.08	
5003.00	2.64	8.17	0.00	8.17	21.38	8.17	
5003.50	3.13	9.14	0.00	9.14	23.91	9.14	9.2
5004.00	3.66	10.01	0.00	10.01	26.19	10.01	
5004.50	4.22		12.73	12.73	28.29	12.73	
5005.00	4.81		28.17	28.17	30.24	28.17	34.0
5005.50	5.45		34.50	34.50	32.07	32.07	
5006.00	6.12		39.84	39.84	33.81	33.81	
5006.50	6.84		44.54	44.54	35.46	35.46	

Figure 13.23 Underground detention basin. (a) Sump intake and (b) pipes for storage volume.

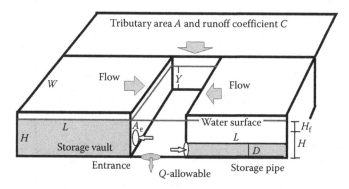

Figure 13.24 Example underground basin for design.

Table 13.17 Calculation of maximized detention volume for underground basin

Duration (min)	Rainfall intensity (in./h)	Inflow volume (acre-ft)	Adjustment factor, m	Outflow volume (acre-ft)	Storage volume (ft³)
70.00	2.33	0.35	0.57	0.11	10,230.63
80.00	2.13	0.36	0.56	0.12	10,253.44
90.00	1.96	0.37	0.56	0.14	10,205.40

The maximization of storage volume is presented in Table 13.17.

For this case, the detention volume is $10,253.44\,ft^3$, which is equivalent to a vault 5 ft deep (H) on a square of 45×45 (W × L) ft^2 or 11 pipes of 5 ft in diameter for a length of 50 ft. Details of the flow entrance into the underground pipes or vaults can be sized using the energy principle to balance the entrance area, A_e, headwater, H, and friction losses, H_f.

13.11 Evaluation of detention effectiveness

The ultimate goal of flood mitigation is to preserve the predevelopment watershed regime after the development. As aforementioned, after development, we face two major alterations in watershed hydrology: increased flow rates (Q-problem) and runoff volumes (V-problem). According to the drainage law, any and all upstream developments are not allowed to transfer drainage problems and/or damage to the downstream properties, even after the drainage easement is granted. The Q-problem is a concern of how to release extreme events at their predevelopment rates. As illustrated in Figure 13.25, curves 1 and 4 represent the predevelopment and postdevelopment flow-frequency relations that provide a basis to quantify the increases of peak flows. The effort of stormwater management is to convert Curve 1 (postdevelopment condition) back to Curve 4 (predevelopment condition).

Since 1980s, stormwater detention has been recognized to be the most effective measure to reduce the flow release. By applying a low-flow orifice and a high-flow weir (as shown in Figure 13.26), the outflow structure can produce curve 2. The extended release with a perforated plate or riser will achieve Curve 3 (Guo, 2013). Although a *conventional detention basin* using orifice and weir is recognized as an effective method, it controls only the 10- to 100-year extreme events, which are approximately 2% of the runoff population. Most frequent events trickle through the low-flow orifice without any detention effect. In the 1990s, the major improvement to stormwater detention was to install a perforated plate or riser to extend the drain time from 24 to 72 h, depending on the local regulation of water rights. As shown in Figure 13.26, an *extended detention basin* using orifice, weir, and perforated plate can produce curve 3, which shows the flow release control from 2- to 100-year events, which are approximately 5% of the runoff population. It means, the practice of stormwater detention is a good approach to mitigate the Q-problem that is related to flooding and safety concerns, but it is not sufficient to provide an overall control for the entire spectrum of runoff flows. How to cope with the second issue—V-problem? We need more effort in the new concept of LID to control frequent events (Guo, 2007, 2009; Guo and Cheng, 2008).

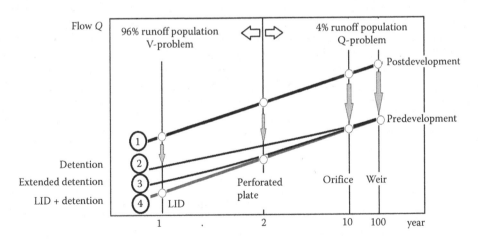

Figure 13.25 Effectiveness of stormwater detention.

Figure 13.26 Outlet box to control flow releases. (a) Orifice only, (b) orifice and weir, and (c) orifice, weir, and riser.

13.12 Maintenance and safety

Detention basins present as an attraction to the public because it is open space and is designed as a neighborhood park. During a heavy storm event, the flash flood can be lethal. It is necessary to have flash flood-warning signs posted around the basin. All inlets must be protected with a trash rack. For public safety, a fence may be used around the shoreline where there are potential hazards.

Upon the completion of the basin, an inspection shall be conducted to ensure that the as-built basin complies with the design. During the first 3 years in service, it is necessary to frequently check on vegetation growth in infiltration bed and on the side slope. Supplemental plantings are added as needed to ensure good cover. On an annual basis, the basin needs a basic maintenance before the wet months. The outlet structure should be inspected after a severe storm event, and debris blockages should be removed. Trash and debris have to be removed regularly.

13.13 Homework

Q13.1 Figure Q13.1 presents the topographic map for a subdivision before and after the development. (A) Mark the flow direction under the predevelopment condition. (B) Identify the flow paths after the development. (C) Identify the location of detention basin to alleviate the increased runoff due to the development.

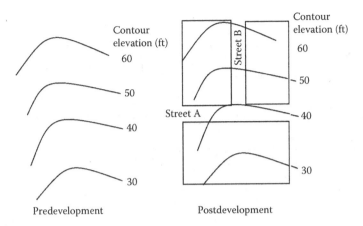

Figure Q13.1 Predevelopment and postdevelopment watershed conditions.

1. Identify the historical stormwater flow paths.
2. Identify the developed stormwater flow paths.
3. Explain why Avenue B becomes inundated after the development.
4. Suggest a measure to mitigate this situation.

Q13.2 A watershed has a drainage area of 50 acres. The design rainfall IDF is given as

$$i\text{(in./h)} = \frac{28.5 \times P_1}{(10 + T_d)^{0.784}}$$

in which $P_1 = 1\,h$ precipitation in inches and T_d = rainfall duration in minutes.
    The time of concentration of the watershed is 25 min.

Tasks for this project are as follows:

1. Knowing $P_1 = 2.6\,\text{in.}$, runoff coefficient $C = 0.66$, and allowable release rate $= 1.0\,\text{cfs/}$ acre for the 100-year event, estimate the stormwater detention volume (*Solution*: 4.56 acre-ft)
2. Knowing $P_1 = 1.6\,\text{in.}$ runoff coefficient $C = 0.45$, and allowable release rate $= 0.30\,\text{cfs/}$ acre for the 10-year event, estimate the stormwater detention volume (*Solution*: 1.78 acre-ft)

Q13.3 A detention pond is designed to control the flow release after the tributary area of 40 acres has been developed. The layout of this pond is shown in Figure Q13.3.

Step 1: Determination of allowable release flows
According to the local design criteria, the allowable release rates are 0.23 cfs/acre for the 10-year event and 0.85 cfs/acre for the 100-year event. What are the allowable release rates for the 10- and 100-year events from this detention basin under design? (*Solution*: Q_{10}-allowable = 9.2 cfs and Q_{100}-allowable = 34 cfs)

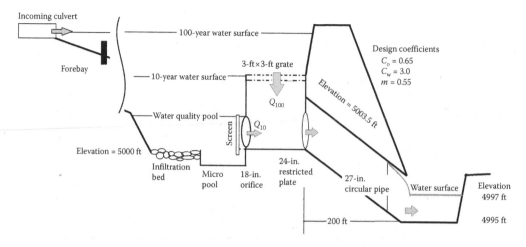

Figure Q13.3 Layout of detention basin.

Step 2: Determination of detention volumes: V-10 and V-100
From the hydrologic study, the 10- and 100-year inflow hydrographs are given as

Time (min)	10-year runoff (cfs)	100-year runoff (cfs)
0	0	0
10	19	17
20	46	64
30	51	108
40	40	103
50	29	78
60	22	56
70	17	37
80	13	22
90	10	13
100	9	10
110	7	7
120	4	4

Apply the hydrograph method to determine the 10- and 100-year detention volumes. (*Solution*: V-10 = 1.4 acre-ft and V-100 = 2.3 acre-ft)

Step 3: Design of basin geometry
Consider a triangular basin with its bottom width = 100 ft and length = 300 ft. The basin's bottom elevation is at 5000 ft, and the freeboard height is set to be 1.0 ft. Shape this triangular basin using the given side slopes as

1. Identify the 10- and 100-year water surface elevations
2. Establish the stage-cross-sectional area curve
3. Construct the stage-storage curve

Water surface elevation (ft)	Basin side slope (ft/ft)	Width of cross-section (ft)	Length of cross-section (ft)	Cross pond (acres)	Sectional area (ft²)	Cumulative pond (acre-ft)	Storage volume (cfs)	Identify design water elevation
5000.00	0.00	100.00	300.00	0.34	15,000.0	0.00	0.0	
5000.50	3.00	103.00	303.00	0.36	15,604.5	0.18	7651.1	
5001.00	3.00	106.00	306.00	0.37	16,218.0	0.36	15,606.8	
5002.00	5.00							
5003.00	5.00							
5003.50	**5.00**	**131.00**	**331.00**	**0.50**	**21,680.5**	**1.44**	**62,854.9**	**WS = 10 year**
5004.00	7.00							
5004.50	7.00							
5005.00	10.00	**155.00**	**355.00**	**0.63**	**27,512.5**	**2.28**	**99,320.4**	**WS = 100 year**
5006.00	10.00	175.00	375.00	0.75	32,812.5	2.97	129,457.9	**Freeboard**

Step 4: Design of flow collection (orifice and grate)

The low-flow outlet for the 10-year release is a vertical orifice of 18-in. in diameter. The center of this vertical orifice is at 9 in. above the floor. The high-flow outlet is another horizontal 3.0×3.0-ft grate set at an elevation of 5003.50 ft. The net opening area ratio is 0.55. Using the foot-sec units, the weir coefficient is 3.0, and the orifice coefficient is 0.65. (*Solution*: At $WS_{10} = 5003.5$ ft, which is the 10-year water surface elevation, the 18-in. orifice can collect $8.41 < 9.2$ cfs; at $WS_{100} = 5005$ ft, which is the 100-year water surface elevation, the 3-ft $\times$ 3-ft grate can collect $31.62 < 34$ cfs.)

Step 5: Design of flow discharge (outfall culvert pipe)

The tailwater elevation is set to be 4997 ft. A 27-in. circular pipe of 200-ft long is laid on a slope of 0.025 ft/ft. The pipe roughness is 0.022. The entrance and exit loss coefficients are 0.2 and 0.5. The restriction plate installed at the pipe entrance has a diameter of 24 in. (*Solution*: 25.9 cfs at $WS_{10} = 5003.5$ ft and 33.42 cfs at $WS_{100} = 5005$ ft.)

Step 6: Construct the SSO curve.

Bibliography

Akan, A.O. (1990). "Single Outlet Pond Analysis and Design," *ASCE Journal of Irrigation and Drainage Engineering*, Vol. 116, No. 4, pp. 527–536.

Aron, G., and Kibler, D.F. (1990). "Pond Sizing for Rational Formula Hydrographs," *Water Resources Bulletin*, Vol. 26, No. 2, pp. 255–258.

ASCE. (1984). "Final Report of the Task Committee on Stormwater Detention Outlet Structures," American Society of Civil Engineers, New York.

ASCE. (1992). "Design and Construction of Urban Stormwater Management Systems," Water Environmental Federation Manual of Practice FD-20, and ASCE Manual and Reports No. 77.

Brate, E.F., and King, H.W. (1976). *Handbook of Hydraulics*, McGraw-Hill Book Company, New York.

EPA. (1983). "Results of the Nationwide Urban Runoff Program," Final Report, U.S. Environmental Protection Agency, NTIS No. PB84-185545, Washington, DC.

EPA. (1986). "Methodology for Analysis of Detention Basins for Control of Urban Runoff Quality," U.S Environmental Protection Agency, EPA440/5-87-001, September 1986.

EPA. (1994). *Urbanization and Water Quality*, prepared by Terrene Institute for U.S. Environmental Protection Agency, Washington, DC.

FAA. (1970). "Airport Drainage," Report 150/5320-5B, U.S. Department of Transportation, Federal Aviation Administration, Washington, DC.

FAA. (1977). "Surface Drainage Facilities for Airfields and Heliports," Technical Manual No. 5-820-1, U.S. Department of the Army and Air Force, Washington, DC.

Guo, J.C.Y. (1999). "Detention Storage Volume for Small Urban Catchment," *ASCE Journal of Water Resources Planning and Management*, Vol. 125, No. 6, pp. 380–382.

Guo, J.C.Y. (2004). "Hydrology-Based Approach to Storm Water Detention Design Using New Routing Schemes," *ASCE Journal of Hydrologic Engineering*, Vol. 9, No. 4, pp. 333–336.

Guo, J.C.Y. (2007). "Stormwater Detention and Retention LID systems," Invited, *Journal of Urban Water Management*.

Guo, J.C.Y. (2009). "Retrofitting Detention Basin for LID Design with a Water Quality Control Pool," *ASCE Journal of Irrigation and Drainage Engineering*, Vol. 135, No. 6.

Guo, J.C.Y. (2012). "Off-Stream Detention Design for Stormwater Management," *ASCE Journal of Irrigation and Drainage Engineering*, Vol. 138, No. 4.

Guo, J.C.Y. (2013). "Green Concept in Stormwater Management," *Journal of Irrigation and Drainage Systems Engineering*, Vol. 2, No. 3, p. 114, doi:10.4172/2168-9768.1000114.

Guo, J.C.Y., and Cheng, J.Y.C. (2008). "Retrofit Stormwater Retention Volume for Low Impact Development (LID)," *ASCE Journal of Irrigation and Drainage Engineering*, Vol. 134, No. 6, pp. 872–876.

Guo, J.C.Y., and Clark, J. (2006). "Rational Volumetric Method for Off-line Storm Water Detention Design," *Journal of PB Network for Water Engineering and Management*, Vol. 21, No. 3, Issue No. 64.

HEC HMS. (2005). "Hydrologic Analysis System," Hydrologic Engineering Center, Corps of Engineers, Davis, CA.

HEC5. (1965). *Hydraulic Charts for the Section of Highway Culverts*, U.S. Department of Commerce, Bureau of Roads, Washington, DC.

Jones, J, Guo, J.C.Y., and Urbonas, B. (2006). "Safety on Detention and Retention Pond Designs," *Journal of Storm Water*.

McCuen, R.H. (1998). *Hydrologic Analysis and Design*, Chapter 8, 2nd ed., Prentice Hall, New York.

Shueler, T.R., and Helfrich, M. (1989). "Design of Extended Detention Wet Pond System," *Design of Urban Runoff Quality Controls*, edited by L.A. Roesner, B. Urbonas, and M.B. Sonnen, American Society of Civil Engineers, New York.

Urbonas, B., and Stahre, P. (1992). *Stormwater Best Management Practices and Detention*, Prentice Hall, Englewood Cliffs, NJ.

US EPA SWMM. (2005). Stormwater Management Model. http://www2.epa.gov/water-research/storm-water-management-model-swmm.

USWSCM. (2010). *Urban Storm Drainage Design and Criteria*, Volumes 1, 2, and 3, Urban Drainage and Flood Control District, Denver, CO.

Chapter 14

Flow diversion

When a flood channel is overloaded with stormwater, it may be directly drained into a widened floodplain as an *on stream detention basin*, or diverted into an *off stream detention* basin, which is located outside of the channel system (ASCE, 1994). Figure 14.1 presents a comparison between on stream and off stream detention systems.

The flow in a channel is accumulated in the downstream direction. When the channel reaches its critical capacity, only the *straight-through discharge*, Q_1, is allowed to go through the channel. The excess flow has to be temporarily stored in a detention basin. The *after-detention flow release*, Q_2, from the basin may be returned to the channel at a downstream point or completely diverted out of the channel system (Guo, 2012; Guo and Clark, 2006). A detention basin shall be sized to comply with the *allowable flow release*, Q-allowable.

14.1 Flow diversion

Flow diversion is an important element in the operation of an off stream basin. In practice, flow diversion (as shown in Figure 14.2) may be a simple overtopping side weir installed on top of a channel bank. The elevation of the weir crest is selected to control the flow diversion into the off stream basin. A flat conduit is installed to act as an *equalizer* between the basin and the channel. An equalizer allows an active flow to balance the water surfaces on the both sides. The *flood gate* is operated by the hydraulic grade line (HGL). During the loading process, the flood gate remains closed because the channel has a higher water stage than that in the basin. During the period of flow recession, the water stage in the channel continues decreasing and the high water stage in the basin remains unchanged. As soon as the water surface differential reaches the *threshold depth*, ΔY, the flood gate would be pushed open. In addition to the flood gate, the basin may have another outlet, Q_2, to release stormwater. Regardless of the details, the basic rule is that the total flow release, Q_1 and Q_2, shall not exceed the allowable release, Q_a, from the tributary watershed. The design criteria of allowable release flow can be found in Chapter 13 (Table 13.1) (UDFCD, 2010).

The off stream detention process is an open-channel flow system. The flow diversion system consists of a downstream culvert on the channel alignment and an upstream side weir on the top of a channel bank. As illustrated in Figure 14.3, the performance of a side weir depends on the reliable headwater depth, h, on top of its crest. It is recommended that the downstream culvert be designed to pass the straight-through capacity under an inlet-control condition (Guo et al., 2007). The inflow channel shall be laid on a mild slope. The headwater at the culvert entrance creates an M-1 backwater profile across the length of the side weir. Usually, an M-1 water surface profile is long and stable. Of course,

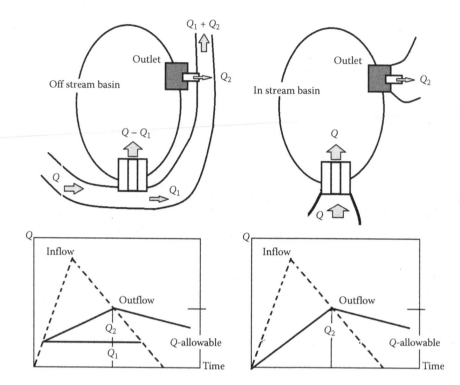

Figure 14.1 Flow diversion into detention basin.

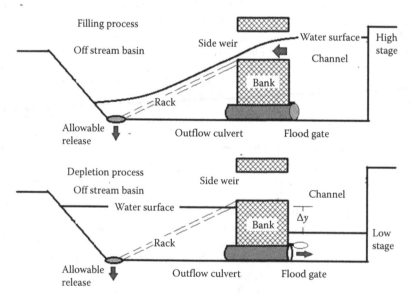

Figure 14.2 Loading process in off stream basin.

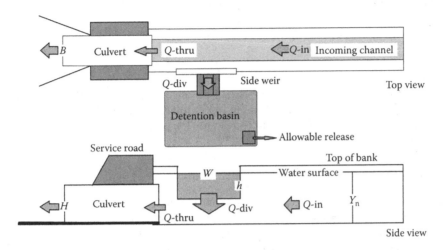

Figure 14.3 Flow diversion system.

it is even better if the normal depth in the channel is close to the required headwater depth for the culvert flow. Under the assumption of normal flow, the peak flow, Q_p, in the channel is carried with its normal flow condition as

$$Q_p = \frac{K}{N} A^{\frac{5}{3}} P^{\frac{-2}{3}} \sqrt{S_o} \tag{14.1}$$

in which Q_p = peak flow in channel in $[L^3/T]$, $K = 1.0$ for meter-second units or 1.486 for feet-second units, N = Manning's roughness coefficient, A = flow area in $[L^2]$, P = wetted parameter in $[L]$, and S_o = channel bottom slope in $[L/L]$. The solution for Equation 14.1 is the normal depth, Y_n, for the peak flow in the channel. This peak flow in the channel is divided into the straight-through flow, Q_1, and the diverted flow, Q_w. The continuity equation is

$$Q_p = Q_w + Q_1 \tag{14.2}$$

Under inlet control, the straight-through flow is described as a culvert flow, which is computed as

$$Q_1 = C_o B H \sqrt{2g\left(Y - \frac{H}{2}\right)} \tag{14.3}$$

in which C_o = discharge coefficient varied from 0.55 to 0.65, B = width of box culvert in $[L]$, H = height of box culvert in $[L]$, Y = required headwater depth in $[L]$, and g = gravitational acceleration in $[L/T^2]$. The diverted flow is a side weir flow that is calculated as

$$Q_w = \frac{2}{3} C_o \sqrt{2g} W (Y - h)^{1.5} \tag{14.4}$$

in which Q_w = diverted flow in $[L^3/T]$, W = length of side weir in $[L]$, and h = headwater depth in $[L]$ on top of weir crest. In practice, set the headwater depth, Y, equal to the normal depth, Y_n, in the channel to solve for three unknowns: B, W, and h.

$$Y = Y_n \tag{14.5}$$

The width of the downstream culvert should be selected to be slightly smaller than or equal to the channel bottom width. Setting the required headwater depth for the culvert flow equal to the normal flow depth in the channel warrants a reliable headwater depth, h, for the side weir flow.

EXAMPLE 14.1

A trapezoidal channel has a bottom width of 10 ft, side slope of 1V:4H, bottom slope of 0.25%, and Manning's $N = 0.035$. The 100-year peak inflow in this channel is 980 cfs. The downstream capacity in this channel is limited to 350 cfs. (1) Use $C_o = 0.65$ to design the box culvert. (2) Determine the headwater depth, h, on the 100-ft side weir.

Applying Equation 14.1 to the design flow of 980 cfs in the given channel yields the normal depth, $Y_n = 6$ ft. Try a concrete box with $B = 7$ ft and $H = 5$ ft. Under a headwater of 6 ft, the inlet-control capacity for this box culvert is

$$Q_1 = C_o HB \sqrt{2g \left(Y - \frac{H}{2} \right)} = 0.65(5 \times 7) \sqrt{2 \times 32.2 \left(6 - \frac{5}{2} \right)} = 341 \text{ cfs close to 350 cfs}$$

The diversion flow overtopping the side weir is

$$Q_w = Q_p - Q_1 = 980 - 341 = 639 \text{ cfs}$$

The headwater for the 100-ft side weir flow is

$$Q_w = \frac{2}{3} C_o \sqrt{2g} W h^{1.5} = \frac{2}{3} \times 0.65 \times \sqrt{2 \times 32.2} \times 100 \times h^{1.5} = 639 \text{ cfs}$$

So, $h = 1.50$ ft below the water surface or 4.5 ft above the channel bottom.

Equations 14.1 through 14.4 were derived based on the peak flow. For a period of 10 min, the diverted water volume is

$$\text{Diverted volume} = \frac{639 \times 10 \times 60}{43,560} = 8.81 \text{ acre-ft}$$

Flow diversion is not solely determined by the peak flow. In fact, the total volume diverted into the basin has to be further confirmed with the numerical simulation to route the inflow hydrograph through the flow diversion structure. The dimension of flow diversion may need minor adjustments till the total diverted water volume is close to the required detention volume.

14.2 Flood gate

To transfer flows between the flood channel and the off stream detention basin, a short conduit (as illustrated in Figure 14.4) is installed on the basin floor. Preferably, this conduit is laid horizontally as an equalizer to balance the difference between

Figure 14.4 Flood gate. (a) Closed gate and (b) open gate.

the water stages. A trash rack should be installed at the conduit entrance on the basin side, while a flood gate is installed at the conduit exit on the channel side (Guo et al., 2010b). For safety, the surface area of the trash rack shall be at least four times the conduit opening area (Guo and Jones, 2010a). During the basin's loading cycle, the high flow in the channel overtops the side weir to fill up the basin. When both the channel and basin are full, the hydrostatic pressure is balanced on the both sides. During the recession of flood flow in the channel, the water depth in the channel depletes, while the high water depth in the basin remains. As a result, the positive force-moment about the top hinge would automatically open the gate to return the stored water volume to the channel (Skipwith et al., 1990). For engineering practice, it is necessary to define the minimum water depth differential that can trigger the gate to be pushed open. This threshold depth can be determined by the balance of moment about the hinge of the gate.

As shown in Figure 14.5, the hydrostatic pressure distribution can be converted into a single force, which is determined by the gate area and the hydrostatic pressure at the centroid of the gate area. Such a single force should be placed at the point of application to represent the distributed pressure diagram. The horizontal hydrostatic forces on the gate are calculated as (Guo, 2012)

$$F_1 = \gamma(Y_1 - R)A_G \tag{14.6}$$

$$F_2 = \gamma(Y_2 - R)A_G \tag{14.7}$$

in which F_1 = hydraulic force from channel in pounds or Newton, Y_1 = water depth in channel in [L], A_G = gate area in [L^2], R = radius for circular gate in [L], F_2 = hydraulic force from basin in pounds or Newton, Y_2 = water depth in basin in [L], and γ = specific weight of water equal to 62.4 pounds/ft^3 or 1000 kg/m^3. The points of application for these two horizontal forces are located as

$$Y_{P1} = Y_{C1} + \frac{I_C}{Y_{C1}A_G} = (Y_1 - R) + \frac{R^2}{4(Y_1 - R)} \text{ for circular gate} \tag{14.8}$$

$$Y_{P2} = Y_{C2} + \frac{I_C}{Y_{C2}A_G} = (Y_2 - R) + \frac{R^2}{4(Y_2 - R)} \text{ for circular gate} \tag{14.9}$$

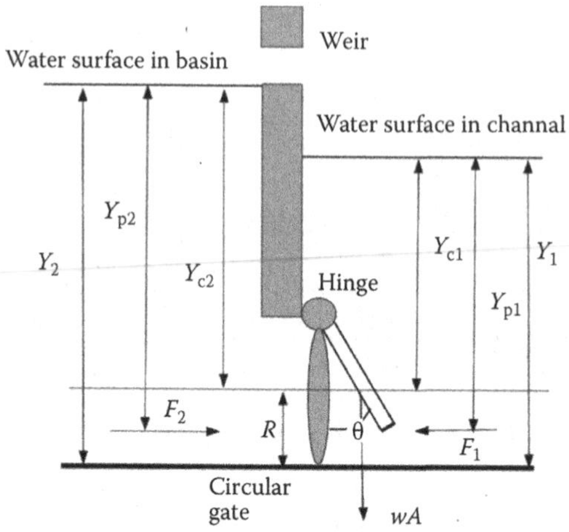

Figure 14.5 Forces applied to flood gate.

For convenience, the weight of the gate is expressed in its unit weight per area as

$$W = wA_G = S\gamma A_G t \tag{14.10}$$

in which W = weight of gate in pounds or Newton, S = specific gravity of flood gate such as 7 to 8 for steel, w = unit weight per area of gate in pounds/ft^2 or N/m^2, and t = thickness of gate in [L]. As illustrated in Figure 14.5, the operation of the flood gate is dynamic, starting from its vertical position and then becoming open at an inclined angle, θ. The turning moment about the hinge is calculated as

$$\gamma(Y_2 - R)\left[R + \frac{R^2}{4(Y_2 - R)}\right] - \gamma(Y_1 - R)\left[R + \frac{R^2}{4(Y_1 - R)}\right] - w\frac{(S-1)}{S}R\sin\theta \geq 0 \tag{14.11}$$

When the basin is full, both Y_2 and Y_1 are numerically much greater than R. As a result, Equation 14.11 is reduced to

$$\Delta Y = Y_2 - Y_1 \geq \frac{w\sin\theta}{\gamma}\frac{(S-1)}{S}\,\text{or} \geq t(S-1)\sin\theta \tag{14.12}$$

and ΔY= water depth difference in [L] and θ = minimal inclined angle to open gate, such as 45°. Equation 14.12 significantly simplifies the complicated force-moment calculation when determining the operation of a flood gate in numerical simulations. At each computing time step, the operation of the flood gate, open or close, can be easily identified by the water depth difference between the basin and the channel.

EXAMPLE 14.2

An 18-in. circular iron gate has a thickness of 0.5 ft. This gate is considered open when the inclined angle is 45° or greater. The specific weight of this gate is 7.5. The unit weight per gate area is

$$w = \gamma St = 62.4 \times 7.5 \times 0.5 = 234 \text{ pound per ft}^2\text{of iron gate}$$

For this case, the minimum depth differential required to open the gate is

$$\Delta Y = (Y_2 - Y_1) \geq 0.5 \times (7.5 - 1) \sin 45° \text{ or } \geq 2.29 \text{ ft}$$

It means that the gate will be pushed open by the gravitational force as soon as the depth differential exceeds 2.29 ft.

14.3 Forebay for sediment settlement

A sediment forebay (Figure 14.6) is a small pool installed immediately downstream of the flow entrance of a detention basin. A forebay is designed to slow down the inflow with a shallow water depth of 2–3 ft, which can be formed with earth, riprap, or concrete berm. The major function of a forebay is to trap and settle down large particles (>1.0 mm in diameter) and heavy pollutants in stormwater. A forebay acts as a pretreatment feature that makes the basin's maintenance easier and less expensive by trapping large particles in a small, confined area. After the event, the sediment deposit can be loaded up and then removed.

A forebay consists of an *energy dissipater* to reduce the flow momentum, *a shallow pool* to diffuse the flow volume, an *overtopping wall* to pass high flows, and a *small on-floor outlet* to pass low flows. During an extreme event, the forebay will be filled up, and the excess water overtops the wall. Low flows will be accumulated in the pool, and then the water depth acts as the headwater to drain the stored water through the on-floor orifice. As shown in Figure 14.7, a low-flow outlet in the forebay should be sized to release

Figure 14.6 Forebay in stormwater basin. (a) Forebay pool and (b) forebay with energy dissipater.

Figure 14.7 Low-flow outlet and trapped sediment in forebay. (a) Vertical slot for low-flow outlet and (b) pipe for low-flow outlet.

1%–2% of the 100-year peak discharge. The low-flow outlet can be constructed using an on-floor pipe or a vertical slot on the wall.

To design a forebay, the flow path through the forebay should be maximized, and the bottom slope should be minimized to encourage particle's settling. A riprap berm should not have a side slope steeper than 1V:3H. The floor of the forebay should be lined with concrete or grouted riprap. As expected, a forebay berm will be overtopped frequently. It is necessary to protect the berm with riprap blankets to avoid erosion.

14.3.1 Forebay Design

The movement of a particle in water is dominated by the *buoyance force* due to displacement of water volume, the *weight of the particle* due to the gravity, and the *drag force* due to the particle's movement (Pemberton and Lara, 1971). Assuming that the particle is spherical in shape, the buoyancy of a falling particle is calculated as

$$F_b = \rho_w g \frac{\pi D_s^3}{6} \qquad (14.13)$$

where F_b = buoyancy force in pounds or Newton, ρ_w = density of water equal to 1.94 slug/ft^3 or 1000 kg/m^3, g = gravitational acceleration in $[L/T^2]$, and D_s = target diameter in $[L]$ of particle such as 1 mm. The body weight is defined as

$$F_b = \rho_s g \frac{\pi D_s^3}{6} \qquad (14.14)$$

where F_b = body weight in pounds or Newton and ρ_s = density of particle. The drag force acts in the opposite direction of the movement. For a settling particle, the drag force acts in opposition to the gravitational force and in concurrence to the buoyant force. The drag force can be computed as

$$F_d = C_d \frac{\pi D_s^2}{4} \frac{\rho_w V_s^2}{2} \qquad (14.15)$$

where F_d = drag force in pounds or Newton, C_d = drag coefficient, and V_s = fall velocity in $[L/T]$ of particle. Referring to Figure 14.8, the aforementioned forces shall be balanced as

$$F_g = F_b + F_d \qquad (14.16)$$

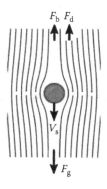

Figure 14.8 Flow Field around Falling Particle in Water.

Aided by Equations 14.14 through 14.16, the fall velocity is solved as

$$V_s = \sqrt{\frac{4gD_s}{3C_d}\left(\frac{\rho_s}{\rho_w} - 1\right)} = \sqrt{\frac{4gD_s}{3C_d}(S_s - 1)} \qquad (14.17)$$

where S_s = specific gravity of soil particle ranging from 2.4 to 2.8. The *drag coefficient* depends on the particle's *Reynolds number*. For a spherical particle, the empirical formula for C_d is

$$C_d = \frac{24}{Re} + \frac{3}{\sqrt{Re}} + 0.34 \approx 0.4 \text{ when } Re > 10,000 \qquad (14.18)$$

$$Re = \frac{V_s D_s}{\nu} \qquad (14.19)$$

where Re = Reynolds number as the ratio of particle's momentum force to the viscous force of water and ν = water kinematic viscosity such as $1.2 \times 10^{-5} \text{ ft}^2/\text{s}$. Using an iterative process, one can start with an estimated value for Re. For example, set Re = 5000 to calculate C_d in Equation 14.18 and V_s in Equation 14.17, and then check on the value of Re in Equation 14.19. Repeat this process until the estimated Re closely agrees with the computed value. As illustrated in Figure 14.9, a particle is considered captured in the forebay if the following condition is satisfied:

$$T_s = \frac{H}{V_s} \qquad (14.20)$$

$$L_s \geq U_s T_s \qquad (14.21)$$

where H = water depth in [L] in forebay, T_s = travel time in [T] or residence time for the target particle with diameter, D_s, L_s = flow length in [L] through forebay, and U_s = horizontal flow velocity in [L/T]. Equation 14.21 implies that the particle has reached the floor before it flows through the basin horizontally (Randle, 1984).

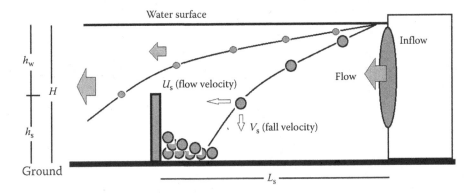

Figure 14.9 Particle settlement in forebay.

The flow horizontal velocity, U, is the cross-sectional average velocity calculated as

$$U_s = \frac{Q}{HW}$$
(14.22)

The required headwater depth on top of the berm crest is determined as

$$Q = \frac{2}{3}C_o\sqrt{2g}W\,h_w^{1.5}$$
(14.23)

The water depth in the basin is summed as

$$H = h_w + h_s$$
(14.24)

where Q = design discharge in $[L^3/T]$, W = width of berm in $[L]$, h_w = headwater depth in $[L]$ for weir flow on top of berm, and h_s = height of berm in $[L]$. In practice, the sediment load is depicted with a particle gradation curve. Therefore, we have to apply Equations 14.14 through 14.24 to a selected particle size, D_s, to examine whether particles $\geq D_s$ would be trapped.

EXAMPLE 14.3

Use the 100-year peak discharge of 100 cfs for design. The 100-year water surface elevation is 5005 ft, base elevation = 5000 ft, and entrance elevation = 5001 ft. Set the width of the berm $W = 25$ ft. Consider $S_s = 2.6$ for the solids in stormwater. Figure 14.10 is the distribution of particle sizes observed in urban stormwater. Size a forebay in Figure 14.11 to capture particles ≥ 1.0 mm or $D_s = 1.0$ mm.
   Solution for this case is summarized in Table 14.1.

EXAMPLE 14.4

Expand the forebay sized in Example 14.3 to the particle's gradation curve in Figure 14.10. Table 14.2 summarizes the incremental and cumulative percentage of sediment removal.

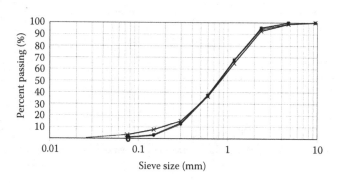

Figure 14.10 Size distributions of particles in urban stormwater.

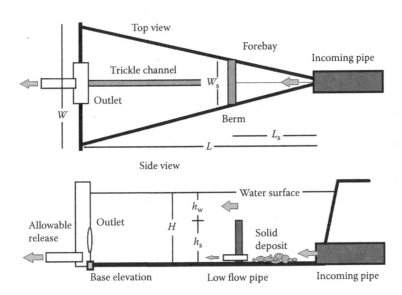

Figure 14.11 Design of forebay.

Table 14.1 Design of forebay to capture particles >0 mm

The design peak flow rate	Q-design = 100.00 cfs
Enter sediment specific gravity	S_s = 2.60
Base elevation at the basin bottom	E-base = 5000.00 ft
Elevation at forebay floor at entrance	E-entrance = 5001.00 ft
Enter design water surface elevation	E-WS = 5005.00 ft
Depth of the basin: $H = h_w + h_s$	H = 4.00 ft
Enter width of forebay or overtopping weir	W_s = 25.00 ft
Enter weir coefficient for feet-second unit	C_w = 3.00
Enter target solid size for sediment removal	D_s = 1.00 mm
Enter fall velocity for target solid size	V_s = 0.165 mps

Particle size (mm)	Reynolds no.	Drag coefficient	Fall velocity (m/s)	Difference in Re
	Guess			Check
1.000	137.23	0.77	0.165	0.00

Sediment residence time or travel time thru forebay	T_s = 7.47 s
Average longitudinal velocity through forebay	U_s = 1.00 fps
Minimum length for forebay	L_s = 7.47 ft
Headwater depth on weir crest (berm)	h_s = 1.21 ft
Height of overtopping weir (berm)	h_s = 2.79 ft

Solution: As shown in Example 14.3, for D_s = 1.0 mm, T_s = 7.47 s. For each particle group, its V_s and T_s are determined for its diameter. As shown in Table 14.1, U_s = 1 fps. If $(U_s \cdot T_s) < L_s$, the particle group would be settled in the forebay. Otherwise, the particle group is washed through the forebay. As summarized in Table 14.2, the sediment removal rate is 40%. Any particles <1.00 mm will be washed into the detention basin for further treatment.

Table 14.2 Sediment removal rate for a given particle gradation curve

Particle size (mm)	Cumulative percentage	Reynolds no.	Drag Coefficient C_d	Fall velocity V_s (m/s)	Residence time T_s (s)	Horizontal distance (m)	Settlement analysis	Sediment trap efficiency (%)
10.000	100	6084.169	0.38	7.30E-01	1.67	1.67	Settled	0.00
5.000	98	2057.966	0.42	4.94E-01	2.47	2.47	Settled	2.00
2.000	90	461.527	0.53	2.77E-01	4.40	4.40	Settled	10.00
1.000	60	135.080	0.78	1.62E-01	7.52	7.52	Settled	40.00
0.500	30	33.530	1.57	8.05E-02	15.15	15.15	Washed	40.00
0.300	12	10.295	3.61	4.12E-02	29.61	29.61	Washed	40.00
0.200	10	3.657	8.47	2.19E-02	55.59	55.59	Washed	40.00
0.100	5	0.591	44.85	6.74E-03	180.89	180.89	Washed	40.00
0.080	3	0.303	84.92	4.38E-03	278.29	278.29	Washed	40.00
0.030	1	0.017	1440.49	6.52E-04	1871.70	1871.70	Washed	40.00

14.4 Micropool for syphon flow

Between two adjacent storm events, the basin remains dry. Urban debris and leaves are built up in the dry basin. As soon as flood water enters the basin, light debris float up with the rising water. As shown in Figure 14.12, floating debris tend to flow around the outflow box. As soon as the outflow devices become clogged, the basin will carry standing water. This prolonged drainage problem is a serious public concern. Often, a screen or rack is installed in front of the perforated plate to reduce the clogging potential, and a syphon device is also built in the micropool to provide a syphon flow in case of clogging.

As shown in Figure 14.13, a micropool is a sunken wet pool that is built in front of the outflow structure to house the submerged syphon device. An up-sloped pipe or submerged perforated plate serves like a syphon that will continually drain water as soon as the hydraulic head is developed due to clogging around the outlet structure. The standing water in the basin produces a sufficient headwater depth to lift water through the gap between the screen and the plate. Such a syphon flow will continue until the basin is emptied out (Guo et al., 2012).

Figure 14.12 Observed clogging on perforated plate. (a) Micropool in front of outflow structure and (b) micropool and syphon flow.

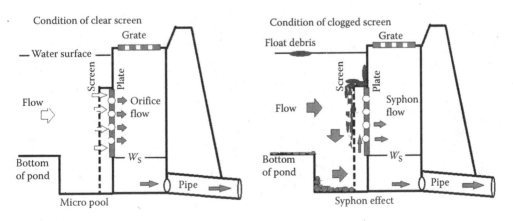

Figure 14.13 Clear and clogging drainage condition.

14.4.1 Micropool Design

To convert the unsteady flow around a moving particle into a steady flow condition, we need to "freeze" the particle. Adding a downward velocity, V_b, which is the float velocity of the particle, to the entire flow field (illustrated in Figure 14.14), the steady flow pattern is a downward flow around the debris particle. This downward velocity is equivalent to the syphon flow that goes through the micropool surface.

The forces around a floating particle include the downward body weight and drag forces that are balanced with the upward buoyance lift. Aided with Equations 14.13 through 14.15, the upward floating velocity for debris particle is derived as

$$F_b = F_g + F_d \tag{14.25}$$

$$V_b = \sqrt{\frac{4gD_b}{3C_d}(1 - S_b)} \tag{14.26}$$

where V_b = float velocity of the particle in [L/T], D_b = minimum size of the float particle that is allowed to flow through the perforated plate, such as 1.0 mm, and S_b = specific gravity of saturated debris float such as 0.8 to 0.9. Figure 14.15 is the evidence of sediment removal

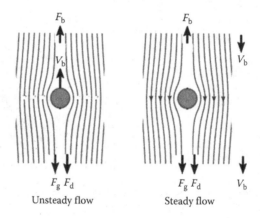

Figure 14.14 Flow field around floating debris particle.

Figure 14.15 Sediment removal at forebay and micropool. (a) Debris cake intercepted by screen and (b) sediment captured in forebay.

in an urban detention basin with a forebay at the entrance and a micropool in front of the outflow structure.

The analyses of particle size (as shown in Figure 14.16) were conducted based on the collected debris cakes accumulated on the clogged screen in front of the perforated plates in several urban detention basins (Mendi, 2015). Considering that the screen intercepts 90% of floating particles, $D_b = 0.3\,\text{mm}$ (as shown in Figure 14.16). Any particles $<D_b$ will flow through the perforated plate to enter the downstream water body.

A micropool is a backup system in case that a clogging situation is developed around the outflow structure. The main purpose of a micropool is to warrant the drain time and flow release rate under the design condition. Usually, a detention basin is not clogged during an extreme event because of its huge flow volume with diffused debris loads. An outflow structure is, in fact, more vulnerable to debris clogging during a series of small events in the magnitude of 3- to 6-month events. The trickle flow continually carries debris and trash into a basin. As the clogging on the outlet box is being developed, the basin will accumulate standing water that tends to become a mosquito bed. A micropool is designed to provide a suction head to produce a syphon flow to continually drain the accumulated runoff volume. Consider the WQCV as the target volume for micropool design. Usually, the drain time for WQCV is 12–24 h. Therefore, the average release rate is defined as

$$q = \frac{\text{WQCV}}{T} \tag{14.27}$$

in which q = average release in $[L^3/T]$, WQCV = water quality capture volume in $[L^3]$, and T = drain time in hours.

The flow release, q, has to go through the micropool's surface area, which can be determined as

$$A_M = \frac{q}{V_b} \tag{14.28}$$

As soon as it rains, the micropool is the first low spot to be filled up, and the syphon device (Figure 14.17) is submerged and primed with its surcharge depth, H_b. As the water

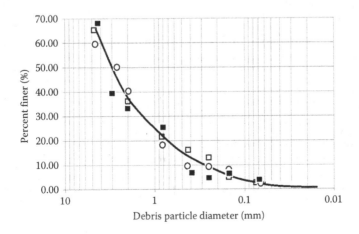

Figure 14.16 Distribution of particle size for float debris.

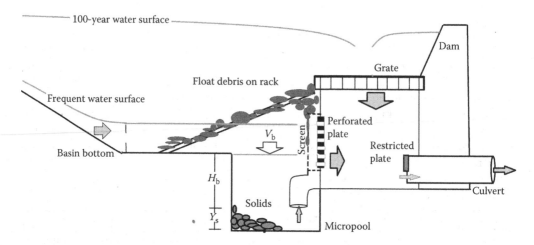

Figure 14.17 Syphon pipes to lift water flow.

surface rises in the detention basin, floating debris will be built up on the screen in front of the perforated plate. After the screen becomes plugged, a standing water depth, Y, is developed in the basin. The suction flow is calculated as

$$q_S = C_o(1 - c\log)A_S\sqrt{2g(H_b + Y)} \tag{14.29}$$

where q_S = suction flow in [L³/T], A_S = cross-sectional area of syphon flow in [L²], c log = area clogging ratio due to algae in pool, g = gravitational acceleration 32.2 ft/s² or 9.81 m/s², Y = standing water depth in [L], and H_b = specified surcharge depth in [L] such as 6 to 12 in. To be conservative, the cross-sectional area of the syphon device is determined with $Y \approx 0$ as

$$A_S = \frac{q}{C_o(1 - c\log)\sqrt{2gH_b}} \tag{14.30}$$

Equation 14.30 will allow $q_S > q$ for an accelerated emptying process. Considering the dead storage for settled solids, and evaporation loss, the depth of the micropool shall not be less than

$$H_M = H_b + Y_S + Y_V \tag{14.31}$$

where H_M = depth of micropool in [L], Y_S = dead storage depth for sediment deposit in [L], and Y_V = evaporation depth in [L]. Evaporation rate, E_V, is local and seasonal. The proper evaporation depth shall be estimated for the wet months in a year. From the local hydrology, the average interevent time, T_I, provides a basis to estimate the required evaporation depth as

$$Y_V = E_V T_I \tag{14.32}$$

EXAMPLE 14.5

A detention basin is designed to serve a tributary area of 20 acres. The WQCV is determined to be 0.19 in. per watershed area with a drain time of 12 h. Design the micropool to intercept 90% of floating debris particles. Design information includes the following: $S_b = 0.83$, $T_l = 7$ days, $E_V = 0.35$ in./day, and $H_b = 0.5$ ft.

$$WQCV = 0.19 \text{ in./watershed area} = (0.19/12) \times 20 = 0.31 \text{ acre-ft}$$

$$q = \frac{WQCV}{T} = \frac{0.31 \times 43,560}{12 \times 3600} = 0.32 \text{ cfs}$$

For an interception rate of 90%, the particle size for floating debris is 0.3 mm from Figure 14.15. Details of computations are as follows:

Specific gravity of floating debris	$S_s = 0.83$
Minimum size of the float particle	$D_b = 0.30$ mm
Water viscosity	Viscosity $= 0.0000012$ m^2/s

Analysis of floating debris velocity

Float size (mm)	Reynolds no. Guess	Drag coefficient	Float velocity (m/s)	Difference in Re Check
0.300	1.481	19.011	0.006	0.00

Float velocity of the particle	$V_b = 0.019$ ft/s

Syphon capacity

Orifice discharge coefficient	$C_o = 0.60$
Clogging factor due to algae	$c \log = 0.10$
Surcharge depth	$H_s = 0.50$ ft
Cross-sectional area of syphon flow	$A_s = 0.10$ ft^2

Geometry of micropool

Evaporation daily rate	$E_V = 0.35$ in./day
Interevent time between adjacent storms	Inter time $= 7.00$ days
Sediment dead storage depth	$Y_S = 1.50$ ft
Micropool surface area	$A_M = 16.34$ ft^2
Micropool depth	$Y_M = 2.20$ ft

The dimension of the micropool for this case is presented in Figure 14.18.

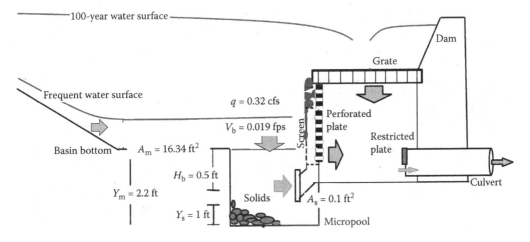

Figure 14.18 Design of micropool with up-sloped syphon.

14.5 Emergency spillway

Spillway is an emergency outlet. Water released from a spillway is modeled by weir hydraulics. A spillway (Figure 14.19) is designed to provide a safe overflow when situations such as the blockage of the primary outlet structures or the occurrence of an event larger than the design capacity of the basin arise. For a large detention basin that has a high embankment >10 ft, and/or a large storage volume >100 acre-ft, the spillway shall be designed to be able to withstand up to a *probable maximum flood* (PMF). For a small detention basin, the embankment is designed to safely pass the 100-year overtopping flows.

The overtopping flow from a spillway drains into either a collector channel or a stilling basin through a concrete chute with baffle blocks. An emergency spillway must be accompanied by adequate erosion control and energy dissipating measures to ensure the stability of the embankment. In an urban area, a labyrinth weir (Figure 14.20) is often considered because it has a high hydraulic efficiency (Tullis et al., 1995).

Design of a labyrinth weir begins with the selection of the height of the weir, P, the length of the apron, B, and the labyrinth angle, θ. The wall thickness, t, is recommended as

$$t = \frac{P}{6} \tag{14.33}$$

The inside apex width, D_1, is recommended to be

$$D_1 = \text{a value between } t \text{ and } 2t \tag{14.34}$$

According to the geometry of the labyrinth weir, the design parameters can be calculated as

$$D_2 = D_1 + \frac{t}{2}\tan\left(45 - \frac{\theta}{2}\right) \text{in which } \theta \text{ is in degrees} \tag{14.35}$$

$$L_1 = \frac{(B-t)}{\cos\theta} \tag{14.36}$$

$$L_2 = L_1 - t\tan\left(45 - \frac{\theta}{2}\right) \tag{14.37}$$

Figure 14.19 Labyrinth weir and straight weir. (a) Overtopping weir and (b) labyrinth weir.

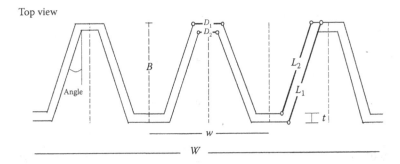

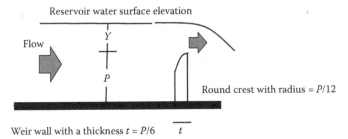

Figure 14.20 Parameters of labyrinth weir.

$$w = 2L_1 \sin\theta + D_1 + D_2 \tag{14.38}$$

$$W = nw \tag{14.39}$$

$$L = n(2L_1 + D_1 + D_2) \tag{14.40}$$

$$L_e = 2n(D_1 + L_2) \tag{14.41}$$

in which L = weir length in [L], L_e = effective weir length in [L], n = number of cycle in the labyrinth weir, and other variables are defined in Figure 14.20. When $n = 1$, a labyrinth weir is reduced to a straight (linear) weir.

To apply weir hydraulics to the labyrinth weir, the discharge coefficient, C_d, is a function of Y/P. The weir height, P, is the difference between the crest elevation and the elevation of the upstream apron. The weir height influences losses in the approach flow and spillway capacity. For a labyrinth weir, the value of C_d continues to decrease as the head increases. Eventually, the labyrinth weir approaches that of a linear weir having the length equal to the apron width. Tullis et al. (1995) reported a set of empirical equations for determining the value of discharge coefficient as

$$C_d = 0.49 + C_1\left(\frac{Y}{P}\right) + C_2\left(\frac{Y}{P}\right)^2 + C_3\left(\frac{Y}{P}\right)^3 + C_4\left(\frac{Y}{P}\right)^4 \tag{14.42}$$

$$Q_L = \frac{2}{3}\sqrt{2g}C_d L_e H^{1.5} \tag{14.43}$$

Table 14.3 Coefficients developed for labyrinth weir used for spillway

Labyrinth angle (°)	C_1	C_2	C_3	C_4	Limitation
6.00	−0.24	−1.20	2.17	−1.03	
8.00	1.08	−5.27	6.79	−2.83	
12.00	1.06	−4.43	5.18	−1.97	
15.00	1.00	−3.57	3.82	−1.38	
18.00	1.32	−4.13	4.24	−1.50	
25.00	1.51	−3.83	3.40	−1.05	
35.00	1.69	−4.05	3.62	−1.10	
90.00	1.46	−2.56	1.44	0.00	$E < 0.70$

in which Q_L = weir flow in [L³/T] and C_d = discharge coefficient. Coefficients C_1, C_2, C_3, and C_4 are listed in Table 14.3. Equation 14.42 was compared with field measurements for $0.1 < Y/P < 0.9$. The standard deviation between the measured and calculated data for a labyrinth angle from 6° to 18° was less than ±3.0%.

EXAMPLE 14.6

A labyrinth weir is designed to have a height, $P = 3.0$ m, length of apron, $B = 10$ m, weir angle, $\theta = 12°$, and number of cycle, $n = 5.0$. Determine the capacity of the spillway under an effective headwater, $y = 1.8$ m.

Firstly, the geometric parameters for this labyrinth weir are

$$t = \frac{P}{6} = \frac{3.0}{6.0} = 0.5 \text{ m}$$

$D_1 = 0.75$ m between t and $2t$.

$$D_2 = D_1 + \frac{t}{2} \tan\left(45 - \frac{\theta}{2}\right) = 0.75 + \frac{0.5}{2} \tan\left(45 - \frac{12}{2}\right) = 0.95 \text{ m}$$

$$L_1 = \frac{(B-t)}{\cos\theta} = \frac{(10-0.5)}{\cos 12°} = 9.71 \text{ m}$$

$$L_2 = L_1 - t \tan\left(45 - \frac{\theta}{2}\right) = 9.71 - 0.5 \times \tan 39° = 9.31 \text{ m}$$

$$L = n(2L_1 + D_1 + D_2) = 5 \times (2 \times 9.71 + 0.75 + 0.95) = 105.6 \text{ m}$$

The outflow capacity of this spillway is calculated as

$$\frac{Y}{P} = \frac{1.8}{3.0} = 0.60$$

$$C_d = 0.49 + 1.60 \times 0.60 - 4.43 \times 0.60^2 + 5.18 \times 0.60^3 - 1.97 \times 0.60^4 = 0.40$$

Substituting the variables into Equation 14.43 yields

$$Q_L = \frac{2}{3} \times 0.40 \times \sqrt{2 \times 9.80} \times 105.60 \times 1.8^{1.5} = 302.5 \ m^3/s$$

Repeating this process, a rating curve, stage-outflow curve, can be constructed for this spill-way. Of course, this emergency capacity can be added to the stage-outflow characteristic curve for the detention basin under design.

14.6 Closing

When on stream detention is not possible, the excess stormwater has to be diverted into an off stream detention basin. The flow diversion system can be designed to have side weirs and flood gates that will be reliably operated by the gravity. The inflow shall be first delivered into a forebay to settle coarse solids such as particles ≥1.0 or 2.0 mm. A functional forebay not only extends the life of the stormwater detention pond but also adds more enhancements to water quality control. The cost of a sediment forebay depends on its design criteria and requirements. Design features including impermeable liners, baffles, and embankments will dictate the final cost of the forebay. In case the outlet is clogged, the water pool in the forebay tends to grow algae. Stormwater generated from the neighborhood often carries excessive fertilizer and chemical, and the abundant nutrients encourage algae to grow. It is recommended that a sediment forebay be cleaned out every 3–5 years or as required when algae become a problem.

All drainage facilities are vulnerable to urban debris that remains dry between events, and becomes wet and floating when water rises along the rack and screen in front of the outflow structure. A micropool is an important element to reduce the risk of standing water in case of clogging developed around the outflow structure. A micropool is not necessary to be large, but it has to be adequate to provide a syphon effect when the outflow system becomes plugged. During the loading process in the basin, a micropool needs to be sufficiently submerged with the rising water. A micropool system shall be sized to warrant a continual syphon flow to empty the standing water in time.

14.7 Homework

Q14.1 A trapezoidal channel in Figure Q14.1 has a bottom width of 10 ft, side slope of 1V:4H, invert slope of 0.25%, and Manning's N = 0.035. The 100-year peak inflow in this channel is 690 cfs. The downstream capacity of this channel is limited to 370 cfs.

1. Verify that a 4-ft × 10-ft box culvert can regulate the straight-through flow to be 370 cfs.
2. Determine the diversion flow into the detention basin using three 24-in. circular conduits that are laid on the channel bottom (Solution: 321 cfs).
3. Determine the threshold depth differential between the channel and the basin in order to open the 24-in. steel flood gate. The specific gravity for steel flood gate is 7.50. The thickness of the flood gate is 6 in. (Solution: Δy = 2.3 ft).

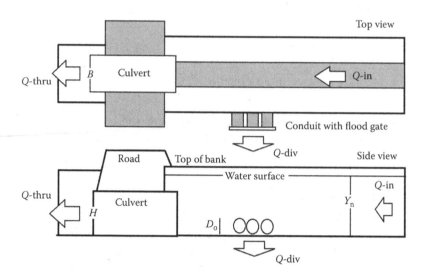

Figure Q14.1 Illustration of flow diversion system using circular conduits.

Q14.2 A detention pond is designed to have a forebay and micropool as shown in Figure Q14.2. The detention system is sized to reduce the 100-year peak inflow from 110 cfs down to 34 cfs. A forebay is required to settle particles ≥2 mm, and a micropool is sized to drain a WQCV of 0.35 in. for a tributary area of 40 acre over a drain time of 12 h.

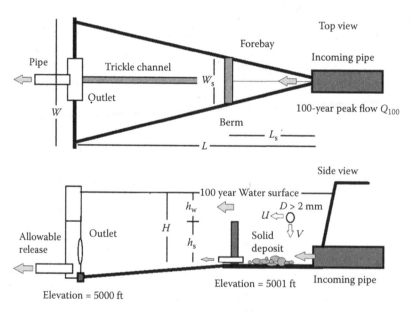

Figure Q14.2 Layout of forebay.

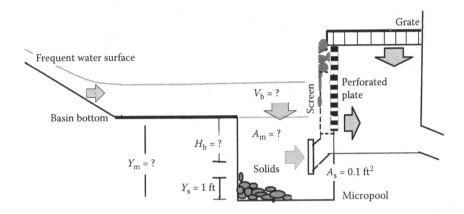

Figure Q14.3 Design of micropool.

1. Use the 100-year peak flow of 110 cfs as the inflow to the forebay that is shaped to capture solids ≥ 2 mm. The width of the weir is set to be 25.0 ft. Determine the height of the weir and the headwater depth on the weir when using the weir coefficient of 3.0.
2. The water quality capture volume for this case is 0.35 in. for a tributary area of 40 acres. The drain time is set to be 12 h. Design the micropool in Figure Q14.3 to intercept 90% of floating debris particles, or $D_b = 0.3$ mm. Given design information includes $S_b = 0.83$, $T_I = 7$ days, and $E_V = 0.35$ in./day.

Bibliography

ASCE. (1994). "Design and Construction of Urban Stormwater Management System," American Society of Civil Engineers, Manuals and Reports of Engineering Practice No. 77, New York.

Guo, J.C.Y. (2012). "Off-Stream Detention Design for Stormwater Management," *ASCE Journal of Irrigation and Drainage Engineering*, Vol. 138, No. 4, pp. 371–376.

Guo, J.C.Y., and Clark, J. (2006). "Rational Volumetric Method for Off-Line Storm Water Detention Design," *Journal of PB Network for Water Engineering and Management*, Vol. 21, No 3, Issue No. 64.

Guo, J.C.Y., and Jones, J. (2010a). "Pinning Force during Closure Process at Blocked Pipe Entrance," *ASCE Journal of Irrigation and Drainage Engineering*, Vol. 136, No. 2, pp. 141–144.

Guo, J.C.Y., Jones, J., and Earles, A. (2010b). "Method of Superimposition for Suction Force on Trash Rack," *ASCE Journal of Irrigation and Drainage Engineering*, Vol. 136, No.11, pp. 781–785.

Guo, J.C.Y., Shih, H.M., and MacKenzie, K. (2012). "Stormwater Quality Control LID Basin with Micropool," *ASCE Journal of Irrigation and Drainage Engineering*, Vol. 138, No. 5.

Guo, J.C.Y., Urbonas, B., Mackenzie, K., and Lloyd, D. (2007). "Stormwater Computer Design Tools," Invited, *Journal of Urban Water Management*.

Jones, J, Guo, J.C.Y., and Urbonas, B. (2006). "Safety on Detention and Retention Pond Designs," *Journal of Storm Water*, Vol. 35.

Mendi, S. (2015). "Design of Micropool for Detention Basin System," Master Report, Department of Civil Engineering, University of Colorado Denver, Denver.

Pemberton, E.L., and Lara, J.M. (1971). *A Procedure to Determine Sediment Deposition in a Settling Basin*, Sedimentation Section, U.S. Bureau of Reclamation, Denver, CO.

Randle, T.J. (1984). *User's Guide to Computer Modeling of Settling Basins*, U.S. Bureau of Reclamation (Sedimentation and River Hydraulics Section, Hydrology Branch), Denver, CO.

Skipwith, W., Denton, M., and Askew, M. (1990). "Closing the Floodgates," *ASCE Civil Engineering*, Vol. 60, No. 7, pp. 54–55.

Tullis, J.P., Amanian, N., and Waldron, D. (1995). "Design of Labyrinth Spillways," *ASCE Journal of Hydraulic Engineering*, Vol. 121, No. 3, pp. 247–255.

UDFCD. (2010). *Urban Stormwater Design Criteria Manual*, Volumes 1 and 2, Urban Drainage and Flood Control District, Denver, CO.

Chapter 15

Grate and rack hydraulics

A water flow has to undertake the processes of deceleration and acceleration when going through a hydraulic structure. Any change in the flow velocity means that a resultant force is acting on the flow and its reaction force is acting on the hydraulic structure. For the sake of safety and maintenance, grates and racks (as shown in Figure 15.1) are always recommended to be installed at an entrance of a conduit (CDOT, 2004; FHWA, 1971). Over the recent years, higher standards were developed to preserve the water environment, and more detention and retention basins were built in the neighborhood. Safety around a stormwater facility has become an increasing concern for the public because urban flood flows are quick, concentrated, and fast. Many forensic studies indicate that a trash rack at the entrance of a hydraulic structure can prevent human body or a small animal from being washed into the closed system such as culverts and sewers. Therefore, grates and racks act not only as trash control but also as life savers (Jones et al., 2006).

15.1 Grate geometry

Grates are often used to collect storm runoff into an outflow structure installed in a storage basin or on a highway median. A type C grate (Figure 15.2) has a standardized square surface area of 1.0 m × 1.0 m, and a type D grate is a doubled type C grate or has a surface area of 1.0 m × 2.0 m. Both type C and D grates are commonly used for inlet designs that are aimed at a high hydraulic efficiency to intercept storm runoff. In practice, a grate is installed at an inclined angle because an inclined grate surface may act as a regulator to the flow release and an inclined grate is not as vulnerable as the flat grate to debris clogging.

In addition to hydraulic natures, the performance of a grate also depends on its surface geometry. As shown in Figure 15.3, a grate is formed with I-beam bars that reduce the flow-through area on the grate surface. The net opening ratio for a grate is first calculated based on the clear opening area for water to flow through the grate surface when the water depth is deep or to overtop the grate sides when the water depth is shallow. Next, taking the debris clogging into consideration, the net opening area ratio is defined as

$$n = (1 - c\log)\frac{LB - L_b B}{LB} = (1 - c\log)\frac{L - L_b}{L} \tag{15.1}$$

where n = net area-opening ratio, c log = clogging factor $0 \leq c\log \leq 1.0$ due to debris, L = grate length in [L], B = grate width in [L], and L_b = cumulative width in [L] of I-beam bars on grate.

Figure 15.1 Examples of rack and grate. (a) Rack in front of box culvert and (b) inclined grate on inflow conduit.

Figure 15.2 Grates used for stormwater drains. (a) Type C grate in sump, (b) type D grate in median, and (c) type C on outflow box.

Equation 15.1 indicates that the grate's *area-opening ratio* for an orifice flow is equal to the *length-opening ratio* for a weir flow overtopping the grate's sides. The selection of clogging factor depends on the surrounding condition. A decay-based clogging factor is recommended for multiple grates in series. Details can be found elsewhere (Guo in 2000, 2006).

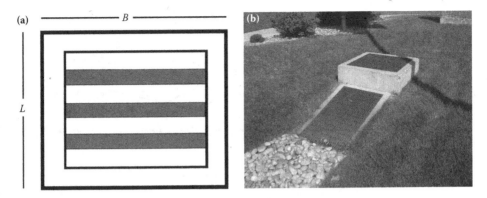

Figure 15.3 Inclined grate. (a) Steel I-beam on grate and (b) inclined grate.

15.2 Grate hydraulics

The hydraulic performance of a grate depends on the water depth on top of the grate. When the water depth is too shallow to submerge the entire grate surface, the grate operates like a weir. When the grate area is completely submerged, the grate operates like an orifice. The transition from weir to orifice flow is a mixing flow (Guo et al., 2008). The hydraulic capacity of a type C grate is quantified based on its flow interception. The integral of flow interception is described as

$$Q = nC_d \int \sqrt{2gh} \, dA \qquad (15.2)$$

where Q = flow rate in [L^3/T], C_d = discharge coefficient, g = gravitational acceleration in [L/T^2], dA = infinitesimal flow area in [L^2], and h = headwater depth in [L] on dA. A grate may operate like a weir or an orifice, depending on the water depth. Two sets of flow interception equations were derived to predict weir and orifice flows; the smaller of the two dominates the grate's hydraulic capacity.

15.2.1 Weir flow capacity

As illustrated in Figure 15.4, the inclined angle is formed by the grate length, L, and its rise, H_b. The coordination system (h, x) is set to describe the flow condition in which h = water depth measured downward from the water surface and x = wetted distance measured along the grate. Under a shallow water depth, only the lower grate surface is submerged and acts like a weir to receive the flow overtopping the three submerged sides: two inclined sides and the lower base width.

The infinitesimal flow area (Figure 15.4) for a weir flow is derived as

$$dA = (H - h) \cot\theta \, dh \quad \text{for } H < H_b \qquad (15.3)$$

$$y = H - h \qquad (15.4)$$

where θ = inclined angle, H = water depth in [L], y = location in [L] of dA above the ground (or $y = 0$), and dh = infinitesimal thickness in [L] for flow area. The weir flow

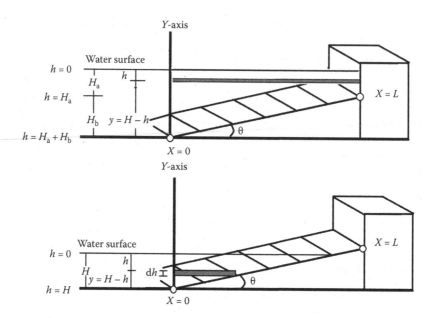

Figure 15.4 Weir flow overtopping submerged side along grate.

overtopping the wetted length along the grate's side is integrated from $h = 0$ to $h = H$. Aided by Equation 15.3, Equation 15.2 yields

$$Q_{WS} = \frac{4}{15} n\, C_d \sqrt{2g}\, \cot\theta\, H^{\frac{5}{2}} \quad \text{for } H < H_b \tag{15.5}$$

where Q_{WS} = side weir flow in $[L^3/T]$. Under a high water depth (as illustrated in Figure 15.4), for mathematical convenience, the integration limit is divided into two zones as follows:

$$H = H_b + H_a \tag{15.6}$$

where H_a = surcharge depth in $[L]$ above the top base of the grate. The infinitesimal areas for the weir flow in these two flow zones are respectively formulated as

$$dA_1 = (H - h)\, \cot\theta\, dh \quad 0 \le h \le H_a \quad \text{for Zone 1} \tag{15.7}$$

$$dA_2 = L\cos\theta\, dh \quad H_a \le h \le H \quad \text{for Zone 2} \tag{15.8}$$

The subscripts 1 and 2 represent the variables in zones 1 and 2. The weir flow overtopping the submerged length is integrated as

$$Q_{WS} = n C_d \int_{h=0}^{h=H_a} \sqrt{2gh}\, L\cos\theta\, dh + n C_d \int_{h=H_a}^{h=H} \sqrt{2gh}\,(H-h)\cot\theta\, dh \tag{15.9}$$

Integrating Equation 15.9 yields

$$Q_{WS} = \frac{4}{15} n C_d \sqrt{2g} \cot\theta (H^{5/2} - H_a^{5/2})$$ (15.10)

Rearranging Equation 15.10 yields

$$Q_{WS} = \frac{4}{15} n C_d \sqrt{2g} L \cos\theta \, H^{\frac{3}{2}} \left[\frac{H^{\frac{5}{2}}}{H^{\frac{3}{2}} H_b} - \frac{(H-H_b)^{\frac{5}{2}}}{H^{\frac{3}{2}} H_b} \right] \text{ for } H \geq H_b$$ (15.11)

At $H = H_b$, Equation 15.11 agrees with Equation 15.5. The total flow collected into the inlet box is the sum of the weir flows overtopping the two wetted sides along the grate and the lower base width of the grate. The weir flow over the lower base is computed as

$$Q_{WB} = \frac{2}{3} n C_d \sqrt{2g} \, B \, H^{\frac{3}{2}}$$ (15.12)

in which Q_{WB} = flow in $[L^3/T]$ overtopping the low base width. The total weir flow is the sum as

$$Q_W = 2Q_{WS} + Q_{WB}$$ (15.13)

in which Q_W = total interception in $[L^3/T]$ for weir flow.

15.2.2 Orifice flow capacity

The grate surface area operates like an orifice (as illustrated in Figure 15.5). As mentioned earlier, the integration of the orifice flow into the inlet box is separately conducted for the low and high water depth conditions.

For $H \leq H_b$, the infinitesimal flow area for orifice flow in Figure 15.5 is defined as

$$dA = n \, B \cos\theta \, dx$$ (15.14)

The head water depth, h, can be related to the submerged length, x, along grate's side as

$$h = \left(1 - \frac{x}{X}\right) H$$ (15.15)

where X = submerged length in $[L]$ that is ranged as $0 \leq X \leq L$ and x = submerged distance in $[L]$ along the grate's side ranged as $0 \leq x \leq X$. Under a low-flow condition, $H \leq H_b$, the orifice flow through the submerged surface area on the grate is integrated from $x = 0$ to $x = X$ as

$$Q_o = \frac{2}{3} n C_d B H \cot\theta \sqrt{2gH} \quad \text{for } H \leq H_b$$ (15.16)

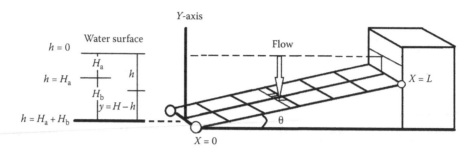

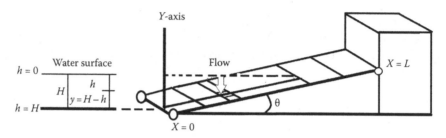

Figure 15.5 Orifice flow through submerged area on grate.

in which Q_o = orifice flow in $[L^3/T]$. When $\theta = 0$, Equation 15.16 is reduced to a horizontal orifice as

$$Q_o = \frac{2}{3} n C_d B L \sqrt{2gH} \quad \text{for } H_b = 0 \text{ and } \theta = 0 \tag{15.17}$$

Under a high-flow condition, the entire grate surface area is submerged. The headwater is related to the submerged length along the grate as

$$h = H - \frac{x}{L}(H - H_a) = H - \frac{x}{L}H_b \tag{15.18}$$

For mathematical convenience, the flow depth is divided into two zones for numerical integration: (1) above the top of the grate and (2) below the top of the grate. The orifice flow under a high water depth is integrated from $x = 0$ to $x = L$ as

$$Q_o = \frac{2}{3} n C_d B L \cos\theta \sqrt{2gH} \left[\frac{H^{\frac{3}{2}}}{H_b \sqrt{H}} - \frac{(H - H_b)^{\frac{3}{2}}}{H_b \sqrt{H}} \right] \quad \text{for } H > H_b \tag{15.19}$$

At $H = H_b$, Equation 15.19 agrees with Equation 15.16.

15.2.3 Discharge coefficients

In practice, grates in Figure 15.6 are laid out with different inclined angles and centerline orientation relative to the direction of incoming flow. Inclined angles are ranged from flat to 30° above the ground. The alignment of a grate may be parallel or normal to the flow line as illustrated in Figure 15.6.

The efficiency of flow interception through a grate is termed *discharge coefficient*, C_d. The laboratory models shown in Figure 15.6 were tested to derive the values of C_d (Comport et al., 2010). Figure 15.7 is the summary of the variation of C_d with respect to inclined angle. As revealed in Figure 15.7, a leveled grate has the highest hydraulic efficiency.

Figure 15.6 Layouts of grate. (a) Flat type C grate, (b) 20° type D grate, and (c) 10° rotated type D.

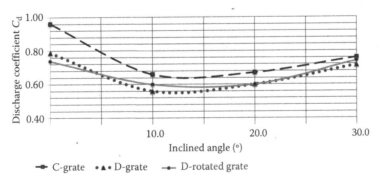

Figure 15.7 Discharge coefficients for type C and D grates.

As the inclined angle increases from 0° to 15°, the grate's discharge coefficient decreases. The grate gradually recovers its hydraulic efficiency as the inclined angle increases from 15° to 30°. The influence of the inclined angle becomes diminished when a horizontal grate is gradually raised toward a vertical.

Comparing with the conventional approach, the orifice and weir coefficients can be related to the discharge coefficient:

$$C_o = \frac{2}{3}C_d \tag{15.20}$$

$$C_w = \frac{4}{15}C_d\sqrt{2g} \tag{15.21}$$

in which C_o = orifice coefficient and C_w = weir coefficient. Using the orifice and weir coefficients, the governing equations for various flow conditions are summarized as follows:

For $H \leq H_b$, the orifice and weir flows, respectively, are estimated as

$$Q_o = nC_o BH\cot\theta\sqrt{2gH} \quad \text{for low head orifice flow} \tag{15.22}$$

$$Q_w = 2n\,C_w\cot\theta\,H^{\frac{5}{2}} + \frac{5}{2}nC_w BH^{\frac{3}{2}} \quad \text{for low head weir flow} \tag{15.23}$$

For $H \geq H_b$, the orifice and weir flows, respectively, are estimated as

$$Q_o = nC_o BL\cos\theta\,\sqrt{2gH}\left[\frac{H^{\frac{3}{2}}}{H_b\sqrt{H}} - \frac{(H-H_b)^{\frac{3}{2}}}{H_b\sqrt{H}}\right] \quad \text{for high head orifice flow} \tag{15.24}$$

$$Q_w = 2nC_w L\cos\theta\,H^{\frac{3}{2}}\left[\frac{H^{\frac{5}{2}}}{H^{\frac{3}{2}}H_b} - \frac{(H-H_b)^{\frac{5}{2}}}{H^{\frac{3}{2}}H_b}\right] + \frac{5}{2}nC_w BH^{\frac{3}{2}} \quad \text{for high head weir flow}$$

$$\tag{15.25}$$

For a given water depth, the flow interception capacity through an inclined grate is dictated as shown by weir or orifice flows, whichever is less (Mays, 2001):

$$Q_c = \min\left(Q_w,\ Q_o\right) \quad \text{for a given water depth} \tag{15.26}$$

in which Q_c = flow interception in $[L^3/T]$ through grate. On the contrary, for a given design flow, the required headwater depth, H, acting on an inclined grate is determined as (HEC-22, 2002)

$$H = \max\left(H_w,\ H_o\right) \quad \text{for a given design flow} \tag{15.27}$$

where H_w = headwater for weir flow in $[L]$, H_o = headwater for orifice flow in $[L]$, and H = design headwater in $[L]$.

EXAMPLE 15.1

An inclined type D grate in Figure 15.8 is installed on the outflow structure in a detention basin. The basic information is provided in Table 15.1. The lower portion of the outflow structure is reserved for a perforated plate from an elevation of 5000–5002 ft. As a result, the low base of the grate is set at an elevation of 5002 ft. Use a net opening ratio of 0.60 to construct the stage-flow collection curve for this grate.

According to Figure 15.7, the discharge coefficient for an inclined angle of 30° is 0.75. Aided with Equations 15.20 and 15.21, $C_o = 0.50$ and $C_w = 1.60$.

As shown in Table 15.2, the grate remains dry until the water stage reaches the lower grate base, i.e., an elevation of 5002 ft for this case. The collection capacity switched from weir to orifice flow when the water stage changes from 5003 to 5004 ft for this case.

For example, when water stage is at 5003 ft, the water depth, $H = 1.0$ ft $< H_b$. So Equations 15.22 and 15.23 are applicable. The outflow is determined as

$$Q_o = nC_oBH\cot\theta\sqrt{2gH} = 0.6\times0.5\times(3\times1)\cot30°\sqrt{2\times32.2\times1} = 12.51 \text{ cfs}$$

$$Q_w = 2n\,C_w\cot\theta\,H^{\frac{5}{2}} + \frac{5}{2}nC_wBH^{\frac{3}{2}} = 2\times0.6\times1.60\times\cot30°\times1.0^{\frac{5}{2}} + \frac{5}{2}\times0.6\times1.60\times3\times1.0^{\frac{3}{2}}$$
$$= 10.56 \text{ cfs}$$

$$Q_c = \min\ (10.56,\ 12.51) = 10.56 \text{ cfs, namely it is dictated by the weir flow.}$$

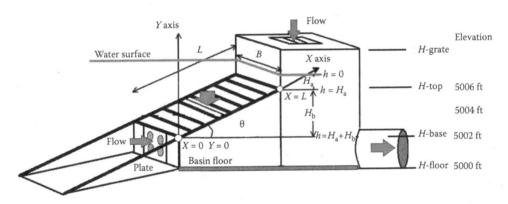

Figure 15.8 Inclined grate on outflow structure for flow detention.

Table 15.1 Design information for inclined gate

Width of inclined grate	$B = 3.00$
Length of inclined grate	$L = 6.00$ ft
Vertical rise of inclined grate	$H_b = 3.00$ ft
Elevation of basin floor	H-floor $= 5000.00$ ft
Elevation of lower grate base	H-base $= 5002.00$ ft
Opening ratio of grate	$n = 0.60$
Grate discharge Coefficient	$C_d = 0.75$
Inclined angle	$\theta = 30.00°$

Table 15.2 Stage-flow collection curve for inclined grate

Water stage (ft)	Water depth (ft)	Submerged side weir length, X (ft)	Inclined left side weir (cfs)	Inclined right side weir (cfs)	Base weir (cfs)	Total weir (cfs)	Total orifice (cfs)	Outflow (cfs)
5000.00	0.00	0.00	0.00	0.00	0.00	0.00	0.00	0.00
5001.00	0.00	0.00	0.00	0.00	0.00	0.00	0.00	0.00
5002.00	0.00	0.00	0.00	0.00	0.00	0.00	0.00	0.00
5003.00	1.00	2.00	1.67	1.67	7.22	10.56	12.51	10.56
5004.00	2.00	4.00	9.44	9.44	20.43	39.30	35.38	35.38
5005.00	3.00	6.00	26.00	26.00	37.53	89.53	65.00	65.00
5006.00	4.00	6.00	51.71	51.71	57.78	161.19	87.57	87.57

Similarly, when water stage is 5006 ft, the water depth, $H = 4$ ft $>H_b$. So Equations 15.24 and 15.25 are applicable.

$$Q_o = nC_oBL\cos\theta\sqrt{2gH}\left[\frac{H^{\frac{3}{2}}}{H_b\sqrt{H}} - \frac{(H-H_b)^{\frac{3}{2}}}{H_b\sqrt{H}}\right]$$

$$= 0.6\times0.5\times3\times6\times\cos(30°)\sqrt{2\times32.2\times4}\left[\frac{4^{\frac{3}{2}}}{3\sqrt{4}} - \frac{(4-3)^{\frac{3}{2}}}{3\sqrt{4}}\right] = 87.57 \text{ cfs}$$

$$Q_w = 2nC_wL\cos\theta\,H^{\frac{3}{2}}\left[\frac{H^{\frac{5}{2}}}{H^{\frac{3}{2}}H_b} - \frac{(H-H_b)^{\frac{5}{2}}}{H^{\frac{3}{2}}H_b}\right] + \frac{5}{2}nC_wBH^{\frac{3}{2}}$$

$$= 2\times0.6\times1.6\times6\times\cos(30°)\times4^{\frac{3}{2}}\left[\frac{4^{\frac{5}{2}}}{4^{\frac{3}{2}}\times3} - \frac{(4-3)^{\frac{5}{2}}}{4^{\frac{3}{2}}\times3}\right] + \frac{5}{2}\times0.6\times1.6\times3\times4^{\frac{3}{2}} = 161.19 \text{ cfs}$$

$Q_c = \min\ (87.57, 161.19) = 87.57$ cfs, namely it is dictated by the orifice flow

Table 15.2 presents the stage-outflow relationship for this inclined grate with an angle of 30°.

As observed in field and laboratory, an inclined angle does not elevate the floating debris to the water surface, but it decreases the grate's performance. Any float landed onto the grate surface becomes nailed by the hydrostatic suction forces. It is suggested that straw or sand bags be used around the grate as a debris control.

15.3 Rack geometry

A rack is simply a bar-screen that intercepts tree stumps, twigs, and floating debris in stormwater flows. As illustrated in Figure 15.9, a prefabricated culvert is often attached with its flare-end section (FES). The trash rack is directly laid on top of the FES with an inclined angle between 30° and 40° (USWDCM, 2001). Figure 15.10 presents examples of trash racks. A trash rack must have a surface area more than *four times* the opening

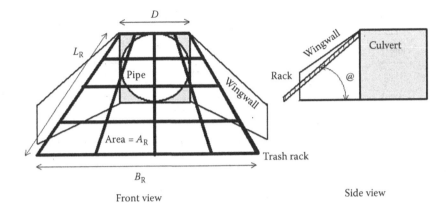

Figure 15.9 Flare-end section for culvert unit.

Figure 15.10 Examples of rack in front culvert. (a) Sufficient rack in front of culvert and (b) overloaded rack.

area of the culvert. A small trash rack may be effective to intercept debris, but it can be easily clogged or even plugged with a large amount of urban trash.

15.4 Rack hydraulics

The force analysis on a partially blocked trash rack is complicated because of the FES geometry at the culvert entrance. The major external forces acting on the water flow include *the reaction force* from the wing walls, *the contraction force* through the trash rack, and *the pinning force* on the blocking object (Weisman, 1989, Allred-Coonrod, 1994). It is a challenge to solve the three unknown forces simultaneously using the principle of flow momentum. The momentum principle is a vector approach. To apply the momentum principle to the FES, it is reasonable to assume that the vertical force components are balanced by the ground support, and the horizontal forces are balanced in the flow direction (Guo et al., 2010). As illustrated in Figure 15.11, the control volume of water flow is set to be between sections 1 and 2. Section 1 is the

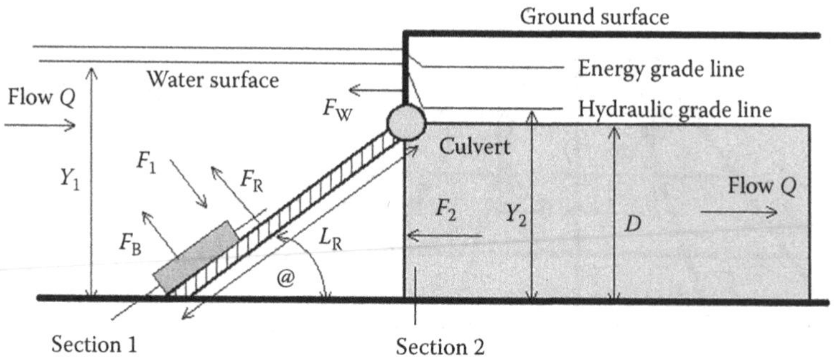

Figure 15.11 Flow through trash rack at culvert entrance.

FES, and section 2 represents the culvert entrance section. The balance of forces in the flow direction is formulated as (Guo and Jones, 2010)

$$F_1 \sin\theta = F_2 + F_W + F_R \sin\theta + F_B \sin\theta \quad \text{in flow direction} \tag{15.28}$$

$$F_1 = \gamma \bar{Y}_1 A_1 + \rho Q V_1 \quad \text{in flow direction perpendicular to rack surface} \tag{15.29}$$

$$F_2 = \gamma \bar{Y}_2 A_2 + \rho Q V_2 \quad \text{in flow direction} \tag{15.30}$$

in which F = force acting on water flow, F_W = reaction force from wing walls, F_R = contraction force through rack, F_B = pinning force on block, γ = water specific weight, $\bar{Y}$ = depth to centroid of flow area, V = cross sectional average velocity, A = flow area, Q = flow in pipe, ρ = water density, and θ = inclined angle of rack. The subscripts 1 and 2 represent the variables of sections 1 and 2, respectively.

Equation 15.28 has three unknown forces—F_R, F_W, and F_B. It is suggested that Equation 15.28 be solved using the *method of superimposition*, starting with (1) solving the unknown force, F_W, when the system has no rack, (2) solving the unknown force, $(F_R \sin\theta + F_W)$ when the rack is added to the system, and (3) solving the unknown force, $(F_R \sin\theta + F_B \sin\theta + F_W)$ when the rack is partially blocked. Between sections 1 and 2 in Figure 15.11, the flow velocities and areas are calculated as

$$A_1 = A_R \tag{15.31}$$

$$V_1 = \frac{Q}{A_R} \tag{15.32}$$

$$A_2 = \frac{\pi D^2}{4} \tag{15.33}$$

$$V_2 = \frac{4Q}{\pi D^2} \tag{15.34}$$

$$\bar{Y}_1 = Y_1 - \frac{L_R}{2} \sin\theta \tag{15.35}$$

$$\bar{Y}_2 = Y_2 - \frac{D}{2} \tag{15.36}$$

where A_R = rack's surface area on FES, D = culvert diameter or height, Y = hydraulic gradient line (HGL) relative to the culvert invert, and L_R = length of rack on FES. Let $F_B = F_R = 0$ in Equation 15.28. Substituting Equations 15.33 through 15.36 into Equation 15.28 yields the reaction force, F_W, from the wing walls.

15.4.1 Force balance under clear rack condition

A rack is formed with steel bars. The area-opening ratio is used to calculate the net opening area for water to flow through the rack. The flow velocity at section 1 is then calculated as

$$V_1 = V_R = \frac{Q}{nA_R} \qquad (15.37)$$

where V_R = flow velocity through clear area on the rack and n = area-opening ratio of rack surface, depending on the number and size of steel bars used to form the rack. The rest of the variables remain the same as the base condition. Because the rack is not blocked, $F_B = 0$. Aided by Equation 15.37, Equation 15.28 can be solved for the sum of $(F_R \sin\theta + F_W)$. The pinning force on the rack is the difference between $(F_R \sin\theta + F_W)$ and F_W, or it can be derived as

$$F_R = \rho\frac{Q^2}{nA_R}(1-n) = \rho V_R Q(1-n) \qquad (15.38)$$

Equation 15.38 provides a direct solution to the force on the clear rack. It implies that the presence of a rack does not change the hydrostatic force because the rack is submerged.

15.4.2 Force balance under blocked rack condition

A rack intercepts debris in the water flow. For simplicity, the blockage on the rack is represented by the clogged surface area that can be approximated by the projected area on the rack. The blocked area to rack surface area ratio, m, is defined as

$$m = \frac{A_B}{A_R} \qquad (15.39)$$

in which A_B = blocked area on rack surface and m = blocked area to rack surface area ratio. As a result, the flow velocity through the blocked rack is calculated as

$$V_1 = V_B = \frac{Q}{A_R(n-m)} \qquad (15.40)$$

where V_B = flow velocity through the clear portion of the partially blocked rack. The pinning force on the rack is solely proportional to the change of the flow momentum or linearly varied with respect to the area-opening ratio on the rack surface. Aided by Equations 15.38 through 15.40, the pinning force on the block landed on the rack surface is derived as

$$F_B = \rho\frac{Q^2}{nA_R}\left[\frac{1}{(1-m/n)}-1\right] = \rho V_R Q\left[\frac{1}{(1-m/n)}-1\right] \qquad (15.41)$$

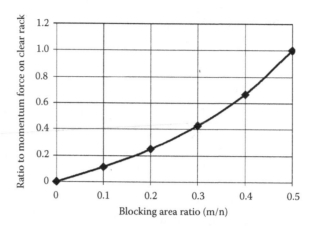

Figure 15.12 Force acting on block landed on rack surface.

Equation 15.41 indicates that the pinning force on the block is solely dominated by the change in the flow momentum force, $\rho V_R Q$. As illustrated in Figure 15.12 the force on the clogging block exponentially increases as m/n increases. It implies that the larger the rack surface area, the less resultant pinning force acting on the block. As a common practice, the rack surface area is recommended to be at least four times the pipe-opening area (USWDCM, 2001).

EXAMPLE 15.2

In a forensic study, it is frequent to ask how the pinning force acts on the clogging block, such as human body landed on the rack. For instance, a 120-cm circular culvert is equipped with a trash rack that is laid on the FES. The opening area of the FES is 3.35 m². The inclined angle of the trash rack is 30° relative to the streambed. A trash rack installed on FES at the entrance has a inclined length, L_R, of 2.4 m. During the storm event, the trash rack was partially clogged by a block that had a projected area of 0.32 m² on the rack surface area. As illustrated in Figure 15.13, the HGL and energy gradient line through the system were analyzed using a flow rate of 3.2 cms. The headwater depth immediately upstream of the rack was 1.7 m, and the hydraulic head immediately downstream of the culvert entrance was 1.26 m.

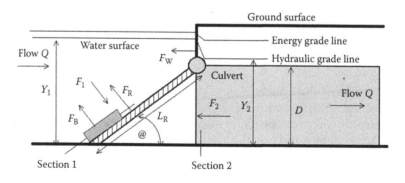

Figure 15.13 Flow condition for force analysis.

To quantify the pinning force, the *method of superimposition* begins with the base condition, i.e., no trash rack on the FES, as

$$A_1 = A_R = 3.35 \text{ m}^2$$

$$V_1 = \frac{Q}{A_R} = \frac{3.20}{3.35} = 0.96 \text{ m/s}$$

$$A_2 = \frac{\pi D^2}{4} = \frac{\pi (1.2)^2}{4} = 1.13 \text{ m}^2$$

$$V_2 = \frac{4Q}{\pi D^2} = \frac{4 \times 3.20}{\pi (1.2)^2} = 2.83 \text{ m/s}$$

$$\overline{Y}_1 = Y_1 - \frac{L_R}{2} \sin\theta = 1.70 - \frac{2.40}{2} \sin 30° = 1.10 \text{ m}$$

$$\overline{Y}_2 = Y_2 - \frac{D}{2} = 1.26 - \frac{1.20}{2} = 0.66 \text{ m}$$

$$F_1 = \gamma \overline{Y}_1 A_1 + \rho Q V_1 = 9.8 \times 1.10 \times 3.35 + 1.0 \times 3.2 \times 0.96 = 39.13 \text{ kN}$$

$$F_2 = \gamma \overline{Y}_2 A_2 + \rho Q V_2 = 9.8 \times 0.66 \times 1.13 + 1.0 \times 3.2 \times 2.83 = 16.33 \text{ kN}$$

$F_B = F_R = 0$ for the base condition. Substituting the values of F_1 and F_2 into Equation 15.28 yields

$$F_W = F_1 \sin\theta - F_2 = 39.13 \sin 30° - 16.33 = 3.23 \text{ kN or 727 lbs}$$

The force, F_w, is the reaction force from the wing wall. Having the trash rack installed, the area-opening ratio for the rack is 0.77%, or 23% of the rack surface area is occupied by steel bars. The flow velocity through the rack surface is

$$V_1 = V_R = \frac{Q}{nA_R} = \frac{3.20}{0.77 \times 3.35} = 1.24 \text{ m/s}$$

$$F_1 = \gamma \overline{Y}_1 A_1 + \rho Q V_1 = 9.8 \times 1.10 \times 3.35 + 1.0 \times 3.2 \times 1.24 = 40.05 \text{ kN}$$

The force, F_2, remains the same as that without a rack. Under the clear rack condition, Equation 15.28 is solved as

$$F_W + F_R \sin\theta = F_1 \sin\theta - F_2 = 40.05 \sin 30° - 16.33 = 3.69 \text{ kN or 830 lbs}$$

The force on the trash rack is found to be $F_R = 0.91$ kN, which is equal to 23% of the flow momentum through the clear rack. Of course, this force can be directly calculated using Equation 15.38 as

$$F_R = \rho \frac{Q^2}{nA_R}(1-n) = 1.0 \times \frac{3.20^2}{0.77 \times 3.35} \times [1 - 0.77] = 0.91 \text{ kN or 205 lbs}$$

Next, the clogging block is landed on the trash rack. As mentioned earlier, the additional blocking area on the rack is $0.32\,m^2$ or $m = 0.096$. Applying Equation 15.41 to the blocked rack, the flow velocity is calculated as

$$V_1 = V_B = \frac{Q}{A_R(n-m)} = \frac{3.20}{3.35 \times (0.77 - 0.096)} = 1.42 \text{ m/s}$$

Substituting the flow velocity at section 1 into Equation 15.29 yields

$$F_1 = \gamma \bar{Y}_1 A_1 + \rho Q V_1 = 9.8 \times 1.10 \times 3.35 + 1.0 \times 3.2 \times 1.42 = 40.60 \text{ kN}$$

The difference between F_1 and F_2 represents the total external force as

$$F_W + F_R \sin\theta + F_B \sin\theta = F_1 \sin\theta - F_2 = 40.60 \sin 30° - 16.33 = 3.97 \text{ kN or 893 lbs}$$

The pinning force, F_B, on the block is found to be $0.56\,kN$, which is 14% of the flow momentum force through the clear rack as shown in Figure 15.13. Of course, this force can be directly calculated using Equation 15.41 as

$$F_B = \rho \frac{Q^2}{nA_R}\left[\frac{1}{(1-m/n)} - 1\right] = 1.0 \times \frac{3.20^2}{0.77 \times 3.35}\left[\frac{1}{(1-0.096/0.77)} - 1\right] = 0.56 \text{ kN or 125 lbs}$$

In this case study, the reaction force from the wing walls is as high as $3.23\,kN$ (727 pounds) in the flow direction. This reaction force is essentially resulted from the hydrostatic force due to the headwater at the culvert entrance. The pinning force on the clear trash rack is $0.91\,kN$ (205 pounds) perpendicular to the rack surface. With a clogged area of 9.6%, the pinning force on the block is $0.56\,kN$ (125 pounds) perpendicular to the block surface. The trash rack is submerged in the water flow. Consequently, the pinning force on the rack is mainly resulted from the change in the flow momentum force that is much smaller than the hydrostatic force.

15.5 Closing

A pinning force is the normal force perpendicular to the rack surface. The effort to re-lease a person from being pinned on the trash rack surface is not to overcome the pinning force itself. Rather, it is to overcome the friction force along the rack surface. Therefore, a friction coefficient needs to be considered. The method of force superimposition derived in this chapter is limited to the steady flow condition, or the blockage on the rack is not so severe as to reduce the flow through the system. This approach is applicable to approx-imate the pinning force acting on a person or a small animal that is washed and pushed against the trash rack.

15.6 Homework

Q15.1 The photo in Figure Q15.1 shows a sheet of 4-ft × 4-ft pry wood laid in front of dual 24-in. culverts. As soon as it rains, the sheet of pry wood may be floated and then may block the 24-in. pipe entrance. Under a water depth of 3 ft, your tasks are as follows:

Figure Q15.1 Force analysis on blocked pipe.

1. Determine the design flow for this 24-in. culvert using the orifice flow formula.
2. Estimate the force on the headwall when the culvert entrance is clear. (The force is equal to the momentum change of the design flow from the upstream pool into the pipe.)
3. Calculate the force acting on the sheet of pry wood as soon as the lower half of the culvert entrance is blocked.
4. Determine the hydrostatic force acting on the sheet of pry wood as soon as the culvert entrance is completely blocked.

Bibliography

Allred-Coonrod, J.E. (1994). "Safety Grates in Supercritical Channels," *Journal of Irrigation and Drainage Engineering*, Vol. 120, No. 1, pp. 218–224.

CDOT. (2004). *Drainage Design Manual*, Colorado of Department of Transportation, Denver, CO.

Comport, B.C., Cox, A.L., and Thornton, C. (2010). "Performance Assessment of Grate Inlets for Highway Median Drainage," research report submitted to Urban Drainage and Flood Control District, Denver, CO.

FHWA (Federal Highway Administration). (1971). "Debris Control Structures," Hydraulic Engineering Circular No. 9 (HEC-9), FHWA-EPD-106.

Guo, J.C.Y. (2000). "Design of Grate Inlets with a Clogging Factor," *Advances in Environmental Research*, Vol. 4, pp. 181–186.

Guo, J.C.Y. (2006). "Decay-Based Clogging Factor for Curb Inlet Design," *ASCE Journal of Hydraulic Engineering*, Vol. 132, No. 11.

Guo, J.C.Y., and Jones, J. (2010). "Pinning Force during Closure Process at Blocked Pipe Entrance," *ASCE Journal of Irrigation and Drainage Engineering*, Vol. 136, No. 2, pp. 141–144.

Guo, J.C.Y., Jones, J., and Earles, A. (2010). "Method of Superimposition for Suction Force on Trash Rack," *ASCE Journal of Irrigation and Drainage Engineering*, Vol. 136, No. 11, pp. 781–785.

Guo, J.C.Y., McKenzie, K., and Mommandi, A. (2008). "Sump Inlet Hydraulics," *ASCE Journal of Hydraulic Engineering*, Vol. 135, No. 1.

HEC-22. (2002). *Urban Drainage Design Manual*, US Department of Transportation, Federal Highway Administration, Washington, DC.

Jones, J, Guo, J.C.Y., and Urbonas, B. (2006). "Safety on Detention and Retention Pond Designs," *Journal of Storm Water*.

Mays, L.W. (2001). *Stormwater Collection Systems Design Handbook*, McGraw Hill Publication Company, New York.

Weisman, R.N. (1989). "Model Study of Safety Grating for Culvert Inlet," *Journal of Transportation Engineering*, Vol. 115, No. 2, pp. 130–138.

USWDCM. (2001). *Urban Stormwater Design Criteria Manual*, Urban Drainage and Flood Control District, Denver, CO.

Chapter 16

Stormwater quality capture volume

A rural watershed is characterized with its hydrologic losses including interception, infiltration, and depression losses. In comparison, interception losses due to bushes and trees are negligible in an urban area. Depression loss depends on the storage volume associated with the depressed area. Infiltration loss depends on the type of soils, and it applies to the entire watershed's surface. Developments of an urban area result in more impervious surfaces and fills of depressed areas. As illustrated in Figure 16.1, an urban drainage system is a replica of the natural drainage network, including the underground storm sewers sized to carry the minor event and the street gutter designed to deliver the major event. Such a double-deck flow system is to mimic the natural waterway that consists of the low-flow main channel and the overbank floodplains.

In an urban catchment, the source of storm runoff is the impervious areas. Before the overland flows become concentrated, the *increased runoff volume per unit area* is the cause of water quality problems or the V-problem. After the overland flows are collected into street gutters, sewers, and channels, the *increased runoff flow* is the cause of flooding problems or the Q-problem. The conventional stormwater design has been focused on how to reduce peak flows using stormwater detention, while the latest development in the concept of low-impact development (LID) is to integrate various infiltrating and filtering designs to reduce the increased runoff volume. An LID design is to apply a filtering process to enhance stormwater quality and an infiltration process to reduce stormwater volume. Because the LID designs are aimed at the runoff source control, they are applicable to a small tributary area (2–5 acres or 1–2 ha). The latest developments in LID designs include infiltration beds, rain gardens, bioswales, and porous pavements. As shown in Figure 16.2, stormwater LID designs are classified into the following:

1. *Flow-over conveyance type*, such as porous pavements using flat infiltration beds
2. *Flow-in storage type*, such as rain gardens using shallow infiltration basins.

Obviously, the effectiveness of an LID design depends on how much the surface runoff volume can be intercepted. To differ from the storm*water detention storage volume* (WDSV) for extreme events, the storage volume used to design an LID device is termed the *water quality capture volume* (WQCV). A WQCV shall be in the same magnitude as the natural depression volume that was filled and leveled during the urbanization process.

Since the 1990s, the urban stormwater management has rapidly changed from the conventional concerns on flood flow mitigation to a new focus on urban runoff quality enhancement. The 1987 *Federal Water Quality Act* is a reflection of the public's support for improvements in urban water environment, and such legislation gives a new direction to renovate urban drainage systems. In the Unites States, governments and

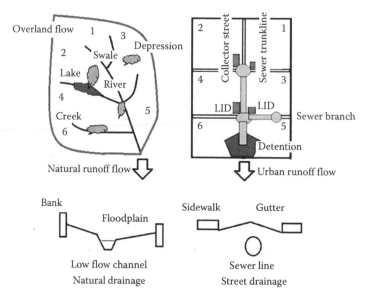

Figure 16.1 Comparison between natural and street drainage networks.

Figure 16.2 Conveyance and storage low-impact designs. (a) Conveyance type—porous pavement and (b) storage type—rain garden.

industries have a mandate from Congress to minimize the discharge of pollutants to receiving waters. As a result, stormwater *best management practices* (BMPs) have been developed to offer practical alternatives to address these problems. Stormwater BMPs are the best-known practical techniques available and affordable. Over 20 years of learning, stormwater BMPs have reached the conclusion that the concept of LID is the best approach to integrate both flood mitigation and water quality enhancement altogether.

16.1 Rainfall and runoff distributions

It is important to realize that the conventional criteria used to design stormwater detention basins for extreme events can no longer be used to design *stormwater quality-control*

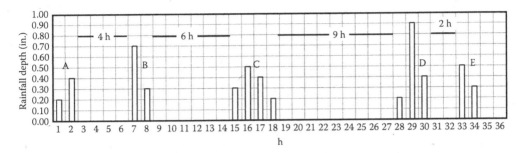

Figure 16.3 Event separation by interevent time.

basins (WQCB). It is simply because the goal of a WQCB is to capture frequent runoff events, not the extreme.

Frequent events must be identified from a continuous rainfall or runoff record. Between two adjacent rainfall events in a continuous record (as shown in Figure 16.3), the *interevent time* represents the period of time in which there is no rain. As illustrated in Figure 16.3, the analysis of a continuous record begins with delimiting individual rainfall events using a preselected *event separation time* or minimum interevent time of no rain (Tucker et al., 1989). As depicted in Figure 16.3, considering an event separation time of 6 h, there are three individual events identified because groups A and B are lumped into a single event, so are groups D and E. After each individual event is identified, the *event rainfall depth* and *duration* can be further calculated for statistical analyses.

Although the selection of event separation time is somewhat subjective, the EPA's general guidance is to apply an event separation time of 6 h to identify the number of events in a continuous record. Of course, for other purposes, the selection of event separation time shall depend on the basin's operation. For instance, an investigation of runoff volume captured by a WQCB, it is advisable that the event separation time be equal to the basin's drain time. In doing so, every rainfall event introduced into the basin will begin with an empty basin. No two events would be overlapped. Similarly, the study on sediment trap ratio shall set the event separation time equal to the particle's residence time. In doing so, before the next event to come, the particles in the current event have already been settled in the basin. The 1986 EPA study reported that about 80%–90% of solids were removed if a 12-h drain time was applied to a wet pond or a 24-h drain time was applied to a dry pond (EPA, 1986). In practice, the drain time of a WQCB shall be slightly longer than the particle's residence time (Guo and Urbonas, 1996).

After a continuous rainfall record is divided into individual events, Figure 16.4 is the distribution of the event rainfall depths observed at the City of Denver, CO. Although a 2-year storm event is often considered a small storm for flood-control projects, a 2-year event, in fact, has a rainfall depth >95% of the rainfall population. Figure 16.5 shows the distribution of the event rainfall depths segregated from the 30-year rainfall record observed at the City of San Diego, CA. It shows that 97% of the events have a depth less than the local 2-year rainfall depth. Although the skewness of the event rainfall depth distribution varies with respect to the meteorological region, it is generally true that the number of smaller rainfall events absolutely dominate the rainfall population.

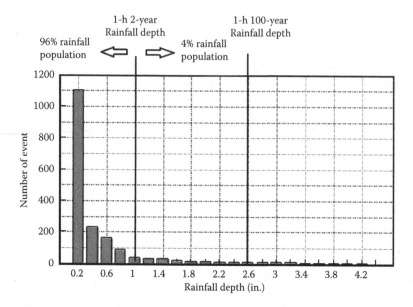

Figure 16.4 Rainfall depth distribution at Denver, CO.

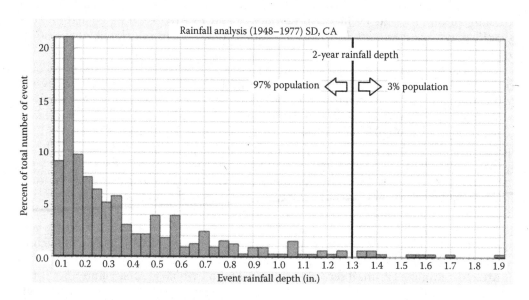

Figure 16.5 Rainfall depth distribution at San Diego, CA.

Raindrops do not reach the ground until the event rainfall depth is greater than the interception loss. An on-ground rainfall depth is the difference between the recorded rainfall depth and the interception loss:

$$d_i = D_i - I_s \tag{16.1}$$

in which d_i = on-ground rainfall depth in [L] for the ith event, D_i = observed rainfall depth in [L], and I_s = interception loss in [L] such as 0.05–0.1 in. (Guo and Urbonas, 1996). Having the continuous rainfall record divided into individual storms, the statistics for event depth, duration, and interevent time can further be calculated as

$$D_m = \frac{1}{N} \sum_{i=1}^{i=N} d_i \tag{16.2}$$

$$S_D = \left[\frac{1}{(N-1)} \sum_{i=1}^{i=N} (d_i - D_m)^2 \right]^{\frac{1}{2}} \tag{16.3}$$

$$C_s = \frac{1}{S_D^3 N(N-1)(N-2)} \left[\sum_{i=1}^{i=N} (d_i - D_m)^3 \right] \tag{16.4}$$

$$T_m = \frac{1}{N} \sum_{i=1}^{i=N} t_i \tag{16.5}$$

in which D_m = average event rainfall depth in [L], N = total number of events in the record, S_D = standard deviation in [L], C_s = skewness coefficient, t_i = time interval to the next event in [T], and T_m = average interevent time in [T]. The abovementioned approach was employed to analyze the continuous rainfall records observed in seven metropolitan areas (Guo and Urbonas, 1996). Approximately 1000–1500 individual events were identified from each continuous rainfall record using an event separation time of 6, 12, or 24 h. The rainfall statistics in inches and average interevent time in hours are summarized in Table 16.1. As expected, the distributions of rainfall depth are skewed by the number of small events in all cities. Figure 16.6 presents the mean event rainfall depth derived from the study using a 6-h event separation time and 0.1 in. as the interception loss (Driscoll et al., 1989).

16.2 Runoff capture analysis

For a single event, the effectiveness of a WQCB is determined by the percentage of runoff volume captured. Over a long period of time, the performance of the WQCB is evaluated with the cumulative percentage of runoff volume or the number of events captured.

16.2.1 Runoff volume capture analysis for single event

For convenience, all volumes used in the calculation of runoff capture are converted to the same unit as rainfall depth, namely inches or millimeters per watershed. Applying the point rainfall-to-runoff volumetric approach, an on-ground event rainfall depth is converted to its runoff depth:

$$P_R = C(d_i - I_s) \tag{16.6}$$

Table 16.1 Rainfall statistics using 1-, 6-, and 24-h separation times

City	6-h				12-h				24-h			
	D_m (in.)	S_D (in.)	C_s	T_m (h)	D_m (in.)	S_D (in.)	C_s	T_m (h)	D_m (in.)	S_D (in.)	C_s	T_m (h)
Seattle, WA	0.48	0.49	2.75	53.5	0.60	0.64	2.67	72.7	0.78	0.90	3.06	98.1
Sacramento, CA	0.61	0.62	2.96	166.7	0.72	0.76	3.50	208.8	0.82	0.92	3.44	251.6
Phoenix, AZ	0.42	0.36	2.59	261.3	0.45	0.40	2.41	300.1	0.48	0.44	2.57	341.8
Denver, CO	0.44	0.48	3.59	106.4	0.46	0.51	3.47	121.4	0.51	0.56	3.30	144.2
Cincinnati, OH	0.58	0.55	3.03	65.2	0.66	0.64	2.76	81.1	0.73	0.71	2.51	97.8
Tampa, FL	0.66	0.78	4.40	71.4	0.71	0.83	4.46	79.6	1.01	1.10	2.89	114.7
Boston, MA	0.70	0.79	4.98	70.7	0.73	0.81	4.60	82.1	0.78	0.84	4.28	94.8

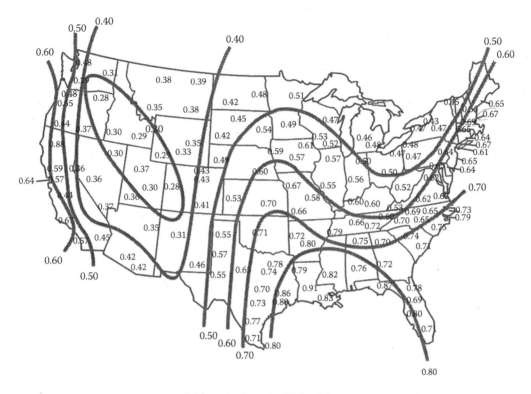

Figure 16.6 Average event rainfall depths for the United States.

in which P_R = event runoff volume in inch or millimeter per watershed, I_S = incipient runoff depth in [L] such as 0.1 inch due to evaporation and interception loss, and C = runoff coefficient. During a long event, the WQCB is loaded and then overtopped. The runoff volume captured and treated by the WQCB is equal to the basin storage volume plus the runoff volume flowing through the basin during the storm duration:

$$P_T = D_o + qT_d \tag{16.7}$$

in which P_T = potential capacity in [L/watershed] that can be captured and treated, D_o = basin's storage volume in [L/watershed], and T_o = basin's drain time in [T] such as hours. The average flow release is determined as

$$q = \frac{D_o}{T_o} \tag{16.8}$$

in which q = *average flow release* from WQCB in [L/T] and T_o = basin's drain time in [L] such as hours. The product, qT_d, represents the runoff volume flowing through the WQCB during the event duration. Because not every event can overtop the WQCB, there are two possibilities: (1) If the event runoff volume, P_R, is greater than the potential capacity, P_T, the excess runoff volume, $(P_R - P_T)$, is the overflow without any treatment, and (2) If the runoff volume $P_R \le P_T$, the event is completely captured and treated. As a result, the numerical procedure for determining the runoff capture volume, P_A, and overflow volume, P_o, is

$$P_A = \min(P_R, P_T) \tag{16.9}$$

$$P_o = P_R - P_A \quad \text{if } P_o \geq 0 \text{ or } P_o = 0 \tag{16.10}$$

For this single event, the *runoff volume capture ratio* (RVCR), R_V, is defined as

$$R_V = \frac{P_A}{P_R} \tag{16.11}$$

EXAMPLE 16.1

An urban catchment has a tributary area of 2.0 acres and a runoff coefficient of 0.85. Its WQCB is designed to have a storage volume of 0.07 acre-ft and a drain time of 6 h. Determine the percentage of captured volume for the event of 1.0 in. over a duration of 3.0 h.

Solution: With $I_S = 0.1$ inch, the given event produces a runoff volume as

$$P_R = 0.85(1.0 - 0.1) = 0.765 \text{ in./watershed}$$

Let us convert the basin storage volume from acre-ft to inch/watershed as

$$D_o = \frac{0.07 \times 12}{2.0} = 0.42 \text{ in./watershed}$$

The average release from this WQCB is

$$q = \frac{0.42}{6.0} = 0.07 \text{ in./h per watershed}$$

The basin's potential capacity for this case is

$$P_T = 0.42 + 0.07 \times 3.0 = 0.63 \text{ in.} < 0.765 \text{ in.}$$

Because $P_R > P_T$, the basin was fully loaded and then spilled. The RVCR is calculated as

$$P_A = \min(0.765, 0.63) = 0.63 \text{ in./watershed}$$

$$P_o = 0.765 - 0.63 = 0.132 \text{ in./watershed}$$

$$R_V = 0.63 / 0.762 = 82\%$$

EXAMPLE 16.2

Continued with Example 16.1, determine the percentage of captured volume for the event of 0.5 in. over a duration of 3.0 h.

Solution: The given event produces a runoff volume as

$$P_R = 0.85(0.5 - 0.1) = 0.34 \text{ in./watershed}$$

$$P_A = \min(0.765, 0.34) = 0.34 \text{ in./watershed}$$

$P_o = 0.34 - 0.34 = 0.0$ in./watershed

$R_V = 0.34 / 0.34 = 100\%$

For both cases, the average $R_V = 91\%$. Examples 16.1 and 16.2 can be repeated for a long-term continuous rainfall record to determine the long-term runoff captured volume for a given WQCV.

16.2.2 Runoff volume capture analysis for continuous record

The long-term performance for a selected storage volume, P_D, and drain time, T_o, can be evaluated by the continuous rainfall record. Therefore, the total runoff volume generated from the watershed is summed up as

$$P_{RT} = \sum_{i=1}^{i=n} P_{R_i} \tag{16.12}$$

in which $i = i$th event, $n =$ number of events, and $P_{RT} =$ total runoff volume in the record. The total runoff volume captured by the basin is

$$P_{AT} = \sum_{i=1}^{i=n} P_{A_i} \tag{16.13}$$

in which $P_{AT} =$ total runoff volume captured through the period of the record. The overflow runoff volume, P_{OT}, is calculated as

$$P_{OT} = \sum_{i=1}^{i=n} (P_{R_i} - P_{A_i}) \quad \text{if } P_{R_i} \geq P_{C_i}, \text{ otherwise zero} \tag{16.14}$$

The over-all RVCR for the period of the record is defined as

$$R_V = \frac{P_{AT}}{P_{RT}} \tag{16.15}$$

in which, $R_V =$ runoff volume capture ratio ranging between zero and unity.

16.2.3 Runoff event capture ratio (RECR)

By following the same approach, set the counter on the number of events that were completely intercepted without any overflow. The over-all *runoff event capture ratio (RECR)*, R_E, is defined as

$$R_E = \frac{N_C}{n} \tag{16.16}$$

in which $N_C =$ number of runoff events that were completely captured. In comparison with RVCR, RECR is a preferable approach when outliers exist in the record. As indicated in Equations 16.7 through 16.11, both RVCR and RECR depend on event separation

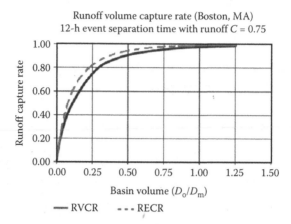

Figure 16.7 Comparison of runoff volume capture ratio and runoff event capture ratio curves for Boston, MA.

time, T_o, and runoff coefficient, C. For a pair of (T_o, C), both RVCR and RECR can be produced. Figure 16.7 is an example of RVCR and RECR prepared for the City of Boston when using an event separation time of 12 h and runoff coefficient of 0.75. As expected, the RECR gives a slightly higher capture ratio than the RVCR.

16.3 Optimal water quality capture volume

The main objective in the design of a WQCB is to maximize the runoff volume captured. However, this objective leads to the conclusion that the larger the basin, the more the runoff volume captured. Consequently, the runoff capture amount fails to serve as a basis to choose the optimal basin size. However, the marginal benefit, which is defined by the tangent on the curve (as illustrated in Figure 16.8) or the ratio between the incremental runoff capture rate and the incremental basin size, provides a clue. It is noted that for a small basin, its marginal benefit is on the trend of increasing return, or upsizing this basin is encouraged. By the same token, for a very large basin, its marginal benefit is on a diminishing return, or downsizing the basin is encouraged. In between, there is a break-even point for the optimal design.

Figure 16.8 was prepared with the basin sizes normalized by the one that can intercept 99% of the total runoff volume in the record. In doing so, any outliers (depth > 99 percentile value) in the record are purged out. For a pair of (T_o, C), a normalized curve is constructed, and each curve provides an optimal basin volume based on the break-even point. A matrix of similar curves can be produced for $C = 0.3, 0.5, 0.7,$ and 0.9 and $T_o = 12,$ 24, and 48 h. This method has been applied to 30-year continuous hourly rainfall data recorded at *Seattle, WA, Sacramento, CA, Cincinnati, OH, Boston, MA, Phoenix, AZ, Denver, CO,* and *Tampa, FL,* to find the optimal runoff capture volumes. Findings from these seven gages form a database for regression analyses using the model as

$$\frac{D_o}{D_m} = aC + b \tag{16.17}$$

in which a and b = coefficients derived from regression analysis and listed in Table 16.2. The values for variable, b, are numerically negligible for practice, or $b = 0$ is acceptable. For the seven metropolitan cities, the regression equations show excellent correlation

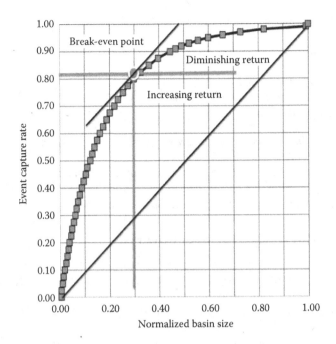

Figure 16.8 Determination of optimal basin volume.

Table 16.2 Coefficients for determining optimal basin sizes

Drain time	Volume ratio			Event ratio		
	a	*b*	r^2	*a*	*b*	r^2
12-h	1.36	−0.034	0.80	1.1.96	0.010	0.97
24-h	1.62	−0.027	0.93	1.256	0.030	0.91
48-h	1.98	−0.021	0.84	1.457	0.063	0.85

coefficients, r^2, ranging from 0.80 to 0.97, depending on drain time. Generally, the equation for RECR has a higher correlation.

The optimal basin size defined by Equation 16.17 is termed WQCV (USWDCM, 2001). The runoff capture rate for WQCV is between 82.0% and 88.0% for the seven cities studied. A RVCR is similar to, but not the same as, its RECR. In comparison, RECR is much less sensitive to the outliers in the database than the RVCR. In design, the use of RECR is similar in concept to that which is used in *combined sewer overflow control strategy* (CSOCS). Namely, it is indicative of the annual average number of CSO.

EXAMPLE 16.3

The tributary watershed of 2.0 acres is located in the City of Denver, CO. The watershed runoff coefficient is 0.70. The WQCB is designed to have a drain time of 24 h. Determine the WQCV.

Solution: The coefficients for Equation 16.17 are $a = 1.62$ and $b = -0.027$ from Table 16.2 for the RVCR. The WQCV to average event rainfall depth ratio is

$$\frac{D_o}{D_m} = 1.62 \times 0.70 - 0.027 = 1.11$$

From Figure 16.6, the average event rainfall depth at the City of Denver is $D_m = 0.41$ in. As a result, the WQCV for this case is

$$D_o = 0.41 \times 1.11 = 0.45 \text{ in./watershed or a volume of } 0.45/12 \times 2.0 = 0.076 \text{ acre-ft}$$

The WQCB shall be designed to have a storage volume of 0.076 acre-ft. A WQCB is sized to mimic the natural depression losses. The suggested WQCV of 0.45 in. is to make up the depression storage capacity under the predevelopment condition.

16.4 Exponential model for overflow risk

As aforementioned, the WQCV is in the same magnitude as the natural depression loss. As expected, a WQCB will be overtopped frequently. The overflow risk for a given WQCB can be formulated by the joint probability determined by the interevent time and rainfall amount for the next incoming event. Considering that the arrival of a rainfall event is a random process, and only one arrival can occur at an instant in time, the chance of occurrence is a typical *Poisson process* whose *probability density function* (PDF) is formulated as

$$f(t) = me^{-mt} \quad \text{for } t > 0 \tag{16.18}$$

in which $f(t)$ = PDF in Figure 16.9, m = constant, and t = elapsed time.

The probability *cumulative density function* (CDF) is an integration of PDF as

$$P(T_1 \le t \le T) = \int_{T_1}^{T} me^{-mt}\, dt = e^{-mT_1} - e^{-mT} \tag{16.19}$$

in which $P(T_1 \le t \le T)$ = PDF from time T_1 to time T. When $T_1 = 0$, Equation 16.19 becomes

$$P(0 \le t \le T) = \int_{0}^{T} me^{-mt}\, dt = 1 - e^{-mT} \tag{16.20}$$

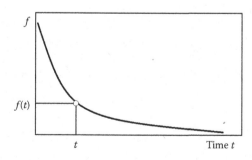

Figure 16.9 PDF for Poisson process.

When T becomes infinite, the integration of Equation 16.19 becomes

$$P(T_1 \le t \le \infty) = e^{-mT_1} \tag{16.21}$$

Of course, for the range from $T_1 = 0.0$ to $T = \infty$, Equation 16.19 becomes unity. The *mean*, $E(t)$, and *variance*, $\mathrm{Var}(t)$, of Equation 16.18 are

$$E(t) = \frac{1}{m} \tag{16.22}$$

$$\mathrm{Var}(t) = \frac{1}{m^2} \tag{16.23}$$

The Poisson distribution has a standard deviation, $S_D = 1.0$, and a skewness coefficient, $C_s = 2.0$. With the known mean event rainfall depth (Figure 16.6), the PDF of the rainfall depth distribution is

$$f(D) = \frac{1}{D_m} e^{\frac{-D}{D_m}} \tag{16.24}$$

in which D = rainfall event depth and D_m = average rainfall event depth in Figure 16.6. Similarly, the PDF of the interevent time distribution is

$$f(t) = \frac{1}{T_m} e^{\frac{-t}{T_m}} \tag{16.25}$$

in which t = interevent time and T_m = average interevent time. Their cumulative probability function (CPF) functions are similar to Equations 16.19 through 16.23. During an event, the operation of a basin is a cycle of filling and depletion. Between two events, the WQCB undergoes a waiting period for the next event. Consider that the arrival of a rainfall event is a random process, and there is only one arrival at an instant in time. The cumulative probability for an event to occur during a period is an integration of Equation 16.25 as

$$P(T_1 \le t \le T) = \int_{T_1}^{T} \frac{1}{T_m} e^{\frac{-t}{T_m}} \, dt = e^{\frac{-T_1}{T_m}} - e^{\frac{-T}{T_m}} \tag{16.26}$$

in which $P(T_1 \le t \le T)$ = probability to have an event during the time period from T_1 to T. When $T_1 = 0$, Equation 16.26 is reduced to the probability to have an event during the elapsed time, T, as

$$P(0 \le t \le T) = 1 - e^{\frac{-T}{T_m}} \tag{16.27}$$

Also, Equation 16.27 is the nonexceedance probability that is the chance to have the next event not to exceed a waiting time, T. Of course, the exceedance probability for the interevent time is

$$P(T \le t \le \infty) = e^{\frac{-T}{T_m}} \tag{16.28}$$

Equation 16.28 represents the probability to have the next event after a time period of T. Similar probability formulas can be derived for rainfall depth. For instance, the nonexceedance probability for the next event not to exceed a depth, D, is

$$P(0 \le d \le D) = 1 - e^{\frac{-D}{Dm}} \qquad (16.29)$$

Its exceeding probability is

$$P(D \le d \le \infty) = e^{\frac{-D}{Dm}} \qquad (16.30)$$

in which d = next incoming rainfall depth in [L] and D = basin's storage capacity in [L]. When selecting the storage capacity for a WQCB, Equation 16.29 provides an important basis to estimate the overtopping risk for the next rainfall event to come.

16.5 Runoff capture curve

Although both Equations 16.29 and 16.30 are formulated for rainfall depths, the design of WQCB depends on the runoff depth. The basic relationship between rainfall and runoff depths is

$$D_o = C(D - I_S) \qquad (16.31)$$

in which D_o = basin's storage volume equal to the design storage volume for WQCB in mm/watershed, D = design rainfall depth in mm/watershed, C = runoff coefficient, and I_S = incipient runoff depth such as 0.1 in. Substituting Equation 16.31 into Equation 16.29, the nonexceedance probability is derived as

$$C_V = P(0 \le d \le D_o) = 1 - ke^{\frac{-D_o}{CDm}} \qquad (16.32)$$

$$k = e^{\frac{-I_S}{Dm}} \qquad (16.33)$$

in which $P(0 \le d \le D_o)$ = nonexceedance probability and k = a constant representing initial loss. Equation 16.32 is the *runoff capture volume rate*, C_V, for a given WQCB. In theory, $C_V = R_V$ and $D_o = WQCV$. In practice, we shall expect some minor differences because both C_V and D_o were derived from a continuous model based on the rainfall distribution, whereas R_V and $WQCV$ were derived from a discrete field database representing the runoff distribution. In the flow simulation, any event that produced a runoff volume more than D_o will overload the basin. Therefore, the overflow risk is calculated as

$$R_o = P(D_o \le d \le \infty) = ke^{\frac{-D_o}{CDm}} \qquad (16.34)$$

The plot of C_V versus D_o using Equation 16.32 is termed the *runoff capture curve* for the basin site with a specified runoff coefficient. With $C = 0.5$, Figure 16.10 presents a comparison between Equation 16.33 and the runoff capture curves generated by the long-term records observed in several major cities in the United States.

The runoff capture curve is the required information for designing an LID facility. Figure 16.11 presents a set of generalized runoff capture curves produced using Equation 16.32 for runoff coefficients of 0.2, 0.4, 0.6, 0.8, 0.90, and 1.0. It is noticed that the

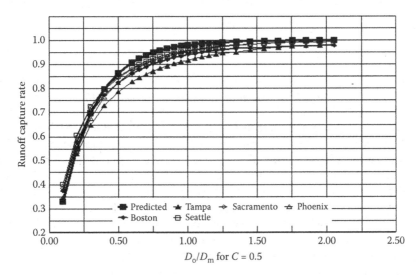

Figure 16.10 Runoff capture curves for several cities in the Unites States (C = 0.5).

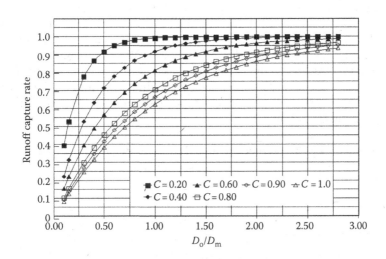

Figure 16.11 Runoff capture curves for various runoff coefficients.

curvature of runoff capture curve increases when the runoff coefficient decreases. The runoff capture curve becomes almost a linear response between rainfall and runoff amount when $C = 1.0$. This tendency reflects the fact that the higher the imperviousness in a catchment, the lesser the surface detention. As a result, the response of a catchment to rainfall is quick and direct.

EXAMPLE 16.4

The tributary watershed of 2.0 acres is located in the City of Denver, CO. The runoff coefficient is 0.70. Considering that a runoff incipient depth is 0.1 in. and the target runoff capture rate is 82%, find the design basin storage volume.

Solution: From Figure 16.6, $D_m = 0.41$ in. for the Denver area. Equations 16.33 and 16.32 become:

$$k = e^{\frac{-0.1}{0.41}} = 0.784$$

$$C_v = 1 - 0.784 \times e^{\frac{-D_o}{0.7 \times D_m}} = 0.82 \quad \text{Solution: } D_o/D_m = 1.11 \text{ or } D_o = 0.45 \text{ in.}$$

The basin volume in the case is found to be $D_o = 0.45$ in./watershed. Recall that WQCV $= 0.45$ in. was derived for Example 16.3 using a drain time of 24 h. This case indicates that Equation 16.32 is similar to Equation 16.17. It is noted that Equation 16.32 is a continuous model applicable to all drain times, while Equation 16.17 is a regression model tailored for various drain times. In practice, we can produce a synthetic runoff capture curve using Equation 16.32 for the Denver area. Because the runoff capture curve is asymptotic to unity when D_o/D_m becomes infinite, it is advisable that the runoff capture curve be constructed for a target range. For example, Figure 16.12 was developed for runoff capture rates from 50.0% to 90.0% of runoff events. Within this range, the engineer takes all design factors into consideration to make the final selection.

In comparison, a flood-control detention system is designed for a preselected recurrence interval using a *flood-frequency curve*, while a WQCB is designed to capture a target percentage of a complete rainfall series described by the *runoff capture curve (volume or event)*. Both flood-frequency curve and runoff capture curve define the *inherent overflow risk* for the selected design event. The overflow risk of a flood-control detention basin is referenced to the recurrence interval of the design flood, while the overflow risk of a WQCB is defined by the runoff capture percentage. In addition to the *inherent overflow risk*, a basin is also subject to the *operational overflow risk* through a cycle of basin operation, including the following:

1. Overflow caused by the current event, which is greater than the basin's capacity
2. Overflow caused by the next incoming event during the draining process

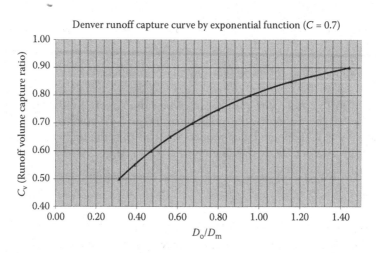

Figure 16.12 Synthetic runoff volume capture curve for Denver area.

16.6 Overflow risk

As expected, a WQCB designed for *microevents* will be overtopped several times in a year. The operational cycle of a basin includes the *dynamic (loading) period* during an event and the *quiescent (waiting) period* between two events. The overflow risk is separately derived for both periods as described in the following sections.

16.6.1 Inherent overflow risk

During a quiescent period, the overflow risk for an empty basin consists of two conditions: (1) A rainfall event will come within the waiting time, T, and (2) its rainfall depth of the next event exceeds the basin's storage volume. Such a joint probability during a quiescent period is calculated as

$$R_e(0 \le t \le T) = P_T(0 \le t \le T) \times P_D(D_o \le d \le \infty) \tag{16.35}$$

in which R_e = overflow risk when the basin is empty, T = waiting time, P_T = probability for interevent time, P_D = probability for rainfall depth, C = runoff coefficient, and d = rainfall depth. Aided by Equations 16.27 and 16.34, Equation 16.35 becomes

$$R_e(0 \le t \le T) = \left(1 - e^{\frac{-T}{Tm}}\right) k e^{\frac{-D_o}{CDm}} \tag{16.36}$$

When the waiting time becomes long enough, Equation 16.36 is reduced to

$$R_e = k e^{\frac{-D_o}{CDm}} \tag{16.37}$$

Equation 16.37 is the same as Equation 16.34. Both represent the *inherent overflow risk for the basin*. As indicated in Equation 16.37, the inherent overflow risk depends on the basin storage capacity relative to the local average rainfall event depth.

16.6.2 Operational overflow risk

Once the basin is loaded by a large event, the overflow risk of the basin is subject to the magnitude of the next event. During the draining process, the available storage volume in the basin increases as the elapsed time, T, increases, and can be calculated as

$$V(T) = qT \quad \text{for } T < T_d \tag{16.38}$$

in which $V(T)$ = available storage volume in mm/watershed at elapsed time T, q = release rate from the basin in terms of mm/watershed, and T_d = drain time. When $T = T_d$, the available storage volume in the basin is equal to the design capacity

$$V_o = qT_d \tag{16.39}$$

The overflow risk during the draining period from T to T_d depends on the two probabilities: (1) The next event will come between T and T_d, and (2) the rainfall depth will exceed the available storage volume. Such a joint probability can be formulated as

$$R_D(T \le t \le T_d) = P_T(T \le t \le T_d) \times P_D(V(T) \le d \le \infty) \tag{16.40}$$

in which R_D = overflow risk during the draining process. Aided by Equations 16.28 and 16.30, Equation 16.40 becomes

$$R_D(T \le t \le T_d) = \left(e^{\frac{-T}{T_m}} - e^{\frac{-T_d}{T_m}} \right) ke^{\frac{-qT}{CD_m}} \tag{16.41}$$

Equation 16.41 describes the *operational overflow risk* that is caused by a sequential rainfall event during the basin's draining process. As expected, the operational overflow risk decreases as the elapsed time increases, and vanishes when $T = T_d$.

16.6.3 Overflow risk for a cycle of operation

Considering a cycle of operation, the basin is loaded by a large event and then subject to the next event. The total overflow risk is equal to the sum of Equations 16.36 and 16.41:

$$R(T) = R_e + R_D \tag{16.42}$$

in which $R(T)$ = total overflow risk at elapsed time $T < T_d$. Substituting Equations 16.36 and 16.41 into Equation 16.42 yields

$$R(T) = ke^{\frac{-D_o}{CD_m}} + \left(e^{\frac{-T}{T_m}} - e^{\frac{-T_d}{T_m}} \right) ke^{\frac{-qT}{CD_m}} \quad \text{for } 0 \le T \le T_d \tag{16.43}$$

Equation 16.43 indicates that $R(T)$ has its highest value at $T = 0$ and the lowest value at $T = T_d$. Substituting $T = 0$ into Equation 16.43 yields

$$R(0) = ke^{\frac{-D_o}{CD_m}} + k\left(1 - e^{\frac{-T_d}{T_m}} \right) \tag{16.44}$$

Equation 16.44 is the total overflow risk at the beginning of the draining process. Care has to be taken because $R(0)$ may become greater than unity when D_o/CD_m is unreasonably small or T_d/T_m is unreasonably large. Substituting $T = T_d$ into Equation 16.43 yields

$$R(T_d) = ke^{\frac{-D_o}{CD_m}} \quad \text{for } T_d \le T \tag{16.45}$$

Equation 16.43 begins with $R(0)$ prescribed by Equation 16.44 and then converges to $R(T_d)$ by Equation 16.45 as the elapsed time, T, increases. Because the design capacity in the basin becomes available after T_d, the overflow risk is reduced to the inherent risk of the basin, i.e., $R(T_d)$.

EXAMPLE 16.5

A WQCB is located in Boston, MA. The tributary watershed has a drainage area of 8098 m^2 (2.0 acres) with a runoff coefficient of 0.5. At Boston, the average rainfall event depth is 17.78 mm, and the average interevent time is 70.65 h. Considering a runoff incipient depth of

2.5 mm, the value of k is 0.86 for the Boston area. Based on the characteristics of sediments found in the local stormwater runoff, the WQCV is determined to be 13.2 mm/watershed and the drain time is set to be 24 h. Evaluate the overflow risk for this basin.

Solution: With $D_o = 13.2$ mm and a drain time $T_o = 24$ h, the average release rate from the basin is determined as

$$q = D_o / T_o = 13.2 / 24.0 = 0.55 \text{ watershed mm/h}$$

According to Equation 16.37, the inherent overflow risk for an empty basin is

$$R_e = 0.86 e^{\frac{-13.20}{0.50 \times 17.78}} = 0.195 \quad \text{for } T \geq T_d$$

Substituting $D_m = 17.78$ mm and $T_m = 70.6$ h into Equation 16.43 yields

$$R(T) = 0.195 + 0.86 \left(e^{\frac{-T}{70.65}} - e^{\frac{-24.0}{70.65}} \right) e^{\frac{-0.55T}{0.50 \times 17.78}} \quad \text{for } 0 \leq T \leq T_d$$

According to Equation 16.44, the upper limit of Equation 16.43 is defined by $T = 0$ as

$$R(0) = 0.195 + 0.86 \left(1 - e^{\frac{-24.0}{70.65}} \right) = 0.447$$

During the operation through an event, as the elapsed time increases, the overflow risk for this basin decreases from 0.447 when the basin was full to 0.195 when the basin becomes empty. From the aspect of sedimentation, the longer the residence time, the more the particles captured. On the other hand, it also introduces a higher overflow risk to the basin's operation. In practice, a range of drain times shall be selected by the sedimentation requirements. Each drain time can then be evaluated with its associated overflow risk. This process assists the engineer in making a final selection of drain time based on the tradeoff between the overflow risk and the amount of sediment captured. For example, using WQCV of 13.2 mm/watershed and $C = 0.5$, Table 16.3 presents the variations of the overflow probabilities for drain times of 12-, 24-, 48-, 72-, and 96-h. For a selected drain time, the overflow risk begins with its highest level when the basin is full and then gradually reduces through the emptying process. After the basin becomes empty,

Table 16.3 Overflow risk versus various drain times

Elapsed time (h)	Overflow risk				
	12-h	24-h	48-h	72-h	96-h
0.00	0.332	0.447	0.626	0.752	0.841
6.00	0.226	0.320	0.493	0.624	0.718
12.00	0.195	0.250	0.397	0.523	0.617
24.00	0.195	0.195	0.280	0.380	0.466
36.00	0.195	0.195	0.222	0.294	0.365
48.00	0.195	0.195	0.195	0.242	0.297
60.00	0.195	0.195	0.195	0.212	0.253
72.00	0.195	0.195	0.195	0.195	0.224
84.00	0.195	0.195	0.195	0.195	0.206
96.00	0.195	0.195	0.195	0.195	0.195

the overflow risk converges to the inherent risk level determined by the basin's storage capacity. Among various drain times, as expected, the longer the drain time, the higher the overflow risk. In design, it is important to know the associated overflow risk for the selected drain time.

16.7 Retrofitting of detention basin

Often, a detention basin designed for flood control (as shown in Figure 16.13a through c) would only focus on flow releases for the 10- to 100-year events. With the latest development on stormwater quality enhancement, an existing basin needs to be retrofitted in order to reshape the bottom portion of the basin to accommodate the WQCV. Essentially, it is how to incorporate a perforated plate in Figure 16.13b through d into the outflow structure to provide a comparable flow control to the frequent, small flows.

The following example is employed to illustrate the retrofitting procedure applied to the existing detention basin. The basin has been constructed to control the 10- and 100-year outflows. The 10- and 100-year water depths are 4 and 8 ft above the basin floor, respectively. The existing outlet system for this basin includes a 2-ft × 2-ft concrete box culvert that is 100 ft long on a slope of 2%. This basin needs to be modified for full-spectrum runoff control. As illustrated in Figure 16.14, the tasks include (1) determination of WQCV, (2) reshaping the bottom of the basin to accommodate the WQCV, (3) design of the concrete vault, and (4) adding a micropool.

Figure 16.13 Outlets for flow control. (a) Outlet for extreme events, (b) outlet for all events, (c) outlets for10- and 100-year events and (d) perforated plate for all flow events.

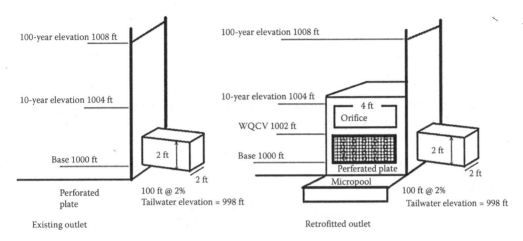

Figure 16.14 Retrofitted outlet system using concrete vault.

The tributary watershed to this basin has a drainage area of 50 acres (20.0 ha) and a runoff coefficient of 0.6. At the site, the average rainfall event depth is 0.41 in. (10.4 mm), and the incipient runoff depth is 0.1 in. (2.54 mm). Aided by Equation 16.32 with an RVCR of 80%, the on-site WQCV is determined as

$$C_v = 1 - e^{-\left(\frac{0.1}{0.41} \times \frac{D_o}{0.6 \times 0.41}\right)} = 1 - 0.78 e^{-4.2 D_o} \tag{16.46}$$

To target a runoff capture rate at 80%, the required WQCV, D_o, is found to be 0.32 in. (8.1 mm) per watershed or $WQCV = 1.3$ acre-ft (1604 m³) for a tributary area of 50 acres (20 ha). The bottom of this basin is reshaped and widened to accommodate the proposed WQCV with a depth of 2 ft.

To retrofit the outfall system for the design WQCV, a concrete vault (as shown in Figure 16.14) is added to the entrance of the existing box culvert. Considering a drain time of 24 h, the perforated plate, 2-ft (60 cm) high and 4-ft (120 cm) wide with five rows and five columns of 1-in. holes, is selected for this basin. The perforated plate is protected with a trash screen. This perforated plate creates a pool 2 ft (60 cm) deep for the required WQCV. Immediately above the perforated plate, a vertical orifice, 1.5 ft (45 cm) high and 4 ft (120 cm) wide, is installed on the vault to control the extreme-event releases. For this case, the top horizontal grate is not needed.

The concrete vault is connected to the existing box culvert. The total flow collection capacity into the concrete vault is the sum of the inflows through the perforated plate and the vertical orifice. For a specified water surface elevation in the basin, the flow through the vertical orifice is calculated as

$$Q_o = C_o A_o \sqrt{2g(H - h_o)} \tag{16.47}$$

in which Q_o = vertical orifice flow in [L³/T], C_o = orifice coefficient such as 0.65, A_o = flow area in [L²], h_o = central elevation in [L] of orifice opening area, g = gravitational acceleration in $[L/T^2]$, and H = water surface elevation in $[L]$ in detention basin. Similarly, the flow through a row of holes on the perforated plate is determined as

$$Q_p = C_o M A_p \sqrt{2g(H - h_p)} \qquad (16.48)$$

in which Q_p = flow in $[L^3/T]$ collected by the holes with their center elevation at h_p in $[L]$, M = number of holes, and A_p = unit hole area in $[L^2]$. For the given water surface elevation, H in $[L]$, the discharge capacity through the outfall culvert is calculated as

$$Q_c = A_c \sqrt{\frac{1}{K+1}} \sqrt{2g(H - h_t)} \qquad (16.49)$$

in which Q_c = culvert discharge capacity in $[L^3/T]$, A_c = culvert opening area in $[L^2]$, h_t = tailwater elevation in $[L]$, and K = sum of loss coefficients determined as

$$K = K_e + K_x + \beta \frac{N^2 L}{R^{4/3}} \qquad (16.50)$$

in which R = hydraulic radius of box pipe in $[L]$, K_e = entrance loss coefficient such as 0.3, K_x = exit loss coefficient between 0.5 and 1.0, β = 29 for foot-second units or 19.5 for meter-second units, and N = Manning's roughness coefficient such as 0.014 for a concrete pipe. For the given water surface elevation, the outflow from the detention basin is dictated by the smaller one between the collection capacity into the concrete vault and the discharge capacity through the outfall box culvert as

$$Q = \min(Q_o + Q_p, Q_c) \quad \text{for a given headwater depth, } H \qquad (16.51)$$

where Q = outflow from the detention basin under water surface elevation, H. The detailed calculation procedure for outflow can be found in Chapter 13.

Table 16.4 summarizes the calculations of the two stage-outflow curves: one for the existing box culvert and another for the new outflow vault. The WQCV for this case is 2 ft deep. The

Table 16.4 Stage–outflow curves for existing and retrofitted conditions

Water depth (ft)	Existing condition			Retrofitted condition				
	Collection capacity (cfs)	Discharge capacity (cfs)	Existing outflow (cfs)	Vertical orifice (cfs)	Perforated plate (cfs)	Collection capacity (cfs)	Discharge capacity (cfs)	Retrofitted outflow (cfs)
0.0	0.00	0.00	0.00	0.00	0.00	0.00	0.00	0.00
1.0	6.79	16.19	6.79	0.00	0.24	0.24	16.19	0.24
2.0	23.18	22.89	22.89	0.00	0.62	0.62	22.89	0.62
3.0	32.77	28.04	28.04	9.17	0.99	10.16	28.04	10.16
4.0	40.14	32.37	32.37	31.30	1.23	32.53	32.37	32.37
5.0	46.35	36.20	36.20	44.26	1.41	45.67	36.20	36.20
6.0	51.82	39.65	39.65	54.21	1.60	55.81	39.65	39.65
7.0	56.77	42.83	42.83	62.59	1.75	64.34	42.83	42.83
8.0	61.32	45.78	45.78	69.98	1.89	71.87	45.78	45.78

Note: $K_e = 0.5$, $K_x = 1.0$, $K_b = 0$, $N = 0.015$, and $L = 100\,\text{ft}$.

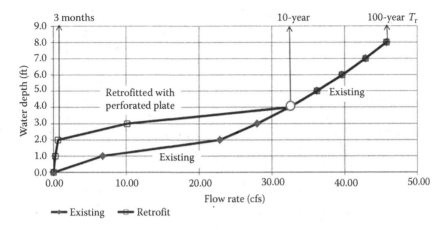

Figure 16.15 Existing and retrofitted stage-outflow curves for detention basin.

new perforated plate is used to control frequent, small events, whereas the orifice and culvert are used to control the 10- and 100-year flow releases. As shown in Figure 16.15, the retrofitted stage-outflow curve is merged into the existing for water depths above the WQCV pool.

The micropool for this basin can be designed for the given WQCV over a drain time of 12 h using the procedure outlined in Chapter 14. The ultimate goal for a detention process is to preserve the watershed regime. The slow release through the perforated plate can simulate the base flow in the receiving stream, and the 10- and 100-year peak flows are released through the orifice and the culvert. In doing so, the after-detention flows can mimic the predevelopment hydrologic condition.

16.8 Homework

Q16.1 A WQCB is located in Tampa, FL. This basin is designed to treat storm runoff from a watershed of 1.5 acre. The watershed area has an imperviousness of 60%.

1. Use Equation 16.17 to estimate the WQCV in acre-ft for this basin.
2. Use Equation 16.32 to estimate the WQCV in inch/watershed with a capture rate at 80%.
3. Consider that the basin has a drain time of 6 h. Based on the basin size in (B), evaluate the runoff capture rate for the first event having a depth of 1.5 in. for a duration of 3 h and then the second event having a depth of 1.0 in. for a duration of 2 h. The interevent time between these two events was 8 h.
4. Calculate the inherent overflow risk for the basin in (B).
5. Calculate the operational risk for a drain time of 48 h for the basin in (B).

Q16.2 Equation Q16.1 represents the runoff capture curve. Its first derivative, Equation Q16.2, represents the slope on the runoff capture curve. In practice, Equation Q16.1 is applied to a preselected range such as the RVCRs from 50% to 95%.

$$C_v = P(0 \le d \le D) = 1 - ke^{\frac{-D}{CD_m}} \qquad \text{(Q16.1)}$$

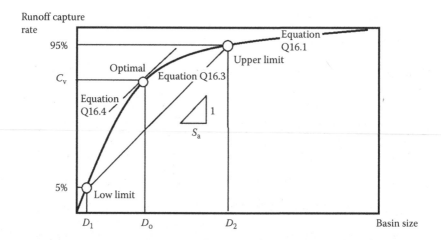

Figure Q16.2 Illustration of optimal capture volume.

$$S_o = \frac{dC_v}{dD} = \frac{1}{CD_m} e^{\frac{-D}{CD_m}}$$
(Q16.2)

The optimal basin volume can then be identified by the average slope for the preselected range. Prove that the average slope for the selected range is determined as Equation Q16.3.

$$S_a = \frac{e^{\frac{-D_1}{CD_m}} - e^{\frac{-D_2}{CD_m}}}{D_2 - D_1}$$
(Q16.3)

in which S_a = average return or slope, D_1 = storage volume for the lower runoff capture rate such as 50%, and D_2 = storage volume for the upper runoff capture rate such as 90%. Prove that the optimal storage volume, D_o, between D_1 and D_2 can be determined as Equation Q16.4 (Figure Q16.2).

$$D_o = CD_m \ln\left\{ \frac{CD_m}{(D_2 - D_1)} \left[e^{\frac{-D_1}{CD_m}} - e^{\frac{-D_2}{CD_m}} \right] \right\}$$
(Q16.4)

Q16.3 The 10- and 100-year peak flows from the tributary watershed were 45 and 150 cfs, respectively, before the development. A detention basin is built to control the 100-year flow release. As illustrated in Figure Q16.3a, the 10- and 100-year water surface elevations are 5008 and 5012 ft, respectively. The 2-ft × 4-ft rectangular orifice on the outlet concrete box has a discharge coefficient of 0.65. The trash rack has an area-opening ratio of 0.6. The outfall system has a restricted circular plate of 3.5 ft in diameter installed at the entrance of the 4-ft × 4-ft box culvert. The culvert pipe is 400 ft long.

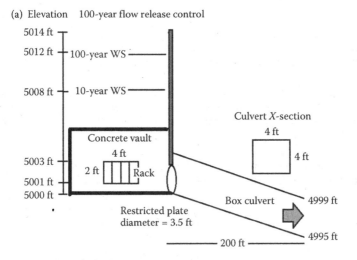

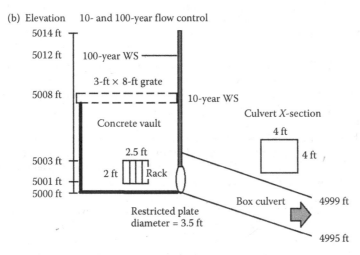

Figure Q16.3 (a) As-built outlet box for 100-year peak flow control. (b) Modified outlet box for both 10- and 100-year peak flow control.

1. Verify that the 100-year peak flow released from the detention basin <150 cfs.
2. Determine the 10-year peak flow, Q10, released from the detention basin. (Is Q10 <45 cfs?)

To improve the as-built condition, the outlet box in Figure Q16.3b is modified with a low-flow orifice and a high-flow grate opening on the top of the concrete box.

1. Verify that the 100-year peak release <150 cfs.
2. Verify that the 10-year peak release <45 cfs.

Bibliography

ASCE WEF. (1998). "Urban Runoff Quality Management," Joint Task Force of the Water Environment Federation and the ASCE, WEF Manual of Practice No. 23 Alexandria VA and Reston, VA.

Athayde, D.N. (1976). "Best Management Practices (BMP)," Proceedings: Urban Stormwater Seminars, Water Quality Management Guide, WPD 03-76-04, U.S. Environmental Protection Agency.

Driscoll, E.D., DiToro, D., Gaboury, D., and P. Shelley (1989), "Methodology for Analysis of Detention Basins for Control of Urban Runoff Quality," U.S. Environmental Protection Agency, Washington, DC, Report No. EPA440/5-87-001.

Einstein, H.A. (1965), "Final Report Spawning Grounds," University of California, Hydraulic Engineering Lab.

EPA. (1983). "Results of the Nationwide Urban Runoff Program, Final Report," U.S. Environmental Protection Agency, NTIS No. PB84-185545, Washington, DC.

EPA. (1986). "Methodology for Analysis of Detention Basins for Control of Urban Runoff Quality," U.S Environmental Protection Agency, EPA440/5-87-001, September 1986.

Flaxman, E.M. (1972). "Predicting Sediment Yield in Western United States." ASCE Journal of the Hydraulics Division, Vol. 98, No. HY12, pp. 2073–2085.

Grizzard, T.J., Randall, C.W., Weand, B.L., and Ellis, K.L. (1986). "Methodology for Analysis of Detention Basins for Control of Urban Runoff Quality," U.S Environmental Protection Agency, Washington, DC, Repot No. EPA440/5-87-001.

Guo, J.C.Y. (2009). "Retrofitting Detention Basin for LID Design with a Water Quality Control Pool," *ASCE Journal of Irrigation and Drainage Engineering*, Vol. 135, No. 6, pp. 671–675.

Guo, J.C.Y. (2013). "Green Concept in Stormwater Management," *Journal of Irrigation and Drainage Systems Engineering*, Vol. 2, No. 3, p. 114, doi:10.4172/2168-9768.1000114.

Guo, J.C.Y., and Cheng, J.Y.C. (2008). "Retrofit Stormwater Retention Volume for Low Impact Development (LID)," *ASCE Journal of Irrigation and Drainage Engineering*, Vol. 134, No. 6, pp. 872–876.

Guo, J.C.Y., Shih, H.M., and MacKenzie K. (2012). "Stormwater Quality Control LID Basin with Micropool," *ASCE Journal of Irrigation and Drainage Engineering*, Vol. 138, No. 5.

Guo, J.C.Y., and Urbonas, B. (1996). "Maximized Detention Volume Determined by Runoff Capture Rate," *ASCE Journal of Water Resources Planning and Management*, Vol. 122, No. 1, pp. 33–39.

Guo, J.C.Y., and Urbonas, B. (2002). "Runoff Capture and Delivery Curves for Storm Water Quality Control Designs," *ASCE Journal of Water Resources Planning and Management*, Vol. 128, No. 3, pp. 208–215.

Guo, J.C.Y., Urbonas, B., and MacKenzie, K. (2011). "The Case for a Water Quality Capture Volume for Stormwater BMP," *Journal of Stormwater.*

Guo, J.C.Y., Urbonas, B., and MacKenzie K. (2014). "Water Quality Capture Volume for LID and BMP Designs," *ASCE Journal of Hydrologic Engineering*, Vol. 19, No. 4, pp. 682–686

Pemberton, E.L., and Lara, J.M. (1971). "A Procedure To Determine Sediment Deposition in a Settling Basin," Sedimentation Section, U.S. Bureau of Reclamation, Denver.

Tucker, L.S., Urbonas, B., Guo, J.C.Y. (1989). *"Sizing a Capture Volume for Stormwater Quality Enhancement."* Flood Hazed News, Vol. 19, No.1.

Urbonas, B., Guo, J.C.Y., and Tucker, L.S. (1989). "Sizing a Capture Volume for Stormwater Quality Enhancement," *Flood Hazed News*, Vol. 19, No. 1, pp. 1–9.

Urbonas, B., Rosener, L.A., and Guo, J.C.Y. (1996). "Hydrology for Optimal Sizing of Urban Runoff Treatment Control System." *Journal of Water Quality International*, Vol. 1996, No. 1, pp. 30–33.

USWDCM. (2001). *Urban Storm Water Design Criteria Manual*, Urban Drainage and Flood Control District, Denver, CO.

Wischmeier, W.H., and Smith, D.D. (1960). "A Universal Soil-Loss Equation to Guide Conservation Farm Planning," 7th International Congress of Soil Science, Madison, Wisconsin.

Chapter 17

Low-impact development facilities

It takes a long time for a watershed to establish the hydrologic equilibrium that involves the long-term stability between surface erosion and stream morphology. The changes in the spatial and temporal distributions of surface runoff will induce a new hydrologic balance between runoff flows and landscapes in the watershed (Booth and Jackson, 1997). The concept of low-impact development (LID) was evolved from the goal of minimizing the urban negative impacts on the water environment. Flood mitigation and stormwater management are a regional effort in planning and designs. In general, this regional strategy consists of two elements: (1) *stormwater extended detention basin (EDB)* placed at the watershed's outlet as a point control to reduce the flow releases from extreme events and (2) *stormwater LID layout and devices* placed throughout the watershed as non-point runoff source control to reduce the runoff volumes from frequent events.

17.1 LID site plan

Figure 17.1 presents a comparison between the conventional *distributed drainage system* and the innovative *cascading flow system* under the concept of LID. A distributed drainage system consists of two separate flow paths: storm drains for impervious areas and swales for pervious areas. An LID layout applies a cascading flow system to minimize the direct connectivity between adjacent impervious areas. An LID layout is to spread runoff flows generated from the upper impervious surface onto the lower pervious area, such as vegetation beds and porous landscaping areas, for additional infiltration benefits and water quality enhancement.

It is critically important that an LID unit is placed at the source of runoff, such as roof downspout, outfall point of a parking lot, entrance pool upstream of a street inlet, etc. As recommended, an LID site should be developed using the following measures at the source of runoff (USWDCM, 2010):

1. To decrease the impervious areas at the project site.
2. To minimize the directly connected impervious area (MDCIA).
3. To decentralize runoff flows and volumes.
4. To dispose runoff into hydrologically functional landscape such as rain gardens (RGs), biodetention systems, filter/buffer strips, grassed swales, and infiltration trenches (Figure 17.2) (Guo et al., 2014).

In an urban setting, the *Q-problem* of stormwater is directly related to the extreme events for increased peak flow, high-flow velocity, and long inundation. Q-problems are considered more as a public safety issue that can be alleviated by flood mitigation control. The

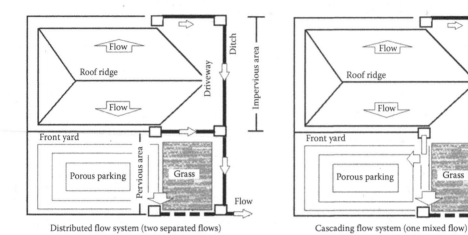

Figure 17.1 LID cascading flow system.

Figure 17.2 Examples of LID designs. (a) Rain garden and (b) infiltration swale.

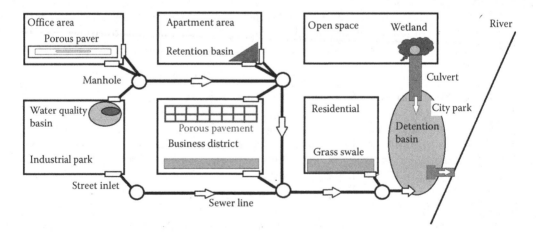

Figure 17.3 Micro–minor–major (3M) cascading flow systems.

Figure 17.4 Examples of LID sites. (a) Cascading detention and WQ basins and (b) grass swale nested in WQ basin.

V-problems are more a concern of public health issue that is directly related to the increased runoff volumes from frequent events. V-problems can be improved by stormwater quality control through infiltration and filtering processes. As illustrated in Figure 17.3, the existing drainage system consists of (1) *storm sewers* to collect the minor events (2- to 5-year events) and (2) *street gutters* to pass the major events (10- to 100-year events). The outfall point for the entire watershed is located at the city park. The urban renewal plan shall improve the existing drainage systems with (1) a regional EDB placed at the city park to reduce the flow releases from extreme events and (2) LID features placed immediately upstream of street inlets to reduce runoff volumes up to microevents (3- to 6-month events). As shown in Figure 17.4, these micro–minor–major (3M) cascading flow systems provide a full-spectrum runoff release control on both runoff flow (Q-problem) and volume (V-problem) (Guo, 2013).

17.2 Effective imperviousness for LID site

An LID site is a flow path–dependent layout. As shown in Figure 17.5, the receiving pervious area in a cascading flow system is covered with native soils, grass, and/or

Figure 17.5 Receiving porous areas. (a) Structured porous pavement and (b) sandy infiltration bed.

structured porous pavers. For example, the use of modular block porous pavement or reinforced turf in low-traffic zones, such as parking areas and infrequently used service drives such as fire lanes, alleys, and sidewalks, can significantly reduce the site imperviousness. Before detailed designing of LID units, it is important that the site be laid with cascading flows onto porous surfaces. This practice can significantly reduce the sizes of the downstream storm sewers and detention basins.

17.2.1 Area-weighted imperviousness

An urban catchment comprises impervious and pervious areas. For modeling convenience, an irregular catchment shall be converted into its equivalent rectangular sloping plane. A distributed flow system (as shown in Figure 17.6) is to divide the catchment into left-impervious and right-pervious planes. Both planes are under the same rainfall, but they produce overland flows separately and independently (SWMM5, 2009; Rossman, 2009).

At the outfall point, the resultant hydrograph is the sum of these two overland flows. The relationship among plane widths is

$$W_T = W_I + W_P \tag{17.1}$$

where W_T = total width in [L], W_I = left-impervious width in [L], and W_P = right-porous width in [L]. Normalizing Equation 17.1 with W_T, we have

$$I_a = \frac{W_I}{W_T} = \frac{A_I}{A_T} \tag{17.2}$$

$$\frac{W_P}{W_T} = \frac{A_P}{A_T} = 1 - I_a \tag{17.3}$$

$$A_T = A_I + A_P \tag{17.4}$$

where I_a = area imperviousness percentage, A_I = impervious area in [L^2], A_P = pervious area in [L^2], and A_T = total area in [L^2].

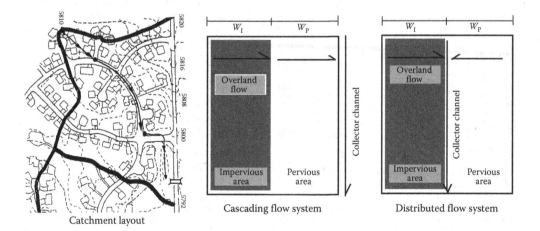

Figure 17.6 Flow systems at project site.

17.2.2 Volume-weighted imperviousness

Equations 17.2 and 17.3 are valid only for a distributed flow site. Under a cascading flow system, the additional infiltration benefit can be added to the flow path. Aided with Equations 17.2 and 17.4, the relationship between the receiving pervious area and the source impervious area is derived as

$$I_a = \frac{A_I}{A_I + A_P} = \frac{A_r}{A_r + 1} \tag{17.5}$$

$$A_r = \frac{A_I}{A_P} \tag{17.6}$$

where A_r = area ratio between upstream impervious area and downstream porous area. At an LID site, the *area imperviousness* is replaced with the *effective imperviousness, I_E,* which is defined by the runoff volume-weighted ratio as (Guo, 2008)

$$V_T = I_E V_I + (1 - I_E) V_P \tag{17.7}$$

$$V_I = P A_T \tag{17.8}$$

$$V_P = (P - F) A_T \tag{17.9}$$

in which V_T = runoff volume in $[L^3]$ from the cascading plane, V_I = runoff volume in $[L^3]$ as if the entire plane was impervious, V_P = runoff volume in $[L^3]$ as if the entire plane was pervious, I = rainfall depth in $[L]$, and F = soil infiltration amount in $[L]$. Rearranging Equation 17.7 yields

$$I_E = \frac{V_T - V_P}{V_I - V_P} \tag{17.10}$$

With an additional infiltration loss on the receiving pervious area, the effective imperviousness, I_E, must be numerically less than the area imperviousness, I_a. Let the pavement area reduction factor (PARF), K, be defined as (Blackler and Guo, 2012, 2013)

$$I_E = K I_a \tag{17.11}$$

In engineering practice, PARF serves as an indicator of the effectiveness of a cascading flow system. PARF can be used as a basis to evaluate various alternatives of LID designs. PARF is proportional to the ratio of soil infiltration rate to rainfall intensity and the ratio of impervious to pervious area (Guo, 2008).

$$K = fct\left(\frac{F}{P}, A_r\right) = fct\left(\frac{F}{P}, I_a\right) \tag{17.12}$$

Under the assumption that the pervious area would have a 100% interception of the runoff flow generated from the upstream impervious area, Figure 17.7 was produced based on numerical simulations using the EPA SWMM computer model (Guo, 2008). As expected, the higher the F/P ratio, the lower the PARF, and the higher the A_r ratio, the higher the PARF.

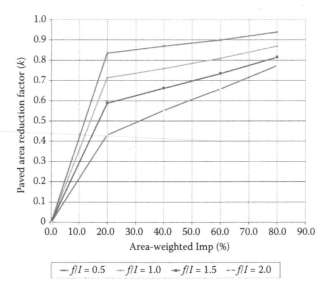

Figure 17.7 Paved area reduction factor (PARF).

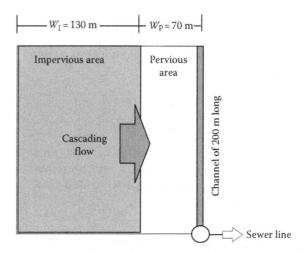

Figure 17.8 Example LID layout.

EXAMPLE 17.1

The catchment in Figure 17.8 has a drainage area of 10 acres (4.0 ha or an equivalent square of 200 m × 200 m). It is estimated that 65% of the catchment area is covered by roofs and driveways, and the rest is covered with soils and turfs. All roofs and driveways are connected to turf surfaces before the stormwater reaches the street inlets.

Solution: For this case, $W_I = 130$ m, and $W_P = 70$ m for numerical simulations.

$$A_r = \frac{A_I}{A_P} = \frac{130 \times 200}{70 \times 200} = 1.86$$

$$I_a = \frac{A_r}{A_r + 1} = \frac{1.86}{1.86 + 1} = 0.65$$

At the project site, the 2-year event has a ratio of $F/P = 1.0$. From Figure 17.7, $K = 0.88$. As a result, this project site has an effective imperviousness of $I_E = 0.88 \times 0.65 = 0.57$. The 100-year event has a ratio of $F/P = 0.5$. The corresponding $K = 0.99$ or no impact on the 100-year event. This is a typical characteristic of LID effect. The effect of LID cascading flow is significant for frequent events, but it diminishes for extreme events.

17.3 Location of LID unit

As shown in Figure 17.9, an LID device is sized to intercept up to the water quality capture volume (WQCV) generated from the tributary area. The size of the LID device depends on its tributary area. As a rule of thumb, the area ratio between the tributary and the LID unit is 5–10, depending on the cascading layout. Often, not the entire impervious area can be intercepted by the LID device. A small portion of the tributary area will bypass the LID device to become a *directly connected impervious area* (DCIA) that drains into the street without water treatment.

As illustrated in Figure 17.9, an LID device is equipped with a level spreader to receive storm runoff and an overtopping weir (another level spreader) to release the excess stormwater. The intercepted water volume will go through the filtering and infiltration media underneath the LID device.

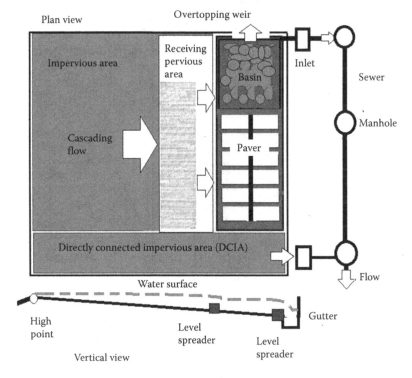

Figure 17.9 Layout of LID device in tributary area.

There are two major types of LID devices: (1) *flow-over porous pavement* and (2) *flow-in porous basin*. A porous pavement shall be laid on a flat surface to enhance the flow interception, whereas a porous basin shall be designed with a pool of 30–40 cm deep (12–18 in.) to store the WQCV. Underneath, the porous bottom is built with multiple layers of filtering media to filter the water flow for better quality. Ideally, an LID unit shall have a drain time of 12–24 h, depending on the local regulations of water rights on storm runoff release.

17.4 Porous basins

Applications of porous basin include *bioretention basin* (BRB), RG, and *porous landscaping detention basin* (PLDB). Although the general structure among these porous basins is similar, there are some minor differences. A BRB (Figure 17.10) comprises a porous basin with a water depth of up to 12–15 in. (30–38 cm) and multiple sublayers of filtering media. Beneath the porous basin bottom are the upper layer of 15–18-in. (38–45 cm) sand-mix for filtering and the lower layer of 8–12-in. (20–30 cm) gravel serving as a subsurface reservoir (Hsieh and Davis, 2005). An RG does not have the gravel layer. A geotextile fabric is used to wrap the RG unit if no leak into the native soils is allowed; otherwise, a perforated fabric is preferred. A PLDB should have thicker layers of sand-mix and gravel to accommodate the growth of vegetation roots. Care must be taken when choosing bushes and plants for a PLDB.

Stormwater that is intercepted by the surface basin will be infiltrated into the subsurface reservoir where the seepage flow is filtered, stored, and gradually released into the perforated subdrain pipes. The subdrain systems are tied into the downstream sewer manhole. The infiltrating rate on the porous bottom represents the inflow to the subsurface RG system, whereas the seepage rate into the subdrain pipe represents the outflow. The operation of a RG is controlled by either the infiltrating rate or the seepage rate, whichever is smaller (Guo, 2012). During an event, all the aggregate voids are filled up with infiltrating water before the seepage flow can be fully developed through the saturated medium (Davis, 2007). If the subsurface seepage flow cannot sustain the infiltrating

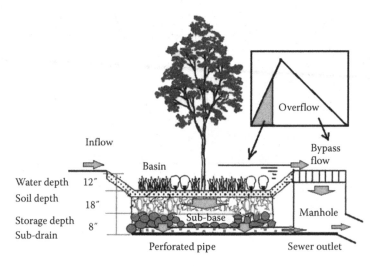

Figure 17.10 Layout of bioretention (rain garden).

flow, the water mounding will be built up to balance the inflow and outflow rates. This phenomenon is manifested by standing water remaining in the surface basin. Therefore, it is advisable that the subsurface geometry beneath the RG be designed to provide an adequate hydraulic gradient in order to sustain the continuity of flow.

17.5 Porous pavement

A porous pavement system consists of permeable surfaces with multiple sublayers to store and to infiltrate runoff into a subdrain pipe or the native soils for groundwater recharge. Porous pavements are suitable for low-traffic areas, including patios, walkways, driveways, fire lanes, and parking spaces. As illustrated in Figure 17.11, a porous pavement is formed with 4- to 6-in. permeable asphalt, concrete, tiles, or pavers on a gently sloped ground (<1%). The effectiveness of a porous pavement system highly depends on the interception of surface runoff. A level spreader using a 1- to –2-in. berm shall be installed at the lower end of the pavement surface. Excess stormwater overtops the berm.

Underneath the porous pavement, it is preferred to have two layers of filtering medium. The upper layer is built with a 4- to 8-in. gravel layer to serve as a subsurface reservoir, while the lower layer is filled with sand-mix for filtering and infiltrating processes. A porous pavement unit may be wrapped with a geotextile fabric if no water leak is preferred. A subdrain system using flexible pipes of 2- to 4-in. diameter shall be installed to collect and deliver the clean water to the downstream manhole. Otherwise, a perforated geotextile fabric is used to allow the local groundwater recharge at a rate of 0.5–1.0 in./h.

17.6 Surface storage basin

An RG is designed as an on-site stormwater disposal facility. The storage volume for an RG is often sized for the WQCV or equivalent for the natural depression storage volume ranging from 0.25 to 1.0 in. As mentioned in Chapter 16, WQCV is directly related to the local *average event rainfall depth*, D_m, and runoff capture rate. The runoff volume capture rate (*RVCR*) is preferred if a continuous probability model is used to portray the rainfall distribution, or the runoff event capture rate (*RECR*) shall be employed if the runoff discrete database is used. Applying the exponential distribution to the rainfall database, the RVCR is derived as (Guo and Urbonas, 2002)

$$C_v = 1 - \alpha\, e^{\frac{-D_0}{CD_m}} \tag{17.13}$$

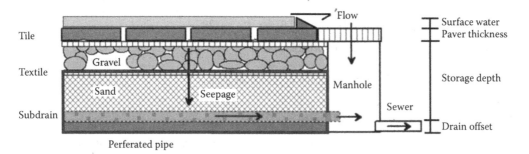

Figure 17.11 Porous pavement system.

$$\alpha = e^{\frac{-D_i}{D_m}} \tag{17.14}$$

in which C_v = RVCR based on exponential distribution, numerically $0 \le C_v \le 1.0$, D_m = local average rainfall event-depth in [L], D_o = WQCV in [L]/watershed determined by exponential distribution, C = runoff coefficient of the tributary catchment to RG, D_i = interception loss in [L] such as 0.05 to 0.1 in., and α = constant related to rainfall interception loss (Guo and Urbonas, 2002). For a preselected C_v, the value of D_o can be determined by Equation 17.13. Next, the storage volume for the RG's surface storage basin is

$$V_o = D_o A \tag{17.15}$$

where V_o = WQCV in [L^3] and A = catchment area in [L^2] tributary to RG's basin. Safety is always a concern when designing an RG. Often, the water depth in an RG is set to be 12–15 in. (30–38 cm). With a preselected basin depth, the basin's cross-sectional area is determined as

$$A_B = \frac{V_o}{Y_o} \tag{17.16}$$

where A_B = average cross section equal to porous bed's bottom area in [L^2] and Y_o = basin depth in [L] of 12–15 in. To enhance the infiltrating process, the basin bottom shall be on a flat to mild slope ($\le 1.0\%$).

EXAMPLE 17.2

The catchment in Figure 17.8 has a drainage area of 10 acres (4.0 ha). The catchment is located at the project site where $D_m = 0.41$ in. The effective imperviousness for this catchment is determined to be $I_e = 57\%$ in Example 17.1. According to Chapter 5, the runoff coefficient for imperviousness of 57% is $C = 0.52$. Design the surface basin for this RG unit.

Solution: Consider $D_i = 0.1$ in., $D_m = 0.41$ in. The value of α is calculated as

$$\alpha = e^{\frac{-D_i}{D_m}} = e^{\frac{-0.1}{0.41}} = 0.784$$

Setting $C_v = 0.8$, the value for D_o is calculated as

$$C_v = 1 - \alpha e^{\frac{-D_o}{CD_m}} = 1 - 0.784 e^{\frac{-D_o}{0.52 \times 0.41}} = 0.8 \quad \text{So,} \quad D_o = 0.29 \text{ in.}$$

$$V_o = D_o A_B = (0.29/12) \, \text{ft} \times 10 \, \text{acre} = 0.24 \, \text{acre-ft}$$

Let $Y_o = 12$ in. The basin's cross-sectional area is 0.24 acre.

17.7 Subbase storage volume

Most soil properties are related to the *soil moisture content* that represents the percentage of the pore volume in the sand layer that has been filled up with water. The layer of sand becomes saturated when the moisture content is equal to its porosity. During a storm event, the available pore volume in a sand layer is defined as

$$D_1 = H_1(\theta_1 - \theta_0) \qquad (17.17)$$

where D_1 = equivalent water depth in the sand layer in [L], H_1 = thickness in [L] of upper sand layer, θ_1 = saturated moisture content in sand layer such as 0.30–0.35 for sand, and θ_0 = initial moisture content. It is noted that the saturated moisture content cannot exceed the porosity of the sand layer, and the initial moisture content cannot be below the wilting point as shown in Figure 17.12.

Similarly, the equivalent water depth in the gravel layer is calculated as

$$D_2 = H_2(\theta_2 - \theta_0) \qquad (17.18)$$

where D_2 = equivalent water depth in [L] in the gravel layer, H_2 = thickness in [L] of lower gravel layer, θ_2 = saturated moisture content in gravel layer such as 0.40–0.45 for gravel, and θ_0 = initial moisture content such as 0.05–0.1. The total excavated depth for the LID unit is the sum of:

$$D_T = Y_o + H_1 + H_2 \qquad (17.19)$$

where D_T = excavated depth in [L]. If the excavated depth reaches the local groundwater table, the design has to be changed to reduce the storage volume. For instance, two smaller LID units are used for a shallower excavated depth.

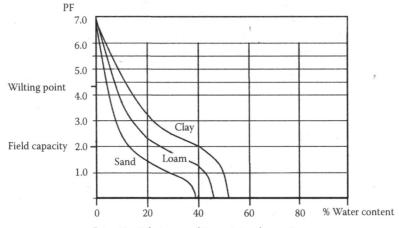

Saturation is between wilting point and porosity.
PF = Log (soil suction head in cm).
PF = 2 (at field capacity) = Log (suction head) or suction head = 100 cm.

Figure 17.12 Soil properties and water contents.

EXAMPLE 17.3

A RG to be added to the catchment in Figure 17.8 has Y_o = 12 in. The local groundwater table is 10 ft below the ground. Check whether the subsurface dimension in Figure 17.10 is acceptable.

Solution: The subsurface storage volume for the two layers of sand and gravel is summed as

$$H_1(\theta_1 - \theta_0) + H_2(\theta_2 - \theta_0) = 18 \times (0.35 - 0.1) + 8.0 \times (0.45 - 0.10) = 7.3 \text{ in.}$$
$$D_T = Y_o + H_1 + H_2 = 12 + 18 + 8 = 38 \text{ in.} < 10 \text{ ft.}$$

During the water loading process, the subbase layers act as a reservoir to be filled with a water volume of 7.3 in. over a vertical distance of 26 in. below the porous bed. As soon as the layer becomes saturated, the porous bed is hydraulically connected to the subdrain pipe through the saturated subbase layers. As a result, the subbase layers become a vertical tube to transmit water from the surface basin into the subdrain. Such a seepage flow is dominated by either the seepage process through the filtering layers or the orifice flow at the subdrain exit, whichever is smaller.

17.8 Seepage flow and drain time

Drain time is critically important for the operation of an LID unit because it controls the enhancement of water quality. Based on the characteristics of urban pollutants, a drain time for an RG is usually set to be 12–24 h (USWDCM, 2010). The average infiltration rate is

$$f = \frac{Y_o}{T_o} \tag{17.20}$$

where f = water infiltrating rate in [L/T] such as mm/h or in./h and T_o = design drain time in [T] such as hour. As a dual flow system above and below the porous bed, the flow movement through a porous bed can be analyzed using the principle of continuity between the infiltrating water flow on the porous bed and the seepage flow through the subbase media (Guo et al., 2009). As illustrated in Figure 17.13, under the assumption of steady state, the infiltrating flow rate described by Horton formula must be equal to the seepage flows determined by Darcy's law as

$$Q = fA_B = K_s I_s A_B = K_g I_g A_B \tag{17.21}$$

in which Q = flow released from LID unit [L^3/T], A_B = porous bed's bottom area [L^2], K_s = hydraulic conductivity coefficient in [L/T] for sand-mix layer such as 2.5 in./h, I_s = hydraulic gradient through sand layer, K_g = hydraulic conductivity coefficient in [L/T] for gravel layer such as 25 in./h, and I_g = hydraulic gradient through gravel layer.

From the laboratory test (Kocman et al., 2012), the hydraulic conductivity coefficients for sand-mix and gravel layers are summarized in Table 17.1.

Referring to Figure 17.13, the flow is driven by the available head in the system as

$$H = Y_o + H_1 + H_2 \tag{17.22}$$

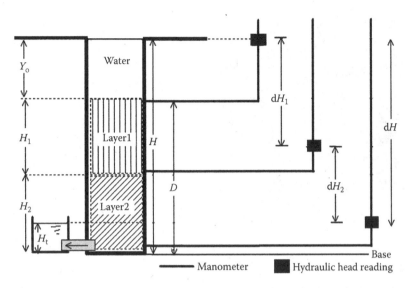

Figure 17.13 Seepage flow through filtering media.

Table 17.1 Hydraulic conductivity coefficients for subbase filtering media

Subbase filtering material	Hydraulic conductivity under fresh condition (in./h)	Hydraulic conductivity under clogged condition (in./h)
Sand-mix	2.50	1.0
Gravel	25.0	1.0

Note: 1 in. = 25.4 mm.

in which H = total hydraulic head in [L], H_1 = thickness of upper layer in [L], and H_2 = thickness of lower layer in [L]. It is noted that the upper layer for a RG is filled with sand-mix, whereas the porous pavement requires the upper layer to be gravel. The energy losses for the seepage flow through the sand and gravel layers are calculated as

$$\Delta H_1 = I_1 H_1 \tag{17.23}$$
$$\Delta H_2 = I_2 H_2 \tag{17.24}$$

As a result, the residual head applied to the underdrain pipe is

$$\Delta H = H - \Delta H_1 - \Delta H_2 \tag{17.25}$$

in which ΔH_1 = energy loss in upper layer, ΔH_2 = energy loss in lower layer, and ΔH = total loss through two filtering layers. Design of RG or Paver involves the uncertainty of how to select design parameters. For instance, the infiltration rate through the sand-mix layer varies from 10 to 15 in./h for a newly constructed basin, 3–5 in./h for a matured porous bed, and 1.0 in./h or less for a clogged basin (Kocman et al., 2012). An infiltration

Figure 17.14 Outlet of perforated pipe with and without a cap orifice.

below 1.0 in./h will result in such a prolonged inundation that the infiltrating bed needs a replacement. For design, a moderate infiltration rate, 1.0–3.0 in./h, is selected to meet the design criteria for both water quality and quantity control. Consequently, how to mimic the predevelopment flow release during the early years of a RG's operation becomes a challenge. In practice, a cap orifice (Figure 17.14) is employed to regulate the flow release at the exit of the underdrain pipe. In comparison, the gravel layer in an RG has a much higher seepage capacity than the sand layer. The perforated underdrain pipe through the gravel layer may be in a saturated or unsaturated condition, depending on the operation of the cap orifice that can be turned down to produce a tailwater effect to the flow system.

17.8.1 Case 1: Without a cap orifice

Without a cap orifice, the perforated underdrain pipe is directly connected to the downstream manhole. At the underdrain exit, the flow pressure drops to the atmospheric pressure. As a result, the infiltration rate in Equation 17.25 must satisfy the balance of energy as

$$\Delta H = H - \Delta H_1 - \Delta H_2 = 0 \text{ without a cap orifice} \tag{17.26}$$

It implies that without a cap orifice, the available headwater in the system dictates the flow capacity. As a result, the RG may be drained at a release rate higher than the predevelopment condition. Its operation may have a drain time shorter than the required residence time for stormwater filtering and solid settlement.

17.8.2 Case 2: With a cap orifice

To regulate the flow release, a cap orifice can be installed at the exit of the perforated underdrain pipe. A cap orifice backs up the flow system to cause saturation in the lower layer. In doing so, the flow release is regulated with the cap orifice. To satisfy the principle of energy, the friction loss through the underdrain pipe is computed as

$$\Delta H_N = kL \frac{N^2 Q^2}{D^{(16/3)}} \tag{17.27}$$

in which ΔH_N = friction loss in [L] through circular underdrain pipe, L = pipe length in [L], D = diameter in [L] of underdrain pipe, N = Manning's roughness coefficient, and k = 4.65 for unit of feet-second or 10.28 for unit of meter-second. The cross-sectional area for the required cap orifice is calculated as

$$A_o = \frac{Q}{C_d\sqrt{2g(H - \Delta H_1 - \Delta H_2 - \Delta H_N)}} \quad \text{with a cap} \tag{17.28}$$

in which A_o = opening area of cap orifice in [L^2], C_d = discharge coefficient, and g = gravity acceleration in [L/T^2]. In practice, the cap orifice must have a diameter smaller than the underdrain pipe.

EXAMPLE 17.4

A RG is designed to have an infiltration basin and two-layered filtering system. The infiltration bed for an RG has a flat area of A_B = 500 ft^2. Referring to Figure 17.10, the dimensions of filtering system are as follows: Y = 12 in., H_1 = 18 in., and H_2 = 8 in. The infiltration rate for the filtering media is estimated to decay from 10.0 to 1.0 in./h. The hydraulic conductivity is 2.5 in./h for the upper sand layer and 25.0 in./h for the lower gravel layer. Without a cap orifice, the flow rate released from this RG is determined using a trial-and-error procedure. Let us start with a guessed infiltrating rate of 5.0 in./h.

$$Q = fA_B = \frac{5.0}{12 \times 3600} \times 500 = 0.058 \, \text{cfs}$$

The energy gradients through the two filtering layers are computed as

$$I_1 = \frac{f}{K_1} = \frac{5.0}{2.5} = 2.0 \quad \text{for the sand layer}$$

$$I_2 = \frac{f}{K_2} = \frac{5.0}{25.0} = 0.2 \quad \text{for the gravel layer}$$

The energy losses through the sand and gravel layers are calculated as

$$\Delta H_1 = I_1 H_1 = 2.0 \times 18 = 36.0 \, \text{in.}$$
$$\Delta H_2 = I_2 H_2 = 0.2 \times 8 = 1.6 \, \text{in.}$$
$$H = Y_o + H_1 + H_2 = 12.0 + 18.0 + 8.0 = 38.0 \, \text{in.}$$
$$\Delta H = H - \Delta H_1 - \Delta H_2 = 38.0 - 36.0 - 1.6 = 0.4 \, \text{in. close to zero}$$

With f = 5.0 in./h, the total energy loss is 37.6 in. in comparison with the total available energy of 38 in. in the system. Therefore, the unregulated flow rate through the RG is determined to be 5.1 in./h after the second iteration.

Based on the predevelopment condition at the project site, the flow release from this RG is not allowed to exceed 3.0 in./h. The task is to design a cap orifice that will reduce the flow release from 5.1 to 3.0 in./h. Repeating the above procedure with f = 3.0 in./h, the cap orifice is determined as

$$Q = fA_B = \frac{5.0}{12 \times 3600} \times 500 = 0.058 \, \text{cfs}$$

The energy gradients through the two filtering layers are computed as

$$I_1 = \frac{f}{K_1} = \frac{3.0}{2.5} = 1.2 \text{ for the sand layer}$$

$$I_2 = \frac{f}{K_2} = \frac{3.0}{25.0} = 0.12 \text{ for the gravel layer}$$

The energy losses through the sand and gravel layers are calculated as

$$\Delta H_1 = I_1 H_1 = 1.2 \times 18 = 21.6 \text{ in.}$$

$$\Delta H_2 = I_2 H_2 = 0.12 \times 8 = 0.96 \text{ in.}$$

Considering that the underdrain pipe is described as follows: $D = 4$ in., $L = 25$ ft, and $N = 0.012$, the friction loss through the underdrain pipe is

$$\Delta H_N = 4.62 L \frac{N^2 Q^2}{D^{(16/3)}} = 4.62 \times 25 \times \frac{0.012^2 \times 0.035^2}{(4/12)^{(16/3)}} = 0.007 \text{ ft} = 0.084 \text{ in.}$$

With $C_d = 0.70$, the cross-sectional area for the cap orifice is calculated as

$$A_o = \frac{0.035}{0.70\sqrt{2 \times 32.2(38 - 21.6 - 0.96 - 0.084)/12}} = 0.0055 \text{ ft}^2 \text{ or one in. in-diameter.}$$

The design example reveals that ΔH_2 for the lower gravel layer and ΔH_N through the short subdrain pipe are numerically negligible in comparison with ΔH_1 for the sand-mix layer dictates because the conductivity of sand-mix is much smaller than that of gravel. As a result, Equation 17.28 is reduced to and normalized as

$$\frac{A_o}{A_B} = \frac{F_f}{C_d\sqrt{2\left(1 - \frac{\Delta H_1}{H}\right)}} \tag{17.29}$$

$$F_f = \frac{f}{\sqrt{gH}} \tag{17.30}$$

$$\frac{\Delta H_1}{H} = \frac{f}{K_1}\frac{H_1}{H} \tag{17.31}$$

where F_f = infiltration Froude Number, and the subscript of 1 represents the variable for the sand layer. Equation 17.29 indicates that this system is characterized with the infiltration flow Froude number. Considering that $C_d = 0.7$, Equation 17.29 is converted into a design chart in Figure 17.15.

For instance, the design example has an infiltration Froude number as

$$F_f = \frac{f}{\sqrt{gH}} = \frac{3.0/(12 \times 3600)}{\sqrt{32.2 \times 38/12}} = 6.88E - 06$$

$$\frac{\Delta H_1}{H_t} = \frac{f}{K_1}\frac{H_1}{H} = \frac{3.0}{2.5}\frac{18}{38} = 0.57$$

From Figure 17.15, the area ratio is found to be $10.5E - 06$ or $A_o = 0.0055 \text{ ft}^2$ for the cap orifice.

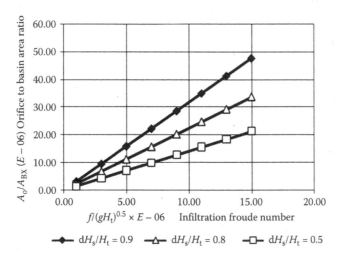

Figure 17.15 Cap orifice for rain garden.

17.9 Dry time of subbase

As discussed earlier, the *drain time* is defined as the period of time to deplete the water depth in the storage basin. The *dry time* is the period of time to drip water out of the sand and gravel layers by the gravity. In comparison, the dry time is dominated by the sand layer. The depletion process in a basin is an unsteady flow that is subject to varied headwater. Under the assumption that the seepage flow is faster than the orifice flow, the depletion process is then dominated by the orifice hydraulics. Referring to Figure 17.16, the continuity between the volume in the storage basin and the flow released through the orifice is

$$A_B \Delta D = Q \Delta t = C_d A_o \sqrt{2g(D-h)} \Delta t \qquad (17.32)$$

Where A_B = porous bed's bottom area in $[L^2]$, ΔD = recession depth in $[L]$, Q = flow release in $[L^3/T]$ through subdrain, Δt = time step, C_d = orifice coefficient for subdrain outlet, A_o = opening area in $[L^2]$ as the cross-sectional area of the sub-drain, g = gravitational

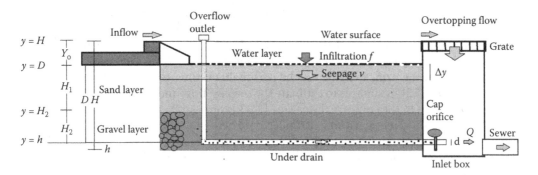

Figure 17.16 Depletion of water volume through rain garden.

acceleration in $[L/T^2]$, H = total headwater in $[L]$, Y_o = water depth in storage basin in $[L]$, D = total saturated depth in $[L]$ in filtering layers, including H_1 in $[L]$ for sand-mix layer and H_2 in $[L]$ for gravel layer, and h = height in $[L]$ at the center line of subdrain.

Integrating Equation 17.32 from H to D yields the drain time, T_{DW}, for the storage basin as

$$T_{DW} = \frac{2A_B\left[(H-h)^{1/2} - (D-h)^{1/2}\right]}{C_d A_o \sqrt{2g}} / 3600 \text{ (h)} \tag{17.33}$$

After empting the storage basin, the LID unit continues depleting water content from the saturated filtering layer. The water volume in the sand-mix layer depends on the porosity. As a result, Equation 17.32 is modified to

$$n_S A_B \Delta D = Q \Delta t = C_d A_o \sqrt{2g(D-h)} \Delta t \tag{17.34}$$

in which n_S = porosity of sand-mix. Integrating Equation 17.34 from D to H_2 yields the dry time, T_{DS}, for the sand layer

$$T_{DS} = \frac{2n_S A_B[(D-h)^{1/2} - (H_2-h)^{1/2}]}{C_d A_o \sqrt{2g}} / 3600 \text{ (h) for sand layer} \tag{17.35}$$

After the sand layer becomes dry, the dry time for the gravel layer, T_{DG}, is calculated from H_2 to h using the porosity of gravel, n_G, as

$$T_{DG} = \frac{2n_G A_B (H_2-h)^{1/2}}{C_d A_o \sqrt{2g}} / 3600 \text{ (h) for gravel layer} \tag{17.36}$$

For a simple case, the filtering system has only a layer of sand-mix. Grouping all system parameters together, Equation 17.35 is reduced to

$$C_f = \frac{C_d A_o \sqrt{2g}}{n_S A_B} = \frac{2(D-h)^{1/2}}{T_{DS}} \text{ (in.}^{0.5}/\text{h)} \tag{17.37}$$

where C_f = flow coefficient, which is a required input parameter when using the EPA SWMM-LID computer model. Equation 17.37 converts the dry time into a depth-based ratio.

It is important to understand that Equation 17.20 is applicable during the peaking period of the runoff event that keeps the storage basin filled up. Therefore, both the seepage flow and the orifice release can be reasonably estimated under a constant headwater. Equation 17.20 is suitable to estimate the seepage flow and orifice release under the design condition. On the contrary, Equation 17.33 is only applicable during the depletion process. The drain time is estimated under a varied headwater after the rain ceases. Equations 17.33 through 17.37 tend to underestimate the drain and dry times because the friction losses through the sand and gravel layers are completely ignored. Of course, a calibrated orifice coefficient can be introduced to compensate the friction loss if sufficient data is available. As a rule of thumb, the orifice coefficient for Equations 17.33

through 17.37 is reduced to the range of 0.2–0.3 without a cap orifice. When the RG is operated with a cap orifice, the entire sand and gravel layers remain so saturated that the friction losses can be ignored. Such a depletion process is similar to a water tank. As a result, with a cap orifice, no reduction on the orifice coefficient is recommended.

EXAMPLE 17.5

Referring to Figure 17.16, an infiltration system has a water depth of 12 in. in the storage basin and a subsurface filtering sand layer of 26 in. The design parameters are as follows: $A_B = 1500\,ft^2$, $n_S = 0.35$, $H = 38$ in., $D = 26$ in., $d = 2.0$ in., $h = 1.0$ in., and $C_d = 0.70$. Determine the drain time for the 12-in. water and the dry time for the 26-in. sand layer.

Solution: The drain time is determined for water to recede from $H = 38$ in. to $D = 26$ in. Applying Equation 17.33 to the storage basin yields

$$T_{DW} = \frac{2 \times 1500 \times \left[\left(\frac{38}{12} - \frac{1}{12} \right)^{1/2} - \left(\frac{26}{12} - \frac{1}{12} \right)^{1/2} \right]}{0.70 \times \left[3.14 \times \left(\frac{2.0}{12} \right)^2 \right] / 4 \times \sqrt{2 \times 32.2}} / 3600 = 2.13\,h$$

After the storage basin is emptied, the saturated filtering depth is 26 in. The dry time is determined to dry the sand layer from $D = 26$ in. to $h = 1.0$ in. as

$$T_{DS} = \frac{2 \times 0.35 \times 1500 \times \left(\frac{26}{12} - \frac{1}{12} \right)^{1/2}}{0.7 \times [3.14 \times (2.0/12)^2/4] \times \sqrt{2 \times 32.2}} / 3600 = 3.44\,h$$

The corresponding flow coefficient for the sand layer is

$$C_f = \frac{2(D-h)^{1/2}}{T_{DS}} = \frac{2 \times (26-1)^{1/2}}{3.44} = 2.91\,in.^{0.5}/h$$

EXAMPLE 17.6

For the purpose of rain harvest, a 48-in. circular tank in Figure 17.17 is used to store storm runoff to a depth of 60 in. The underdrain on this tank is controlled by an orifice with $d = 1.0$ in. and $h = 0.5$ in. Determine the drain time.

Solution: Emptying a rain tank is a process of unsteady flow that does not involve any filtering medium. Thus, Equation 17.33 shall be used with $A_B = 12.6\,ft^2$, $C_d = 0.7$, and $d = 1.0$ in. to calculate the drain time from $H = 60$ in. to $h = 0.5$ in. as

$$T_{DW} = \frac{2 \times 1.0 \times 12.6 \times \left\{ [(60-0.5)/12]^{1/2} - (0.5-0.5)/12]^{1/2} \right\}}{0.7 \times \left[3.14 \times (1/12)^2/4 \right] \times \sqrt{2 \times 32.2}} / 3600 = 0.51\,h$$

Figure 17.17 Rain tank. (a) Industrial rain tank and (b) household rain tank.

The operation of a rain tank can be synchronized with the local weather forecasting. The tank shall carry a slow flow release during dry days and then get reloaded during wet days. A storage tank can be emptied much faster than an RG. The time of delay in the operation of rain harvest is an important factor that shall observe the local water right regulation.

17.10 Clogging effect and life cycle

Over years, clogging effect will be developed in an infiltration-based LID unit. Sediment deposit in an LID unit is accumulated on top of the porous bed and gradually forms a layer of hardened cake that clogs the pores on the porous bed. Clogging effect may also migrate into the upper filtering layer to cause a reduction in infiltration rate. With a reduced infiltration rate, the drain time becomes prolonged. As recommended, the RG needs to be repaired or replaced after the clogged infiltration rate becomes less than 1.0 in./h (UDFCD, 2010). Such a low infiltration rate results in inundation of longer than 12 h in the infiltration bed. Prolonged standing water is considered as a hazard to the public. To be conservative, the design infiltration rate for LID units is set to be the minimum infiltration rate of 1.0 in./h. In operation, the LID unit needs to be replaced when the infiltration rate decays to such a minimal rate.

In the laboratory, the clogging effect was studied by continuously adding sediment-laden storm water onto the porous bed. After many cycles of wet-dry applications, the decay of infiltration rate versus the sediment load was investigated and plotted as Figure 17.18. An empirical equation was derived from the data regression analysis as (Kocman et al., 2012)

$$\frac{f_T}{f} = f_1 e^{-0.1369 L_S} \tag{17.38}$$

where f_T = clogged infiltration rate [L/T] measured after sediment load, L_s, in [10^3 kg/m²] applied to a unit-area of porous bed, f_1 = first-year infiltration rate in [L/T] such as 10–12 in./h, and f = design infiltration rate in [L/T] that is the minimal rate acceptable for operation such as 1.0 in./h. The first-year infiltration rate can be greater than 12 in./h and then gradually decays to the design infiltration rate of 1.0 in./h over years of service.

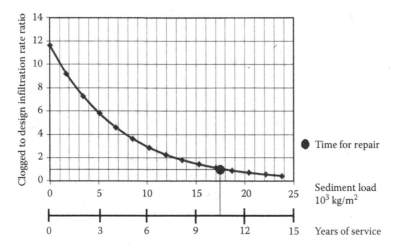

Figure 17.18 Clogging effect through life cycle of LID unit.

In practice, it is more meaningful if the reduction in infiltration rate can be converted into the years of service for the LID unit under design. The annual sediment yield generated from the tributary area to the LID unit is estimated by the annual event mean concentration of sediment and the annual runoff volume as

$$V_o = CP_oA \tag{17.39}$$
$$L_o = C_SV_o \tag{17.40}$$

in which V_o = annual runoff volume in $[L^3/year]$, C = runoff coefficient representing the land use in tributary area, P_o = annual rainfall depth in $[L]$, A = tributary area in $[L^2]$, C_S = mean event sediment concentration $[M/L^3]$, and L_o = annual sediment load generated from the tributary area $[M/year]$. It is important that the annual snowfall depth in the high country is excluded from the annual rainfall depth because the sediment yield is more related to the rainfall-runoff in the thunderstorm season rather than the snowmelt runoff in the early spring. In practice, the RG is designed to intercept the stormwater generated from the tributary area. Aided by Equations 17.39 and 17.40, the annual sediment load added to the RG is estimated as

$$L_B = \frac{L_o}{A_B} = C_S CP_o \frac{A}{A_B} \tag{17.41}$$

in which A_B = porous bed's area or basin's bottom area in $[L^2]$ and L_B = annual unit-area sediment load in RG $[M/L^2/year]$. Aided by Equation 17.41, a cumulative sediment load, L_S, can be converted into RG's years of service as

$$N_Y = \frac{L_S}{L_B} \tag{17.42}$$

where N_Y = RG's years of service, and L_S = cumulative sediment load into RG $[M/L^2]$.

Figure 17.19 Repair of clogged porous bed. (a) Plugged porous bed ($f < 1$ in./h) and (b) vacuum clean to remove cake layer.

EXAMPLE 17.7

A PLDB is built at the outfall point of a parking lot. The ratio of area of the parking lot to that of the PLDB is 20 to 1. The event mean sediment concentration in urban runoff is approximately 240 mg/L. The annual precipitation at the project site is 400 mm (15.7 in.). Aided by Equation 17.41 with $C_S = 240$ mg/L (equivalent to mg/kg), $A/A_B = 20$, $C = 0.9$ for parking lot, and $P_o = 0.4$ m, the annual unit-area sediment load, L_B, is calculated as

$$L_B = (240\,\text{mg/kg}) \times (1000\,\text{kg/m}^3) \times (0.9 \times 20 \times 0.4\,\text{m}) = 1.728 \times 10^3\,\text{kg/m}^2$$

According to Figure 17.18, a PLDB needs a replacement when the clogged infiltration rate decays to 1.0 in./h or the cumulative unit-area sediment load (L_S) is close to 17.5 (10^3 kg/m^2). For this case, this PLDB is expected to have a service of approximately 10 years before an overall repair or replacement. Figure 17.19 represents a plugged porous bed due to tremendous sediment load generated from an adjacent construction site. The cake layer on the bed was removed with a turbo-power vacuum cleaner.

17.11 Evaluation of LID performance

The ultimate goal of LID is to preserve the watershed hydrologic regime. From the hydrologic point of view, it is an effort to mimic the watershed hydrologic condition under the predevelopment condition in terms of the distributions and patterns of flow rate, runoff volume, frequency of event, duration of flow, etc. As mentioned in Chapter 13, stormwater detention alleviates the Q-problem from the extreme events, but it does not effectively control the frequent events that lead to the V-problem. The green concept in stormwater management is to convert curve 1 into curve 4 in Figure 17.20. A flood mitigation approach is to implement an EDB at the watershed outlet that will preserve the flow releases from extreme events (curve 3), whereas a watershed management approach is to recover the natural depression storage volume

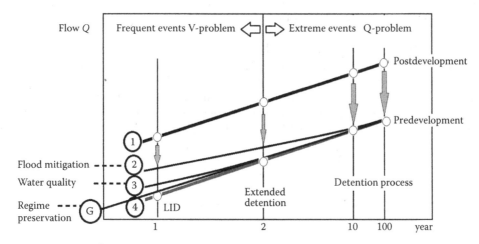

Figure 17.20 Preservation of flow–frequency relationship.

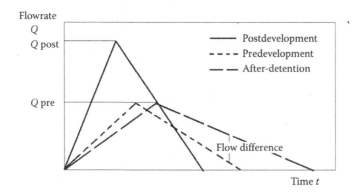

Figure 17.21 Prolonged high flows after detention.

throughout the watershed. The latter is to install LID infiltrating facilities that intercept WQCV at the source of runoff such as roof downspouts, parking lots, business districts, industrial parks, etc. It is important to understand that the LID effort deals with 90%–96% of runoff events, whereas the detention basin only controls 4%–10% of runoff events.

The Q-problem is represented by the changes in the *flow–frequency curve,* which depicts the relationship between the flow and its probability of recurrence. Preservation of a flow–frequency curve implies that the development does not change the expected flood damage. In practice, a well-designed stormwater detention basin may reduce the peak flow (as illustrated in Figure 17.21), but the stored water volume will have to be released at a high rate for a long time. The prolonged high flows do induce erosions to the stream bed and scours to the channel banks. These are typical changes in stream morphology. It is important to recognize the fact that a detention basin provides a solution to Q-problems by reducing the peak flow, but it does not reduce the runoff volume at all. Therefore, the V-problems can only be alleviated with infiltration basins that are designed to reduce runoff volumes (EPA, 1983, 2006).

The V-problem can be quantified by the *flow–duration curve*, which describes the percentage of time to have flows exceeding a given magnitude. The area under a duration curve represents the expected runoff volume. Therefore, on top of the *flow–frequency curve*, it is suggested that the flow–duration curve be preserved as well. Figure 17.22 presents a case study in which a 1-acre lot was developed into residential use with an impervious percent of 75%. The changes in the flow–duration relationship

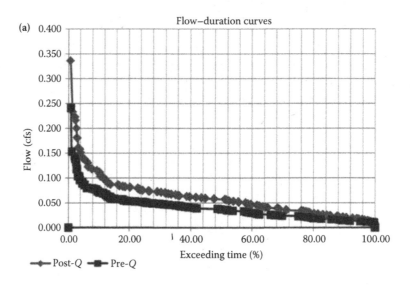

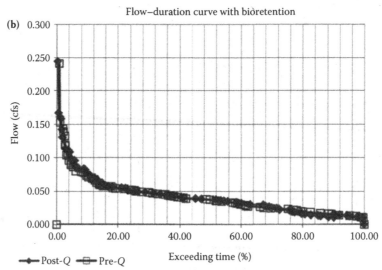

Figure 17.22 Preservation of flow–duration curve. (a) No LID units and (b) with bioretention pond.

are quantified by the runoff simulation using the 30-year 1-h continuous precipitation record. The gap between the pre- and post-flow–duration curves represents the increased runoff volume. After adding a 300-ft^2 bioretention pond to the site, the computer simulation indicates that the infiltration capacity of the bioretention pond is sufficient to preserve the flow–duration curve.

This BRB intercepts runoff volume up to 300 ft^3 and delays the flow release by 10–11 h. Figure 7.23 shows that the inflow is intercepted up to the WQCV, and then the stored volume is divided into a subdrain flow to the downstream manhole and an infiltrating flow to recharge the local groundwater table.

With an adequate LID infiltrating and filtering process at the source of runoff and a slow detention process at the watershed outlet, the watershed may behave as curve G in Figure 17.20 that mimics the predevelopment hydrologic condition and is termed the *Green Stormwater Management* (Guo, 2013).

An urban renewal project is often involved adding LID facilities as the *microdrainage system* that intercepts 3- to 6-month events. With the infiltration capability in an LID facility, the amount of WQCV is transferred into the soil media that can be either for the purpose of groundwater recharge or a delayed release into the storm sewer system as a base flow. Storm drains and sewers are designed to serve as the *minor drainage system* that conveys 2- to 5-year peak runoff flows. During the 10- to 100-year events, the excess stormwater will overflow into street gutters as the *major drainage system*. The 3-M *(micro, minor, and major) drainage system* shall be laid out as a cascading flow system that begins with LID devices for upstream source control and ends with drainage facilities for downstream flood control. As illustrated in Figure 17.24, this continuous overtopping flow system allows runoff flows to become run-on flows for more benefits of infiltration, filtering, and settlement (Guo 2013).

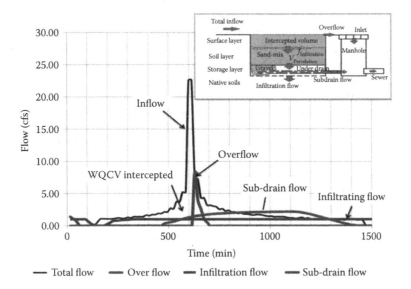

Figure 17.23 Flows through bioretention system.

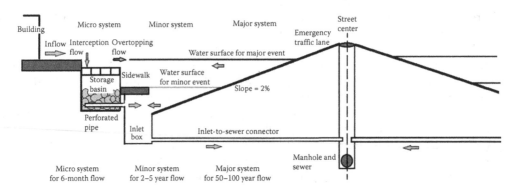

Figure 17.24 Micro–minor–major drainage system for stormwater green approach.

17.12 Closing

Effectiveness of a stormwater LID site is flow path dependent. It is necessary to develop cascading flows to drain storm runoff from the upstream impervious surface onto the downstream pervious areas for more infiltration. Based on the impervious to pervious area ratio, the effective imperviousness is determined and then used to estimate the required WQCV for the project site. Selections of LID units can be porous pavements, porous basins, or both. A porous LID unit shall be equipped with an inlet to intercept the surface runoff and an overtopping weir to release the overflow.

A porous basin such as BRB, RG, and PLDB shall be designed using the concept of hydrologic system to take both the on-surface and subsurface flows into consideration. The filtering layers beneath the porous bed shall be structured to completely consume the hydraulic head available in the system. In practice, it is important that the infiltration rate on the porous bed and the seepage rate through the sand layer are properly evaluated for the design condition to avoid undesirable prolonged standing water in the basin. The drain time for the LID unit is defined as how fast the water depth in the basin can be emptied, while the dry time is calculated as to how fast the subbase becomes unsaturated again. The drain time is approximated by the infiltration rate through the porous bed, and the dry time is calculated based on the seep flow rate through the subdrain pipes. Use the local annual runoff amount and event mean concentration to predict the decay of infiltration rate when the LID unit is getting clogged. When the infiltration rate is reduced to the design rate such as 1 in./h, the LID unit needs to be replaced.

As aforementioned, LID units are designed to reduce the runoff volumes from frequent events. As a result, any large events will overtop the LID unit. As shown in Figure 17.25, the reduction percentage on peak flows depends on the area ratio of the LID unit to its tributary catchment, and its effectiveness is diminished from small to extreme events. The more the LID units, the less the storm drains. In design, it is a challenge to determine the tradeoff between the LID unit and the storm sewer. Nevertheless, an LID unit will reduce the size of the storm drain.

On top of flow reduction, an LID unit captures many small events and reduces the infiltrating flows into combined sewer systems. This effect is a significant improvement

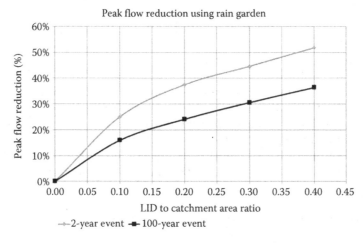

Figure 17.25 Reduction on peak flow using rain garden as LID unit.

to the combined sewer overflow (CSO) system when facing how to reduce the number of overflows from a combined sewer into the downstream water body. Numerically, the reduction on the number of overflows from a CSO system can be well modeled with a long-term rainfall-runoff simulation using the EPA SWMM5.

17.13 Homework

Q17.1 As illustrated in Figure Q17.1, the site of 40 m × 40 m is designed to have an LID layout. The LID basin has a depth of 0.225 m on an area of 10 m × 40 m. Based on the given layout, determine the WQCV in mm per watershed. Verify that the sand and gravel layers are adequate to absorb the given WQCV.

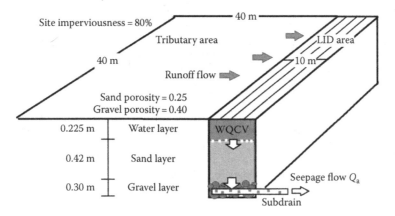

Figure Q17.1 Layout of LID infiltration basin.

Q17.2 The RG in *Q17.1* is built at the outfall point of a parking lot. The ratio of the area of parking lot to that of the RG is 4:1. The event mean sediment concentration in the local runoff flows is approximately 200 mg/L. The annual precipitation at the project site is 600 mm (23.6 in.). (A) Determine the annual sediment amount loaded into the RG. (B) Estimate the lifetime of service for this RG.

Q17.3 A catchment is drained with a CSO system that is designed to divert the low flow up to 2.0 cfs into a wastewater treatment plant. Any high-flow events producing more than 2.0 cfs will result in overflows into the downstream lake. Figure Q17.3 presents the distribution of 144 runoff flows observed in a period of 3 years. After placing an RG to intercept up to 3 cfs from the catchment, verify that the number of runoff event into the sewer line is reduced to 19 events, and the number of overtopping events is reduced to 5 times.

Q17.4 Referring to Figure Q17.4, the RG is built with a basin that has a storage depth of 12-in. on a porous area of 1500 ft^2 ($Y_o = 12$ in.), sand layer ($H_s = 18$ in.), gravel layer ($H_g = 8$ in.), and subdrain pipe (4-in. in diameter for a length of 300-ft). Consider that the conductivity coefficients are 0.25 in./h for the sand layer and 25.0 in./h for the gravel layer. Determine the infiltration rate when the basin is loaded with a depth of 12 in.

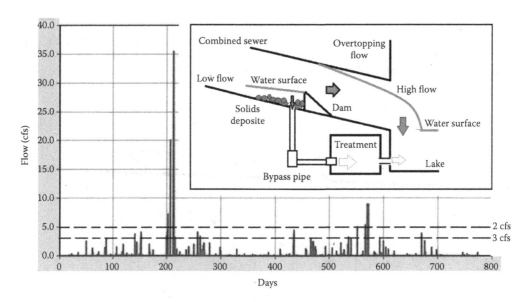

Figure Q17.3 Distribution of runoff flows for CSO system.

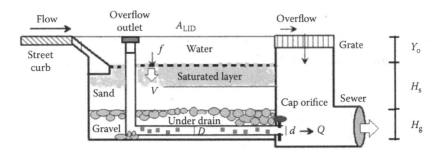

Figure Q17.4 Illustration of rain garden for seepage flow analysis.

Solution:

1. Rain garden flow system

Surface area for LID unit   A-LID = 1500.0 ft^2
Water depth in the surface basin Y_o = 12.00 in.

2. Subbase geometry for two-layered LID basin

Enter initial soil moisture content θ_0 = 10.00%
Enter thickness of upper sand layer H_s = 18.00 in.
Enter hydraulic conductivity of sand layer K_s = 2.50 in./h
Enter porosity for upper sand layer θ_s = 35.00%
Enter thickness of lower gravel layer H_g = 8.00 in.
Enter conductivity of lower gravel layer K_g = 25.00 in./h
Enter porosity for lower gravel layer θ_g = 40.00%
Enter subdrain pipe diameter D = 4.00 in.
Enter subdrain Manning's roughness N = 0.025
Enter the length of subdrain pipe L = 300.0 ft
Enter diameter for cap orifice D_o = 4.00 in.
Enter discharge coefficient for cap orifice C_d = 0.60
Total storage depth = $Y_o + H_s (\theta_s - \theta_0) + H_g (\theta_g - \theta_0)$ H-storage = 18.90 in.

3. Enter the design infiltration rate

Equivalent water depth = $Y_o + H_g \times \theta_g + H_s \times \theta_s$ D-water = 21.47 in.
                                                                             Guess f = 4.07 in./h
Seepage flow through surface area = $f \times$ A-LID   Q-in = 0.1413 cfs
Total headwater depth available = $Y_o + H_g + H_s$ H = 38.00 in.
Energy loss through upper layer = $I_s \times H_s = f/K_s \times H_s$ ΔH_s = 29.30 in. >0
Energy loss through lower layer = $I_g \times H_g = f/K_g \times H_g$ ΔH_g = 1.30 in. >0
Subdrain pipe flowing full velocity = Q/A V = 1.619 fps
Energy slope for flowing full = $(NV)^2/(2.22R^{4/3})$ S_e = 0.0201 ft/ft
Energy headwater for orifice = $H - \Delta H_g - \Delta H_s - LS_e$ ΔH_o = 1.36 in. >0
Cap orifice flow = $C_d \, \varphi \, D_o^2/4 \times (2g \, \Delta H_o)^{0.5}$ Q-out = 0.1414 cfs
Check if $\Delta q = Q_{in} - Q_{out}$ Δq = 0.000 close to 0
Drain time for the basin on surface = Y_o/f T-drain = 2.95 h
Dry time for soil layers = $(\theta_s \times H_s + \theta_g \times H_g)/f$ T-dry = 2.33 h
Sum of drain and dry times = T-drain + T-dry   T-total = 5.28 h

For this case, the system drains very fast. It takes a drain time of 2.95 h to deplete the storage water depth of 12 in.

Q17.5 Referring to Figure Q17.3, the RG was designed to have a drain time of 12 h. A cap orifice with a diameter of 0.92 in. ($d = 0.92$ in.) is installed at the exit of the subdrain pipe. Prove that the drain time is close to 12 h.

Solution:

1. Rain garden flow system

Surface area for LID unit	A-LID = 1500.0 ft^2
Water depth in the surface basin	Y_o = 12.00 in.

2. Subbase geometry for two-layered LID basin

Enter initial soil moisture content	θ_0 = 0.10
Enter thickness of upper sand layer	H_s = 18.00 in.
Enter hydraulic conductivity of sand layer	K_s = 2.50 in./h
Enter porosity for upper sand layer	θ_s = 35.00%
Enter thickness of lower gravel layer	H_g = 8.00 in.
Enter conductivity of lower gravel layer	K_g = 25.00 in./h
Enter porosity for lower gravel layer	θ_g = 40.00%
Enter subdrain pipe diameter	D = 4.00 in.
Enter subdrain Manning's roughness	N = 0.025
Enter the length of subdrain pipe	L = 300.0 ft
Enter diameter for cap orifice	D_o = 0.92 in.
Enter discharge coefficient for cap orifice	C_d = 0.60
Total storage depth = $Y_o + H_s (\theta_s - \theta_0) + H_g (\theta_g - \theta_0)$	H-storage = 18.90 in.

3. Enter the design infiltration rate

Equivalent water depth = $Y_o + H_g \times \theta_g + H_s \times \theta_s$	D-water = 21.47 in.
	Guess f = 1.01 in./h
Seepage flow through surface area = $f \times$ A-LID	Q-in = 0.0351 cfs
Total headwater depth available = $Y_o + H_g + Hs$	H = 38.00 in.
Energy loss through upper layer = $I_s \times H_s = f/K_s \times H_s$	ΔH_s = 7.27 in. >0
Energy loss through lower layer = $I_g \times H_g = f/K_g \times H_g$	ΔH_g = 0.32 in. >0
Subdrain pipe flowing full velocity = Q/A	V = 0.402 fps
Energy slope for flowing full = $(NV)^2/(2.22R^{4/3})$	S_e = 0.0012 ft/ft
Energy headwater for orifice = $H - \Delta H_g - \Delta H_s - LS_e$	ΔH_o = 30.03 in. >0
Cap orifice flow = $C_d \varphi D_o^2/4 \times (2g \Delta H_o)^{0.5}$	Q-out = 0.0352 cfs
Check if $\Delta q = Q_{in} - Q_{out}$	Δq = 0.000 Close to 0
Drain time for the basin on surface = Y_o/f	T-drain = 11.88 h
Dry time for soil layers = $(\theta_s \times H_s + \theta_g \times H_g)/f$	T-dry = 9.41 h
Sum of drain and dry times = T-dry + T-drain	T-total = 21.29 h

For this case, the proposed cap orifice of 0.92 in. in diameter will reduce the flow release down to $f = 1.01$ in./h and extend the drain time to 12 h. Throughout the lifetime of service, the infiltration rate continues decaying from its highest rate down to the clogged rate. It is necessary to operate a cap orifice to maintain such a low release rate that the water quality enhancement is warranted.

Bibliography

Blackler, G., and Guo, J.C.Y. (2012). "Field Test of Paved Area Reduction Factors Using a Storm Water Management Model and Water Quality Test Site," *ASCE Journal of Irrigation and Drainage Engineering*, Vol. 17, No. 8.

Blackler, G., and Guo, J.C.Y. (2013). "Paved Area Reduction Factors under Temporally Varied Rainfall and Infiltration," *ASCE Journal of Irrigation and Drainage Engineering*, Vol. 139, No. 2, pp. 173–179.

Booth, D.B., and Jackson, C.R. (1997). "Urbanization of Aquatic Systems: Degradation Thresholds, Stormwater Detention, and the Limits of Mitigation," *Journal of the American Water Resources Association*, Vol. 22, No. 5, pp. 1–20.

Davis, A.P. (2007). "Field Performance of Bioretention: Water Quality," *Environmental Engineering Science*, Vol. 24, No. 8, p. 1048.

Earles, T., Guo, J., MacKenzie, K., Clary, J., and Tillack, S. (2010). "A Non-Dimensional Modeling Approach for Evaluation of Low Impact Development from Water Quality to Flood Control," *Proceedings of the 2010 International Low Impact Development Conference, San Francisco, CA*, pp. 362–371.

EPA. (1983). "Results of the Nationwide Urban Runoff Program," Final Report, U.S. Environmental Protection Agency, NTIS No. PB84-185545, Washington, DC.

EPA. (2006). "National Recommended Water Quality Criteria," U.S. Environmental Protection Agency, Office of Water, Washington, DC.

Guo, J.C.Y. (2007). "Stormwater Detention and Retention LID Systems," Invited, *Journal of Urban Water Management*.

Guo, J.C.Y. (2008). "Runoff Volume-Based Imperviousness Developed for Storm Water BMP and LID Designs," *ASCE Journal of Irrigation and Drainage Engineering*, Vol. 134, No. 2.

Guo, J.C.Y. (2010). "Preservation of Watershed Regime for Low Impact Development Using (LID) Detention," *ASCE Journal of Engineering Hydrology*, Vol. 15, No. 1.

Guo, J.C.Y. (2012). "Cap-Orifice as a Flow Regulator for Rain Garden Design," *ASCE Journal of Irrigation and Drainage Engineering*, Vol. 138, No. 2.

Guo, J.C.Y. (2013). "Green Concept in Stormwater Management," *Journal of Irrigation and Drainage Systems Engineering*, Vol. 2, No. 3, p. 114.

Guo, J.C.Y., and Urbonas, B. (1996). "Maximized Detention Volume Determined by Runoff Capture Rate," *ASCE Journal of Water Resources Planning and Management*, Vol. 122, No. 1.

Guo, J.C.Y., and Urbonas, B. (2002). "Runoff Capture and Delivery Curves for Storm Water Quality Control Designs," *ASCE Journal of Water Resources Planning and Management*, Vol. 128, No. 3.

Guo, J.C.Y., Blackler, E.G., Earles, A., and MacKenzie, K. (2010). "Effective Imperviousness as Incentive Index for Stormwater LID Designs," *ASCE Journal of Irrigation and Drainage Engineering*, Vol. 136, No. 12.

Guo, J.C.Y., Kocman, S., and Ramaswami, A. (2009). "Design of Two-layered Porous Landscaping LID Basin," *ASCE Journal of Environmental Engineering*, Vol. 145, No. 12.

Guo, J.C.Y., Urbonas, B., and MacKenzie, K. (2011). "The Case for a Water Quality Capture Volume for Stormwater BMP," *Journal of Stormwater*.

Guo, J.C.Y., Urbonas, B., and MacKenzie, K. (2014). "Water Quality Capture Volume for LID and BMP Designs," *ASCE Journal of Hydrologic Engineering*, Vol. 19, No. 4, pp. 682–686

Hsieh, C.H., and Davis, A.P. (2005). "Evaluation and Optimization of Bioretention Media for Treatment of Urban Storm Water Runoff," *Journal of Environmental Engineering*, Vol. 131, No. 11, pp. 1521–1531.

Hunt, W.F., Jarrett, A.R., Smith, J.T., and Sharkey, L.J. (2006). "Evaluating Bioretention Hydrology and Nutrient Removal at Three Field Sites in North Carolina," *Journal of Irrigation and Drainage Engineering*, Vol. 132, No. 6, pp. 600–608.

Kim, H., Seagren, E.A., and Davis, A.P. (2003). "Engineered Bioretention for Removal of Nitrate from Stormwater Runoff," *Water Environmental Research*, Vol. 75, No. 4, pp. 335–367.

Kocman, S.M., Guo, J.C.Y., and Ramaswami, A. (2012). "Waste-Incorporated Subbase for Porous Landscape Detention Basin," *ASCE Journal of Environmental Engineering*, Vol. 137, No. 10.

Lee, J.G., and Heaney, J.P. (2003). "Estimation of Urban Imperviousness and its Impacts on Storm Water Systems," *Journal of Water Resources Planning and Management*, Vol. 129, No. 5, pp. 419–426.

Rossman, L.A. (2009). Storm Water *Management Model User's Manual Version 5.0*, U.S. Environmental Protection Agency, Water Supply and Water Resources Division National Risk Management Research Laboratory, Cincinnati, OH.

SWMM5. (2009). *Storm Water Management Model*, US EPA, Cincinnati, OH.

Thompson, A.M., Paul, A.C., Balster, N.J. (2008). "Physical and Hydraulic Properties Engineered Soil Media for Bioretention Basins," *Transactions of the American Society of Agricultural and Biological Engineers*, Vol. 51, No, 2, pp. 499–514.

UDFCD. (2010). *Urban Storm Drainage Criteria Manual*, Volume 3, Best Management Practices, Urban Drainage and Flood Control District, Denver, CO.

USWDCM. (2010). *Urban Storm Drainage Criteria Manual*, Volume 3, Best Management Practices, Urban Drainage and Flood Control District, Denver, CO.

Chapter 18

Design of infiltration basin

Onsite disposal of storm runoff is the best policy to reduce the increased runoff volumes through soil infiltrations. Examples of onsite stormwater infiltration practice include vegetation beds, sand filters, dry wells, bioswales, infiltration ponds, rock-filled trenches, and permeable pavements. An *infiltration bed* (Figure 18.1) is often constructed as part of the landscaping and is then blended into a vegetation area. Runoff from a roof or parking area shall first be diverted into infiltration beds for onsite stormwater quality enhancement. *Rock-filled trenches* (Figure 18.1) are an effective measure to control the quality of highway runoff. Trenches are built with grass banks and a riprap bottom to increase the infiltration capacity.

A *sand filter* (Figure 18.2) is a shallow, wide, flat infiltration bed that is located at the outfall point of a parking lot, automobile service station, business area, etc. A *dry well is* a vertical tube with backfilled gravels and rocks to store and then to transfer the intercepted stormwater into the groundwater table. A dry well shall be deep enough to effectively recharge the local groundwater table.

A *roof drain basin* (Figure 18.3) is a common practice as an upstream runoff source control device that collects roof runoff and spreads the stored water onto grass areas. On the other hand, a bioretention system (shown in Figure 18.3) is often installed at the downstream outfall point for continuous groundwater recharge through its permanent pool. On top of the permanent pool, a surcharged detention volume may be added for the purpose of flood mitigation. If the water pool is sustainable by the local runoff or groundwater table, wetland vegetation may healthily grow around the wet pool. Therefore, a bioretention pond is operated for multiple benefits, including stormwater disposal, water quality enhancement, and peak flow reduction for flood mitigation.

18.1 Layout of infiltration basin

An *infiltration basin* (Figure 18.4) is designed to remove pollutants from a tributary area up to 5–10 acres (2–4 ha). At the basin site, the minimum distance from the basin's bottom to the groundwater table shall be greater than 5–10 ft. At the entrance of an infiltration basin, an energy dissipation system is always recommended to protect the banks from local scours. A trickle channel shall be implemented to transport frequent, small flows from the entrance to the infiltration bed next to the outlet structure. To increase the infiltration capacity, a shallow, large, and flat porous bed is preferred. Layers of high infiltrating sand-mix material shall be placed underneath the porous bed. Preferably, dry wells are also installed to provide deep percolation. An infiltration bed is very susceptible to sediment clogging. All construction activities involving ground disturbances will become a sediment source from the areas upstream of the basin. It is critically important to

(a)   (b)

Figure 18.1 Examples for infiltration bed and trench. (a) Infiltration bed and (b) rock-backfilled trench.

(a)   (b)

Figure 18.2 Examples for sand filter and dry well. (a) Sand filter and (b) dry well.

(a)   (b)

Figure 18.3 Examples for bioretention and roof drain basins. (a) Roof drain basin and (b) bioretention pond.

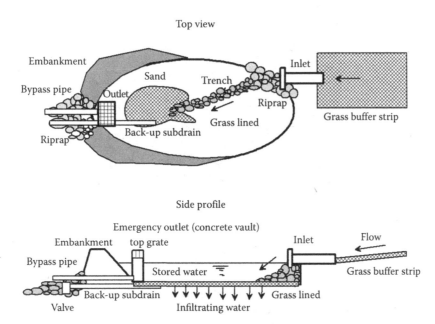

Figure 18.4 Layout of infiltration basin.

Figure 18.5 Construction stormwater BMP measures for sediment control. (a) Without construction stormwater BMP and (b) with construction stormwater BMP.

apply stormwater construction *best management practices* (BMPs) measures (as shown in Figure 18.5) to reduce the surface erosion. Protective measures at street inlets must be implemented to prevent sedimentation of solids from entering the basin. For instance, grass buffer strips shall be used at places where overland flows enter the basin, whereas a sediment forebay shall be installed where concentrated flows enter the basin. As shown in Figure 18.4, it is advisable that a basin be equipped with a backup subdrain system in case of standing water developed due to clogging. A bypass overflow structure is also required to pass major events.

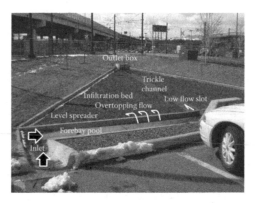

Figure 18.6 Example of layout of infiltration basin.

As an example in Figure 18.6, an infiltration basin consists of four basic elements:

1. Level spreader inlet to spread storm runoff into the porous basin
2. Porous basin to store runoff volume
3. Outlet structure to release excess storm water
4. Emergency bypass facility

Storm runoff enters the basin through a series of filtering grass turf strips and energy dissipaters. The concrete trickle channel runs through the basin from the level spreader to the outlet structure. The porous area is wide and flat to promote the infiltration capacity.

18.2 Design consideration

18.2.1 Basin site

At a site, it is necessary to examine the construction limits, including the topographic slope, the nature of the soil, the proximity of building foundations, and any documented groundwater pollution. The topographic slope for porous pavements and vegetated swales shall be relatively flat, not exceeding 0.5%–1.0%. It is advisable to restrict the use of infiltration practices on steep slopes, because a steep slope increases the chance of downstream water seepage from the subgrade and reduces the amount of infiltrating runoff into the soil. Infiltration practice is not suitable on fill areas because infiltrating water tends to create creeping motions between soil layers. Such an instability condition may become more aggravated to lead to a failure after the fill material on a slope becomes saturated.

18.2.2 Infiltrating devices

The type of pollutants dictates the selection of the infiltration device. For instance, oil and grits in storm water can be more efficiently removed by a *sand filter* than *vegetative beds*. An *infiltration trench* is often installed adjacent to roadways to intercept highway debris. Storm runoff shall have gone through grass buffer strips before entering the trench. The buffer strips shall have a slope less than 1%–3% and shall maintain a flow depth of less than 2–3 in. An infiltrating trench is backfilled with gravel aggregate of 1–3 in. in diameter. The aggregate void ratio ranges from 30%–40%, depending on the gravel size. A filter fabric is required around the walls, or a 6-in. sand layer shall be installed at the bottom of the trench to prevent the migration of fine soil particles from entering into the gravel voids. The challenge in the design of an *infiltration basin* is to ensure that the infiltration rate through the basin bottom can be sustained by the seepage flow through the soil media beneath the basin. The basic design information for an infiltration basin includes the following:

1. Hydrologic parameters of the tributary watershed
2. Predominant pollutants and their residence times for sedimentation
3. Storm water storage volume
4. Feasibility and safety factors at the basin site
5. Surface soil texture and infiltration capacity
6. Subsurface soil seepage capacity
7. Groundwater information

Design information of the tributary catchment includes the design rainfall pattern, tributary area, runoff coefficient representing the level of development, watershed-borne pollutants, and solids from surface erosion. For water quality enhancement, an infiltration basin shall be designed for the frequent storm events. On top, a surcharged layer can be added for the purpose of peak flow reductions during an extreme event.

18.2.3 Soil infiltration rate

The primary factors affecting soil infiltration are *soil type, antecedent soil moisture* (AMC), *vegetative cover,* and *the soil surface texture* such as crusted or frozen. Soils are hardly homogeneous. Adequate field samples shall be surveyed to estimate the infiltration capacities at a site. Table 18.1 presents the recommended infiltration rates for design.

In general, soil textures with infiltration rates less than 0.50 in./h are not suitable for infiltration practices. Soils with a poor drainage capacity are also susceptible to frost heaving and swelling expansion that may cause possible structural instability. Soil infiltration rates can be significantly reduced by clogging effects because of the accumulation of pollutants in soil pores. If an infiltration device is not properly designed, it has the tendency to become clogged rapidly due to sediment entry during and after construction. Therefore, it is necessary to implement surface soil erosion controls in the tributary

Table 18.1 Soil infiltration rates

Soil group	Infiltration rate (in./h)
Sand, open-structured	0.50–1.00
Loam	0.10–0.50
Clay, dense-structured	0.01–0.10
Sand and gravel mix material	0.80–1.00
Silty gravels and silty sands	0.30–0.60
Silty clay sand to sand clay	0.20–0.30
Clays, inorganic or organic	0.10–0.20
Bare rock, not highly fractured	0.00–0.10

catchment and to protect the basin site from clogging debris. A constant standing water pool implies possible clogging due to sediment deposit on the porous bottom. Over the years of service, it is recommended that an infiltration device be adequately maintained, and the filtering layers are backwashed to warrant the functional integrity.

The movement of seepage flows in soils depends on the *soil porosity, hydraulic conductivity,* and *energy gradient. Darcy's law* describes a seepage flow through soil medium as

$$q = KiA \tag{18.1}$$

Where i = hydraulic gradient in [L/L], q = seepage flow in [L^3/T], A = flow area in [L^2], and K = hydraulic conductivity in [L/T] of the soil medium. The *hydraulic conductivity* is also referred to as *permeability coefficient,* which reflects how fast water flows through the soil. The values of hydraulic conductivity for various porous media can be found in Chapter 3. *Hydraulic gradient, i,* is defined as the energy loss per unit length along the flow path.

$$i = \frac{\Delta h}{\Delta l} \tag{18.2}$$

in which Δh = energy loss in terms of headwater depth in [L] and Δl = flow distance in [L]. The flow pattern of infiltrating water underneath the basin can be depicted by the *flow net* that consists of *streamlines* and *equal-potential lines.* A flow net forms a network of meshes that can be used in the finite difference numerical schemes to simulate the flow movement (McDonald and Harbaugh, 1984). Streamlines are tangent to the local flow velocity and divide the flow field into flow tubes. Equal-potential lines are normal to streamlines and represent the energy grade line through the flow field.

18.3 Volume and shape of infiltration basin

The basic design considerations for an infiltration basin include (1) *storage volume* based on the catchment hydrologic requirements, (2) *basin geometry,* and (3) *soil medium* to form the porous bed. At the basin site, the required storage volume and soil infiltration rate on the porous bed dictate the size of the basin. Second, the porosity of the selected soil medium underneath the porous bed sets up the required vertical distance to the local groundwater table. The design procedure begins with the determination of storage volume and then the selection of basin geometry.

18.3.1 Storage volume in basin

An infiltration basin is often placed at the exit of the waterway through a small, highly paved catchment such as a parking lot or business strip. The storage volume in the basin is equal to the difference between the inflow and outflow volumes. The inflow volume to the basin depends on the local *intensity–duration–frequency (IDF) formula* as

$$I_d = \frac{a}{\left(b + T_d\right)^n} \tag{18.3}$$

in which I_d = rainfall intensity in in./h or mm/h and a, b, and n = constants on the IDF formula. The outflow volume depends on the soil infiltration rate, which is described by Horton's formula as

$$f(t) = f_c + (f_0 - f_c)e^{-kt} \tag{18.4}$$

in which $f(t)$ = infiltration rate in in./h or mm/h at elapsed time t in hours, f_0 = *initial infiltration rate* in in./h or mm/h, f_c = *final infiltration rate* in in./h or mm/h, and k = decay coefficient in 1/h. Integration of Equation 18.4 yields

$$F(t) = f_c t + \frac{(f_0 - f_c)}{k}\left(1 - e^{-kt}\right) \tag{18.5}$$

in which $F(t)$ = cumulative infiltration depth in in. or mm at elapsed time t. When time, t, becomes large, the exponential term in Equation 18.5 becomes negligible.

$$F(t) = f_c t + \frac{(f_0 - f_c)}{k} \tag{18.6}$$

For a specified storm duration, the storage volume in the basin is computed as

$$V_d = \alpha C I_d A T_d - \beta A_b F(T_d) \tag{18.7}$$

in which V_d = storage volume in ft^3 or m^3, C = runoff coefficient, A = tributary watershed area in acres or hectares, T_d = rainfall duration in minutes, A_b = basin's porous bottom area in acres or hectares, and α and β are unit conversion factors.

EXAMPLE 18.1

At the project site, the tributary area is a residential subdivision of 2.1 acres with a runoff coefficient of 0.65. The local rainfall IDF curve is described with a = 45.92, b = 10.0, and n = 0.786 using Equation 18.3. Storm runoff produced from the tributary area drains into an infiltration trench of 180 ft long by 20 ft wide. The trench is designed to have infiltration rates: f_0 = 6.50 in./h, f_c = 1.80 in./h, and k = 6.50/h. Determine the storage volumes for the range of $300 \le T_d \le 360$ min.

Solution: Using the units of in./h and acres in computation, the unit conversion factors are α = 60 and β = 1/12, according to Equation 18.7.
   The porous bottom area of the trench is

$$A_b = \frac{180.0 \times 20.0}{43,560} = 0.083\,\text{acre}$$

Set the range of rainfall duration from 300 to 360 min. Table 18.2 summarizes the calculations of storage volumes. It is noted that the infiltration rate has decayed to the final value. The maximal storage volume occurs when T_d = 340 min.

Table 18.2 Storage volume for specified rainfall duration

Duration (min)	Rainfall intensity (in./h)	Inflow volume (acre-ft)	Infiltration rate (in./h)	Infiltration depth (in.)	Outflow volume (acre-ft)	Storage volume (acre-ft)
300.0	0.506	0.288	1.800	9.646	0.066	0.221
320.0	0.481	0.292	1.800	10.246	0.071	0.221
340.0	**0.460**	**0.296**	**1.800**	**10.846**	**0.075**	**0.222**
360.0	0.440	0.300	1.800	11.446	0.079	0.221

The *volume-based method* is a numerical procedure to identify the critical rainfall duration by which the storage volume can be maximized. The maximal value of Equation 18.7 can also be achieved by setting its first derivative with respect to T_d equal to zero, and it results:

$$\frac{dV_d}{dT_d} = \left\{ CA\alpha \left[\frac{-nT_d}{(T_d+b)^{n+1}} + \frac{1}{(T_d+b)^n} \right] - \beta A_b f(T_d) \right\} = 0 \quad \text{when } T_d = T_m \tag{18.8}$$

in which T_m = design rainfall duration in minutes described by Equation 18.8. The solution of Equation 18.8 is

$$T_m = \frac{1}{n} \left[(T_m+b) - (T_m+b)^{n+1} \frac{\beta A_b}{a\alpha CA} f(T_m) \right] \tag{18.9}$$

For most cases, T_m is so long that the infiltration rate, $f(T_m)$, is decayed to the final rate. When the value of b is numerically negligible, Equation 18.9 is reduced to

$$T_m = \left[\frac{2a\alpha CA(1-n)}{\beta A_b f_c} \right]^{\frac{1}{n}} \tag{18.10}$$

Equation 18.10 provides the first approximation to the solution for Equation 18.9 during the trial-and-error procedure. Correspondingly, the maximum storage volume, V_m, is computed as

$$V_m = \alpha Cl_m A T_m - \beta A_b F(T_m) \quad \text{at } T_d = T_m \tag{18.11}$$

EXAMPLE 18.2

Apply the volume-based method to repeat Example 18.1.

Solution: Substituting the design variables into Equations 18.4 and 18.9 yields

$$f(T_m) = 1.80 + (6.50 - 1.80)e^{\frac{-T_m}{60 \times 6.5}}$$

$$T_m = \frac{1}{0.786} \left[(T_m+10) - (T_m+10)^{0.786+1} \frac{\frac{1}{12} \times 0.083 \times f(T_m)}{45.92 \times 60 \times 0.65 \times 2.1} \right]$$

The above two equations are solved for two unknowns, T_m and $f(T_m)$. The design storm duration was found to be $T_m = 340.0$ min. It is important to understand that the uniform rainfall intensity is the average value to represent the nonuniform rainfall time distribution for a period of 340 min. With $T_m = 340.0$ min, the required storage volume is 0.22 acre-ft.

An infiltration basin need not necessarily be a low-impact development facility. For this case, although the basin is sized for the selected design event, not for the water quality capture volume (WQCV), its operation needs to be evaluated with a long-term simulation to warrant a balance among inflows and outflows (Ferguson, 1990).

18.3.2 Water depth in basin

Without taking the subsurface geometry into consideration, the abovementioned procedure yields a storage volume based on the catchment's hydrology. If the *soil infiltration rate* at the land surface is higher than the *underground seepage rate*, the system may become backed up and even cause a failure in the operation. To be conservative, the water storage volume in the soil pores must be examined because the saturated distance in the soil layer will serve as the hydraulic gradient to drive the seepage flows. According to the *diffusive theory* discussed in Chapter 3, the seepage flow through the soil medium (as shown in Figure 18.7) is described as follows (Green and Ampt, 1911):

$$\frac{\partial \theta}{\partial t} + \frac{\partial f}{\partial z} = 0 \tag{18.12}$$

in which θ = soil moisture content, t = elapsed time in [T], f = infiltration rate in [L/T], and z = vertical distance in [L] below the basin. Consider the soil medium between the porous bottom and groundwater table as the control volume of water flow. The finite difference form of Equation 18.12 is reduced to

$$\Delta\theta = \frac{\Delta f \Delta t}{\Delta z} \tag{18.13}$$

As illustrated in Figure 18.7, the value of $\Delta\theta$ is the difference between the soil's initial and saturated moisture contents. The value of Δz is the saturated depth of the soil medium beneath the basin. The value of Δf is equal to the infiltration rate from the porous bottom

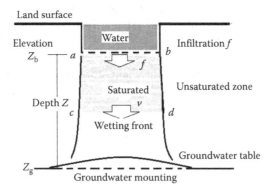

Figure 18.7 Illustration of soil saturation beneath a basin.

because there is no recharge to the groundwater table before the wetting front reaches the groundwater table. As a result, Equation 18.13 becomes (Guo and Hughes, 2001)

$$(\theta_s - \theta_o) = \frac{(f - 0)(T_d - 0)}{(Z_b - Z_g)} = \frac{T_d f}{Z} \quad (18.14)$$

in which θ_s = soil porosity, θ_o = initial soil water content, Z_b = elevation at basin's porous bottom in [L], Z_g = elevation of groundwater table in [L], T_d = drain time in [T], and Z = distance to groundwater table in [L].

Rearranging Equation 18.14, the drain time at the basin site is derived as

$$T_d = \frac{Z(\theta_s - \theta_o)}{f} \quad (18.15)$$

Equation 18.15 indicates that the drain time of an infiltration basin is dictated by the storage capacity in the soil pores and the infiltration rate. The water storage volume in the soil pores is

$$d = Z(\theta_s - \theta_o) \quad (18.16)$$

in which d = equivalent water depth in soil pores in [L]. The seepage flow is driven by the hydraulic gradient, which is produced by the saturated depth underneath the porous bottom. Equation 18.16 sets the minimum saturated depth required to sustain the incoming infiltration rate from the porous bottom. Aided with Equation 18.16, the porous bottom area for the basin under design is

$$A_o = \frac{V_m}{Z(\theta_s - \theta_o)} \quad (18.17)$$

Equation 18.11 defines the required *storage volume* in the basin; Equation 18.16 sets the *minimal saturated water depth* required to sustain the infiltration rate; Equation 18.17 defines the *minimal bottom's area* of the porous basin; and Equation 18.15 calculates the *drain time* to release the stored volume.

EXAMPLE 18.3

At the project site in Example 18.1, the soil porosity is 0.35. The initial soil water content is 0.15. The distance to the local groundwater table is 10 ft. Design the basin geometry for a storage volume of 0.22 acre-ft.

Solution: Under the saturated condition, the water storage volume in the 10-ft soil column is

$$d = 10 \times (0.35 - 0.15) = 2.0 \, ft \, of \, water$$

Assuming that the basin is designed to have the brim-full depth of 2 ft, the basin area is determined as

$$A_o = \frac{0.22}{2.0} = 0.11 \, acre$$

The infiltration rate is 1.8 in./h (see Example 18.1). Therefore, the drain time is

$$T_d = \frac{2.0 \times 12}{1.8} = 13.3h$$

So, the area of the basin is 0.11 acre. According to Example 18.1, the tributary area to this basin is 2.1 acres. The area of this basin is 4.7% of its tributary area.

18.4 Sustainability of infiltration rate

After the soil media become saturated, the infiltrating water directly recharges the groundwater table. If the seepage flow through the soil column is slower than the infiltration rate on the land surface, the excess inflow will cause *water mounting effect*. As a result, the operation of the basin under a saturated soil condition may be rather controlled by the subsurface seepage rate than the on-surface infiltration rate. Therefore, when selecting a basin site for groundwater recharge, it is important to know whether the subsurface geometry at the site sustains the design infiltration rate.

18.4.1 Seepage flow under porous trench

Trenches (Figure 18.8) are used to collect roadway runoff. The filtering process through the porous bottom removes solids and heavy metal. The infiltrating water flow underneath a trench can be modeled by a *two-dimensional potential flow model* (Griffin and Warrington, 1988; Guo, 1998). Under a saturated soil condition, Figure 18.9 shows a unit-width section of an infiltration trench. The infiltrating water moves vertically downward underneath the trench. The diffusive nature of the *wetting front* results in flow movements in both vertical and lateral directions.

In practice, the vertical distance from the porous bed to the local groundwater table is an important factor when determining the seepage capacity. The flow pattern of a seepage flow through soil medium can be depicted by the potential flow using the stream function. After the downward seepage flow reaches the groundwater table, the water flow will be diffused into the top *effective layer* in the groundwater table (Hantuch, 1967).

(a)

(b)

Figure 18.8 Examples of trenches. (a) Trench for highway runoff and (b) trench for roof runoff.

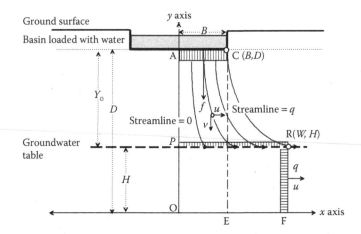

Figure 18.9 Potential flow pattern for infiltrating water under trench.

The thickness of the effective layer, H, (in Figure 18.9) depends on the amount of seepage flow. Using point O in Figure 18.9 as the reference point, a two-dimensional potential flow model is developed to describe the vertical and lateral flow movements beneath the trench as (Guo, 1998)

$$\psi = \frac{f}{D} xy \qquad (18.18)$$

in which ψ = stream function representing the flow rate per linear foot of trench in $[L^2/T]$, f = infiltration rate on porous bed in $[L/T]$, D = saturated vertical depth in $[L]$, x = horizontal distance from the central axis in $[L]$, and y = vertical distance in $[L]$ above Point O in Figure 18.9. The value of stream function represents the cumulative flow rate from the base line. For this case, the base line, which is the line APO in Figure 18.9, has $\psi = 0$. Based on the continuity principle, the value of streamline, ψ_c, from point C to R, must be equal to the total infiltrating flow as

$$\psi_c = q = fB \qquad (18.19)$$

in which q = infiltration volume rate per linear foot of trench in $[L^2/T]$ and B = half width of the trench in $[L]$. Aided with Equations 18.18 and 18.19, the width of the wetting front at a distance, y, is

$$w = \frac{BD}{y} \qquad (18.20)$$

in which w = width of the wetting front in $[L]$ and y = vertical distance in $[L]$ above point O in Figure 18.9. Derivatives of Equation 18.18 with respect to y and x represent the velocity components as

$$u = \frac{\partial \psi}{\partial y} = \frac{f}{D} x \qquad (18.21)$$

$$v = -\frac{\partial \psi}{\partial x} = -\frac{f}{D} y \qquad (18.22)$$

Under a steady-state condition, the flow rate remains constant through sections AC, PR, and RF in Figure 18.9. On section PR, the hydraulic gradient, i, can be approximated by unity, $i = -1$ (downward) when the soil suction is ignored (Bouwer et al., 1999; Guo, 1998). According to the Darcy's law, the vertical velocity of the recharge flow at section PR is

$$v = \frac{q}{W} = \frac{K_y i W}{W} = -K_y \qquad (18.23)$$

in which W = recharging width in [L] and K_y = hydraulic conductivity in [L/T] in the vertical direction. Equation 18.23 must agree with Equation 18.22 at $y = H$. As a result, it is concluded that

$$D = \lambda_y H \qquad (18.24)$$

$$\lambda_y = \frac{f}{K_y} \qquad (18.25)$$

Streamlines of recharging flow near the mound are concentrated in the upper or active thickness of the aquifer, with almost stagnant water in the deeper portion of the aquifer. Such an active flow depth, H, below the groundwater table can be determined by Dupuit-Forchheimer equation as (Brooks and Corey, 1964)

$$q = \psi_c = \frac{K_x}{2} \frac{\left(D^2 - H^2\right)}{(W - B)} \qquad (18.26)$$

Equation 18.26 divided by the effective flow depth, H, yields a sectional velocity, which must be equal to Equation 18.21 at $x = W$. As a result, the effective thickness of aquifer is derived as

$$\frac{H}{B} = \sqrt{\frac{2\lambda_x}{\lambda_y + 1}} \qquad (18.27)$$

$$\lambda_x = \frac{f}{K_x} \qquad (18.28)$$

Aided by Equation 18.24, the vertical saturation depth, D, is

$$\frac{D}{B} = \lambda_y \sqrt{\frac{2\lambda_x}{\lambda_y + 1}} \qquad (18.29)$$

Aided by Equations 18.27 and 18.29, the minimum saturated depth, Y_o, required to sustain the continuity of flow is calculated as

$$\frac{Y_o}{B} = \frac{D - H}{B} \qquad (18.30)$$

Equations 18.27, 18.29, and 18.30 define the required subsurface geometry in order to sustain the infiltration rate from the porous bed. At the basin site, if the vertical distance to the groundwater table is less than Y_o, the design infiltration rate shall be reduced because the site does not have an adequate hydraulic head to sustain such an infiltration rate. Equation 18.30 provides a basis to quantify the subsurface seepage capacity.

EXAMPLE 18.4

A trench has a width of 20 ft and a length of 180 ft. The soil infiltration rate and coefficient of permeability at the site are 1.80 and 1.00 in./h, respectively. The distance to the groundwater table is 10.50 ft. Assuming that the soil medium at the site is homogenous, can this subsurface geometry under the trench sustain the infiltration rate?

The bottom porous area, A, of the trench is

$$A = 20 \times \frac{180}{43,560} = 0.083 \, \text{acres}$$

The infiltration flow rate is

$$Q_o = f \times A = \frac{1}{12 \times 3600} \times 1.80 \times (0.083 \times 43,560) = 0.151 \, \text{cfs}$$

The half width for the trench is $B = 10$ ft. The unit flow rate for half of the basin width is

$$q_o = \frac{1}{12 \times 3600} \times 1.80 \times 10 = 0.00042 \, \text{cfs/ft}$$

Assume that the subsurface medium is isotropic, i.e., $K_x = K_y = K$. At the project site, we have

$$\lambda_x = \lambda_y = \frac{1.8}{1.0} = 1.8$$

$$\frac{H}{B} = \sqrt{\frac{2.0 \times 1.8}{1.8+1}} = 1.13 \quad \text{or } H = 1.13 \, \text{ft}, \, B = 11.3 \, \text{ft}$$

$$\frac{D}{B} = 1.8 \sqrt{\frac{2.0 \times 1.8}{1.8+1}} = 2.04 \quad \text{or } D = 2.04 \, \text{ft}, \, B = 20.4 \, \text{ft}$$

$$Y_o = D - H = 9.10 \, \text{ft}$$

The required saturated depth is 9.1 ft. The existing condition is 10.5 ft. Therefore, this basin site is sufficient to sustain the design infiltration rate. This is a case of direct recharge to the groundwater table.

Table 18.3 presents a comparison among cases with different f/K ratios. When the infiltration rate is slightly greater than the permeability coefficient, the required saturation depth,

Table 18.3 Required saturation depths for various f/K ratios for trench

f/K	D/B	H/B	Y_o/B	D ft	H ft	Y_o ft
1.1	1.13	1.02	0.10	11.26	10.24	1.02
1.5	1.64	1.10	0.55	16.43	10.95	5.48
2.0	2.31	1.15	1.15	23.09	11.55	11.55
2.5	2.99	1.2	1.79	29.88	11.95	17.93
3.0	3.67	1.22	2.45	36.74	12.25	24.49

Note: The case study was performed for $K = 1.0$ in./ft and $B = 10.0$ ft.

Y_o, which is 1.02 ft, is relatively shallow, but it sharply increases as the f/K ratio increases. For instance, the required saturation depth becomes 24.49 ft in order to produce a sufficient hydraulic gradient when $f/K > 3.0$.

18.4.2 Seepage flow underneath circular porous basin

Infiltration practice is recommended for low land drainage in places where the positive hydraulic gradient is not available. Figure 18.10 presents examples of an infiltration basin with a porous bottom and a retention pond for groundwater recharge through a permanent pool.

The flow pattern underneath the circular porous basin is illustrated in Figure 18.11. Such a three-dimensional axially symmetric flow can be described by the stream function as (Guo, 1999a,b; Ortiz et al., 1979)

(a)   (b)

Figure 18.10 Examples of infiltration facility. (a) Infiltration basin and (b) infiltration pond.

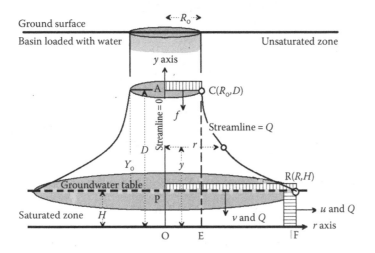

Figure 18.11 Flow pattern under circular basin.

$$\psi = \pi \frac{f}{D} r^2 y \qquad (18.31)$$

in which r = radius in [L] at distance y in [L] above the reference point O in Figure 18.11. The streamlines below the porous bed are distributed as concentric circles with $\Psi = 0$ along the y-axis and Ψ = infiltration volume rate in [L³/T] at the circumference of the basin bottom, i.e., point $C(r, y) = (R_o, D)$ in Figure 18.11. The infiltration volume rate released from the circular porous basin is

$$Q = f\pi R_o^2 \qquad (18.32)$$

in which Q = infiltration volume rate in [L³/T], f = soil infiltration rate in [L/T], and R_o = radius in [L] of circular basin. For this flow field, the stream function is

$$\psi = \frac{\pi f}{D} r^2 y \qquad (18.33)$$

For a specified stream function between zero and Q, its streamline can be plotted through points (r, y) defined by Equation 18.33 on a vertical plane. To maintain the continuity of the flow, the radius, r, of a horizontal cross-section at specified y is obtained by equating Equation 18.32 to Equation 18.33 to yield

$$r = \sqrt{\frac{D}{y}} R_o \qquad (18.34)$$

At section PR in Figure 18.11, aided with Equation 18.34, the radius of the recharge area, R, at $y = H$ is

$$R = \sqrt{\frac{D}{H}} R_o \qquad (18.35)$$

The velocity components in the flow field can be depicted by the derivatives of Equation 18.33 with respect to r and y as

$$u = \frac{1}{2\pi r} \frac{\partial \psi}{\partial y} = \frac{f}{2} \frac{r}{D} \qquad (18.36)$$

and

$$v = \frac{1}{2\pi r} \frac{\partial \psi}{\partial r} = -f \frac{y}{D} \qquad (18.37)$$

According to Darcy's law, the seepage rate across section PR is

$$Q = K_y i \left(\pi R^2 \right) \qquad (18.38)$$

in which i = vertical hydraulic gradient in [L/L] and K_y = coefficient in [L/T] of vertical permeability. The downward velocity, v, at section PR is

$$v = \frac{Q}{\pi R^2} = K_y i \tag{18.39}$$

The downward hydraulic gradient in Equation 18.39 is $i = -1$. Because Equation 18.39 must be equal to Equation 18.37 at $y = H$, it yields

$$D = \lambda_y H \tag{18.40}$$

and

$$\lambda_y = \frac{f}{K_y} \tag{18.41}$$

At the outflow section, section RF, the seepage rate is dictated by the hydraulic gradient between sections CE and RF. Therefore, we have

$$Q = K_r \left(2\pi r y\right)\left(-\frac{\partial y}{\partial r}\right) \tag{18.42}$$

in which K_r = radial hydraulic conductivity. Integrating Equation 18.42 from $y = D$ to $y = H$ and from $r = R_o$ to $r = R$ yields

$$Q = \frac{K_r \pi}{\ln\left(\dfrac{R}{R_o}\right)}\left(D^2 - H^2\right) \tag{18.43}$$

in which H = active thickness in [L] of aquifer to be affected by recharge. Aided by Equations 18.35, 18.40, and 18.43, the cross-sectional average velocity, u, at section PF is

$$u = \frac{Q}{2\pi R H} = K_r \frac{H}{R}\frac{\left(\lambda_y^2 - 1\right)}{\ln \lambda_y} \tag{18.44}$$

Similarly, Equation 18.44 must be consistent with Equation 18.36 at $r = R_o$. Aided by Equations 18.35 and 18.40, the ratio of H/R_o can be derived by setting Equation 18.44 equal to Equation 18.36 as

$$\frac{H}{R_o} = \sqrt{\frac{\lambda_r \ln \lambda_y}{2\left(\lambda_y^2 - 1\right)}} \tag{18.45}$$

and

$$\lambda_r = \frac{f}{K_r} \tag{18.46}$$

in which K_r = radial hydraulic conductivity in [L/T] in the radial direction. Substituting Equation 18.45 into Equation 18.35 yields

$$\frac{D}{R_o} = \lambda_y \sqrt{\frac{\lambda_r \ln \lambda_y}{2\left(\lambda_y^2 - 1\right)}} \tag{18.47}$$

Aided by Equations 18.46 and 18.47, the required saturated depth to sustain the flow continuity is calculated as

$$\frac{Y_o}{R_o} = \frac{D-H}{Y_o} \tag{18.48}$$

As mentioned earlier, Equation 18.48 defines the minimum vertical distance required from the basin bottom to the groundwater table. If the site satisfies Equation 18.48, the infiltrating water will directly recharge the groundwater table; otherwise, the water mounding effect will be developed to reduce the infiltrating efficiency (Morel-Seytous et al., 1990).

EXAMPLE 18.5

Application of the abovementioned model is illustrated by the design of a circular infiltration basin with a diameter of 68.0 ft. The distance to the groundwater table at the site is 16.5 ft. The basin will have a layer of loamy sand lining. From pumping well tests, the hydraulic conductivity is found to be 1.55 in./h for the seepage flow. Recommend a design infiltration rate for the site.

Solution: Under the assumption of an isotropic environment, $K_r = K_y = K$. Let us assume that the soil infiltration rate is 3.6 in./h. The design infiltration volume rate from the basin is estimated as

$$Q = f\pi R_o^2 = \frac{3.60}{12.0 \times 3600.0} \times 3.1416 \times 34.0^2 = 0.30 \, cfs$$

The required subsurface geometry is calculated as

$$\lambda = \frac{3.60}{1.55} = 2.32$$

$$\frac{H}{R_o} = \sqrt{\frac{2.32 \times \ln(2.32)}{2(2.32^2 - 1)}} = 0.471 \quad or \ H = 16.1 \, ft$$

$$\frac{D}{R_o} = 2.32\sqrt{\frac{2.32 \times \ln(2.32)}{2(2.32^2 - 1)}} = 1.09 \quad or \ D = 37.2 \, ft$$

$$Y_o = 37.2 - 16.1 = 21.1 \, ft$$

The required saturated distance is greater than the available (21.1 > 16.5 ft). Therefore, the design infiltration rate must be reduced. Table 18.4 presents the cases with different f/K ratios.

Table 18.4 Design infiltration rates for various f/K ratios

f/K	H/R_o	D/R_o	Y_o/R_o	$H \, ft$	$D \, ft$	$Y_o \, ft$
1.20	0.50	0.60	0.10	16.95	20.34	3.39
1.50	0.49	0.74	0.25	16.77	25.15	8.38
2.00	0.48	0.96	0.48	16.34	32.69	16.34
2.50	0.47	1.17	0.70	15.88	39.7	23.82
3.00	0.45	1.36	0.91	15.34	46.29	30.86

When $f/K = 2.0$, the required saturated depth is 16.34 ft. Therefore, it is suggested that the basin lining material be designed to have infiltration rate: $f = 2K = 2 \times 1.55 = 3.1$ in./h.

18.5 Closing

An infiltration basin is a hydrologic system that is operated with the inflow determined by the design rainfall and the outflow determined by the design infiltration rate. The storage volume in the basin shall be maximized through the selection of rainfall duration. Usually, the design event for an infiltration basin is much longer than the time of concentration, which is often used for peak flow predictions.

Similar to the culvert capacity that has to be examined by its inlet and outlet controls, the flow system in a porous infiltration basin is also examined by the infiltration rate on the porous bed and the seepage rate through the soil medium. The operation of the basin relies on the soil conductivity and the subsurface hydraulic gradient. The potential flow model developed in this chapter provides a basis to estimate the flow field underneath the porous bed, and the minimum saturated depth required to provide a sufficient hydraulic gradient. When the basin site does not have an adequate vertical depth from the porous bottom to the groundwater table, the design infiltration rate has to be reduced to avoid water mounding (Guo, 1998, 1999a,b).

As always, the operation of an infiltration basin is sensitive to sediment clogging. Most of the sediment loads are generated from the construction sites in the tributary area. Care has to be taken when using the construction BMPs for sediment control. Figure 18.12 presents examples of clogged porous bed in an infiltration basin. For this case, the upstream construction site was not adequately equipped with stormwater BMPs for sediment control. In a couple of months, the downstream infiltration basin is clogged owing to a large amount of sediment load.

Regular maintenance of an infiltration basin and enforcement of construction BMPs through the tributary area are critically important to the efficiency of an infiltration basin and its life cycle economics.

(a)   (b)

Figure 18.12 Plugged pond without upstream construction BMP. (a) Plugged infiltration porous bed and (b) interception of urban debris.

18.6 Homework

Q18.1 At the project site, the tributary area is a commercial subdivision of 2.5 acres with a runoff coefficient of 0.80. The local rainfall IDF curve is described with $a = 56$, $b = 10.0$, and $n = 0.786$. Storm runoff produced from the tributary area drains into a circular infiltrating basin with a radius of 50 ft. The porous bottom is designed to have infiltration rates: $f_0 = 3.50$ in./h, $f_c = 1.50$ in./h, and $k = 6.50$/h. Determine the storage volume for this basin. (Solution: $T_m = 332$ min, $V_m = 0.24$ acre-ft)

Q18.2 Continue with *Q18.1*. At the site, the vertical distance to the groundwater table is 10 ft. The coefficient of soil permeability underneath the basin is 1.0 in./h. What is the maximum infiltration rate without causing water mounding at the site? (Solution: 1.41 in./h)

Bibliography

Bouwer, H., Back, J.T., and Oliver, J.M. (1999). "Predicting Infiltration and Ground-water Mounds for Artificial Recharge," ASCE Journal of Hydrologic Engineering, Vol. 4, No. 4, pp. 350–357.

Brooks, R.H., and Corey, A.T. (1964). "Hydraulic Properties of Porous Media," Hydrologic Paper No. 3, Colorado State University, Ft. Collins, CO.

Ferguson, B.K. (1990). "Role of the Long-term Water Balance in Management of Storm Water Infiltration," Journal of Environmental Management, Vol. 30, No. 3, pp. 221–233.

Griffin, D.M. Jr., and Warrington, R.O. (1988). "Examination of 2-D Groundwater Recharge Solution," ASCE Journal of Irrigation and Drainage Engineering, Vol. 114, No. 4, pp. 691–704.

Green, W.H., and Ampt, G.A. (1911). "Studies of Soil Physics, I: The Flow of Air and Water Through Soils," Journal of Agriculture Science, Vol. 4, No. 1, pp. 1–24.

Guo, J.C.Y. (1998). "Surface-Subsurface Model for Trench Infiltration Basins," ASCE Journal of Water Resources Management and Planning, Vol. 124, No. 5, pp. 280–284.

Guo, J.C.Y. (1999a). "Design of Circular Infiltration Basin under Water Mound Effects," ASCE Journal of Water Resources Management and Planning.

Guo, J.C.Y. (1999b). "Application of Potential Flow Model to Water Mound under Infiltration Basin," Journal of Water International, American Water Resources Association.

Guo, J.C.Y., and Hughes, W. (2001). "Runoff Storage Volume for Infiltration Basin," ASCE Journal of Irrigation and Drainage Engineering, Vol. 127, No. 3, pp. 170–174.

Hantuch, M.S. (1967). "Growth and Decay of Groundwater-Mounds in Response to Uniform Percolation," Water Resources Research, Vol. 3, No. 1, pp. 227–234.

McDonald, M.G., and Harbaugh, A. (1984). "A Modular Three-Dimensional Finite-Difference Groundwater Flow Model," U.S. Department of the Interior, USGS, National Center, Reston, VA.

Morel-Seytous, H.J., Miracapillo, C., and Abdulrazzak, M.J. (1990). "A Reductionist Physical Approach to Unsaturated Aquifer Recharge from a Circular Spreading Basin." Water Resources Research, Vol. 26, No. 4, pp. 771–777.

Ortiz, N.V., Zachmann, D.W., McWhorter, D.B., and Sunada, D.K. (1979). "Effects of In-Transit Water on Groundwater Mounds beneath Circular and Rectangular Recharge Areas," Water Resources Research, Vol. 15, No. 3, pp. 577–582.

Rastogi, A.K., and Pandy, S.N. (1998). "Modeling of Artificial Recharge Basins of Different Shapes and Effect on Underlying Aquifer System," ASCE Journal of Hydrologic Engineering, Vol. 123, No. 3, pp. 62–68.

Shansai, A., and Sitar, N. (1991). "Method for Determination of Hydraulic Conductivity in Unsaturated Porous Media," ASCE Journal of Irrigation and Drainage Engineering, Vol. 117, No. 1, pp. 64–78

Sumner, D.M., Rolston, D.E., and Marino, M.A. (1999). "Effects of Unsaturated Zone on Groundwater Mounding," ASCE Journal of Hydrologic Engineering, Vol. 4, No. 1, pp. 65.

Swanee, P.K., and Ojha, C.S.P. (1997). "Ground-Water Mound Equation for Rectangular Recharge Area," ASCE Journal of Irrigation and Drainage Engineering, pp. 215–217.

Chapter 19

Hydraulic routing

When the flow rate in a water system varies with respect to time, the flow is termed *unsteady flow*. An unsteady flow in a river, reservoir, and an estuary is propagated in the form of long waves, including tidal waves, flood waves, storm surges, etc. The motion of a long wave in a river system is best described by Saint-Venant equations.

From the hydraulic point of view, the propagation of flood waves in a river is a typical example of *spatially varied flow*. In a well-defined channel, the unsteady flow can be numerically simulated by the principles of momentum and continuity. The mathematical algorithm solving the governing equations of flood wave propagation is termed *flood routing*. A flood routing scheme is aimed at predicting or forecasting the properties of flood wave, including flow rate, wave speed, and height. There are many flood routing methods developed to solve unsteady flows. Using the concept of hydrologic system, the motion of flood wave is modeled with three elements: *input, output, and throughput*. Input parameters to a flow system include the inflow hydrograph at the entrance. Throughput parameters are associated with the hydraulic characteristics of the river system such as conveyance geometry, floodplain roughness, bridge structures, etc. The operation of flood routing scheme is to predict the outflow hydrograph at the exit of the reach. Although many flood routing methods have been developed, they can be classified into two categories as follows:

1. Hydrologic routing
   A *hydrologic routing method* is formulated to solve the mass balance of inflow, outflow, and volume of storage using the continuity equation. Hydrologic routing is suitable for a water system that has a significant storage volume. A typical example is to transport a flood wave through a reservoir. A hydrologic routing method requires the *stage–storage–discharge* relation to determine the outflow at each time step. Applications of hydrologic routing methods include simulation of overland flows from a watershed, prediction of flood flows through a stormwater detention basin, operations of a reservoir, etc.
2. Hydraulic routing
   Hydraulic flood routing often applies to a water system without a significant storage volume, such as a channel reach or river network. A hydraulic flood routing method is to solve the governing equations of continuity and momentum simultaneously by numerical schemes. A hydraulic routing method applies either the explicit or the implicit numerical method to solve the finite difference equations that govern the flow motion. The selection of numerical method depends on the flow regime and the level of accuracy.

This chapter summarizes the concepts and procedures used in *hydraulic flood routing*. Overland flows are employed as an example to illustrate the derivations of both analytical and numerical solutions.

19.1 Continuity principle

An open-channel flow is three-dimensional in nature. When the longitudinal velocity dominates the flow characteristics, an open-channel flow can be simplified by the one-dimensional approach. The continuity principle for an unsteady flow in a channel has to count for the storage change between two adjacent channel sections. The water surface rises when the inflow is greater than the outflow, or vice versa. The *continuity principle* states

$$\frac{\partial A}{\partial t} + \frac{\partial Q}{\partial x} = q_i \tag{19.1}$$

in which Q = flow in channel in $[L^3/T]$, A = flow area in $[L^2]$, q_i = lateral inflow per unit length of the reach in $[L^2/T]$, x = distance in the flow direction in $[L]$, and t = time in $[T]$. Equation 19.1 applies to an open system. In a closed system such as a sewer pipe between two manholes, any change in the flow rate can only be added to or diverted from the entrance and exit. Therefore, the lateral inflow, q, in Equation 19.1 becomes zero as

$$\frac{\partial A}{\partial t} + \frac{\partial Q}{\partial x} = 0 \tag{19.2}$$

For a steady and uniform flow, Equation 19.2 is reduced to

$$\frac{\partial A}{\partial t} = 0 \ \text{ and } \ \frac{\partial Q}{\partial x} = 0 \tag{19.3}$$

The solution for Equation 19.3 is that Q = constant everywhere and A = constant all the time.

19.2 Momentum principle

The challenge in the numerical modeling of unsteady flow is the formulation to solve the governing equations of motion. Applying different levels of simplification, four types of unsteady long waves have been identified in channel flows, including *dynamic, quasi-dynamic, diffusive, and kinematic waves*. Among them, dynamic wave (DW) is the most complex flow, whereas kinematic wave (KW) is the simplest. DW lies between KW and DW.

1. DW model

    The equation of motion for DW in a channel states

$$\frac{\partial V}{\partial t} + V\frac{\partial V}{\partial x} + g\frac{\partial y}{\partial x} = g(S_o - S_f) \tag{19.4}$$

in which V = cross-sectional average flow velocity in $[L/T]$, g = gravitational acceleration in $[L/T^2]$, S_o = channel bottom slope in $[L/L]$, and S_f = friction slope in $[L/L]$.

Equation 19.4 depicts the DW movement in a channel that has both local and convective accelerations. The *local acceleration*, $\partial V/\partial t$, is determined by the velocity difference at a specified section over a time interval. It is closely related to the increase or decrease of inflow rate with respect to time. The *convective acceleration*, $\partial V/\partial x$, depicts the velocity difference between two adjacent sections at a specified time. For instance, the variation of the channel width from one section to another can induce a significant convective acceleration to the flow. The hydrostatic term, $\partial y/\partial x$, reflects the backwater effects. The slopes of S_o and S_f represent the *gravitational force* and *friction force*, respectively.

2. **Quasi-dynamic wave model**

    Most flood channels are designed to pass the peak design discharge. During the peak time, the flow rates through the reach remain almost constant. Consequently, the *local acceleration* becomes numerically insignificant. Under this circumstance in which the local acceleration vanishes, Equation 19.4 is reduced to *quasi-dynamic wave* as

    $$V\frac{\partial V}{\partial x} + g\frac{\partial y}{\partial x} = g(S_o - S_f) \tag{19.5}$$

3. **Diffusive wave model**

    When the channel cross sections are nearly uniform in a narrow floodplain, the convective acceleration becomes negligible. As a result, the diffusive wave model is derived as

    $$\frac{\partial y}{\partial x} = (S_o - S_f) \tag{19.6}$$

    Diffusive wave is essentially the same as the *gradually varied flow* in which the flow is steady without a significant acceleration. There are 13 water surface profiles identified by Equation 19.6 as gradually varied flows. For instance, the *backwater profile* on a mild slope is termed *the M1 curve*. The M2 curve is a *drawdown profile* at the end of a mild slope, and S2 curve is the *drawdown profile* at the beginning of a steep slope.

4. **KW model**

    Without much influence from upstream and downstream sections, an open-channel flow behaves like a uniform flow that is dominated by the gravity force and the flow friction. Under the assumption that the body force in the flow direction is balanced by the friction in the flow, Equation 19.5 is further reduced to KW as

    $$S_o = S_f \tag{19.7}$$

    Equation 19.7 implies that the flow rate and depth satisfy a single-valued rating curve as

    $$Q = \alpha A^m \tag{19.8}$$

    in which α = empirical constant and m = empirical exponent depending on the channel cross-sectional geometry. The *normal flow condition* described by *Manning's formula* is a special case of KW in which the flow rate is constant. Because Equation 19.8 does not reflect the downstream backwater effects, solutions for a KW model provide an instantaneous uniform flow for the given discharge at the specified time. At each time step, the two unknowns of *flow depth* and *flow discharge* in a KW model must satisfy both Equations 19.1 and 19.8.

19.3 Applicability limit of kinematic waves

The *monoclinal flood wave* (Figure 19.1) is the simplest wave form that can be produced by a sudden lift of a sluice gate. A sudden increase in the discharge creates a monoclinal wave on the base flow. This is a good analogue of the runoff flow generated from an impervious surface as a quick response to rainfall. An intense rainfall results in a sudden increase of runoff flow. For every time interval, the overland flow is best depicted by a KW.

After adding a negative kinematic wave speed, V_w, to the flow field, the flow is transformed to a steady flow. The continuity principle becomes

$$(V_2 - V_w) A_2 = (V_1 - V_w) A_1 \tag{19.9}$$

Rearranging Equation 19.9 yields

$$V_w = \frac{A_2 V_2 - A_1 V_1}{A_2 - A_1} = \frac{Q_2 - Q_1}{A_2 - A_1} = \frac{dQ}{dA} \tag{19.10}$$

Equation 19.10 implies that a KW flow can be described by a single-valued rating curve as shown in Figure 19.2. In order to have a comparison between dynamic and KWs, Equation 19.10 is further simplified using a wide rectangular channel as

$$C = \frac{dQ}{dA} \approx V_w = \frac{1}{B} \frac{dQ}{dy} \text{ for rectangular channel} \tag{19.11}$$

in which C = flood wave celerity in [L/T], A = flow area in [L^2], B = width for rectangular channel in [L], and V_w = KW speed in [L/T]. The *flood wave celerity* is reduced to the *kinematic wave speed* in a rectangular channel. Substituting Equation 19.8 into Equation 19.11 yields

$$V_w = \alpha \frac{m}{B} y^{m-1} \tag{19.12}$$

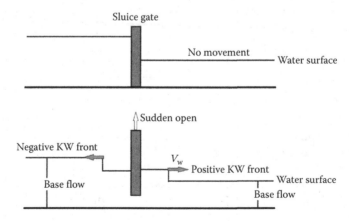

Figure 19.1 Movement of kinematic (monoclinal) wave.

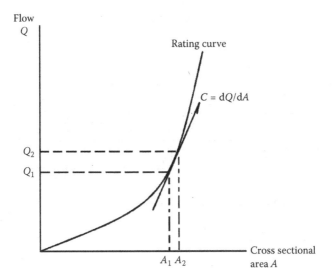

Figure 19.2 Rating curve for kinematic (monoclinal) wave.

The value of m is empirical. For instance, Manning's formula suggests that $m = 2/3$. Chezy's formula suggests that $m = 1/2$. Aided by Equation 19.12, the cross-sectional average velocity, V, is

$$V = \frac{Q}{A} \approx \frac{\alpha y^m}{By} = \frac{V_w}{m} \tag{19.13}$$

in which $V =$ cross-sectional average velocity in $[L^3/T]$. Rearranging Equation 19.13 yields

$$V_w = mV \tag{19.14}$$

Equation 19.14 presents the basic relationship between *KW speed* and *cross-sectional average velocity* when a single-valued rating curve is applicable to the flow. However, the DW speed, V_d, in a wide rectangular channel is

$$V_d = V \pm C \tag{19.15}$$

$$C = \sqrt{gy} \tag{19.16}$$

in which $V_d =$ DW speed in $[L/T]$, "+" = forward wave speed, and "−" = backward wave speed. When a flood wave propagates through a channel, both KW and DW coexist. As illustrated in Figure 19.3, the middle of the flood wave may move at the KW speed, and the wave front may move at the DW speed. In fact, solutions from KW and DW can be similar for the reach that is not significantly affected by backwater effects.

As illustrated in Figure 19.3, the DW propagates in both directions. The DW front in Figure 19.3 is faster than the KW speed.

$$V_d \geq V_w \tag{19.17}$$

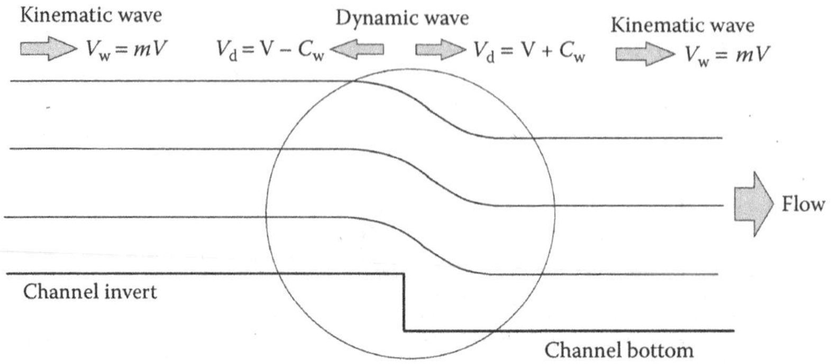

Figure 19.3 Illustration of kinematic and dynamic waves.

Aided with Equations 19.15 and 19.16, Equation 19.17 is converted to

$$\left(V + \sqrt{gy}\right) \geq mV \tag{19.18}$$

Rearranging Equation 19.18 yields

$$F_r = \frac{V}{\sqrt{gy}} \leq \frac{1}{(m-1)} \tag{19.19}$$

Equation 19.19 indicates that the similarity between the DW and the KW in an open-channel flow depends on the flow Froude number. The *flow Froude number* is ≤ 2.0 when $m = 3/2$ as suggested in Chezy's formula, or the *flow Froude number* is ≤ 1.5 when $m = 5/3$ as suggested in Manning's formula (Henderson and Wooding, 1964). Most natural rivers or channels are flat enough to satisfy Equation 19.19. Therefore, it is concluded that the movement of a long flood wave in a river is kinematic in nature when the backwater effect is negligible. For the purpose of runoff modeling, the KW approach is recommended to predict the overland flows generated from watershed and to simulate the motion of channel flow without significant backwater effect.

19.4 Finite difference approach for kinematic flow

To conduct a numerical simulation of rainfall and runoff process, an irregular watershed is converted to its equilibrium rectangular watershed (as shown in Figure 19.4). The rectangular area is further divided into the impervious and pervious plans, according to the watershed's impervious percent, I_a (Details can be found in Chapter 3). The overland runoff hydrograph is produced from a unit-width area under the design rainfall. The channel hydrograph is then produced with the unit-width overland runoff hydrograph as the input to the collector channel. For a unit-width overland flow, Equations 19.1 and 19.8 are converted to

$$\frac{\partial y}{\partial t} + \frac{\partial q}{\partial x} = I_e \tag{19.20}$$

$$q = \alpha_y y^m \tag{19.21}$$

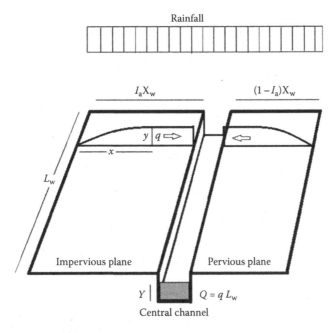

Figure 19.4 Overland and channel flow.

in which y = flow depth for unit-width flow in [L], q = unit-width discharge in [L²/T], α_y = empirical coefficient for unit-width flow, and I_e = net rainfall intensity in [L/T]. For the channel flow, Equations 19.1 and 19.8 are converted to

$$\frac{\partial A}{\partial t} + \frac{\partial Q}{\partial x} = q \tag{19.22}$$

$$Q = \alpha_A A^m \tag{19.23}$$

in which A = flow area in channel in [L²], α_A = empirical coefficient for channel flow, and Q = flow in channel in [L³/T].

The KW model is often employed for overland runoff and channel flow routing. As illustrated in Figure 19.5, a numerical mesh network is formed with I-axis and J-axis. The reach length is divided into segments along the I-axis with a finite distance, dx (or Δx), and the flow time is advanced with an increment of dt (or Δt) to be used in the J-axis. All grids are defined using the coordinates of (i, j).

A finite difference approach provides discrete solutions, $y(i, j)$, at all grids, according to the user-defined initial boundary condition such as dry bed or base flow at all grids at $t = 0$, the downstream boundary condition such as a preknown tailwater condition, and the inflow as inputs to the upstream boundary. The solution of $y(i, j)$ represents the water depth at the jth time step and at the ith station.

19.4.1 Explicit method

At $t = 0$, the initial condition must be provided. At $t = \Delta t$, the inflow is introduced at the upstream boundary. A numerical process is to propagate the inflow at the upstream

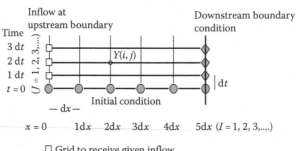

Figure 19.5 Numerical grids and network for flow field.

boundary into the flow field, according to the flow-governing equations. Using the explicit method, the finite difference form of Equation 19.20 is derived as

$$y(i,j+1)=y(i,j)+\frac{\Delta t}{2\Delta x}\Big[q(i-1,j)-q(i+1,j)\Big]+0.5\Big[I_e(i,j+1)+I_e(i,j)\Big]\Delta t \qquad (19.24)$$

in which $i = i$th reach on the plane and $j = j$th time step. Equation 19.24 must be solved in conjunction with the specified rating curve relationship as stated in Equation 19.21. Similarly, Equation 19.22 for channel flow is converted into

$$A(i,j+1)=A(i,j)+\frac{\Delta t}{2\Delta x}\Big[Q(i-1,j)-Q(i+1,j)\Big]+0.5\Big[q(i,j+1)+q(i,j)\Big]\Delta t \qquad (19.25)$$

in which $i = i$th reach along channel. The initial condition for the channel flow can be a dry bed condition, i.e., $y(i, 1) = 0.0$, everywhere when $j = 1$, or a known base flow.

As illustrated in Figure 19.6, Equations 19.24 and 19.25 provide direct solution using the known $y(i, j)$ at the previous time step. An explicit method may become numerically unstable because the gradient in the I-direction, $[y(i + 1, j + 1) - y(i - 1, j + 1)]$, at the $(j+1)$th time step is approximated by $[y(i + 1, j) - y(i - 1, j)]$, which are the solutions at the jth time step. In other words, there is a time lag in the computation convective term.

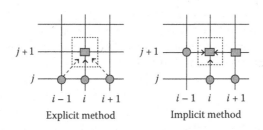

Figure 19.6 Explicit and implicit methods.

19.4.2 Implicit method

The implicit method is an improvement to the explicit method by incorporating the current values into the gradient calculations in the I-direction as

$$y(i,j+1)_{n+1} = y(i,j) + \frac{\Delta t}{2\Delta x}\left[q(i-1,j+1)_n - q(i+1,j+1)_n\right] + 0.5\left[I_e(i,j+1) + I_e(i,j)\right]\Delta t$$

(19.26)

$$A(i,j+1)_{n+1} = A(i,j) + \frac{\Delta t}{2\Delta x}\left[Q(i-1,j+1)_n - Q(i+1,j+1)_n\right] + 0.5\left[q(i,j+1) + q(i,j)\right]\Delta t$$

(19.27)

in which $n = n$th iteration. As illustrated in Figure 19.6, Equation 19.26 does not provide direct solutions because the value at *grid* $(i + 1, j + 1)$ is not yet known when calculating the value at *grid* $(i, j + 1)$. The relaxation method can be applied to Equation 19.26 using successive iterations. To establish the iterative process, the solutions at *grid* $(i + 1, j)$ can serve as the initial guess for the value at *grid* $(i + 1, j + 1)$. With the new input and boundary condition at the $(j + 1)$th time step, Equation 19.26 shall be repeated for all grids. Having the new values calculated by Equation 19.26, all the values marked with a subscript of "n" in Equation 19.26 will be updated with the new values for the next iteration. This process shall be repeated until all values at the $(j + 1)$th time step satisfy the convergence criterion as

$$\frac{\left|y(i,j+1)_{n+1} - y(i,j+1)_n\right|}{y(i,j+1)_n} \leq E$$

(19.28)

in which $n = n$th successive iteration, and $E =$ tolerance for numerical error such as 0.5%. The downstream boundary condition for the channel flow can be one of the following: (1) a known rating curve, (2) normal flow, (3) critical flow, or (4) a fixed tailwater depth.

EXAMPLE 19.1

Referring to Figure 19.7, the watershed has a drainage area of 3.10 acres that can be converted into its equivalent rectangular KW plane with an overflow length of 250 ft. The overland flow rating curve for this watershed is described by $m = 1.70$ and $a = 5.0$. Formulate the explicit finite difference method to determine the overland runoff hydrograph under a rainfall excess of 6.0 in./h for a duration of 20 min.

The rainfall excess of 6.0 in./h is equal to 0.000138 ft/s. Let Δt be 30 s and Δx be 50 ft. A length of 250 ft is divided into five reaches and six stations at $X = 0.0$, 50, 100, 150, 200, and 250 ft. The range of index I increased from I to 6 in Equation 19.24.

Predictions by explicit method

Substituting the rating curve relationship into Equation 19.24, the *explicit finite difference equation* for the case is derived as

$$y(i,j+1) = y(i,j) + 1.50\left[y(i-1,j)^{1.70} - y(i+1,j)^{1.70}\right] + 0.00414$$

(19.29)

Computational mesh network

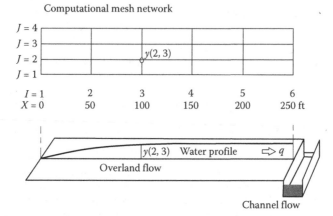

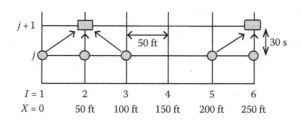

Figure 19.7 Equivalent rectangular watershed for case study.

Figure 19.8 Boundary condition for explicit method.

The initial condition includes $y(i, 1) = 0.0$ everywhere. The upstream boundary condition is

$$y(1, j) = 0.0 \text{ at } x = 0.0 \tag{19.30}$$

The downstream boundary condition is determined by two points as illustrated in Figure 19.8

$$y(6, j+1) = y(6, j) + 3.0\left[y(5, j)^{1.70} - y(6, j)^{1.70}\right] + 0.00414 \text{ at } x = 250 \text{ ft or } i = 6 \tag{19.31}$$

After the rain ceases at 20 min, Equations 19.29 and 19.31 become

$$y(i, j+1) = \max\left\{0, y(i, j) + 1.50\left[y(i-1, j)^{1.70} - y(i+1, j)^{1.70}\right]\right\}$$
$$\text{after } t > 20 \text{ min} \tag{19.32}$$

$$y(6, j+1) = \max\left\{0, y(6, j) + 3.0\left[y(5, j)^{1.70} - y(6, j)^{1.70}\right]\right\}$$
$$\text{after } t > 20 \text{ min} \tag{19.33}$$

The relationship among the grids used in the explicit method is illustrated in Figure 19.8.

Table 19.1 presents the iterative process using the explicit finite difference method for this case. The time of equilibrium for this case is approximately 6 min. As shown in Figure 19.9, during the first 6 min, the flow on the rising hydrograph increases. From 6 to 20 min, the flow is peaking at the equilibrium discharge of 0.0347 cfs/ft. The predicted

Table 19.1 Predicted overland runoff hydrograph by explicit method

Time min	Upstream X = 0.0	Station (ft) 50.0	100.0	150.0	200.0	Downstream 250.0	Overland Flow
		Flow depth × 1000 (ft)					
	$y(t)$ (0)	$y(t)$ (1)	$y(t)$ (2)	$y(t)$ (3)	$y(t)$ (4)	$y(t)$ (5)	$q(t) \times 1000$ cfs/ft
0.00	0.00	0.00	0.00	0.00	0.00	0.00	0.00
0.50	0.00	4.17	4.17	4.17	4.17	4.17	0.53
1.00	0.00	8.17	8.33	8.33	8.33	8.33	1.70
1.50	0.00	11.83	12.48	12.50	12.50	12.50	3.33
2.00	0.00	15.00	16.56	16.66	16.67	16.67	5.39
2.50	0.00	17.57	20.47	20.81	20.83	20.83	7.81
3.00	0.00	19.46	24.06	24.91	25.00	25.00	10.59
3.50	0.00	20.64	27.16	28.88	29.14	29.17	13.69
4.00	0.00	21.16	29.59	32.60	33.25	33.33	17.10
4.50	0.00	21.12	31.22	35.86	37.23	37.47	20.79
5.00	0.00	20.69	31.99	38.46	40.95	41.57	24.71
5.50	0.00	20.06	31.97	40.19	44.22	45.55	28.78
6.00	0.00	19.45	31.33	40.93	46.76	49.30	32.84
6.50	0.00	18.99	30.36	40.70	48.30	50.98	34.72
7.00	0.00	18.76	29.38	39.74	49.20	50.98	34.72
7.50	0.00	18.77	28.64	38.25	49.83	50.98	34.72

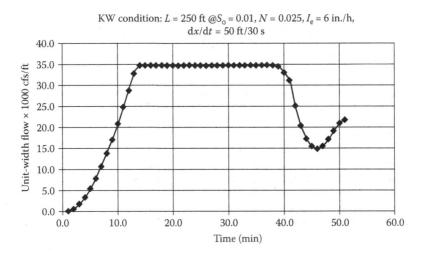

KW condition: $L = 250$ ft @$S_0 = 0.01$, $N = 0.025$, $I_e = 6$ in./h, $dx/dt = 50$ ft/30 s

Figure 19.9 Predicted overland hydrograph by explicit method.

peak flow agrees with the rational method applied to a unit-width area of 250 ft with a runoff coefficient, C_i,

$$q_p = C_i IL = 1.0 \times 6.0 \times \frac{1}{12 \times 3600} \times (1 \times 250) = 0.347\,\text{cfs/ft} \tag{19.34}$$

After the rainfall stops, the numerical term of rainfall excess is removed from the governing equations. Equations 19.33 and 19.34 are used to calculate the flow depth in the recession flow. Numerical instability is introduced to the tail of the hydrograph when the water depth is very shallow.

Predictions by implicit method

Substituting Equation 19.21 into Equation 19.20, the *implicit finite difference equation* for this case is derived as

$$y(i, j+1)_{n+1} = y(i, j) + 1.50 \left[y(i-1, j+1)_{n+1}^{1.70} - y(i+1, j+1)_n^{1.70} \right] + 0.00414 \tag{19.35}$$

in which $n = n$th iteration. At the boundary, $x = 250$ ft, the backward finite difference equation is written as

$$y(6, j+1)_{n+1} = y(6, j) + 3.0 \left[y(5, j+1)_{n+1}^{1.70} - y(6, j+1)_n^{1.70} \right] + 0.00414 \tag{19.36}$$

Boundary and initial conditions for the implicit approach are the same as stated in the explicit method. Both Equations 19.35 and 19.36 have two unknowns: $y(i, j + 1)$, and $y(i + 1, j + 1)$. The solution can be obtained if we start with substituting $y(i + 1, j)$ for $y(i + 1, j + 1)$. We conduct numerical sweeping from $i = 1$ to 6. At the end of each numerical sweeping, we update the value of $y(i + 1, j + 1) = y(i + 1, j)_n$ in which $n = n$th sweeping. This iterative process will be ended till the convergence criterion in Equation 19.28 is satisfied. Table 19.2 presents an example of the iterative procedure for the implicit method. The solutions for flow depth, $y(i, j)$, at $t = 4.5$ min serve as the initial value in

Table 19.2 Iteration procedure used in implicit method

No of iteration	B.C. X = 0.0	1 50.0	2 100.0	3 150.0	4 200.0	5 250.0	ID Number of station / Locations of stations (ft)
	y(t)	y(t)	y(t)	y(t)	y(t)	y(t)	Flow depth × 1000 ft
0.0	0.000	19.086	27.344	32.281	34.894	36.578	Solution of y(i,j) at t = 4.5 min
1	0.000	19.456	28.657	34.852	38.583	41.318	
2	0.000	19.147	27.918	33.638	36.868	39.437	
3	0.000	19.322	28.276	34.214	37.587	40.206	
4	0.000	19.237	28.107	33.970	37.292	39.894	
5	0.000	19.278	28.179	34.071	37.413	40.021	
6	0.000	19.260	28.149	34.029	37.364	39.969	
7	0.000	19.267	28.162	34.046	37.384	39.990	
8	0.000	19.265	28.157	34.039	37.376	39.982	
9	0.000	19.266	28.159	34.042	37.379	39.985	
10	0.000	19.265	28.158	34.041	37.378	39.984	Solution of y(i,j) at t = 5.0 min

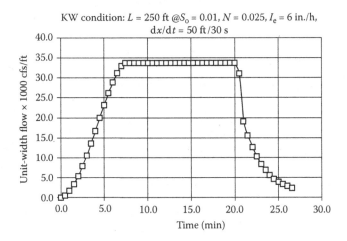

KW condition: $L = 250$ ft @$S_o = 0.01$, $N = 0.025$, $I_e = 6$ in./h, $dx/dt = 50$ ft/30 s

Figure 19.10 Overland hydrograph predicted by implicit method.

Equation 19.35. It takes 10 iterations to update the solution at $t = 4.5$ min to achieve the solution at $t = 5$ min.

Figure 19.10 presents the computed hydrograph by the implicit method. The $\Delta x/\Delta t = 50/30$ fps. Both the explicit and implicit methods produce a similar rising hydrograph and peak flow at the time of equilibrium. In comparison, the implicit method is consistently more stable from the rising to the recession. For this case, the recession hydrograph is a smooth drawdown curve from the peak flow to the dry bed. But the tradeoff is the additional computational efforts. The implicit method generally provides reliable and stable solutions with good agreements to the rational method.

19.5 Characteristic wave method for dynamic wave flows

Although flood flows are kinematic in nature, the KW method does produce erroneous predictions when the downstream condition is controlled by a dam. Both backwater and surge effects must be modeled by the DW approach. The *characteristic wave method* is an algorithm derived to provide numerical DW solutions. To simplify the mathematical details, a rectangular cross section is employed as an example to explain how the characteristic wave method is derived. Conclusions from a rectangular section can be expanded into channels with different cross sections. In a rectangular channel, the continuity equation for a unit-width is written as

$$\frac{\partial y}{\partial t} + \frac{\partial q}{\partial x} = \frac{\partial y}{\partial t} + y\frac{\partial V}{\partial x} + V\frac{\partial y}{\partial x} = 0 \tag{19.37}$$

The momentum equation is

$$\frac{\partial V}{\partial t} + V\frac{\partial V}{\partial x} + g\frac{\partial y}{\partial x} = g\left(S_o - S_f\right) \tag{19.38}$$

Adding Equation 19.38 to Equation 19.37 multiplied by $\sqrt{g/y}$ yields the following:

$$\left[\frac{\partial V}{\partial t}+(V+C)\frac{\partial v}{\partial x}\right]+\sqrt{\frac{g}{y}}\left[\frac{\partial y}{\partial t}+(V+C)\frac{\partial y}{\partial x}\right]=H \tag{19.39}$$

in which H is defined as

$$H=g(S_o-S_f) \tag{19.40}$$

The wave celerity in a rectangular channel is defined as

$$C=\sqrt{gy} \tag{19.41}$$

Equation 19.39 is the sum of two total derivatives for variables V and y as

$$\frac{dV}{dt}+\sqrt{\frac{g}{y}}\frac{dy}{dt}=H \tag{19.42}$$

and

$$\frac{dx}{dt}=V+C \tag{19.43}$$

Aided by Equations 19.41, Equation 19.42 can further be converted to

$$\frac{d}{dt}(V+2C)=H \tag{19.44}$$

Equations 19.43 and 19.44 represent the *forward characteristic wave*. Similarly, Equation 19.38 subtract Equation 19.37 multiplied by $\sqrt{g/y}$ yields the *backward characteristic wave* as

$$\frac{dx}{dt}=V-C \tag{19.45}$$

$$\frac{d}{dt}(V-2C)=H \tag{19.46}$$

Equations 19.43 through 19.46 are the governing equations for characteristic wave method. To apply the characteristic wave method to a channel flow, the friction loss is computed by the flow velocity and hydraulic depth as

$$S_f=\frac{N^2V|V|}{(k_n)^2R^{\frac{4}{3}}} \tag{19.47}$$

in which S_f = friction slope in [L/L], N = Manning's roughness coefficient, V = flow velocity in [L/T], R = hydraulic radius, and k_n = 1.486 in ft-s or 1.0 in m-s. As illustrated in Figure 19.11, two waves exist at a grid: *forward* and *backward* waves. Both

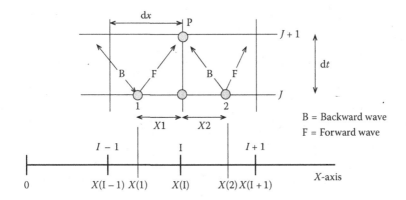

Figure 19.11 Travel distance for characteristic waves.

waves propagate upstream and downstream independently at their own wave speeds, which are varied with respect to the local flow depth and velocity. When the two waves cross each other at an intersection, numerically both waves produce exactly the same flow depth and velocity at the intersection point. As illustrated in Figure 19.11, the characteristic wave method produces discrete solutions at the preselected meshes and grids. For instance, let grid P be a wave-crossing point where the backwater wave from point 2 meets the forward wave from point 1. Points 1 and 2 are not exactly located on the grids; therefore, an interpolation is needed to determine the flow parameters at point 1 between grids $(i-1, j)$ and (i, j) and similarly at point 2 between grids (i, j) and $(i+1, j)$. The schemes of interpolation and extrapolation are employed for numerical calculations.

During each time step, Δt, X1 is the travel distance of the forward wave. Considering the average velocity between grids $(i-1, j)$ and (i, j) as the flow velocity, Equation 19.43 is converted into

$$X1 = X(i) - X(1) = 0.5 \times \left[V(i,j) + C(i,j) + V(i-1,j) + C(i-1,j) \right] \Delta t \qquad (19.48)$$

in which X1 = travel distance upstream of grid (i, j), $X(1)$ = location of point 1, and $X(i)$ = location of grid (i, j).

Similarly, Equation 19.45 is converted to

$$X2 = X(2) - X(i) = 0.5 \times \left[V(i,j) - C(i,j) + V(i+1,j) - C(i+1,j) \right] \Delta t \qquad (19.49)$$

in which X2 = travel distance downstream of grid (i, j) and $X(2)$ = location of point 2. The values of flow variable, such as velocity, depth, or discharge, on points 1 and 2 can also be calculated by interpolation, using the distance as the weighting factor as

$$\Phi(1) = \Phi(i-1,j) \frac{X1}{\Delta x} + \Phi(i,j) \frac{\Delta x - X1}{\Delta x} \qquad (19.50)$$

$$\Phi(2) = \Phi(i+1,j) \frac{X2}{\Delta x} + \Phi(i,j) \frac{\Delta x - X2}{\Delta x} \qquad (19.51)$$

in which Φ = flow variable. Having identified locations of points 1 and 2, the finite difference equations for the forward and backward wave fronts are derived from Equations 19.44 and 19.46 by the explicit method as

$$\{[V+2C](i,j+1)-[V+2C](1,j)\}=H(1,j)\Delta t \tag{19.52}$$

$$\{[V-2C](i,j+1)-[V-2C](2,j)\}=H(2,j)\Delta t \tag{19.53}$$

in which $[V+2C](i,j+1)$ = the value of $(V+2C)$ at grid $(i,j+1)$, etc. The two unknowns, $V(i,j+1)$ and $C(i,j+1)$, can be solved as

$$C(i,j+1)=\frac{1}{4}\Big[(V+2C)(1,j)-(V-2C)(2,j)+H(1,j)\Delta t+H(2,j)\Delta t\Big] \tag{19.54}$$

$$V(i,j+1)=\frac{1}{2}\Big[(V+2C)(1,j)+(V-2C)(2,j)+H(1,j)\Delta t-H(2,j)\Delta t\Big] \tag{19.55}$$

Care must be taken during the numerical computations when using the characteristic wave method, because wave speeds change with the flow regime. For instance, in a supercritical flow, both backward and forward waves will travel downstream as illustrated in Figure 19.12.

During computations, Equations 19.48 and 19.49 shall be used to detect if the wave travel distance becomes greater than Δx. Figure 19.13 shows several possible interpolations

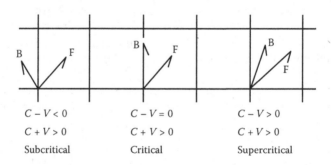

$C-V<0$	$C-V=0$	$C-V>0$
$C+V>0$	$C+V>0$	$C+V>0$
Subcritical	Critical	Supercritical

Figure 19.12 Characteristic waves and flow regime.

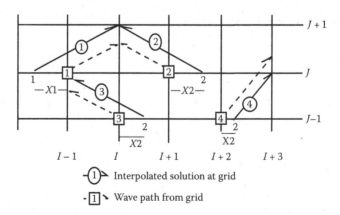

-①⤻ Interpolated solution at grid

-☐⤻ Wave path from grid

Figure 19.13 Interpolation between grid points.

or extrapolations. Over a period of Δt, as shown in case 1 in Figure 19.13, the forward wave issued from grid $(i-1, j)$ did not reach grid $(i, j+1)$. As a result, we need a distance of $X1$ to locate point 1. Use the flow properties at point 1 to send a forward wave that will be able to reach grid $(i, j+1)$. In case 2, we need a distance of $X2$ to locate point 2. Apply Equations 19.54 and 19.55 to points 1 and 2 to determine the flow variables: U and C at grid $(i, j+1)$. Similarly, cases 3 and 4 will need a distance of $X2$ to have two waves meet at grids.

EXAMPLE 19.2

A wide rectangular channel is 3000 ft long. The channel slope is 0.005 ft/ft and has a Manning's roughness of 0.02. To model the flow by the characteristic wave method, this channel is evenly divided into three reaches, i.e., $\Delta x = 1000$ ft in Figure 19.14. Node 1 represents the upstream boundary where the inflow comes into the channel. Node 4 is the downstream outlet for the channel. The unit-width flow condition at time t is listed in Table 19.3. The upstream inflow and downstream outflow at $t + \Delta t$ are also given in Table 19.3. Determine the flow conditions at $t + \Delta t$ and $\Delta t = 30$ s.

Solution: As shown in Table 19.3, the solutions at time t are already known. At $t + \Delta t$, the upstream inflow is given at node 1 and the downstream outflow is also specified by a rating curve at node 4. For this case, the task is to determine the flow depths and discharges at nodes 2 and 3. As a wide channel flow, the hydraulic radius is replaced with the flow depth. At node 2, the variable H is calculated as

$$S_f = \frac{0.02^2 \times 10.15^2}{1.486^2 \times 2.75^{\frac{4}{3}}} = 0.00482 \ \left(\text{see Equation 19.47}\right)$$

$$H = 32.2 \times (0.005 - 0.00482) = 0.00583 \, \text{ft} \ \left(\text{see Equation 19.40}\right)$$

The locations of points $X1$ and $X2$ are calculated as

$$X1 = 0.5 \times [21.64 + 19.56] \times 30 = 618.01 \, \text{ft} \ \left(\text{see Equation 19.48}\right)$$

$$X2 = 0.5 \times [0.74 + 0.48] \times 30 = 18.30 \, \text{ft} \ \left(\text{see Equation 19.49}\right)$$

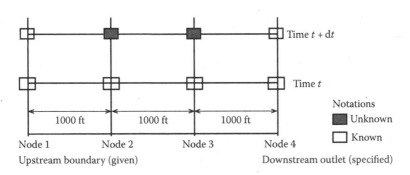

Figure 19.14 Example nodes and links for characteristic wave method.

Table 19.3 Example for characteristic wave method

Variable	Node 1 Upstream inflow (given)	Node 2 X1	Reach 1	X2	Node 3 X1	Reach 2	X2	Node 4 Downstream outflow (specified)
Y (ft)	3.25		2.75			2.45		2.25
q (cfs/ft)	37.08		27.92			22.94		19.85
V (fps)	16.41		10.15			9.36		8.82
C (fps)	10.23		9.41			8.88		8.51
S_f (ft/ft)	0.00487		0.00482			0.00478		0.00476
H (ft/s²)	0.00409		0.00583			0.00702		0.00789
V + C (fps)	21.64		19.56			18.24		17.33
V − C (fps)	1.18		0.74			0.48		0.31
Distance (ft)		618.01		18.32	273.67		7.21	
V + 2C (fps)	31.87	30.76	28.97		27.63	27.13		25.84
V − 2C (fps)	−9.05		−8.67	−8.66			−8.41	−8.21
Weighted H		0.00476		0.00585	0.00669		0.00703	
Solutions:								
V (fps)	16.41		16.05			9.61		8.82
C (fps)	10.23		9.86			9.11		8.51
Y (ft)	3.25		3.02			2.58		2.25

For this case, the forward wave speed is issued at the point where 618.01 ft is upstream of node 2, and the backward wave is issued at the point where 18.3 ft is downstream of node 2. The weighted value for $(V + 2C)$ at $X1$ is determined as

$$(V + 2C) = 31.87 \times \frac{618.01}{1000} + 28.97 \times \frac{(1000 - 618.01)}{1000} = 30.76 \, \text{fps} \quad (\text{see Equation 19.52})$$

The weighted value for $(V - 2C)$ at $X1$ is determined as

$$(V - 2C) = -8.41 \times \frac{18.32}{1000} + (-8.67) \times \frac{(1000 - 18.32)}{1000} = -8.66 \, \text{fps} \quad (\text{see Equation 19.53})$$

Therefore, at $t + 30$ s, the flow condition at node 2 is

$$C = \frac{1}{4} \left[30.76 - (-8.66) + 0.00476 + 0.00585 \right] = 9.86 \, \text{fps} \quad (\text{see Equation 19.54})$$

$$V = \frac{1}{2} \left[30.76 - (-8.66) + 0.00476 + 0.00585 \right] = 11.05 \, \text{fps} \quad (\text{see Equation 19.55})$$

$$Y = \frac{C^2}{g} = \frac{9.86^2}{32.2} = 3.02 \, \text{ft}$$

Applying the same procedure to node 3, the solutions for node 3 are listed in Table 19.3. As expected, both water depths and discharges at nodes 2 and 3 are increased as the rising flood wave enters the reach.

As illustrated in the example, the characteristic wave method converts the partial differential governing equations into a set of algebraic equations. Mathematically, this method is rather straight forward for a subcritical flow. With a high Froude number flow, the selections of Δx and Δt can be critically important for determining the positions of $X1$ and $X2$. When a hydraulic jump occurs in a reach, the characteristic wave method may not provide converged solutions. Often, empirical formula or momentum approach can be an alternative to reach reasonable predictions.

19.6 Numerical weighting method

The finite difference method for the local term in a partial differential equation is often expressed with a *forward difference* as $[y(i, j + 1) - y(i, j)]$, while the convective term can be translated into *backward difference* as $[y(i, j) - y(i - 1, j)]$, *forward difference* as $[y(i + 1, j) - y(i, j)]$, or *central difference* as $[y(i + 1, j) - y(i - 1, j)]$. Referring to Figure 19.15, a cell in a numerical mesh network is formed by four grids. A cell in the flow field is a control volume confined by the four corners. The solution for this control volume shall be weighted with the factors e and r, which are applied to Δx and Δt used in computations. Both weighting factors follow the rule: $0 \le r \le 1$ and $0 \le e \le 1$. Using the weighting factors, both the implicit and explicit numerical methods can be merged into a *finite difference equation for control volume* as (Preissmann, 1961)

$$\frac{r(A_2 - A_1) + (1 - r)(A_4 - A_3)}{\Delta t} + \frac{e(Q_3 - Q_1) + (1 - e)(Q_4 - Q_2)}{\Delta x} = q \qquad (19.56)$$

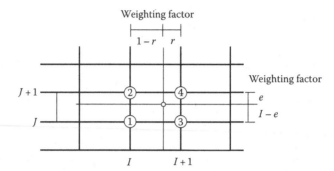

Figure 19.15 Computation cell for weighting finite difference scheme.

Equation 19.56 is the generalized finite difference equation that can be converted into

1. Implicit method when $e = 0.0$,

$$\frac{r(A_2 - A_1) + (1-r)(A_4 - A_3)}{\Delta t} + \frac{(Q_4 - Q_2)}{\Delta x} = q \qquad (19.57)$$

2. Explicit method when $e = 1.0$,

$$\frac{r(A_2 - A_1) + (1-r)(A_4 - A_3)}{\Delta t} + \frac{(Q_3 - Q_1)}{\Delta x} = q \qquad (19.58)$$

3. Backward finite difference when $r = 0.0$,

$$\frac{(A_4 - A_3)}{\Delta t} + \frac{e(Q_3 - Q_1) + (1-e)(Q_4 - Q_2)}{\Delta x} = q \qquad (19.59)$$

4. Forward finite difference when $r = 1.0$,

$$\frac{(A_2 - A_1)}{\Delta t} + \frac{e(Q_3 - Q_1) + (1-e)(Q_4 - Q_2)}{\Delta x} = q \qquad (19.60)$$

5. Central finite difference when $e = 0.50$ and $r = 0.50$,

$$\frac{0.5(A_2 - A_1) + 0.5(A_4 - A_3)}{\Delta t} + \frac{0.5(Q_3 - Q_1) + 0.5(Q_4 - Q_2)}{\Delta x} = q \qquad (19.61)$$

In comparison, the central finite difference offers an overall best estimate that represents a cell unless a special arrangement is required to balance the numerical diffusive effects.

19.7 Numerical stability

There are two major concerns in a numerical study: *numerical stability* and *numerical diffusion*. Both are directly related to the wave movement. To see the role of wave celerity in the flow movement, we shall rearrange Equation 19.22 to include the wave celerity. Referring to Figure 19.2 and aided with Equation 19.11, the first term in Equation 19.22 can be converted into

$$\frac{\partial A}{\partial t} = \frac{dA}{dQ}\frac{\partial Q}{\partial t} = \frac{1}{C}\frac{\partial Q}{\partial t} \tag{19.62}$$

Substituting Equation 19.62 into Equation 19.23 yields the following:

$$\frac{\partial Q}{\partial t} + C\frac{\partial Q}{\partial x} = Cq \tag{19.63}$$

Referring to Equation 19.61, the central finite difference form for Equation 19.63 is as follows:

$$[0.5(Q_2 - Q_1) + 0.5(Q_4 - Q_3)] + \frac{C\Delta t}{\Delta x}[0.5(Q_3 - Q_1) + 0.5(Q_4 - Q_2)] = Cq\Delta t \tag{19.64}$$

Equation 19.64 reveals that the numerical stability of Equation 19.64 is directly related to the ratio of wave celerity and $\Delta t/\Delta x$. The *Courant Number*, C_r, was derived as a criterion for numerical stability and defined as

$$C_r = \frac{C\Delta t}{\Delta x} \le 1.0 \tag{19.65}$$

Figure 19.16 presents a numerical test on various Courant Numbers used in Equation 19.64. Using the ratio of the maximum flow predicted by the finite difference method to the peak flow by the rational method as the basis, the explicit computation begins to become numerically diverged upon the Courant Number >0.6, while the implicit computation remains stable until the Courant Number >1.0.

In practice, the spacing, Δx, used in the numerical mesh network is recommended to be the following:

$$\Delta x \ge \varepsilon\, C\Delta t \tag{19.66}$$

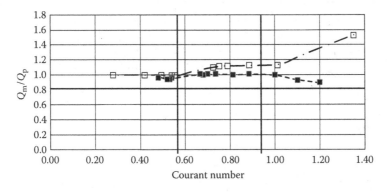

−□− Explicit method −■− Implicit method

Figure 19.16 Comparisons between implicit and explicit methods.

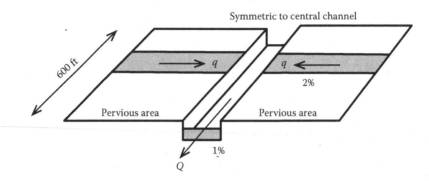

Figure Q19.1 Watershed for developing kinematic wave flow hydraulic models.

Basically, the Courant criterion suggests that the numerical speed, $\Delta x/\Delta t$, shall be faster than the wave celerity. The recommended value of ε is between 1.0 and 1.2 for the implicit method, or between 2.0 and 2.5 for the explicit method.

19.8 Homework

Q19.1 An urban watershed in Figure Q19.1 has a drainage area of 10.0 acres. The central channel has a length of 600 ft on a slope of 0.01 ft/ft. The overland flow planes are on a slope of 0.02 ft/ft. The Manning's N of 0.05 is applied to both overland and channel flows.

The 20-min uniform rainfall intensity of 8.0 in./h is applied to the watershed. The soil infiltration rate is described as

$$f(t) = 0.5 + 4.5e^{-0.12t}$$

in which $f(t)$ = soil infiltration rate (in./h) at time, t, which is the elapsed time in minute.

1. Determine the overland flow length.
2. Establish the constants of α and m on the rating curve for the overland flow.
3. Develop the explicit finite difference model to generate the overland flow hydrograph.
4. Test the numerical stability with Δt = 30, 60, 120, and 240 s.
5. Develop implicit numerical solutions, and repeat the stability test on the $\Delta x/\Delta t$ ratios.
6. Develop the numerical solutions to route the overland hydrographs through the central channel.

Bibliography

Amein, M., and Fang, C.S. (1970). "Implicit Flood Routing in Natural Channels," *ASCE Journal of Hydraulic Engineering Division*," Vol. 96, No. HY12, pp. 2481–2500.

Bedient, P.B., and Huber, W.C. (1992). *Hydrology and Floodplain Analysis*, 2nd ed., Addison Wesley Publishing Company, New York.

Beuter, E.L., Gaebe, R.R., and Horton, R.E. (1940). "Sprinkled-Plat Runoff and Infiltration Experiments on Arizona Desert Soils," *Eos, Transactions American Geophysical Union*, Vol. 21, pp. 550–558.

Chow, V.T. (1964). *Handbook of Applied Hydrology*, Chapters 17 and 21, McGraw-Hill Book Company, London.

Courant, R., Friedricks, O.H., and Lewy, H. (1928). "Uber die partiellen Differentialgleichungen der Mathematishcen Phisik," *Mathematical Annalen*, Vol. 100, pp. 32–74.

DeVries, J.J., and MacArthur, R.C. (1979). "Introduction and Application of Kinematic Wave Routing Techniques Using HEC-1," Training Document 10, Hydrologic Engineering Center, U.S. Army Corps of Engineers, Davis, CA.

Eagleson, P.S. (1970). *Dynamic Hydrology*, McGraw Hill, New York, pp. 344–346.

Fread, D.L. (1978). "National Weather Service Operation Dynamic Wave Model," Verification of Mathematical and Physical Models in Hydraulic Engineering, *Proceedings of 26th Annual Hydraulics Division, Special Conference*, ASCE, College Park, MA.

Guo, J.C.Y. (1998). "Overland Flow on a Pervious Surface," *IWRA Journal of International Water*, Vol. 23, pp. 91–96.

Henderson, F.M., and Wooding, R.A. (1964). "Overland Flow and Groundwater from a Steady Rainfall of Finite Duration," *Journal of Geophysical Research*, Vol. 69, No. 8, pp. 39–67.

Holden, A.P., and Stephenson, D. (1988). "Improved Four-Point Solution of the Kinematic Equations," *Journal of Hydraulic Research*, Vol. 26, No. 4, pp. 413–423.

Holden, A.P., and Stephenson, D. (1995). "Finite Difference Formulations of Kinematic Equations," *Journal of Hydraulic Engineering, ASCE*, Vol. 121, pp. 423–426.

Horton, R.E. (1938). "The Interpretation and Application of Runoff Plot Experiments with Reference to Soil Erosion Problems," *Soil Science Society of America Journal*, Vol. 3, pp. 340–349.

Huang, Y.H. (1978). "Channel Routing by Finite Difference Method," *ASCE Journal of Hydraulics Division*, Vol. 104, No. 10, pp. 1379–1393.

Izzard, C.F. (1944). "The Surface Profile of Overland Flow," *Transactions, American Geophysical Union*, Vol. 25, No. 6, pp. 959–968.

Izzard, C.F. (1946). "Hydraulics of Runoff from Developed Surfaces," Proceedings of the Twenty-Sixth Annual Meeting of the Highway Research Board Held at Washington, DC. December 5–8, 1946, Highway Research Board Proceedings, Vol. 26, pp. 129–146.

Katz, D.M., Watts, F.J., and Burroughs, E.D. (1995). "Effects of Surface Roughness and Rainfall Impact on Overland Flow," *Journal of Hydraulic Engineering, ASCE*, Vol. 121, No. 7, pp. 546–553.

Liggett, J.A., and Woohiser, D.A. (1967). "Finite Difference Solution for the Shallow Water Equations," *ASCE Journal of Engineering Mechanics Division*, Vol. 93, No. 2, pp. 39–71.

Lighthill, M.H., and Whitham, G.B. (1955). "On Kinematic Waves, I, Flood Movement in Long Rivers," *Proceedings of the Royal Society of London, Series A*, Vol. 229, pp. 281–316.

McCuen, R.H., Wong, S.L., and Rawls, W.J. (1984). "Estimating Urban Time of Concentration," *Journal of Hydraulic Engineering, ASCE*, Vol. 110, No. 7, pp. 887–904.

Morgali, J.R. (1970). "Laminar and Turbulent Overland Flow Hydrographs," *Journal of Hydraulic Engineering, ASCE*, Vol. 96, No, HY2, pp. 441–360.

Morgali, J.R., and Linseley, R.K. (1965). "Computer Analysis of Overland Flow," *Journal of Hydraulic Engineering, ASCE*, Vol. 91, No. HY3, pp. 81–100.

Overton, D.E., and Meadows, M.E. (1976). *Stormwater Modeling*, Academic Press, New York.

Ponce, V.M., and Yevjevivh, V. (1978). "Muskingum-Cunge Method with Variable Parameters," *ASCE Journal of Hydraulics Division*, Vol. 104, No. 2, pp. 1663–1667.

Preissmann, A. (1961). *Propogation des intumescences dans les canaux et les riveres*, I'e Congres de l'association Française de Culcule, Grenoble, France, pp. 433–442.

Schaake, J.C. Jr., Geyer, J.C., and Knapp, J.W. (1967). "Experimental Examination of the Rational Method," *ASCE Journal of Hydraulic Engineering*, Vol. 93, No. HY6, pp. 353–370.

Singh, V.P., and Cruise, J.F. (1992). "Analysis of the Rational Formula Using a System Approach," *Catchment Runoff and Rational Formula*, edited by B.C. Yen, Water Resources Publication, Littleton, CO, pp. 39–51.

Urban Highway Storm Drainage Model. (1983). "Inlet Design Program," Volume 3, Federal Highway Administration, Report No. FHWA/RD-83/043.

USGS Open File Report 82-873. (1983). "Rainfall-Runoff Data from Small Watersheds in Colorado, October 1977 through September 1980," USGS, Lakewood, CO.

Wooding, R.A. (1965). "A Hydraulic Model for a Catchment-Stream Problem," *Journal of Hydrology*, Vol. 3, pp. 254–267.

Woolhiser, D.A., and Liggett, J.A. (1967). "Unsteady One-dimensional Flow over a Plane—The Rising Hydrograph," *Water Resources Research*, Vol. 3, No. 3, pp. 753–771.

Yen, B.C., and Chow, V.T. (1974). *Experimental Investigation of Watershed Surface Runoff*, Hydraulic Engineer Series, No. 29, Department of Civil Engineering, University of Illinois at Urbana–Champaign.

Chapter 20

Hydrologic routing

Hydraulic routing is a numerical scheme to transport a hydrograph through a well-defined channel. In practice, the change in the channel storage volume has a negligible effect on peak flow attenuation. However, as the storage capacity increases, the flow velocity and energy are diffused into the storage volume. As the convective term in the Saint-Venant's equation becomes numerically negligible, the flow routing process is termed *hydrologic routing*, which is a numerical process to focus on the balance of water volumes among inflow, outflow, and change in storage volume.

20.1 Hydrologic routing

Hydrologic routing is also called *reservoir routing* because it is a numerical procedure to solve the conservation among water volumes flowing through a reservoir. The governing equation for hydrologic routing is essentially the general continuity principle as

$$I - O = \frac{dS}{dt} \tag{20.1}$$

in which I = inflow in [L³/T], O = outflow in [L³/T], S = volume storage in [L³], and t = time in [T]. Applying the finite difference scheme to Equation 20.1 yields

$$\frac{I(t) + I(t + \Delta t)}{2} - \frac{O(t) + O(t + \Delta t)}{2} = \frac{S(t + \Delta t) - S(t)}{\Delta t} \tag{20.2}$$

in which t = at the beginning of the time interval, $t + \Delta t$ = the end of the time interval, and Δt = time interval used in computation. Reservoir routing is a procedure to systematically apply a numerical approach to solve Equation 20.2 between flow rates and storage volumes. All numerical methods for Equation 20.2 require the *stage–storage–outflow (S–S–O) curves*, which can be developed from the reservoir geometry and outlet hydraulics.

EXAMPLE 20.1

As shown in Figure 20.1, a constant inflow of 2.0 cfs is introduced into a circular tank of 10 ft in diameter. The initial water depth in the tank was 2 ft. The outlet is a 6-in. circular pipe laid at the bottom of the tank. Determine the water depth in 5 min.

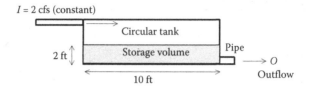

$I = 2$ cfs (constant)

Figure 20.1 Hydrologic routing for tank flow.

Solution: The initial water volume, $S(t)$, in the tank is

$$S(t) = 78.5 \times 2 = 157.0\,\text{ft}^3$$

Beginning with 2 ft of water in the tank, the initial release, O_1, is

$$O(t) = 1.03\sqrt{2} = 1.45\,\text{cfs}$$

With the time interval of $\Delta t = 5\,\text{min}$, $I(t) = I(t + \Delta t) = 2\,\text{cfs}$, Equation 20.2 becomes

$$\frac{2+2}{2} - \frac{1.45 + O(t+\Delta t)}{2} = \frac{S(t+\Delta t) - 157.0}{5 \times 60}\,(\text{continuity equation})$$

Considering that the discharge coefficient is 0.65, the orifice formula is applied to the outlet to establish the *stage–outflow (S–O) curve* as

$$O = 0.65A\sqrt{2gH} = 0.65\frac{\pi \times 0.5^2}{4}\sqrt{2.0 \times 32.2H} = 1.03\sqrt{H}\,(\text{stage–outflow curve})$$

The *stage–storage volume (S–S) curve* in the tank is a function of water depth as

$$S = HA = H\frac{\pi \times 10^2}{4} = 78.5H\,\text{ft}^3\,(\text{stage–storage curve})$$

Combine the S–O and S–S curves into the storage–outflow curve as

$$O = 1.03\sqrt{\frac{S}{78.5}} = 0.116\sqrt{S}$$

At $t + \Delta t$, the two unknowns, $O(t + \Delta t)$ and $S(t + \Delta t)$, shall satisfy

$$O(t+\Delta t) = 0.116\sqrt{S(t+\Delta t)}\,(\text{storage–outflow equation})$$

Now we have two equations for two unknowns. The solutions are $O(t+\Delta t) = 1.87$ cfs and $S(t+\Delta t) = 259.2\,\text{ft}^3$. The water depth, H, at $t + \Delta t$ is found to be 3.3 ft.

The S–S–O curve for an irregular reservoir is more complicated than a simple circular tank. The S–S–O curve has to be obtained by integrating the incremental storage volume between two adjacent topographic contours as

$$\Delta S_{i+1} = 0.5(A_{i+1} + A_i) \times (h_{i+1} - h_i) \tag{20.3}$$

$$S(h) = \sum_{h_0}^{h} \Delta S_i \tag{20.4}$$

in which A = topographic contour area in $[L^2]$, ΔS = incremental storage volume in $[L^3]$, $S(h)$ = accumulated storage volume in $[L^3]$ at stage, h in $[L]$, and i = ith contour at stage, h in $[L]$. Outflow rates at various stages can be calculated by the orifice, weir, and culvert flow formulas. Because both storage volumes and outflows are directly related to stages, the storage–outflow curve can be written as

$$S = a_n O^n + a_{n-1} O^{n-1} + \cdots + \text{constant} = \sum_{m=0}^{m=n} a_{mn} O^m \tag{20.5}$$

where a_n = nth constants derived from regression analysis and m = mth order of polynomial equation. In practice, the third-order polynomial equation, $m = 3$, will be sufficient to describe the storage–outflow curve.

EXAMPLE 20.2

The Belleview Detention Basin (shown in Figure 20.2) is located near the Interstate Highway 470 and W. Belleview Avenue in the City of Denver, CO. This detention system is designed

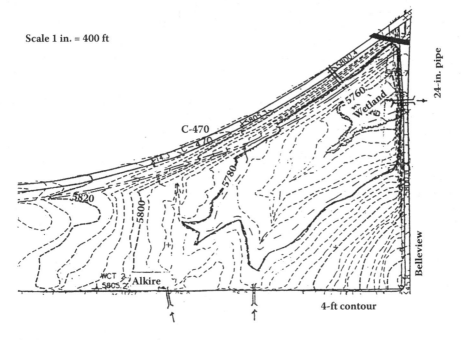

Figure 20.2 Belleview detention basin.

Table 20.1 Stage–storage–outflow relationship for Belleview Basin

Elevation stage (S) (ft)	X-section area (acre)	Cumulative volume (S) (acre-ft)	Water depth (ft)	Orifice flow (cfs)	Weir flow (cfs)	Total outflow (O) (cfs)
5760.0	1.8	0.0	0.0	0.0	0.0	0.0
5764.0	3.5	10.4	4.0	56.8	0.0	56.8
5768.0	5.8	29.0	8.0	86.7	0.0	86.7
5772.0	8.0	56.7	12.0	108.7	0.0	108.7
5774.0	9.6	74.3	14.0	118.2	84.9	203.0
5776.0	11.3	95.2	16.0	126.9	240.0	366.9

to have two 24-in. pipes installed at the bottom of the basin at an elevation of 5760 ft, and another 10-ft weir is installed overtopping the road at an elevation of 5772 ft. The maximum water depth is 16.0 ft. Consider the orifice coefficient of 0.65 and the weir coefficient of 3.0. Determine the S–S–O curve.

Solution: Table 20.1 summarizes the calculations of the S–S–O relationship for this basin. After having the S–S–O curves developed, the selection of a routing method depends on the availability of design information and numerical convenience.

Conduct a regression analysis to develop the empirical equation for the S–O curve defined in Table 20.1 as

$$S = -0.00060^2 + 0.46760 \text{ (correlation coefficient} = 0.94)} \tag{20.6}$$

For this case, $m = 2$. Equation 20.6 is the regression equation to represent the S–O curve for this detention basin. The first term in Equation 20.6 is numerically negligible. As a result, it can be further simplified as a linear equation:

$$S(\text{acre} \times \text{ft}) = 0.46760 \text{ (cfs)} \tag{20.7}$$

Equation 20.7 is useful to provide an approximation during the preliminary design when the detailed information about the reservoir under design is available yet.

20.2 Linear reservoir routing method

A linear reservoir routing method is to assume that the storage and outflow relationship can be linearly related by a constant, K, as

$$S = KO \tag{20.8}$$

Where $K =$ constant in [T]. Aided by Equation 20.8, we have

$$S(t) = KO(t) \tag{20.9}$$

and

$$S(t + \Delta t) = KO(t + \Delta t) \tag{20.10}$$

Substituting Equations 20.9 and 20.10 into Equation 20.2 yields

$$O(t+\Delta t) = C_1\left[I(t+\Delta t)+I(t)\right]+C_2 O(t) \tag{20.11}$$

in which C_1 = inflow coefficient and C_2 = outflow coefficient. They are defined as

$$C_1 = \frac{\Delta t / K}{2+\dfrac{\Delta t}{K}} \tag{20.12}$$

and

$$C_2 = \frac{2-\dfrac{\Delta t}{K}}{2+\dfrac{\Delta t}{K}} \tag{20.13}$$

It is noted that C_1 and C_2 are the weighting factors, and when $\Delta t/K = 0.5$ we have $C_2 = 0$. Aided by Equations 20.12 and 20.13, the outflow can be directly solved by Equation 20.11.

EXAMPLE 20.3

An inflow hydrograph is provided in Table 20.2. Assuming that the Belleview Detention Basin in Figure 20.2 can be described by a linear S–O curve, determine the outflow.

Solution: Set $\Delta t = 5$ min. The value of K in Equation 20.7 needs to be modified on the basis of flow unit in cfs as

$$S\left(\text{ft}^3\right) = 0.4676 \times 43,560\, O\,(\text{cfs}) = 20,368.65\, O\,(\text{cfs})$$

For this case, the value of K is 20,368.65 s. The linear reservoir routing for this case is summarized in Table 20.2.

The calculations in Table 20.2 are verified by the continuity principle in Equation 20.2. For this case, the incoming peak flow of 452 cfs occurs at $t = 35$ min. After the detention process, the peak flow is reduced to 51.99 cfs at $t = 50$ min.

During the final design, all design information is readily available. The performance of the proposed reservoir shall be examined by a more refined routing scheme. For convenience, Equation 20.2 is rearranged as

$$\frac{2S(t+\Delta t)}{\Delta t}+O(t+\Delta t) = \left[I(t)+I(t+\Delta t)\right]+\left[\frac{2S(t)}{\Delta t}-O(t)\right] \tag{20.14}$$

There are two unknowns in Equation 20.14: outflow, $O(t + \Delta t)$, and storage, $S(t + \Delta t)$. At time $t + \Delta t$, both unknowns are related to stage, $h(t + \Delta t)$. There are several methods developed for reservoir routing (Puls, 1928). In this chapter, two new routing methods are introduced to solve Equation 20.14: *storage routing* and *outflow routing* numerical schemes.

Table 20.2 Linear reservoir routing using linear storage–outflow curve

$K = 20,368.000\,s$
$dt = 5.000\,min$
$\Delta t/K = 0.015$
$C_1 = 0.00731$
$C_2 = 0.985$

Time (min)	Inflow (cfs)	Outflow (cfs)	Storage (ft^3)	$I - O$ (cfs)	dS/dt (cfs)
t	$I(t)$	$O(t)$	$S(t)$		
Δt	$I(t + \Delta t)$	$O(t + \Delta t)$	$S(t + \Delta t)$		
	Given	Equation 20.11	Equation 20.10	Left side Equation 20.2	Right side Equation 20.2
0.00	0	0	0.00	0.00	0.00
5.00	3	0.022	446.71	1.49	1.49
10.00	16	0.161	3,269.34	9.41	9.41
15.00	48	0.626	12,751.36	31.61	31.61
20.00	136	1.962	39,963.14	90.71	90.71
25.00	296	5.092	103,705.10	212.47	212.47
30.00	424	10.281	209,399.24	352.31	352.31
35.00	452	16.535	336,776.93	424.59	424.59
40.00	427	22.719	462,738.90	419.87	419.87
45.00	383	28.308	576,584.81	379.49	379.49
50.00	337	33.158	675,364.84	329.27	329.27
55.00	298	37.316	760,043.78	282.26	282.26
60.00	265	40.886	832,763.56	242.40	242.40
65.00	228	43.892	893,996.85	204.11	204.11
70.00	193	46.328	943,613.77	165.39	165.39
75.00	161	48.239	982,528.71	129.72	129.72
80.00	132	49.676	1,011,791.55	97.54	97.54
85.00	108	50.704	1,032,734.65	69.81	69.81
90.00	85	51.373	1,046,373.08	45.46	45.46
95.00	70	51.755	1,054,153.76	25.94	25.94
100.00	60	51.949	1,058,098.10	13.15	13.15
105.00	50	51.994	1,059,006.69	3.03	3.03
110.00	40	51.891	1,056,923.94	-6.94	-6.94
115.00	30	51.644	1,051,893.56	-16.77	-16.77
120.00	20	51.255	1,043,958.67	-26.45	-26.45
125.00	10	50.725	1,033,161.73	-35.99	-35.99
130.00	0	50.056	1,019,544.59	-45.39	-45.39

20.3 Storage routing method

The *storage routing method* is to seek the solutions at time $t + \Delta t$ based on the sum of the storage and the outflow flow volumes at time t. For convenience, Equation 20.14 is rearranged using the unit of flow volumes as

$$\left[I(t) + I(t + \Delta t) - O(t)\right]\Delta t + 2S(t) = O(t + \Delta t)\Delta t + 2S(t + \Delta t) \qquad (20.15)$$

Because the storage volumes and outflow rates of a reservoir are related to its stages, the S–S and S–O curves can be combined into a storage–outflow curve. Let us define the *storage routing function* (SO-function) as

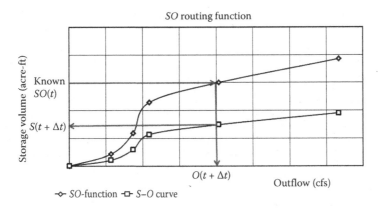

Figure 20.3 Graphic solution using storage routing function.

$$SO = O \times \Delta t + 2S \tag{20.16}$$

The SO-function can be established by the pairs (S, O) described in Table 20.1. Aided by Equation 20.16, Equation 20.15 is divided into the following two portions:

$$SO(t+\Delta t) = \left[I(t) + I(t+\Delta t) - 2O(t)\right]\Delta t + SO(t) \text{(known volume)} \tag{20.17}$$

$$SO(t+\Delta t) = O(t+\Delta t)\Delta t + 2S(t+\Delta t) \text{(solution at } t+\Delta t) \tag{20.18}$$

Figure 20.3 illustrates how solutions can be obtained from the storage routing function and storage–outflow curve. The value of $SO(t + \Delta t)$ is prescribed by the known variables: $I(t)$, $I(t + \Delta t)$, $O(t)$, and $S(t)$. Solutions for the two unknowns, $O(t + \Delta t)$ and $S(t + \Delta t)$, are the pair (S, O) whose SO-function satisfies Equation 20.18. Repeating Equations 20.17 and 20.18 at each time step, the outflow hydrograph can be generated.

20.4 Outflow routing method

The *outflow routing method* is to provide solutions at time, $t + \Delta t$, based on the total outflow rate at time t. Equation 20.14 for this method is rearranged as

$$\frac{2S(t+\Delta t)}{\Delta t} + O(t+\Delta t) = \left[I(t) + I(t+\Delta t)\right] + \left[\frac{2S(t)}{\Delta t} + O(t)\right] - 2O(t) \tag{20.19}$$

Let the *outflow routing function* (OS-function) be defined as

$$OS = \frac{2S}{\Delta t} + O \tag{20.20}$$

It is noted that the outflow routing function has the same unit as flow rate. Aided by Equation 20.20, Equation 20.19 can be divided into two parts as

$$OS(t+\Delta t) = \left[I(t) + I(t+\Delta t) - 2O(t)\right] + OS(t) \text{(know outflow at time } t) \tag{20.21}$$

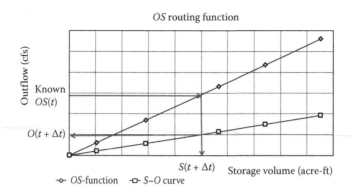

Figure 20.4 Graphic solution using outflow routing function.

and

$$OS(t+\Delta t) = \frac{S(t+\Delta t)}{dt} + O(t+\Delta t)(\text{solution at time } t+\Delta t) \tag{20.22}$$

Figure 20.4 shows that the outflow routing function and storage–outflow curve are related to the storage volume of the reservoir. At each time step, we first compute the value of $OS(t + \Delta t)$ in terms of the variables $I(t)$, $I(t + \Delta t)$, $O(t)$, and $OS(t)$. And the corresponding $S(t + \Delta t)$ and $O(t + \Delta t)$ can then be determined by the functional relationship described in Figure 20.4. Repeating this process, we can advance one Δt at a time until the inflow hydrograph is completely processed.

EXAMPLE 20.4

Considering the time increment of 5 min, derive the SO-function and OS-function for Belleview Basin in Example 20.2.

Solution: Care has to be taken when working on the routing function because the dimensional units have to be consistent. For this case, the two routing functions are formulated using acre-ft and fps as

$$OS = \frac{2S \times 43,560}{5 \times 60} + O \text{ in cfs} \tag{20.23}$$

$$SO = 2S + O \times 5 \times \frac{60}{43,560} \text{ in acre-ft} \tag{20.24}$$

The SO and OS routing functions are developed in Table 20.3.

EXAMPLE 20.5

Repeat Example 20.2 using the OS routing function to determine the outflow hydrograph. The initial condition is at $t = 0$, $O(t) = 0$, and $S(t) = 0$. ·

Table 20.3 SO- and OS-functions for Belleview Basin in Example 20.2

Stage, S (ft)	Storage, S (acre-ft)	Outflow, O (cfs)	SO-function, SO (acre-ft)	OS-function, OS (cfs)
5760.00	0.00	0.00	0.00	0.00
5764.00	10.44	56.77	21.27	3,088.55
5768.00	29.00	86.71	58.60	8,508.31
5772.00	56.70	108.70	114.15	16,574.38
5774.00	74.32	203.02	150.04	21,785.55
5776.00	95.22	366.94	192.97	28,018.83

Solution: Conduct a regression analysis to develop the empirical equation for the S–O curve defined in Equation 20.6. At each time step, the outflow, $O(t + \Delta t)$, is solved by trial and error until the SO-function is balanced. The storage volume, $S(t + \Delta t)$, is determined by Equation 20.16. The hydrologic routing for the given inflow hydrograph is summarized in Table 20.4.

Table 20.4 Hydrologic routing using OS-function

Time (min)	Type-in inflow (cfs)	Guess outflow (cfs)	Storage volume (acre-ft)	OS-function calculated (cfs)	
				$2S \times 43,560/\Delta t + O$	$I(t) + I(t + \Delta t)$
t	$I(t)$	$O(t)$	$S(t)$		$-2O(t) + OS(t)$
Δt	$I(t + \Delta t)$	$O(t + \Delta t)$	$S(t + \Delta t)$		
Given	Given	Guessed	Equation 20.6	Equation 20.23	Equation 20.24
0.00	0	0.00	0.00	0.00	0.00
5.00	3	0.02	0.01	3.00	3.00
10.00	16	0.16	0.08	21.96	21.96
15.00	48	0.63	0.29	85.64	85.64
20.00	136	1.97	0.92	268.38	268.38
25.00	296	5.12	2.38	696.45	696.45
30.00	424	10.42	4.81	1406.20	1406.20
35.00	452	16.89	7.73	2261.37	2261.37
40.00	427	23.40	10.62	3106.59	3106.59
45.00	383	29.38	13.22	3869.78	3869.78
50.00	337	34.65	15.48	4531.01	4531.01
55.00	298	39.21	17.42	5096.73	5096.72
60.00	265	43.17	19.07	5581.31	5581.31
65.00	228	46.52	20.46	5987.97	5987.97
70.00	193	49.25	21.58	6315.93	6315.93
75.00	161	51.39	22.45	6571.43	6571.43
80.00	132	53.00	23.10	6761.64	6761.64
85.00	108	54.13	23.56	6895.65	6895.65
90.00	85	54.85	23.85	6980.39	6980.39
95.00	70	55.24	24.00	7025.69	7025.69
100.00	60	55.40	24.07	7045.22	7045.22
105.00	50	55.39	24.07	7044.41	7044.41
110.00	40	55.22	24.00	7023.62	7023.62
115.00	30	54.87	23.86	6983.19	6983.19
120.00	20	54.37	23.65	6923.44	6923.44
125.00	10	53.70	23.39	6844.71	6844.71
130.00	0	52.88	23.05	6747.31	6747.31
135.00	0	51.98	22.69	6641.56	6641.56
140.00	0	51.11	22.34	6537.59	6537.59

For this case, the incoming peak flow of 452 cfs occurs at $t = 35$ min. After the detention process, the peak flow is reduced to 55.40 cfs at $t = 60$ min.

20.5 Kinematic wave routing approach

The EPA Storm Water Management Model (EPA SWMM, 1983) suggests that a *kinematic wave routing method* be developed for both overland and channel flows. The basic concept is to consider each reach as a reservoir. Over a time interval, the increase or decrease in the flow depth reflects the change in the storage volume due to the inflow and outflow through the reach.

20.5.1 Overland flow routing scheme

An overland flow on a rectangular plane is analyzed as a unit-width flow. The change in the storage volume, V, is computed as

$$\Delta S = S(t + \Delta t) - S(t) = \left[D(t + \Delta t) - D(t)\right]WL \tag{20.25}$$

in which $D(t)$ = flow depth in [L] at time t, $D(t + \Delta t)$ = flow depth in [L] at time $t + \Delta t$ in Figure 20.5, W = width in [L] of the rectangular plane, and L = length in [L] of the reach.

Considering that the overland flow is kinematic, Manning's formula for a wide flow is

$$q(t) = \frac{k}{N} W\left[D(t)\right]^{\frac{5}{3}} \sqrt{S_o} \tag{20.26}$$

$$q(t + \Delta t) = \frac{k}{N} W\left[D(t + \Delta t)\right]^{\frac{5}{3}} \sqrt{S_o} \tag{20.27}$$

in which $q(t)$ = outflow in [L³/T] at time t, $q(t + \Delta t)$ = outflow in [L³/T] at time $t + \Delta t$, N = Manning's roughness (see Table 20.5), $k = 1.486$ for ft-s units or 1.0 for m-s units, and S_o = ground slope in [L/L]. Substituting Equations 20.25 and 20.26 into Equation 20.1 yields

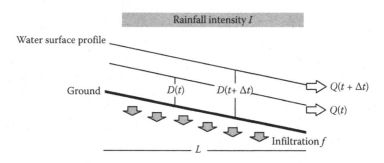

Figure 20.5 Overland flow routing scheme.

Table 20.5 Manning's N for overland flow

Surface texture	Manning's N
Dense growth	0.40–0.50
Pasture	0.30–0.40
Lawn	0.20–0.30
Bluegrass sod	0.20–0.50
Short grass	0.10–0.20
Sparse vegetation	0.05–0.13
Bare clay-loam soil	0.01–0.03
Concrete/asphalt (depth <1/4 in.)	0.10–0.15
Concrete/asphalt (depth >1/4 in.)	0.05–0.10

$$\left[\frac{I(t)+I(t+\Delta t)}{2}-\frac{f(t)+f(t+\Delta t)}{2}\right]-\frac{\left[q(t)+q(t+\Delta t)\right]}{2WL}=\frac{\left[D(t+\Delta t)-D(t)\right]}{\Delta t} \tag{20.28}$$

in which $q(t)$ = inflow in $[L^3/T]$ at time t, $I(t)$ = rainfall intensity in $[L/T]$, $f(t)$ = soil infiltration rate in $[L/T]$, $D(t)$ = water depth in $[L]$, and other variables are at time $t + \Delta t$. Equations 20.27 and 20.28 are simultaneously solved for $Q(t + \Delta t)$ and $D(t + \Delta t)$.

EXAMPLE 20.6

An overland flow is produced under an excess rainfall intensity of 4 in./h. The plane has $N = 0.15$ and $S_o = 0.02\,\text{ft/ft}$. At $t = 3000\,\text{s}$, the flow depth is $D(t) = 0.025\,\text{ft}$ over a length of 1000 ft. Find the unit-width flow at $t = 3300\,\text{s}$.

At $t = 3000\,\text{s}$, the overland flow depth is $D(t) = 0.025\,\text{ft}$, and the flow rate for $W = 1\,\text{ft}$ is

$$q(t) = \frac{1.486}{0.15} \times 1 \times (0.025)^{5/3} \sqrt{0.02} = 0.003\,\text{cfs/ft}$$

At $t = 3300$, the flow depth and unit-width flow rate are calculated as

$$4 \times \frac{1}{(12 \times 3600)} - \frac{q(t+\Delta t)-0.003}{2 \times 1000} = \frac{D(t+\Delta t)-0.025}{300}$$

$$q(t+\Delta t) = \frac{1.486}{0.15} D(t+\Delta t)^{5/3} \sqrt{0.02}$$

We have two equations for two unknowns: $D(t + \Delta t)$ and $q(t + \Delta t)$. Repeat this process to generate the overland flow hydrograph.

20.5.2 Channel flow routing scheme

Consider a typical trapezoidal channel in Figure 20.6. Over a time period, Δt, the reach has an average inflow at the upstream entrance and an average outflow at the downstream exit. The change in the storage volume is the difference between the flow areas over the time interval. As a result, Equation 20.2 is rewritten for the channel flow as

$$\frac{Q(t)+Q(t+\Delta t)}{2} - \frac{O(t)+O(t+\Delta t)}{2} = \frac{\left[A(t+\Delta t)-A(t)\right]L}{\Delta t} \tag{20.29}$$

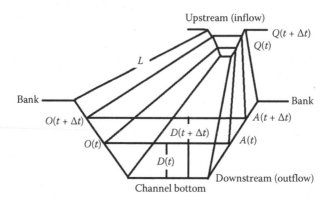

Figure 20.6 Kinematic channel routing scheme.

$$O = \frac{k}{N}\sqrt{S_o}\,P^{-\frac{2}{3}}A^{\frac{5}{3}}$$ (20.30)

in which Q = inflow in $[L^3/T]$ specified by the inflow hydrograph, O = outflow in $[L^3/T]$, A = flow area in $[L^2]$, S_o = channel slope in $[L/L]$, and P = wetted perimeter in $[L]$. Applying Equation 20.29 to the outflows at the beginning and the end over the time interval, the volume differences are calculated as

$$D_V = \frac{Q(t)+Q(t+\Delta t)}{2} - \frac{O(t)+O(t+\Delta t)}{2}$$ (20.31)

$$D_S = \frac{\left[A(t+\Delta t)-A(t)\right]L}{\Delta t}$$ (20.32)

The KW channel routing method is to simultaneously solve Equations 20.31 and 20.32 for $A(t+\Delta t)$ and $O(t+\Delta t)$ at each time step. The error tolerance for the iterative numerical procedure is determined by the criteria as

$$\text{Error} = \left|\frac{D_V - D_S}{D_S}\right| \le 1\%$$ (20.33)

The reliability of KW routing method depends on the Courant criteria. As a rule of thumb, the selection of reach length and time interval shall observe the limits as

$$\frac{L}{2V} \le \Delta t \le \frac{2L}{V}$$ (20.34)

in which V = average flow velocity in $[L/T]$.

EXAMPLE 20.7

The flood channel is described with the parameters as: B (bottom width) = 10 ft, Z (side slope) = 1V:3H, S_o (channel slope) = 0.01 (ft/ft), and N (Manning's roughness) = 0.03. Consider $\Delta t = 300$ s and $L = 2000$ ft. The initial base flow is 120 cfs for this case. Apply the KW routing method to determine the outflow hydrograph from the inflow hydrograph given in Table 20.6.

Table 20.6 Example for kinematic wave channel routing method

Time (s)	Inflow Q_{in} (cfs)	Outflow rate and storage volume					Check	Volume	Balance
		Guess depth Y (ft)	Flow area A (ft^2)	Wetted perimeter P (ft)	Outflow O-out (cfs)	Storage $S(t)$ (acre-ft)	Flow V difference D_V $(Q_{in} - O_{out}) \times \Delta t$ (acre-ft)	Storage V difference D_S $S(t + \Delta t) - S(t)$ (acre-ft)	Difference $D_V - D_S = 0$ (acre-ft)
0.0	120.0	1.53	22.2	19.7	120.0	1.02	0.00	0.00	0.00
300.0	140.0	1.58	23.2	20.0	127.3	1.07	0.04	0.04	0.00
600.0	250.0	1.88	29.5	21.9	178.6	1.35	0.29	0.29	0.00
900.0	500.0	2.62	46.9	26.6	339.7	2.15	0.80	0.80	0.00
1200.0	1100.0	3.89	84.4	34.6	759.6	3.88	1.72	1.72	0.00
1500.0	1200.0	4.72	114.2	39.9	1143.6	5.24	1.37	1.37	0.00
1800.0	900.0	4.54	107.2	38.7	1049.8	4.92	-0.32	-0.32	0.00
2100.0	700.0	4.00	88.1	35.3	804.7	4.04	-0.88	-0.88	0.00
2400.0	550.0	3.57	73.9	32.6	634.1	3.39	-0.65	-0.65	0.00
2700.0	450.0	3.21	63.1	30.3	510.6	2.90	-0.50	-0.50	0.00
3000.0	380.0	2.93	55.2	28.6	424.9	2.53	-0.36	-0.36	0.00
3300.0	310.0	2.68	48.4	27.0	355.1	2.22	-0.31	-0.31	0.00
3600.0	250.0	2.43	41.9	25.3	291.3	1.93	-0.30	-0.30	0.00
3900.0	200.0	2.18	36.1	23.8	236.6	1.66	-0.27	-0.27	0.00
4200.0	150.0	1.93	30.5	22.2	187.4	1.40	-0.25	-0.25	0.00
4500.0	100.0	1.65	24.7	20.4	139.2	1.13	-0.26	-0.27	0.00
4800.0	100.0	1.46	21.0	19.2	110.5	0.96	-0.17	-0.17	0.00
5100.0	100.0	1.41	20.0	18.9	103.0	0.92	-0.05	-0.05	0.00
5400.0	100.0	1.39	19.7	18.8	100.9	0.90	-0.01	-0.01	0.00
5700.0	100.0	1.38	19.6	18.8	100.3	0.90	0.00	0.00	0.00
6000.0	100.0	1.38	19.6	18.8	100.1	0.90	0.00	0.00	0.00

At each time step, the computation is iterated with a guessed water depth until the volume is balanced.

At $t = 1500$ s, the peak outflow, $O(t) = 1143.6$ cfs and the flow area $= 114.2$ ft^2. The average flow velocity is 10.1 ft/s. Substituting these variables into Equation 20.34 yields

$$99.8 \leq \Delta t \leq 399.4 \text{ s}$$

For this case, the time interval of 5 min satisfies the Courant numerical stability criterion. It is noted that the channel storage is so negligible that the peak flow on the inflow hydrograph is slightly attenuated in this case.

20.6 Direct routing method

The direct routing method (Figure 20.7), also called the *time shift method*, was developed to transfer a hydrograph from one design point to the next without any routing process. Therefore, the directing routing approach results in no attenuation to the flood flow at all. The entire hydrograph is shifted along the time axis by the travel time through the reach. The travel time may be estimated by the length of the reach divided by the average flow velocity. The direct routing method is commonly used in watershed modeling for numerical convenience. For instance, to model the hydrograph composition process at a confluence, the inflow hydrographs are combined according to the time sequence. To avoid any numerical attenuation on the resultant hydrograph, the direct routing method is an applicable method to transfer information from one node to another node on a numerical model.

20.7 Closing

Hydrologic routing methods are derived to apply a numerical procedure to verify the performance of a storage facility under design, including water reservoirs, detention basins, storage tanks, etc. Similarly, hydraulic routing methods are developed to confirm the performance of a conveyance facility such as channels, pipes, and culverts. Before the

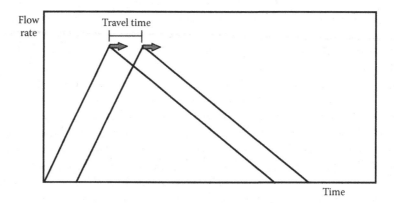

Figure 20.7 Direct routing method.

hydraulic structure is in place for service, hydraulic and hydrologic numerical methods are the only quantifiable basis for making decisions on the selection of design parameters, comparison among alternatives, prediction of performance, and evaluation of cost and damage.

All numerical methods involve empirical variables such as orifice and weir coefficients, Manning's roughness, etc. During the stage of design and sizing, the recommended empirical coefficients are adopted and applied to the numerical procedures. In many cases, the as-built hydraulic structure is equipped with monitoring devices to collect field data. With an adequate field data, the empirical coefficients in the numerical methods can be calibrated and improved for future use.

20.8 Homework

Q20.1 A stormwater detention basin has a S–S–O curve as shown in Table Q20.1. The best-fitted equation for the S–O relationship is

$$S(\text{acre-ft}) = 0.0002O^3 - 0.017O^2 + 0.4661O \quad (O \text{ in cfs})$$

1. Considering the time step of 10 min, derive the SO and OS routing functions for this detention system.
2. The inflow hydrograph is given in Table Q20.2. The initial condition is an empty basin. For each time step, $t + \Delta t$, you may guess the outflow, $O(t + \Delta t)$, and then calculate the corresponding storage volume, $S(t + \Delta t)$, using the best-fitted equation. Next, check on the values of the SO routing function calculated with the known variables at time t, and to-be-found variables at $t + \Delta t$. Iterate this trial-and-error procedure until the value of the SO-function is converged.

Table Q20.1 Stage–storage–outflow curve

Stage, H (ft)	Outflow rate, O (cfs)	Storage volume, OS (acre-ft)	SO-function, SO (acre-ft)	OS-function, OS (cfs)
5000.0	0.00	0.00		
5000.5	0.25	0.41		
5001.0	1.20	0.82		
5001.5	4.09	1.25		
5002.0	5.78	1.70		
5002.5	7.08	2.16		
5003.0	8.17	2.64		
5003.5	9.14	3.13		
5004.0	10.01	3.66		
5004.5	12.73	4.22		
5005.0	28.17	4.81		
5005.5	32.07	5.45		
5006.0	33.81	6.12		

Table Q20.2 Hydrograph routing using SO-function

Time t $t + \Delta t$ (min)	Type-in inflow $I(t)$ $I(t + \Delta t)$ (cfs)	Guess outflow $O(t)$ $O(t + \Delta t)$ (cfs)	Storage volume $S(t)$ $S(t + \Delta t)$ (acre-ft)	SO-function calculated by		Check difference in SO-function (acre-ft)
				$O(t) \times \Delta t/43{,}560 +$ $2 \times S(t)$ (acre-ft)	$[I(t) + I(t + \Delta t) -$ $2O(t)] \times \Delta t/43{,}560$ $+ SO(t)$ (acre-ft)	
10.00	17.0	0.3	0.12	0.23	0.23	0.00
20.00	64.0	1.5	0.66	1.34	1.34	0.00
20.00	64.0	4.6				
30.00	108.0					
40.00	103.0					
50.00	78.0					
60.00	56.0					
70.00	37.0	43.4	4.70	9.99	9.99	0.00
80.00	22.0					
90.00	13.0					
100.00	10.0					
110.00	7.0					
120.00	4.0					

Bibliography

ASCE. (1994). "Design and Construction of Urban Stormwater Management System", American Society of Civil Engineers, Manuals and Reports of Engineering Practice No, 77, Chapter 1.

Bedient, P.B., and Huber, W.C. (1992). *Hydrology and Floodplain Analysis*, 2nd ed., Addison-Wesley Publishing Company, Inc, New York.

EPA Storm Water Management Model. (1983). "Inlet Design Program," Volume 3, Federal Highway Administration, Report No. FHWA/RD-83/043, December.

Guo, J.C.Y. (2004). "Hydrology-Based Approach to Storm Water Detention Design Using New Routing Schemes," *ASCE Journal of Hydrologic Engineering*, Vol. 9, No. 4, pp. 333–336.

McCuen, R.H. (1998). *Hydrologic Analysis and Design*, Chapter 8, 2nd ed., Prentice Hall, New York.

Puls, L.G. (1928). "Construction of Flood Routing Curves," House Document 185, U.S. 70th Congress, First Session, Washington, DC.

US Army Corps of Engineers. (1979). "Introduction and Application of Kinematic Wave Routing Technique Using HEC-1," Training Document 10, May.

Index

Printed in the United States
by Baker & Taylor Publisher Services